AF352885

Pseudorabies Virus

Pseudorabies Virus

Editors

Yan-Dong Tang
Xiangdong Li

MDPI • Basel • Beijing • Wuhan • Barcelona • Belgrade • Manchester • Tokyo • Cluj • Tianjin

Editors
Yan-Dong Tang
State Key Laboratory of
Veterinary Biotechnology
Harbin Veterinary Research
Institute of Chinese Academy
of Agricultural Sciences
Harbin
China

Xiangdong Li
College of Veterinary
Medicine
Yangzhou University
Yangzhou
China

Editorial Office
MDPI
St. Alban-Anlage 66
4052 Basel, Switzerland

This is a reprint of articles from the Special Issue published online in the open access journal *Viruses* (ISSN 1999-4915) (available at: www.mdpi.com/journal/viruses/special_issues/Pseudorabies).

For citation purposes, cite each article independently as indicated on the article page online and as indicated below:

LastName, A.A.; LastName, B.B.; LastName, C.C. Article Title. *Journal Name* **Year**, *Volume Number*, Page Range.

ISBN 978-3-0365-5986-5 (Hbk)
ISBN 978-3-0365-5985-8 (PDF)

Contents

Zongyi Bo and Xiangdong Li
A Review of Pseudorabies Virus Variants: Genomics, Vaccination, Transmission, and Zoonotic
Potential
Reprinted from: *Viruses* **2022**, *14*, 1003, doi:10.3390/v14051003 . . . . **1**

Jing Chen, Gang Li, Chao Wan, Yixuan Li, Lianci Peng and Rendong Fang et al.
A Comparison of Pseudorabies Virus Latency to Other α-Herpesvirinae Subfamily Members
Reprinted from: *Viruses* **2022**, *14*, 1386, doi:10.3390/v14071386 . . . . **17**

Qingyun Liu, Yan Kuang, Yafei Li, Huihui Guo, Chuyue Zhou and Shibang Guo et al.
The Epidemiology and Variation in Pseudorabies Virus: A Continuing Challenge to Pigs and
Humans
Reprinted from: *Viruses* **2022**, *14*, 1463, doi:10.3390/v14071463 . . . . **33**

Hui-Hua Zheng, Peng-Fei Fu, Hong-Ying Chen and Zhen-Ya Wang
Pseudorabies Virus: From Pathogenesis to Prevention Strategies
Reprinted from: *Viruses* **2022**, *14*, 1638, doi:10.3390/v14081638 . . . . **49**

Jiahuan Deng, Zhuoyun Wu, Jiaqi Liu, Qiuyun Ji and Chunmei Ju
The Role of Latency-Associated Transcripts in the Latent Infection of Pseudorabies Virus
Reprinted from: *Viruses* **2022**, *14*, 1379, doi:10.3390/v14071379 . . . . **77**

**Guangqiang Ye, Hongyang Liu, Qiongqiong Zhou, Xiaohong Liu, Li Huang and Changjiang
Weng**
A Tug of War: Pseudorabies Virus and Host Antiviral Innate Immunity
Reprinted from: *Viruses* **2022**, *14*, 547, doi:10.3390/v14030547 . . . . **89**

Yi Wang, Guo-Li Li, Yan-Li Qi, Li-Yun Li, Lu-Fang Wang and Cong-Rong Wang et al.
Pseudorabies Virus Inhibits Expression of Liver X Receptors to Assist Viral Infection
Reprinted from: *Viruses* **2022**, *14*, 514, doi:10.3390/v14030514 . . . . **107**

Xiaohui Yang, Chuanzhao Yu, Qiuyan Zhang, Linjun Hong, Ting Gu and Enqin Zheng et al.
A Nectin1 Mutant Mouse Model Is Resistant to Pseudorabies Virus Infection
Reprinted from: *Viruses* **2022**, *14*, 874, doi:10.3390/v14050874 . . . . **121**

**Qiongqiong Zhou, Longfeng Zhang, Hongyang Liu, Guangqiang Ye, Li Huang and
Changjiang Weng**
Isolation and Characterization of Two Pseudorabies Virus and Evaluation of Their Effects on
Host Natural Immune Responses and Pathogenicity
Reprinted from: *Viruses* **2022**, *14*, 712, doi:10.3390/v14040712 . . . . **133**

**Tong-Yun Wang, Guo-Ju Sang, Qian Wang, Chao-Liang Leng, Zhi-Jun Tian and Jin-Mei Peng
et al.**
Generation of Premature Termination Codon (PTC)-Harboring Pseudorabies Virus (PRV) via
Genetic Code Expansion Technology
Reprinted from: *Viruses* **2022**, *14*, 572, doi:10.3390/v14030572 . . . . **147**

Bing Wang, Hongxia Wu, Hansong Qi, Hanglin Li, Li Pan and Lianfeng Li et al.
Histamine Is Responsible for the Neuropathic Itch Induced by the Pseudorabies Virus Variant
in a Mouse Model
Reprinted from: *Viruses* **2022**, *14*, 1067, doi:10.3390/v14051067 . . . . **159**

Changchao Huan, Jingting Yao, Weiyin Xu, Wei Zhang, Ziyan Zhou and Haochun Pan et al.
Huaier Polysaccharide Interrupts PRV Infection via Reducing Virus Adsorption and Entry
Reprinted from: *Viruses* **2022**, *14*, 745, doi:10.3390/v14040745 . . . . **173**

Xiangbo Zhang, Jingying Xie, Ming Gao, Zhenfang Yan, Lei Chen and Suocheng Wei et al.
Pseudorabies Virus ICP0 Abolishes Tumor Necrosis Factor Alpha-Induced NF-κB Activation by
Degrading P65
Reprinted from: *Viruses* **2022**, *14*, 954, doi:10.3390/v14050954 . . . . **189**

Ningning Zhao, Fan Wang, Zhengjie Kong and Yingli Shang
Pseudorabies Virus Tegument Protein UL13 Suppresses RLR-Mediated Antiviral Innate
Immunity through Regulating Receptor Transcription
Reprinted from: *Viruses* **2022**, *14*, 1465, doi:10.3390/v14071465 . . . . **205**

Panrao Liu, Danhe Hu, Lili Yuan, Zhengmin Lian, Xiaohui Yao and Zhenbang Zhu et al.
Metabolomics Analysis of PK-15 Cells with Pseudorabies Virus Infection Based on
UHPLC-QE-MS
Reprinted from: *Viruses* **2022**, *14*, 1158, doi:10.3390/v14061158 . . . . **219**

Xintan Yang, Shengkui Xu, Dengjin Chen, Ruijiao Jiang, Haoran Kang and Xinna Ge et al.
Proteomic Analysis of Vero Cells Infected with Pseudorabies Virus
Reprinted from: *Viruses* **2022**, *14*, 755, doi:10.3390/v14040755 . . . . **233**

Jianbo Huang, Wenjie Tang, Xvetao Wang, Jun Zhao, Kenan Peng and Xiangang Sun et al.
The Genetic Characterization of a Novel Natural Recombinant Pseudorabies Virus in China
Reprinted from: *Viruses* **2022**, *14*, 978, doi:10.3390/v14050978 . . . . **249**

Jun Yao, Juan Li, Lin Gao, Yuwen He, Jiarui Xie and Pei Zhu et al.
Epidemiological Investigation and Genetic Analysis of Pseudorabies Virus in Yunnan Province
of China from 2017 to 2021
Reprinted from: *Viruses* **2022**, *14*, 895, doi:10.3390/v14050895 . . . . **263**

**Ximeng Chen, Hongxuan Li, Qianlei Zhu, Hongying Chen, Zhenya Wang and Lanlan Zheng
et al.**
Serological Investigation and Genetic Characteristics of Pseudorabies Virus between 2019 and
2021 in Henan Province of China
Reprinted from: *Viruses* **2022**, *14*, 1685, doi:10.3390/v14081685 . . . . **273**

Review

A Review of Pseudorabies Virus Variants: Genomics, Vaccination, Transmission, and Zoonotic Potential

Zongyi Bo [1,2] and Xiangdong Li [2,*]

1   Joint International Research Laboratory of Agriculture and Agri-Product Safety, The Ministry of Education of China, Yangzhou University, Yangzhou 225009, China; zybo@yzu.edu.cn
2   Jiangsu Co-Innovation Center for the Prevention and Control of Animal Infectious Disease and Zoonoses, College of Veterinary Medicine, Yangzhou University, Yangzhou 225009, China
*   Correspondence: 007352@yzu.edu.cn

Abstract: Pseudorabies virus (PRV), the causative agent of Aujeszky's disease, has a broad host range including most mammals and avian species. In 2011, a PRV variant emerged in many Bartha K61-vaccinated pig herds in China and has attracted more and more attention due to its serious threat to domestic and wild animals, and even human beings. The PRV variant has been spreading in China for more than 10 years, and considerable research progresses about its molecular biology, pathogenesis, transmission, and host–virus interactions have been made. This review is mainly organized into four sections including outbreak and genomic evolution characteristics of PRV variants, progresses of PRV variant vaccine development, the pathogenicity and transmission of PRV variants among different species of animals, and the zoonotic potential of PRV variants. Considering PRV has caused a huge economic loss of animals and is a potential threat to public health, it is necessary to extensively explore the mechanisms involved in its replication, pathogenesis, and transmission in order to ultimately eradicate it in China.

Keywords: pseudorabies virus; variant strain; genomics; vaccination; transmission; zoonosis

Citation: Bo, Z.; Li, X. A Review of Pseudorabies Virus Variants: Genomics, Vaccination, Transmission, and Zoonotic Potential. *Viruses* **2022**, *14*, 1003. https://doi.org/10.3390/v14051003

Academic Editor: Julia A. Beatty

Received: 22 April 2022
Accepted: 5 May 2022
Published: 9 May 2022

Publisher's Note: MDPI stays neutral with regard to jurisdictional claims in published maps and institutional affiliations.

1. Introduction

Pseudorabies Virus, also called Aujeszky's disease virus or Suid alphaherpesvirus 1, belongs to genus Varicelloviru, Alphaherpesvirinae subfamily within the family Herpesviridae, and is a double-strand linear DNA with 143 kb and can encode more than 70 proteins [1]. PRV was first documented as the causative pathogen of Aujeszky's disease in 1902 by a Hungarian veterinarian [2]. Different clinical symptoms of pigs can be induced dependent on their housing stages: infected piglets always show fatal and central nervous system disorders, fattening pigs have respiratory symptoms with low mortality, and pregnant sows have abortions with the death of the fetuses [3].

The first case report of PRV in China could be traced back to 1947 in a domestic cat [4]. After that, PRV circulated in many swine herds of China due to the lack of an available PRV vaccine. The Bartha vaccine strain, attenuated through many passages of a virulent strain in culture cells and embryos, was introduced into China in the 1970s, and it provided an ideal protection effect to the early prevalent strains (classical strains) in China [5–7]. In late 2011, PRV occurred in many Bartha K61-vaccinated swine herds [8]. After viral isolation and genome sequencing, the results showed that the newly emerging PRV clustered in an independent branch from the previously isolated strains in China [9,10]. To differentiate it from the classical strains isolated before 2011 in China, the newly isolated PRV was designated as a PRV variant.

Though pigs are the only reservoir of PRV, it can infect many species of animals, including sheep, dogs, foxes, tigers, bears, etc. [11,12]. Previously, experimental studies in nonhuman primates showed that rhesus monkeys and marmosets are susceptible to PRV infection, while other higher-order primates, such as chimpanzees and humans are not susceptible to PRV infection [13]. Most recently, more than 20 human severe pseudorabies encephalitis cases were reported [14]. All these patients infected with PRV variants had close contact with pigs which suggested pigs might be the etiological source of PRV for human infection. So, it is time to pay attention to the public threat induced by PRV variants and the transmission of it between different species of animals and human beings.

In this review, we will briefly summarize the studies of PRV variants from the aspects of its outbreak, genomic characteristics, vaccine development, transmission in different animals, and its significance for public health.

2. The Outbreaks and the Genomic Evolution of PRV Variants

In late 2011, a large-scale occurrence of severe disease with anorexia, neurologic symptoms, high fever, and respiratory distress in piglets, and a high percent of abortion in sows happened in many swine farms of China. Pathological examination showed gross lesions in the lungs, and yellow-white necrosis in the kidneys. Through multiple kinds of diagnostic methods, including ELISA, PCR, viral isolation, immunohistochemical staining, and gene sequencing, the PRV variant was finally recognized as the causative pathogen for these severe clinical diseases [8,15].

After viral isolation and sequencing, there is a large sequence divergence in newly isolated PRV strains from previously classical PRV strains. Here, we constructed a maximum likelihood (ML) phylogenetic tree of 39 strains (Supplemental Table S1) of PRV full-length sequences which have been extracted from NCBI. The results showed that all PRV strains can be phylogenetically clustered into two groups. All foreign strains out of China were clustered into the same group, and nearly all strains isolated from China are located in an independent group with them (Figure 1A). Interestingly, the PRV GD1802 strain, isolated in China in 2018, is clustered into the same group with the strains from foreign countries. We infer that the GD1802 strain might be delivered into China from foreign countries through the introduction of pigs. In addition, when compared with classical strains, such as Ea and Fa, PRV strains isolated after 2011 are located in a relatively dependent branch with them. The gC gene is a major gene which has been commonly used for the phylogenetic analysis of PRV due to its high variability [16,17]. We also analyzed gC gene sequences of these reported PRV strains. As shown in Figure 1B, the gC-based phylogenetic tree reveals that PRV can be phylogenetically divided into two groups, designated as genotype I and genotype II, and genotype II can be further divided into two clades, clade 2.1 and clade 2.2. Clade 2.1 mainly comprises the strains isolated before 2011, and nearly all PRV variants are located at clade 2.2.

Recombination contributes a lot to the genomic divergence of many viruses, such as African swine fever virus [18], porcine reproductive and respiratory syndrome virus [19], classical swine fever virus [20], porcine circovirus [21], and porcine epidemic diarrhea virus [22]. As for PRV, the recombination between different strains has been reported in vivo and vitro [23,24]. So, it is interesting to explore whether the evolution of PRV in China also has a relationship with recombination. As shown in Figure 1B, we found that there are four Chinese-origin strains located at the genotype I cluster, including SC, HLJ-2013, JSY13, and GD1802 strains. Except for GD1802, which might be introduced abroad, the other three strains all have a relationship with recombination. Ye et al. found that SC was a recombinant of an endemic Chinese strain and a Bartha-vaccine-like strain [25]; Bo et al. found that the JSY13 strain was a recombination of the PRV variant JSY7 strain and the Bartha K61 vaccine strain [26]; Liu et al. found that the HLJ-2013 strain is probably a recombination of three origins: a yet unknown parent strain, a European-origin strain, and a Chinese-origin strain [27]. Besides these, Huang et al. found that the FJ62 variant strain was the recombination between the PRV genotype I strain from wild boar and genotype II strain from domestic pig [28]. These reports demonstrated that recombination plays an important role in the evolution of PRV in China.

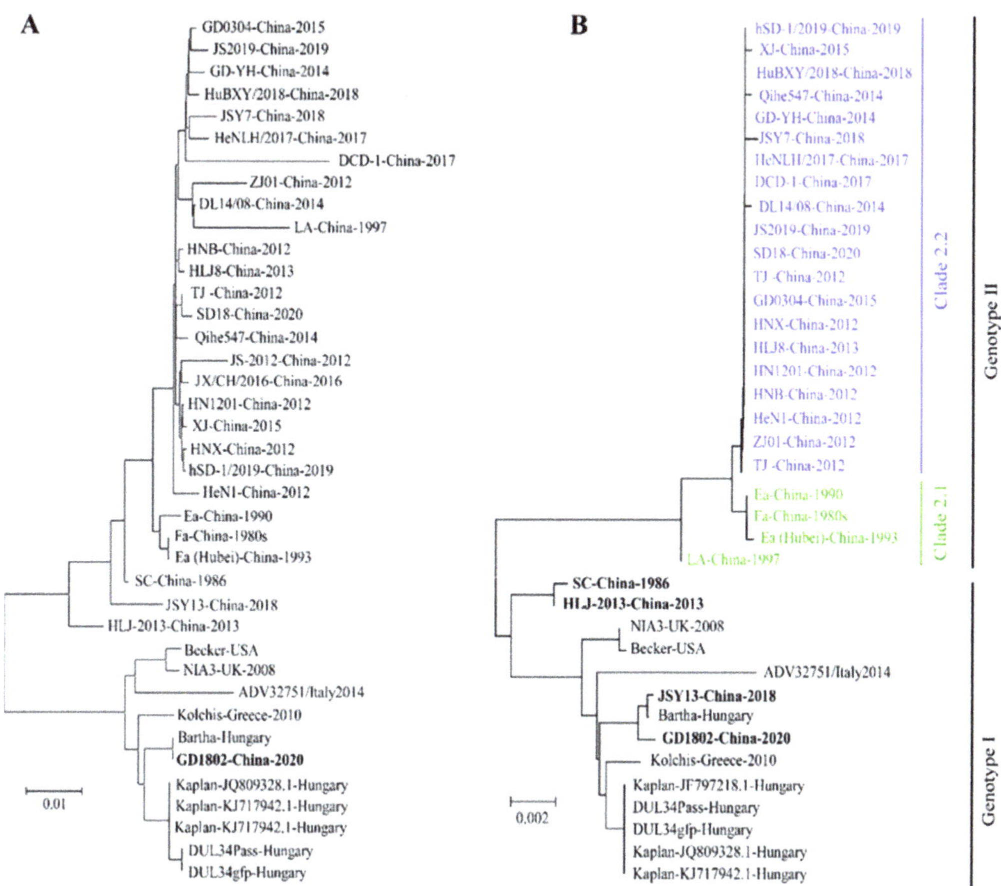

Figure 1. The phylogenetic analysis of PRV full-length genome sequences and gC sequences. (**A**) Phylogenetic analysis of PRV complete genome sequences. (**B**) Phylogenetic analysis of PRV gC gene sequences. Both maximum likelihood (ML) trees were constructed by using MEGA X software.

It was reported that the virulence of PRV variants were higher than the classical strains which were isolated before 2011. So, whether there is a difference in the major virulence-determining genes between PRV variants and classical strains warrants investigation. Therefore, we analyzed gI and gE genes, which are major virulence-determining genes of PRV [29]. The phylogenetic analysis showed that all PRV strains can be divided into two classes, as shown in Figure 2A (gI) and Figure 2B (gE). Compared with PRV from foreign countries or the classical strains isolated in China, there are some typical mutations, insertions, and deletions in the several virulence-determining genes, and non-coding sequences of PRV variants, such as two Aspartate (Asp, D) insertions in gE protein [26,30,31].

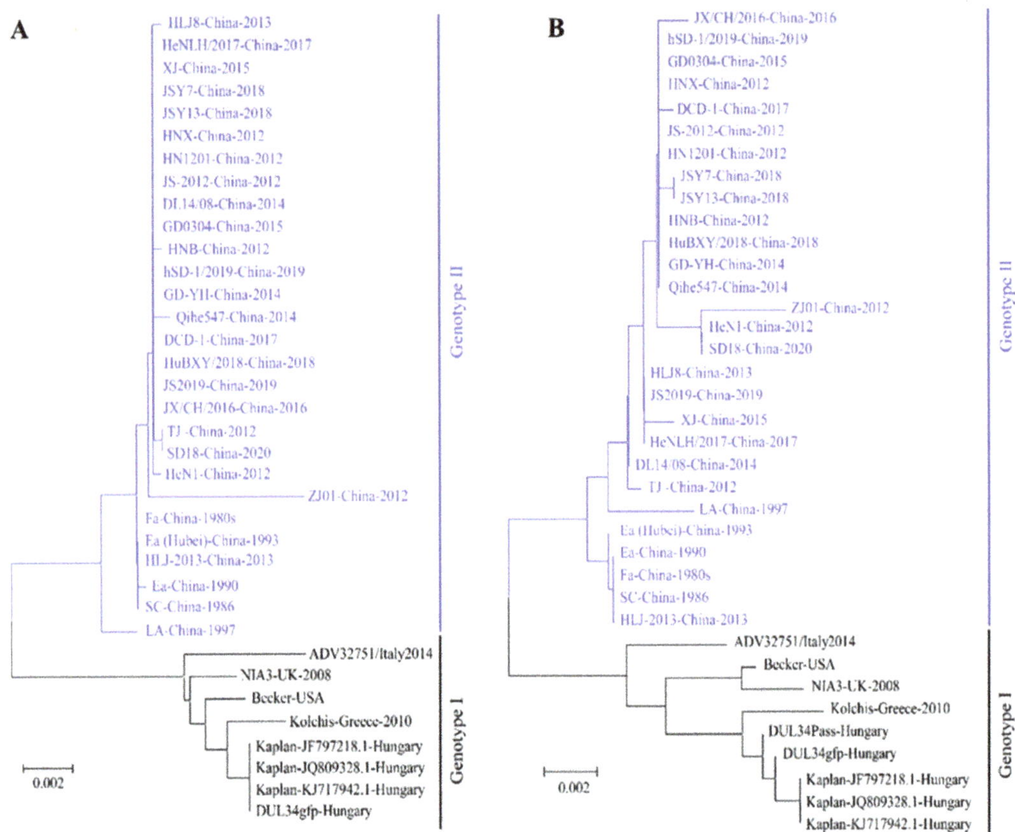

Figure 2. The phylogenetic analysis of PRV gI and gE genes. (**A**) Phylogenetic analysis of PRV gI gene sequences. (**B**) Phylogenetic analysis of PRV gE gene sequences. Both maximum likelihood (ML) trees were constructed by using MEGA X software.

The biological functions and meanings of these insertions or deletions in PRV variant strains are yet to be explored. Previously, a study showed that the exchange of gB genes contributes to an immunogenic difference between PRV variant JS-2012 and Bartha-K61, which indicated the gB protein of the PRV variant has the ability to evade the neutralization antibody induced by the Bartha strain [32]. The gC gene, a main host receptor binding protein of PRV, can bind to heparan sulfate proteoglycans in the extracellular matrix. Whether the mutations in the gC gene influence the cell entry stage of PRV and viral transmission among different kinds of tissues, animals, and even human beings is rarely studied. As the PRV variant is more virulent than classical PRV strains, whether the mutations in TK, gI, and gE virulence-determining genes are related to the increased virulence of PRV variants also needs to be addressed in the further studies.

3. PRV Variant-Based Vaccines Are on the Way

As the PRV variant has been circulating in many Bartha K61-immunized pig farms, researchers tried to explore whether the Bartha K61 strain could provide full protection against PRV variants. An et al. used the PRV variant HeN1 strain to challenge the Bartha K61-immunized pigs and sheep, the results showed that HeN1-challenged pigs showed fever and loss of appetite, while no deaths were found; the experiment on sheep showed that the Bartha K61 strain provided full protection against PRV classical SC strain, but two out of four Bartha K61-immunized sheep challenged with the PRV variant HeN1 strain had clinical signs and died [15]. Luo et al. challenged the Bartha K61-immunized sheep with the PRV variant TJ strain and found that the Bartha K61 vaccine cannot provide complete protection [33]. These experiments demonstrated that the Bartha K61 vaccine can hardly provide full protection against the PRV variant.

Besides the imported Bartha K61 vaccine strain, there are several other licensed PRV live attenuated and inactivated vaccine strains that are based on the PRV strains isolated in China. One of them is the SA125 strain, which is a gE/gI/TK gene-deleted strain based on the cattle-origin Fa strain, and it is the first licensed PRV vaccine in China [34]. Another is the PRV gE/gI/TK gene-deleted HB-2000 strain, which is based on the Ea strain. Besides these two live attenuated vaccines, there is another licensed PRV HB-98 inactivated vaccine which deletes TK/gG genes based on the Ea strain. The existence of the gE gene in the HB-98 vaccine makes it impossible to differentiate the immunized from field strain-infected pigs [35]. Now, there is no direct evidence whether these licensed Chinese-origin PRV strains have better protection than the Bartha K61 vaccine strain to the PRV variant.

The PRV variant has been circulating in China for more than 10 years, and many vaccine candidates that are based on PRV variants have been reported by different research groups. Wang et al. deleted the gE gene of a PRV variant TJ strain and used it to immunize pigs; the pigs had no clinical signs when they were challenged with a PRV variant strain [36]. Wang et al. constructed a TK and gE gene-deleted PRV variant, AH02LA, which could stop the viral shedding, and no clinical signs were observed after being challenged with the PRV variant. By contrast, Bartha K61-immunized pigs showed some mild clinical signs and had viral shedding [37]. Wang et al. used the bacterial artificial chromosome (BCA) manipulation method to construct a PRV variant AH02LA gE gene-deleted strain and found it can provide complete clinical protection against the challenge of the PRV variant [38]. Besides the above-mentioned experimental vaccines, several other PRV variant-based vaccine candidates which have double or triple deletions of TK, gE, or gI genes, also showed complete protection against the challenge of PRV variants [39–41]. Till now, there are two licensed PRV variant-based vaccines. One is the PRV C strain, a gI/gE/Us9/Us2 naturally deleted strain that was isolated in China in 2011 and licensed in 2017 [42]. Another is a gE-deleted inactivated vaccine (HN1201-ΔgE), which was certificated in 2019 and could induce a high neutralization antibody titer against both PRV classical and variant strains.

Despite the inactivated and live attenuated vaccines, several other types of vaccines including a subunit vaccine and nucleic vaccine, could also be the alternative for the development of an effective PRV variant vaccine. Among 11 glycoproteins PRV encoded, gB, gC, and gD are the main immunogenetic antigens that can induce neutralization antibodies [7,43]. Previously, Wang et al. used the baculovirus system to express the PRV variant gB protein and challenged the immunized pigs with PRV variant HN1201 strain. The results showed that the gB-based vaccine can provide full protection for the PRV variant [44]. In another report, gB, gC, and gD proteins of the PRV variant were separately expressed using the baculovirus expression system working as subunit vaccine candidates. Pigs were then immunized with each single protein twice and challenged with the PRV variant HNLH strain. The results showed that the survival rates of gB-, gC-, and gD-vaccinated pigs were 100%, 50%, and 87.5%, respectively [45].

A DNA vaccine can mimic the natural infection in which the immunized antigens could be presented in both major histocompatibility complex classes I and II settings [46]. Previously, E.M.A. van Rooij et al. vaccinated pigs with the plasmids expressing the main immunogenetic antigens gB, gC, and gD, then challenged the immunized pigs with PRV. The results showed that plasmid DNA encoding gB induced the strongest cell-mediated immune responses, whereas plasmid DNA encoding gD induced the strongest neutralizing antibody responses [47]. Furthermore, Hyun A Yoon et al. compared the intramuscular (i.m.) and intranasal (i.n.) immunization routes by immunizing mice with the plasmid that encoded gB. The results showed that immunization can induce the mucosal immunity via the i.n. route. However, it only induced a low IgG response which cannot protect the mice against the challenge of PRV [48].

A messenger RNA (mRNA)-based vaccine that encodes the immunogenetic antigen of pathogens provides a new vaccine developing platform for multiple viruses [49,50]. As PRV can encode several immunogenetic antigens that can induce the neutralization antibodies, mRNA might be a candidate for PRV vaccine development. Jiang et al. developed a PRV-XJ variant strain gD gene-based mRNA vaccine which was formulated via mRNA encapsulated in the liposomes. Compared with the DNA vaccine that encoded PRV gD protein in a recombinant plasmid pVAX-gD, both mRNA and pVAX-gD plasmids induced a high neutralization antibody titer and antigen-specific B and T cell response in mice. Finally, the PRV variant challenge experiment was performed, and the results showed that one mouse in ten which was immunized with the mRNA vaccine died, whereas there was no death in the pVAX-gD plasmid-vaccinated group [51]. These results demonstrated that multiple types of vaccines besides inactivated and live attenuated vaccines, show protection against the challenge of PRV, and gB and gD proteins play an important role among these major immunogenetic proteins.

Despite many reports showing that the Bartha K61 vaccine cannot provide full protection against PRV variants, there are still several studies showing that Bartha K61 can provide complete protection to pigs against the PRV variant [7]. Previously, Zhou et al. used and then challenged Bartha K61-immunized growing pigs with the PRV variant XJ5 strain. The results showed that all immunized/challenged pigs survived, while the unimmunized pigs all died [52]. Another report also showed that Bartha K61-immunized pigs had no death under the challenge of the PRV variant AH02LA strain [53]. Although there were some mild clinical signs in immunized/challenged growing pigs, no death was found in the above two studies, which demonstrates that the Bartha K61 vaccine strain can provide protection for pigs against the PRV variant challenge. However, as the PRV variant could lead to the deaths of newborn piglets, further experiments still need to be conducted to confirm whether piglets are resistant to the challenge of the PRV variant because they are Bartha K61-immunized or because they got maternal antibodies form the immunized sows.

4. The Pathogenicity of PRV Variant to Different Species of Vertebrates

Although the natural reservoir for PRV is the pig, the first case of PRV was reported in cattle, and the disease was described as "mad itch" [54]. PRV has a broad host range, infecting most mammals and even some avian species. PRV infection can cause different clinical signs dependent on the stages of housing pigs. Besides pigs, PRV infection in other species of animals is always fatal due to the neurologic invasion. Till now, it has been reported that there are about 19 species of animals that can be naturally infected by PRV including domestic/wild pigs [55,56], sheep [57], cats [58], coyotes [59], foxes [60], rats [61], deer [62], bears [63], rabbits [64], dogs [65], horses [66], bats [67], wolves [68], raccoons [69], mice [70], ferrets [71], panthers [72], cattle [73], and chickens [74] (Figure 3). Domestic and wild pigs are the reservoirs for PRV, all other species of animals will die after several days upon the onset of PRV. In the last decade, several animal species including dogs, bovine, mink, foxes, goats, and wolves were found dead due to the infection of PRV variants [75–79]. It is reasonable to infer that some other animal species could be susceptible to PRV variants. The wide host range of PRV can also be demonstrated by its incubation in multiple kinds of cells, such as PK15 cells (Porcine kidney), Vero cells (Green monkey kidney cells), HEp-2 (Human Epithelioma cells), DF-1 cells (Chicken embryo fibroblasts cells), and many other cells. This demonstrates that PRV can bind with its receptor in the surface of these cells, which suggests that PRV might have the possibility to infect multiple kinds of animals.

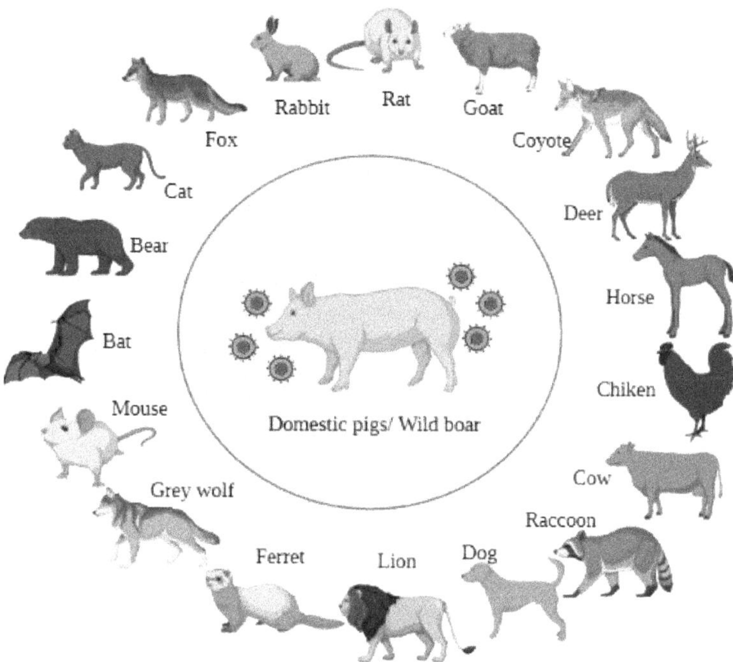

Figure 3. The reported animals that can be infected by PRV.

Besides natural infection, the experimental infections of PRV were also conducted to confirm the susceptibility of different animals, such as domestic pigs [80], wild boars [81], sheep [82], calves [73], dogs [83], cats [84], blue foxes [85], raccoons [86], horses [66], and chickens [74]. The clinical signs after infection mainly include fever, depression, pruritus, self-mutilation, anorexia, neurologic deficits, etc. The viral challenge experiments demonstrated that PRV can be introduced through multiple kinds of routes, including intracranial, intradermal, intramuscular, intranasal, intraocular, intraperitoneal, intratracheal, intravenous, and even foot pads [87]. The animal experiments of PRV can also be performed in some laboratory animals, including mice [88], rats [89], rabbits [90], dogs [90], guinea pigs [91], and rhesus macaques [92]. Due to the wide host range and easy manipulation, PRV can be handled as the model to study many characteristics of herpesvirus, such as molecular biology, pathogenesis, neuroinvasion, and transneuronal spread.

Sometimes the transmission of PRV among different species of animals will lead to the change of PRV pathogenicity. Previously, R E Shope found that PRV was attenuated after passaging through a guinea pig brain [91]. In addition, T F Müller et al. found that PRV strains of wild swine origin might be less virulent than those of domestic pigs [81]. This phenomenon might be explained by the insertions or deletions of multiple segments among different strains, in order to adapt to their host animals.

PRV is an air-borne pathogen. Furthermore, food, water, and excrement can also be intermediate vectors of PRV [93]. The infected pigs or semen are the main origins for PRV infection in swine farms. As for other species of animals, the direct contact of contaminated equipment, or consumption of the infected animals are the main sources for their infection [94]. Thereafter, it is important to keep the animals away from swine farms, especially mice, cats, dogs, etc., in case they are infected with PRV and transmit it to other animals. Furthermore, pig products or byproducts, such as head or offal tissue, which might be polluted by PRV must be kept away from other animals, since they could directly contact or consume them. Finally, surveillance of PRV not only in pigs, but also in other species of animals should be performed, in order to keep PRV out of the susceptible herds.

5. The Zoonotic Potential of PRV Variants

It has been known that PRV has a wide host range, and the duplication and translocation of sequences from the left end of the genome to the UL-US junction plays an important role for the growth of PRV in different hosts [95]. It was reported that Nectin-1, a host cellular receptor that can bind with PRV glycoprotein gD, plays an important role in PRV entry. Previous studies reported that PRV gD glycoprotein showed a similar binding ability to human nectin-1 protein [96]. Not only human nectin-1, the protein that comes from bats, dogs, cats, cows, sheep, and several other animals also showed conserved functional amino acids binding with PRV gD, which provides evidence of cross-species infection for PRV [14].

The first case report of humans infected with PRV was in 1914, in which two laboratory technicians were infected with PRV after they were exposed to an infected cat. Afterwards, several PRV cases in humans were reported after they came into contact with cast, dogs, cows, or other domestic animals (Summarized in Table 1). The typical symptoms of these cases mainly include pruritis, pain, fever, swelling, sweating, dysphagia, and aphthous stomatitis. Though these PRV infected cases were reported, the etiological confirmation was not convincing due to the lack of a definitely etiological and serological diagnosis.

In 2017, a swine herder suffered from an eye disease in China. Through next-generation sequencing (NGS), real-time PCR, and phylogenetic analysis, the PRV variant was recognized as the causative agent of the disease [97]. This is the first PRV case in humans reported in China. In the following years, several other PRV infection cases in human beings were reported in China (Table 1). Different from the clinical symptoms reported from other countries, the clinical signs of Chinese patients in these cases were mainly encephalitis and endophthalmitis. Of note, almost all PRV infected patients have contact with swine or other PRV-susceptible animals. Among these incidences, a milestone case attracted our attention because one PRV variant, designated hSD-1/2019, was firstly isolated from the cerebrospinal fluid of a PRV-infected patient [98]. This is the first case in the world where PRV was successfully isolated from humans, which proved that humans are susceptible to PRV.

Table 1. PRV infection cases in humans.

Year	Numbers	Symptoms	Contacted Animals	Ref.
1914	2	Swelling, reddening, intense itching	Cat	[99]
1940	2	Pruritis, erythema, and pain around the wound	Dog	[99]
1963	2	Throat pain, weakness in the legs,	Dog	[99]
1983	1	Pain in togue, hypersalivation, dysphagia, headache, arthralgia	Cat	[100]
1986	2	Tension in the mouth, nose, and throat; perception of strange smells and taste	Cat, or other domestic animals	[100]
1992	6	Pruritus in the palms, lower and upper arms, shoulders, and back	Cow	[101]
2017	1	Endophthalmitis, fever, headaches	Pig	[97]
2018	4	Encephalitis	Pig	[102]
2019	5	Encephalitis	Pig	[103]
2019	1	Encephalitis	Pig	[104]
2019	1	Encephalitis, fever, headache, seizure	unknown	[105]
2019	1	Encephalitis	Pig	[106]
2020	1	Encephalitis	Pig	[107]
2020	6	panencephalitis	Pig	[108]
2021	1	Retinitis	Pig	[109]
2021	1	Encephalitis, retinal vasculitis, fever	Pig	[110]
2021	1	Encephalitis, Endophthalmitis	Pig	[111]
2022	2	Encephalitis, seizures, endophthalmitis	Pig	[112]

Of note, the cases of humans infected with PRV have been increasing rapidly in China since 2017. Coincidentally, PRV variants have been the dominant strain in Chinese pig farms since then. Whether PRV variants are more sensitive to humans than the classical PRV warrants further investigation. Among these PRV-infected patients, most of them were reported as having swine-related occupations or had close contact with other infected animals. Furthermore, several patients have injuries to their fingers or other places. Therefore, it is necessary to carry out skin protection for people who have close contact with swine. Till now, all infected humans recovered completely, and the clinical and neurological signs disappeared, though sometimes the clinical signs last for one to several months. No patients died from PRV infection in the above cases. However, similar to other herpesviruses, PRV can induce latent and lytic replication in pigs, and the infected pigs will carry PRV for a long time. Whether the recovered patients still carry PRV is unknown.

6. Discussion and Perspectives

In this review, we summarized the biological characteristics of PRV variants, which have antigenic variation and a higher virulence compared with classical PRV strains in China. Since 2011, PRV variants have become the dominant strains in China and caused huge economic losses for the swine industry. There is still dispute over whether the Bartha K61 vaccine can provide full protection against the PRV variant [52,113]. However, PRV variants happened in many Bartha K61-immunized swine farms, and the phylogenetic analysis showed that PRV variants have a large sequence divergence with the Bartha K61 strain. Therefore, more effective vaccines based on local PRV variants need to be developed. Moreover, multiple kinds of vaccines, including a live attenuated vaccine, inactivated vaccine, subunit vaccine, DNA vaccine, and mRNA vaccine have the potential to provide full protection for pigs against PRV variants. Upon the design of PRV vaccines, it is necessary to follow the strategy of DIVA, which has helped to eradicate PRV in many countries [114].

The serological detection of PRV gE antibodies is always used to evaluate the status of PRV field strain infection, due to the application of gE-deleted PRV vaccines. One study reported that PRV positive rates have decreased to 18.12% in 2020 from 38.20% in 2018 after screening 19,292 pig serum samples by using PRV gE ELISA [115]. Another report analyzed the gE antibodies from 256,326 serum samples, which were collected from 29 provinces in China from 2011 to 2021. The results showed that the average positive rate was 29.87% [31]. Zheng et al. collected 4708 pig serum samples from Henan province during 2018–2019 and found the positive rate of gE antibodies was 30.14% (1419/4708) [116]. Lin et at. collected a total of 18,138 serum samples from 808 PRV-vaccinated pig farms during 2016–2020 in Hunan province and detected the presence of gE-specific antibodies. The results showed that 23.55% (4271/18,138) of the samples were positive for PRV gE-specific antibodies [117]. All these results demonstrate that the infection of the PRV field strain is still very common in Chinese swine farms, though nearly all of them have used PRV vaccines. Therefore, strict biosecurity measures, feeding strategies, daily managements, diagnostic methods, and effective vaccination should be jointly performed in swine farms to decrease the infection of the PRV field strain.

Right now, with the increased number of human cases caused by PRV infection, the exact mechanisms involved in its transmission from animals to human beings are still unknown. One explanation is that the "one health" concept is recognized by more and more researchers, and the interdisciplinary communication is more frequent than before, so the human cases with encephalitis and neurological symptoms, which might be induced by PRV, could be jointly diagnosed by doctors and veterinarians. Another possibility is that the evolution of the PRV variant makes it more susceptible to humans as PRV entry-related proteins gD or gC might have a higher binding ability to its receptor in human cells than before. In contrast, a study showed that 455 persons who participated in heroic self-incubation with PRV via intracutaneous and subcutaneous methods showed no clinical signs, which demonstrated that the infection of PRV in humans was occasionally asymptomatic [99]. We hypothesize that the infection of PRV in humans might be related to the immune system of people, and immunocompromised patients might be more susceptible to PRV. Till now, the only consolatory thing is that there is no human-to-human PRV transmission case. More studies about the mechanisms evolved in the transmission of PRV from infected animals to humans still need to be explored.

Due to the lack of effective PRV variant vaccines, many researchers have focused on exploring the available compounds that can inhibit the proliferation of PRV in vitro and in vivo. Among them, two kinds of compounds attract people's attention. One is traditional Chinese herbal medicines, which have a long history of anti-virus effects, but their cellular targets and working mechanisms are unclear. For example, Germacrone, which is extracted from Rhizoma Curcuma, was able to inhibit the proliferation of PRV in the early phase of the PRV replication cycle [118]. Another kind is synthetic chemical compounds, which have a relatively clear cellular target and whose functions are comparatively clear. For example, it was demonstrated that meclizine, a class of H1-antihistamine, could inhibit the replication of PRV at its entry and release stages [119]. The exploration of anti-PRV compounds will contribute to the therapy of PRV infection, especially for the infection of human beings.

Because of the existence of multiple non-essential genes, PRV can be engineered as the vector to express foreign antigens derived from other pathogens. For example, Qiu et al. used PRV as the vector to express the GP5 protein of porcine reproductive and respiratory syndrome virus, and the recombinant virus provided an ideal protection against the challenge of PRRSV [120]. Other proteins, such as the HA gene of the H3N2 subtype of swine influenza virus, and the S protein of porcine epidemic diarrhea virus, were also expressed by PRV, which provided protection to the immunized pigs [121,122]. The advantage of these vectored vaccines are that they provide protection against the infections of other pathogens besides PRV.

In conclusion, as the PRV variant has caused a huge economic loss for the pig industry in China and is a potential threat to public health, it is necessary to pay more attention to the detection and isolation of the PRV variant in different animals, in order to study its epidemiological characteristics. At the same time, it is urgent to develop more safe and efficient DIVA vaccines based on the PRV variants. Furthermore, it is also important to deeply explore the interactions between the PRV infection and host responses, which will not only help to clarify the mechanisms involved in PRV proliferation and pathogenesis, but also contribute to the development of PRV vaccines and antiviral drugs. Most importantly, since PRV has the possibility to infect human beings, researchers need to pay more attention to its cross-species transmission and screen the patients that are suspected of being infected by PRV, especially those who have swine-related occupations.

Supplementary Materials: The following supporting information can be downloaded at: https://www.mdpi.com/article/10.3390/v14051003/s1, Table S1: The genomic information of 39 strains of PRV.

Author Contributions: Conceptualization, X.L. and Z.B.; writing, Z.B. and X.L.; manuscript revision and supervision, X.L. All authors have read and agreed to the published version of the manuscript.

Funding: This work was funded by the National Natural Science Foundation of China (32102635, 32172823), the Natural Science Foundation of Jiangsu Province (BK20210805), and the Project of the Priority Academic Program Development of Jiangsu Higher Education Institutions (PAPD).

Institutional Review Board Statement: Not applicable.

Informed Consent Statement: Not applicable.

Data Availability Statement: All data generated or analyzed during this study are included in the published article.

Conflicts of Interest: The authors declare no conflict of interest.

References

1. Pomeranz, L.E.; Reynolds, A.E.; Hengartner, C.J. Molecular Biology of Pseudorabies Virus: Impact on Neurovirology and Veterinary Medicine. *Microbiol. Mol. Biol. Rev.* **2005**, *69*, 462–500. [CrossRef] [PubMed]
2. Wittmann, G.; Rziha, H.-J. Aujeszk™ s Disease (Pseudorabies) in Pigs. In *Herpesvirus Diseases of Cattle, Horses, and Pigs*; Springer: Boston, MA, USA, 1989.
3. Mettenleiter, T.C. Aujeszky's disease (pseudorabies) virus: The virus and molecular pathogenesis—State of the art, June 1999. *Vet. Res.* **2000**, *31*, 99–115. [CrossRef] [PubMed]
4. Yuan, Q.Z.; Wu, Y.X.; Li, Y.X.; Li, Z.R.; Nan, X. The pseudorabies vaccination research. I: Pseudorabies attenuated vaccine research. *Chin. J. Prev. Vet. Med.* **1983**, *1*, 1–6. (In Chinese)
5. Lomniczi, B.; Kaplan, A.S.; Ben-Porat, T. Multiple defects in the genome of pseudorabies virus can affect virulence without detectably affecting replication in cell culture. *Virology* **1987**, *161*, 181–189. [CrossRef]
6. Sun, Y.; Luo, Y.; Wang, C.-H.; Yuan, J.; Li, N.; Song, K.; Qiu, H.-J. Control of swine pseudorabies in China: Opportunities and limitations. *Vet. Microbiol.* **2016**, *183*, 119–124. [CrossRef]
7. Freuling, C.M.; Müller, T.F.; Mettenleiter, T.C. Vaccines against pseudorabies virus (PrV). *Vet. Microbiol.* **2017**, *206*, 3–9. [CrossRef]
8. Yu, X.; Zhou, Z.; Hu, D.; Zhang, Q.; Han, T.; Li, X.; Gu, X.; Yuan, L.; Zhang, S.; Wang, B.; et al. Pathogenic Pseudorabies Virus, China, 2012. *Emerg. Infect. Dis.* **2014**, *20*, 102–104. [CrossRef]
9. He, W.; Auclert, L.Z.; Zhai, X.; Wong, G.; Zhang, C.; Zhu, H.; Xing, G.; Wang, S.; He, W.; Li, K.; et al. Interspecies Transmission, Genetic Diversity, and Evolutionary Dynamics of Pseudorabies Virus. *J. Infect. Dis.* **2019**, *219*, 1705–1715. [CrossRef]
10. Zhai, X.; Zhao, W.; Li, K.; Zhang, C.; Wang, C.; Su, S.; Zhou, J.; Lei, J.; Xing, G.; Sun, H.; et al. Genome Characteristics and Evolution of Pseudorabies Virus Strains in Eastern China from 2017 to 2019. *Virol. Sin.* **2019**, *34*, 601–609. [CrossRef]
11. Laval, K.; Enquist, L.W. The Neuropathic Itch Caused by Pseudorabies Virus. *Pathogens* **2020**, *9*, 254. [CrossRef]
12. Müller, T.; Hahn, E.C.; Tottewitz, F.; Kramer, M.; Klupp, B.G.; Mettenleiter, T.C.; Freuling, C. Pseudorabies virus in wild swine: A global perspective. *Arch. Virol.* **2011**, *156*, 1691–1705. [CrossRef] [PubMed]
13. Enquist, L.W. Life beyond eradication: Veterinary viruses in basic science. *Arch. Virol. Suppl.* **1999**, *15*, 87–109. [CrossRef] [PubMed]
14. Wong, G.; Lu, J.; Zhang, W.; Gao, G.F. Pseudorabies virus: A neglected zoonotic pathogen in humans? *Emerg. Microbes Infect.* **2019**, *8*, 150–154. [CrossRef]
15. An, T.-Q.; Peng, J.-M.; Tian, Z.-J.; Zhao, H.-Y.; Li, N.; Liu, Y.-M.; Chen, J.-Z.; Leng, C.-L.; Sun, Y.; Chang, D.; et al. Pseudorabies Virus Variant in Bartha-K61–Vaccinated Pigs, China, 2012. *Emerg. Infect. Dis.* **2013**, *19*, 1749–1755. [CrossRef] [PubMed]
16. Ishikawa, K.; Tsutsui, M.; Taguchi, K.; Saitoh, A.; Muramatsu, M. Sequence variation of the gC gene among pseudorabies virus strains. *Vet. Microbiol.* **1996**, *49*, 267–272. [CrossRef]
17. Ye, C.; Zhang, Q.-Z.; Tian, Z.-J.; Zheng, H.; Zhao, K.; Liu, F.; Guo, J.-C.; Tong, W.; Jiang, C.-G.; Wang, S.-J.; et al. Genomic characterization of emergent pseudorabies virus in China reveals marked sequence divergence: Evidence for the existence of two major genotypes. *Virology* **2015**, *483*, 32–43. [CrossRef]
18. Zhu, Z.; Xiao, C.-T.; Fan, Y.; Cai, Z.; Lu, C.; Zhang, G.; Jiang, T.; Tan, Y.; Peng, Y. Homologous recombination shapes the genetic diversity of African swine fever viruses. *Vet. Microbiol.* **2019**, *236*, 108380. [CrossRef]
19. Zhou, L.; Kang, R.; Zhang, Y.; Yu, J.; Xie, B.; Chen, C.; Li, X.; Chen, B.; Liang, L.; Zhu, J.; et al. Emergence of two novel recombinant porcine reproductive and respiratory syndrome viruses 2 (lineage 3) in Southwestern China. *Vet. Microbiol.* **2019**, *232*, 30–41. [CrossRef]
20. Chen, Y.; Chen, Y.-F. Extensive homologous recombination in classical swine fever virus: A re-evaluation of homologous recombination events in the strain AF407339. *Saudi J. Biol. Sci.* **2014**, *21*, 311–316. [CrossRef]
21. Wei, C.; Lin, Z.; Dai, A.; Chen, H.; Ma, Y.; Li, N.; Wu, Y.; Yang, X.; Luo, M.; Liu, J. Emergence of a novel recombinant porcine circovirus type 2 in China: PCV2c and PCV2d recombinant. *Transbound. Emerg. Dis.* **2019**, *66*, 2496–2506. [CrossRef]
22. Chen, N.; Li, S.; Zhou, R.; Zhu, M.; He, S.; Ye, M.; Huang, Y.; Li, S.; Zhu, C.; Xia, P.; et al. Two novel porcine epidemic diarrhea virus (PEDV) recombinants from a natural recombinant and distinct subtypes of PEDV variants. *Virus Res.* **2017**, *242*, 90–95. [CrossRef] [PubMed]
23. Maes, R.K.; Sussman, M.D.; Vilnis, A.; Thacker, B.J. Recent developments in latency and recombination of Aujeszky's disease (pseudorabies) virus. *Vet. Microbiol.* **1997**, *55*, 13–27. [CrossRef]
24. Dangler, C.A.; Henderson, L.M.; Bowman, L.A.; Deaver, R.E. Direct isolation and identification of recombinant pseudorabies virus strains from tissues of experimentally co-infected swine. *Am. J. Vet. Res.* **1993**, *54*, 540–545. [PubMed]
25. Ye, C.; Guo, J.-C.; Gao, J.-C.; Wang, T.-Y.; Zhao, K.; Chang, X.-B.; Wang, Q.; Peng, J.-M.; Tian, Z.-J.; Cai, X.-H.; et al. Genomic analyses reveal that partial sequence of an earlier pseudorabies virus in China is originated from a Bartha-vaccine-like strain. *Virology* **2016**, *491*, 56–63. [CrossRef] [PubMed]
26. Bo, Z.; Miao, Y.; Xi, R.; Gao, X.; Miao, D.; Chen, H.; Jung, Y.S.; Qian, Y.; Dai, J. Emergence of a novel pathogenic recombinant virus from Bartha vaccine and variant pseudorabies virus in China. *Transbound. Emerg. Dis.* **2020**, *68*, 1454–1464. [CrossRef] [PubMed]
27. Liu, H.; Shi, Z.; Liu, C.; Wang, P.; Wang, M.; Wang, S.; Liu, Z.; Wei, L.; Sun, Z.; He, X.; et al. Implication of the Identification of an Earlier Pseudorabies Virus (PRV) Strain HLJ-2013 to the Evolution of Chinese PRVs. *Front. Microbiol.* **2020**, *11*, 612474. [CrossRef]
28. Huang, J.; Zhu, L.; Zhao, J.; Yin, X.; Feng, Y.; Wang, X.; Sun, X.; Zhou, Y.; Xu, Z. Genetic evolution analysis of novel recombinant pseudorabies virus strain in Sichuan, China. *Transbound. Emerg. Dis.* **2020**, *67*, 1428–1432. [CrossRef]

29. Ferrari, M.; Mettenleiter, T.; Romanelli, M.; Cabassi, E.; Corradi, A.; Mas, N.D.; Silini, R. A Comparative Study of Pseudorabies Virus (PRV) Strains with Defects in Thymidine Kinase and Glycoprotein Genes. *J. Comp. Pathol.* **2000**, *123*, 152–163. [CrossRef]
30. Fan, J.; Zeng, X.; Zhang, G.; Wu, Q.; Niu, J.; Sun, B.; Xie, Q.; Ma, J. Molecular characterization and phylogenetic analysis of pseudorabies virus variants isolated from Guangdong province of southern China during 2013–2014. *J. Vet. Sci.* **2016**, *17*, 369–375. [CrossRef]
31. Tan, L.; Yao, J.; Yang, Y.; Luo, W.; Yuan, X.; Yang, L.; Wang, A. Current Status and Challenge of Pseudorabies Virus Infection in China. *Virol. Sin.* **2021**, *36*, 588–607. [CrossRef]
32. Yu, Z.-Q.; Tong, W.; Zheng, H.; Li, L.-W.; Li, G.-X.; Gao, F.; Wang, T.; Liang, C.; Ye, C.; Wu, J.-Q.; et al. Variations in glycoprotein B contribute to immunogenic difference between PRV variant JS-2012 and Bartha-K61. *Vet. Microbiol.* **2017**, *208*, 97–105. [CrossRef] [PubMed]
33. Luo, Y.; Li, N.; Cong, X.; Wang, C.-H.; Du, M.; Li, L.; Zhao, B.; Yuan, J.; Liu, D.-D.; Li, S.; et al. Pathogenicity and genomic characterization of a pseudorabies virus variant isolated from Bartha-K61-vaccinated swine population in China. *Vet. Microbiol.* **2014**, *174*, 107–115. [CrossRef] [PubMed]
34. Ling, Z.; Wan-Zhu, G.; Zhi-Wen, X. Fluctuant Rule of Colostral Antibodies and the Date of Initial Immunization for the Piglet from Sows Inoculated with Pseudorabies Virus Gene-deleted Vaccine SA215. *Chin. J. Vet. Sci.* **2004**, *24*, 320–322. (In Chinese)
35. He, Q.G.; Chen, H.C.; Fang, L.R.; Wu, B.; Liu, Z.F.; Xiao, S.B.; Jin, M.L. The Safety, Stablization and Immunogenicity of Double Gene-negative Mutant of Pseudorabies Virus Strain (PrV HB-98). *Chin. J. Vet. Sci.* **2006**, *26*, 165–168. (In Chinese)
36. Wang, C.-H.; Yuan, J.; Qin, H.-Y.; Luo, Y.; Cong, X.; Li, Y.; Chen, J.; Li, S.; Sun, Y.; Qiu, H.-J. A novel gE-deleted pseudorabies virus (PRV) provides rapid and complete protection from lethal challenge with the PRV variant emerging in Bartha-K61-vaccinated swine population in China. *Vaccine* **2014**, *32*, 3379–3385. [CrossRef] [PubMed]
37. Wang, J.; Song, Z.; Ge, A.; Guo, R.; Qiao, Y.; Xu, M.; Wang, Z.; Liu, Y.; Zheng, Y.; Fan, H.; et al. Safety and immunogenicity of an attenuated Chinese pseudorabies variant by dual deletion of TK&gE genes. *BMC Vet. Res.* **2018**, *14*, 287. [CrossRef]
38. Wang, J.; Guo, R.; Qiao, Y.; Xu, M.; Wang, Z.; Liu, Y.; Gu, Y.; Liu, C.; Hou, J. An inactivated gE-deleted pseudorabies vaccine provides complete clinical protection and reduces virus shedding against challenge by a Chinese pseudorabies variant. *BMC Vet. Res.* **2016**, *12*, 277. [CrossRef] [PubMed]
39. Gu, Z.; Dong, J.; Wang, J.; Hou, C.; Sun, H.; Yang, W.; Bai, J.; Jiang, P. A novel inactivated gE/gI deleted pseudorabies virus (PRV) vaccine completely protects pigs from an emerged variant PRV challenge. *Virus Res.* **2015**, *195*, 57–63. [CrossRef]
40. Tong, W.; Li, G.; Liang, C.; Liu, F.; Tian, Q.; Cao, Y.; Li, L.; Zheng, X.; Zheng, H.; Tong, G. A live, attenuated pseudorabies virus strain JS-2012 deleted for gE/gI protects against both classical and emerging strains. *Antivir. Res.* **2016**, *130*, 110–117. [CrossRef]
41. Hu, R.-M.; Zhou, Q.; Song, W.-B.; Sun, E.-C.; Zhang, M.-M.; He, Q.-G.; Chen, H.-C.; Wu, B.; Liu, Z.-F. Novel pseudorabies virus variant with defects in TK, gE and gI protects growing pigs against lethal challenge. *Vaccine* **2015**, *33*, 5733–5740. [CrossRef]
42. Gao, J.F.; Lai, Z.; Shu, Y.H.; Qi, S.H.; Ma, J.J.; Wu, B.Q.; Gong, J.P. Isolation and identification of porcine pseudorabies virus (PRV) C strain. *Acta Agric. Shanghai* **2015**, *31*, 32–36. (In Chinese)
43. Mettenleiter, T.C. Pseudorabies (Aujeszky's disease) virus: State of the art. August 1993. *Acta Vet. Hung.* **1994**, *42*, 153–177. [PubMed]
44. Wang, Y.; Wang, T.; Yan, H.; Yang, F.; Guo, L.; Yang, Q.; Hu, X.; Tan, F.; Xiao, Y.; Li, X.; et al. Research and development of a novel subunit vaccine for the currently circulating pseudorabies virus variant in China. *Front. Agric. Sci. Eng.* **2015**, *2*, 216–222. [CrossRef]
45. Zhang, T.; Liu, Y.; Chen, Y.; Wang, A.; Feng, H.; Wei, Q.; Zhou, E.; Zhang, G. A single dose glycoprotein D-based subunit vaccine against pseudorabies virus infection. *Vaccine* **2020**, *38*, 6153–6161. [CrossRef] [PubMed]
46. Porter, K.R.; Raviprakash, K. DNA Vaccine Delivery and Improved Immunogenicity. *Curr. Issues Mol. Biol.* **2017**, *22*, 129–138. [CrossRef] [PubMed]
47. Van Rooij, E.M.; Haagmans, B.L.; Glansbeek, H.L.; de Visser, Y.E.; de Bruin, M.G.; Boersma, W.; Bianchi, A.T. A DNA vaccine coding for glycoprotein B of pseudorabies virus induces cell-mediated immunity in pigs and reduces virus excretion early after infection. *Vet. Immunol. Immunopathol.* **2000**, *74*, 121–136. [CrossRef]
48. Yoon, H.A.; Han, Y.W.; Aleyas, A.; George, J.A.; Kim, S.J.; Kim, H.K.; Song, H.J.; Cho, J.G.; Eo, S.K. Protective immunity induced by systemic and mucosal delivery of DNA vaccine expressing glycoprotein B of pseudorabies virus. *J. Microbiol. Biotechnol.* **2008**, *18*, 591–599.
49. Pardi, N.; Hogan, M.J.; Porter, F.W.; Weissman, D. mRNA vaccines—A new era in vaccinology. *Nat. Rev. Drug Discov.* **2018**, *17*, 261–279. [CrossRef]
50. Jackson, N.A.C.; Kester, K.E.; Casimiro, D.; Gurunathan, S.; DeRosa, F. The promise of mRNA vaccines: A biotech and industrial perspective. *NPJ Vaccines* **2020**, *5*, 11. [CrossRef]
51. Jiang, Z.; Zhu, L.; Cai, Y.; Yan, J.; Fan, Y.; Lv, W.; Gong, S.; Yin, X.; Yang, X.; Sun, X.; et al. Immunogenicity and protective efficacy induced by an mRNA vaccine encoding gD antigen against pseudorabies virus infection. *Vet. Microbiol.* **2020**, *251*, 108886. [CrossRef]
52. Zhou, J.; Li, S.; Wang, X.; Zou, M.; Gao, S. Bartha-k61 vaccine protects growing pigs against challenge with an emerging variant pseudorabies virus. *Vaccine* **2017**, *35*, 1161–1166. [CrossRef] [PubMed]
53. Wang, J.; Zeng, R.; Torrents, D.; Martinez, C.; Qiao, Y.; Yiqi, G.U.; Liu, C.J.A.H.; Medicine, V. Protection of pseudorabies vaccine (Bartha K61 strain) against pseudorabies virus variant in pigs. *Anim. Husb. Vet. Med.* **2015**, *47*. (In Chinese)
54. Hanson, R.P. The history of pseudorabies in the United States. *J. Am. Vet. Med. Assoc.* **1954**, *124*, 259–261. [PubMed]

55. Beran, G.W.; Davies, E.B.; Arambulo, P.V., 3rd; Will, L.A.; Hill, H.T.; Rock, D.L. Persistence of pseudorabies virus in infected swine. *J. Am. Vet. Med. Assoc.* **1980**, *176*, 998–1000.
56. Verpoest, S.; Cay, A.B.; De Regge, N. Molecular characterization of Belgian pseudorabies virus isolates from domestic swine and wild boar. *Vet. Microbiol.* **2014**, *172*, 72–77. [CrossRef]
57. Mocsári, E.; Szolnoki, J.; Glávits, R.; Zsák, L. Horizontal transmission of Aujeszky's disease virus from sheep to pigs. *Vet. Microbiol.* **1989**, *19*, 245–252. [CrossRef]
58. Egberink, H.F. Aujeszky's disease in dogs and cats. *Tijdschr. Voor Diergeneeskd.* **1990**, *115*, 349–353.
59. Raymond, J.T.; Gillespie, R.G.; Woodruff, M.; Janovitz, E.B. Pseudorabies in Captive Coyotes. *J. Wildl. Dis.* **1997**, *33*, 916–918. [CrossRef]
60. Murdoch, R.S. Aujeszky's disease in foxhounds. *Vet. Rec.* **1990**, *126*, 226.
61. Dolivo, M.; Beretta, E.; Bonifas, V.; Foroglou, C. Ultrastructure and function in sympathetic ganglia isolated from rats infected with pseudorabies virus. *Brain Res.* **1978**, *140*, 111–123. [CrossRef]
62. Neagari, Y.; Sakai, T.; Nogami, S.; Kaiho, I.; Katoh, C. Incidence of antibodies in raccoon dogs and deer inhabiting suburban areas. *Kansenshogaku Zasshi* **1998**, *72*, 331–334. [CrossRef] [PubMed]
63. Banks, M.; Torraca, L.S.M.; Greenwood, A.G.; Taylor, D.C. Aujeszky's disease in captive bears. *Vet. Rec.* **1999**, *145*, 362–365. [CrossRef] [PubMed]
64. Goto, H.; Burger, D.; Gorham, J., Jr. Quantitative studies of pseudirabies virus in mink, ferrets, rabbits and mice. *Jpn. J. Vet. Sci.* **1971**, *33*, 145–153. [CrossRef] [PubMed]
65. Guillon, J.C.; Chirol, C.; Vallée, A.; Cordaillat, J.C.; Beylot, J.C. A focus of Aujeszky's disease in dogs in the department of Ain. *Bull. L'acad. Vet. Fr.* **1968**, *41*, 177–179.
66. Kimman, T.G.; Binkhorst, G.J.; Ingh, T.S.V.D.; Pol, J.M.; Gielkens, A.L.; Roelvink, M.E. Aujeszky's disease in horses fulfils Koch's postulates. *Vet. Rec.* **1991**, *128*, 103–106. [CrossRef]
67. Reagan, R.L.; Day, W.C.; Marley, R.T.; Brueckner, A.L. Effect of pseudorabies virus (Aujeszky strain) in the large brown bat (Eptesicus fuscus). *Am. J. Vet. Res.* **1953**, *14*, 331–332.
68. Amoroso, M.G.; Di Concilio, D.; D'Alessio, N.; Veneziano, V.; Galiero, G.; Fusco, G. Canine parvovirus and pseudorabies virus coinfection as a cause of death in a wolf (Canis lupus) from southern Italy. *Vet. Med. Sci.* **2020**, *6*, 600–605. [CrossRef]
69. Thawley, D.; Wright, J. Pseudorabies virus infection in raccoons: A review. *J. Wildl. Dis.* **1982**, *18*, 113–116. [CrossRef]
70. Field, H.J.; Hill, T.J. The Pathogenesis of Pseudorabies in Mice following Peripheral Inoculation. *J. Gen. Virol.* **1974**, *23*, 145–157. [CrossRef]
71. Ohshima, K.I.; Gorham, J.R.; Henson, J.B. Pathologic changes in ferrets exposed to pseudorabies virus. *Am. J. Vet. Res.* **1976**, *37*, 591–596.
72. Glass, C.M.; McLean, R.G.; Katz, J.B.; Maehr, D.S.; Cropp, C.B.; Kirk, L.J.; McKeiman, A.J.; Evermann, J.F. Isolation of pseudorabies (Aujeszky's Disease) virus from a florida panther. *J. Wildl. Dis.* **1994**, *30*, 180–184. [CrossRef] [PubMed]
73. McCracken, R.M.; McFerran, J.B.; Dow, C. The Neural Spread of Pseudorabies Virus in Calves. *J. Gen. Virol.* **1973**, *20*, 17–28. [CrossRef] [PubMed]
74. Ramachandran, S.P.; Fraser, G. Studies on the virus of Aujeszky's disease. *J. Comp. Pathol.* **1971**, *81*, 55–62. [CrossRef]
75. Tu, L.; Lian, J.; Pang, Y.; Liu, C.; Cui, S.; Lin, W. Retrospective detection and phylogenetic analysis of pseudorabies virus in dogs in China. *Arch. Virol.* **2021**, *166*, 91–100. [CrossRef]
76. Cheng, Z.; Kong, Z.; Liu, P.; Fu, Z.; Zhang, J.; Liu, M.; Shang, Y. Natural infection of a variant pseudorabies virus leads to bovine death in China. *Transbound. Emerg. Dis.* **2020**, *67*, 518–522. [CrossRef]
77. Lian, K.; Zhang, M.; Zhou, L.; Song, Y.; Wang, G.; Wang, S. First report of a pseudorabies-virus-infected wolf (Canis lupus) in China. *Arch. Virol.* **2020**, *165*, 459–462. [CrossRef] [PubMed]
78. Liu, H.; Li, X.-T.; Hu, B.; Deng, X.-Y.; Zhang, L.; Lian, S.-Z.; Zhang, H.-L.; Lv, S.; Xue, X.-H.; Lu, R.-G.; et al. Outbreak of severe pseudorabies virus infection in pig-offal-fed farmed mink in Liaoning Province, China. *Arch. Virol.* **2017**, *162*, 863–866. [CrossRef]
79. Jin, H.-L.; Gao, S.-M.; Liu, Y.; Zhang, S.-F.; Hu, R.-L. Pseudorabies in farmed foxes fed pig offal in Shandong province, China. *Arch. Virol.* **2016**, *161*, 445–448. [CrossRef]
80. Lin, J.; Li, Z.; Feng, Z.; Fang, Z.; Chen, J.; Chen, W.; Liang, W.; Chen, Q. Pseudorabies virus (PRV) strain with defects in gE, gC, and TK genes protects piglets against an emerging PRV variant. *J. Vet. Med Sci.* **2020**, *82*, 846–855. [CrossRef]
81. Müller, T.F.; Teuffert, J.; Zellmer, R.; Conraths, F.J. Experimental infection of European wild boars and domestic pigs with pseudorabies viruses with differing virulence. *Am. J. Vet. Res.* **2001**, *62*, 252–258. [CrossRef]
82. Schmidt, S.P.; Hagemoser, W.A.; Kluge, J.P.; Hill, H.T. Pathogenesis of ovine pseudorabies (Aujeszky's disease) following intratracheal inoculation. *Can. J. Vet. Res.* **1987**, *51*, 326–333. [PubMed]
83. Zhang, L.; Zhong, C.; Wang, J.; Lu, Z.; Liu, L.; Yang, W.; Lyu, Y. Pathogenesis of natural and experimental Pseudorabies virus infections in dogs. *Virol. J.* **2015**, *12*, 44. [CrossRef]
84. Hagemoser, W.A.; Kluge, J.P.; Hill, H.T. Studies on the pathogenesis of pseudorabies in domestic cats following oral inoculation. *Can. J. Comp. Med.* **1980**, *44*, 192–202. [PubMed]
85. Quiroga, M.I.; Vázquez, S.; López-Peña, M.; Guerrero, F.; Nieto, J.M. Experimental Aujeszky's Disease in Blue Foxes (Alopex lagopus. *J. Vet. Med. Ser. A* **1995**, *42*, 649–657. [CrossRef] [PubMed]
86. Kirkpatrick, C.M.; Kanitz, C.L.; McCrocklin, S.M. Possible Role of wild mammals in transmission of pseudorabies to swine. *J. Wildl. Dis.* **1980**, *16*, 601–614. [CrossRef] [PubMed]

87. Sehl, J.; Teifke, J.P. Comparative Pathology of Pseudorabies in Different Naturally and Experimentally Infected Species—A Review. *Pathogens* **2020**, *9*, 633. [CrossRef]

88. Laval, K.; Vernejoul, J.B.; Van Cleemput, J.; Koyuncu, O.O.; Enquist, L.W. Virulent Pseudorabies Virus Infection Induces a Specific and Lethal Systemic Inflammatory Response in Mice. *J. Virol.* **2018**, *92*, e01614–e01618. [CrossRef]

89. Rassnick, S.; Enquist, L.W.; Sved, A.F.; Card, J. Pseudorabies Virus-Induced Leukocyte Trafficking into the Rat Central Nervous System. *J. Virol.* **1998**, *72*, 9181–9191. [CrossRef]

90. Olander, H.J.; Saunders, J.; Gustafson, D.; Jones, R. Pathologic Findings in Swine Affected with a Virulent Strain of Aujeszky's Virus. *Pathol. Vet.* **1966**, *3*, 64–82. [CrossRef]

91. Shope, R.E. Modification of the pathogenicity of pseudorabies virus by animal passage. *J. Exp. Med.* **1933**, *57*, 925–931. [CrossRef]

92. Hurst, E.W. Studies on pseudorabies (infectious bulbar paralysis, mad itch). *J. Exp. Med.* **1936**, *63*, 449–463. [CrossRef] [PubMed]

93. Li, H.; Liang, R.; Pang, Y.; Shi, L.; Cui, S.; Lin, W. Evidence for interspecies transmission route of pseudorabies virus via virally contaminated fomites. *Vet. Microbiol.* **2020**, *251*, 108912. [CrossRef] [PubMed]

94. Hahn, E.; Page, G.; Hahn, P.; Gillis, K.; Romero, C.; Annelli, J.; Gibbs, E. Mechanisms of transmission of Aujeszky's disease virus originating from feral swine in the USA. *Vet. Microbiol.* **1997**, *55*, 123–130. [CrossRef]

95. Reilly, L.M.; Rall, G.; Lomniczi, B.; Mettenleiter, T.C.; Kuperschmidt, S.; Ben-Porat, T. The ability of pseudorabies virus to grow in different hosts is affected by the duplication and translocation of sequences from the left end of the genome to the UL-US junction. *J. Virol.* **1991**, *65*, 5839–5847. [CrossRef]

96. Zhang, N.; Yan, J.; Lu, G.; Guo, Z.; Fan, Z.; Wang, J.; Shi, Y.; Qi, J.; Gao, G.F. Binding of herpes simplex virus glycoprotein D to nectin-1 exploits host cell adhesion. *Nat. Commun.* **2011**, *2*, 577. [CrossRef] [PubMed]

97. Ai, J.-W.; Weng, S.-S.; Cheng, Q.; Cui, P.; Li, Y.-J.; Wu, H.-L.; Zhu, Y.-M.; Xu, B.; Zhang, W.-H. Human Endophthalmitis Caused By Pseudorabies Virus Infection, China, 2017. *Emerg. Infect. Dis.* **2018**, *24*, 1087–1090. [CrossRef]

98. Liu, Q.; Wang, X.; Xie, C.; Ding, S.; Yang, H.; Guo, S.; Li, J.; Qin, L.; Ban, F.; Wang, D.; et al. A Novel Human Acute Encephalitis Caused by Pseudorabies Virus Variant Strain. *Clin. Infect. Dis.* **2021**, *73*, e3690–e3700. [CrossRef]

99. Skinner, G.; Ahmad, A.; Davies, J. The infrequency of transmission of herpesviruses between humans and animals; postulation of an unrecognised protective host mechanism. *Comp. Immunol. Microbiol. Infect. Dis.* **2001**, *24*, 255–269. [CrossRef]

100. Avak, S.; Bienzle, U.; Feldmeier, H.; Hampl, H.; Habermehl, K.-O. Pseudorabies in man. *Lancet* **1987**, *1*, 501–502. [CrossRef]

101. Anusz, Z.; Szweda, W.; Popko, J.; Trybała, E. Is Aujeszky's disease a zoonosis? *Prz. Epidemiol.* **1992**, *46*, 181–186.

102. Zhao, W.L.; Wu, Y.H.; Li, H.F.; Li, S.Y.; Fan, S.Y.; Wu, H.L.; Li, Y.J.; Lü, Y.L.; Han, J.; Zhang, W.C.; et al. Clinical experience and next-generation sequencing analysis of encephalitis caused by pseudorabies virus. *Zhonghua Yi Xue Za Zhi* **2018**, *98*, 1152–1157. [CrossRef] [PubMed]

103. Yang, X.; Guan, H.; Li, C.; Li, Y.; Wang, S.; Zhao, X.; Zhao, Y.; Liu, Y. Characteristics of human encephalitis caused by pseudorabies virus: A case series study. *Int. J. Infect. Dis.* **2019**, *87*, 92–99. [CrossRef] [PubMed]

104. Wang, Y.; Nian, H.; Li, Z.; Wang, W.; Wang, X.; Cui, Y. Human encephalitis complicated with bilateral acute retinal necrosis associated with pseudorabies virus infection: A case report. *Int. J. Infect. Dis.* **2019**, *89*, 51–54. [CrossRef] [PubMed]

105. Yang, H.; Han, H.; Wang, H.; Cui, Y.; Liu, H.; Ding, S. A Case of Human Viral Encephalitis Caused by Pseudorabies Virus Infection in China. *Front. Neurol.* **2019**, *10*, 534. [CrossRef] [PubMed]

106. Zheng, L.; Liu, X.; Yuan, D.; Li, R.; Lu, J.; Li, X.; Tian, K.; Dai, E. Dynamic cerebrospinal fluid analyses of severe pseudorabies encephalitis. *Transbound. Emerg. Dis.* **2019**, *66*, 2562–2565. [CrossRef]

107. Wang, D.; Tao, X.; Fei, M.; Chen, J.; Guo, W.; Li, P.; Wang, J. Human encephalitis caused by pseudorabies virus infection: A case report. *J. Neuro Virol.* **2020**, *26*, 442–448. [CrossRef]

108. Fan, S.; Yuan, H.; Liu, L.; Li, H.; Wang, S.; Zhao, W.; Wu, Y.; Wang, P.; Hu, Y.; Han, J.; et al. Pseudorabies virus encephalitis in humans: A case series study. *J. Neuro Virol.* **2020**, *26*, 556–564. [CrossRef]

109. Hu, F.; Wang, J.; Peng, X.-Y. Bilateral Necrotizing Retinitis following Encephalitis Caused by the Pseudorabies Virus Confirmed by Next-Generation Sequencing. *Ocul. Immunol. Inflamm.* **2021**, *29*, 922–925. [CrossRef]

110. Ying, M.; Hu, X.; Wang, M.; Cheng, X.; Zhao, B.; Tao, Y. Vitritis and retinal vasculitis caused by pseudorabies virus. *J. Int. Med. Res.* **2021**, *49*, 3000605211058990. [CrossRef]

111. Yan, W.; Hu, Z.; Zhang, Y.; Wu, X.; Zhang, H. Case Report: Metagenomic Next-Generation Sequencing for Diagnosis of Human Encephalitis and Endophthalmitis Caused by Pseudorabies Virus. *Front. Med.* **2021**, *8*, 753988. [CrossRef]

112. Zhou, Y.; Nie, C.; Wen, H.; Long, Y.; Zhou, M.; Xie, Z.; Hong, D. Human viral encephalitis associated with suid herpesvirus 1. *Neurol. Sci.* **2022**, *43*, 2681–2692. [CrossRef] [PubMed]

113. Wang, J.; Cui, X.; Wang, X.; Wang, W.; Gao, S.; Liu, X.; Kai, Y.; Chen, C. Efficacy of the Bartha-K61 vaccine and a gE−/gI−/TK− prototype vaccine against variant porcine pseudorabies virus (vPRV) in piglets with sublethal challenge of vPRV. *Res. Vet. Sci.* **2020**, *128*, 16–23. [CrossRef] [PubMed]

114. Ketusing, N.; Reeves, A.; Portacci, K.; Yano, T.; Olea-Popelka, F.; Keefe, T.; Salman, M. Evaluation of Strategies for the Eradication of Pseudorabies Virus (Aujeszky's Disease) in Commercial Swine Farms in Chiang-Mai and Lampoon Provinces, Thailand, Using a Simulation Disease Spread Model. *Transbound. Emerg. Dis.* **2014**, *61*, 169–176. [CrossRef] [PubMed]

115. Ren, Q.; Ren, H.; Gu, J.; Wang, J.; Jiang, L.; Gao, S. The Epidemiological Analysis of Pseudorabies Virus and Pathogenicity of the Variant Strain in Shandong Province. *Front. Vet. Sci.* **2022**, *9*, 806824. [CrossRef]

116. Zheng, H.-H.; Jin, Y.; Hou, C.-Y.; Li, X.-S.; Zhao, L.; Wang, Z.-Y.; Chen, H.-Y. Seroprevalence investigation and genetic analysis of pseudorabies virus within pig populations in Henan province of China during 2018–2019. *Infect. Genet. Evol.* **2021**, *92*, 104835. [CrossRef]
117. Lin, Y.; Tan, L.; Wang, C.; He, S.; Fang, L.; Wang, Z.; Zhong, Y.; Zhang, K.; Liu, D.; Yang, Q.; et al. Serological Investigation and Genetic Characteristics of Pseudorabies Virus in Hunan Province of China From 2016 to 2020. *Front. Vet. Sci.* **2021**, *8*, 762326. [CrossRef]
118. He, W.; Zhai, X.; Su, J.; Ye, R.; Zheng, Y.; Su, S. Antiviral Activity of Germacrone against Pseudorabies Virus in Vitro. *Pathogens* **2019**, *8*, 258. [CrossRef]
119. Liu, P.; Hu, D.; Yuan, L.; Lian, Z.; Yao, X.; Zhu, Z.; Nowotny, N.; Shi, Y.; Li, X. Meclizine Inhibits Pseudorabies Virus Replication by Interfering with Virus Entry and Release. *Front. Microbiol.* **2021**, *12*, 795593. [CrossRef]
120. Qiu, H.-J.; Tian, Z.-J.; Tong, G.-Z.; Zhou, Y.-J.; Ni, J.-Q.; Luo, Y.-Z.; Cai, X.-H. Protective immunity induced by a recombinant pseudorabies virus expressing the GP5 of porcine reproductive and respiratory syndrome virus in piglets. *Vet. Immunol. Immunopathol.* **2005**, *106*, 309–319. [CrossRef]
121. Tian, Z.-J.; Zhou, G.-H.; Zheng, B.-L.; Qiu, H.-J.; Ni, J.-Q.; Yang, H.-L.; Yin, X.-N.; Hu, S.-P.; Tong, G.-Z. A recombinant pseudorabies virus encoding the HA gene from H3N2 subtype swine influenza virus protects mice from virulent challenge. *Vet. Immunol. Immunopathol.* **2006**, *111*, 211–218. [CrossRef]
122. Zhang, C.; Guo, S.; Guo, R.; Chen, S.; Zheng, Y.; Xu, M.; Wang, Z.; Liu, Y.; Wang, J. Identification of four insertion sites for foreign genes in a pseudorabies virus vector. *BMC Vet. Res.* **2021**, *17*, 190. [CrossRef] [PubMed]

Review

A Comparison of Pseudorabies Virus Latency to Other α-Herpesvirinae Subfamily Members

Jing Chen [1,†], Gang Li [1,†], Chao Wan [1], Yixuan Li [1], Lianci Peng [1], Rendong Fang [1,2,3], Yuanyi Peng [1,*] and Chao Ye [1,*]

1. Joint International Research Laboratory of Animal Health and Animal Food Safety, College of Veterinary Medicine, Southwest University, Chongqing 400715, China; cjing1235@163.com (J.C.); li18438695575@163.com (G.L.); w10241229@163.com (C.W.); lyx03014525@163.com (Y.L.); penglianci@swu.edu.cn (L.P.); rdfang@swu.edu.cn (R.F.)
2. Immunology Research Center, Medical Research Institute, Southwest University, Chongqing 402460, China
3. Chongqing Key Laboratory of Herbivore Science, Chongqing 400715, China
* Correspondence: pyy2002@sina.com (Y.P.); yechao123@swu.edu.cn (C.Y.)
† These authors contributed equally to this work.

Abstract: Pseudorabies virus (PRV), the causative agent of Aujeszky's disease, is one of the most important infectious pathogens threatening the global pig industry. Like other members of alpha-herpesviruses, PRV establishes a lifelong latent infection and occasionally reactivates from latency after stress stimulus in infected pigs. Latent infected pigs can then serve as the source of recurrent infection, which is one of the difficulties for PRV eradication. Virus latency refers to the retention of viral complete genomes without production of infectious progeny virus; however, following stress stimulus, the virus can be reactivated into lytic infection, which is known as the latency-reactivation cycle. Recently, several research have indicated that alphaherpesvirus latency and reactivation is regulated by a complex interplay between virus, neurons, and the immune system. However, with those limited reports, the relevant advances in PRV latency are lagging behind. Therefore, in this review we focus on the regulatory mechanisms in PRV latency via summarizing the progress of PRV itself and that of other alphaherpesviruses, which will improve our understanding in the underlying mechanism of PRV latency and help design novel therapeutic strategies to control PRV latency.

Keywords: pseudorabies virus; latency; miRNA; chromatin; immune regulation

Citation: Chen, J.; Li, G.; Wan, C.; Li, Y.; Peng, L.; Fang, R.; Peng, Y.; Ye, C. A Comparison of Pseudorabies Virus Latency to Other α-Herpesvirinae Subfamily Members. *Viruses* **2022**, *14*, 1386. https://doi.org/10.3390/v14071386

Academic Editors: Yan-Dong Tang and Xiangdong Li

Received: 25 May 2022
Accepted: 22 June 2022
Published: 24 June 2022

Publisher's Note: MDPI stays neutral with regard to jurisdictional claims in published maps and institutional affiliations.

1. Introduction

Pseudorabies virus (PRV) is a member of *Alphaherpesvirinae* under the family *Herpesviridae* that can infect a broad host range of mammals, such as ruminants, carnivores, and rodents. PRV infection mainly causes neurological symptoms and acute death in its non-natural hosts, while causes respiratory disease and reproductive failure in adult pigs and neurological symptoms in piglets, respectively [1]. Recently, it was reported that PRV was isolated from acute encephalitis cases in humans, implying the potential risk of PRV infection from pigs to humans [2]. Although eradication of PRV by the vaccination-DIVA testing strategy has been performed in the United States and several other countries, PRV remains an important pathogen in pig industry in many countries. Particularly, a PRV variant has been reported in China since 2011, which has caused huge economic losses to the Chinese pig industry, making it more difficult for PRV control and eradication [3–5].

Alphaherpesviruses, including human herpesvirus (HSV) 1 and 2, varicella zoster virus (VZV), bovine herpesvirus 1 (BHV-1), and pseudorabies virus (PRV), use either lytic or latent infection strategy to infect their hosts. The replication processes and mechanisms involved in their lytic infection have been reasonably well understood [1]. But for latent infection of PRV and other alphaherpesviruses, it is mainly known that the typical feature of virus latency is the existence of viral genome for long periods but no virus progeny production in the infected hosts. After stressful experiences, the virus can be re-engaged in lytic infection, a process known as reactivation [1]. These stressors for PRV reactivation include but not limited to concomitant disease, long-distance transport, poor animal husbandry, and treatment with immunosuppressive agents such as dexamethasone. Those mechanisms involved in PRV latency establishment, latency maintenance, and reactivation from latency remain largely undefined. This review will address features of PRV and other alphaherpesviruses latency with a focus on the potential contribution of viral latency transcripts and viral proteins, viral non-coding RNAs, host immune system, and chromatin epigenetics to virus latency-reactivation cycle.

2. The Viral Latency-Associated Transcripts in PRV and Other Alphaherpesviruses Latency

During PRV lytic infection, transcription of the viral lytic genes is temporally ordered and known as gene transcription cascade. Viral lytic genes can be subdivided into three classes of successively expressed transcripts, named immediate-early genes, early genes and late genes, respectively [1]. Among them, *IE180* and *EP0* are important regulatory genes during PRV replication. *IE180* gene (homolog of HSV-1 *ICP4*), the only genuine immediate-early gene of PRV, is located in the IRS and TRS repeats and present in two copies in the genome. *IE180* is essential for viral replication in tissue culture, as it is required for the efficient transcription of viral early and possibly late genes. *EP0* is transcribed with early kinetics and functionally homologous to HSV-1 *ICP0*. It is able to activate several viral genes expression such as *IE180*, *UL23*, and *US4*. Although *EP0* is dispensable for viral replication in cell cultures, the viral titers in vitro and virulence in vivo of EP0-defected PRV are significantly reduced [1]. In contrast, when PRV enters a latent infection state, its genome is primarily retained in neurons of the trigeminal ganglion (TG) [6], expression of viral lytic genes is completely inhibited and transcription is restricted to a small region of the viral genome that is named the latency-associated transcript (LAT) locus [7]. The LAT is located at the strand complementary to the *EP0* and *IE180* genes and overlaps the internal repeat sequence of PRV genome [8–10]. As reported, various sizes of LATs transcribed from the LAT region can be detected during PRV latency in the infected swine TG [6,11]. The 8.4 kb large latency transcript (LLT) is the largest transcript in the LAT locus, which is then spliced into different sizes to yield a 4.6 kb intron [9,12]. Although the LLT transcript has the potential of encoding proteins (exon 1 and 2), there is no evidence of any protein product in PRV latent infection. In the 4.6 kb region, a cluster of 11 miRNA genes is identified by deep sequencing in porcine dendritic cell and in PK-15 cell line infected by PRV, which suggests a role of the 4.6 kb intron as a primary miRNA precursor [13,14] (Figure 1). Additionally, transcription from the LAT locus can occur either in latent or lytic infection, but a different set of transcripts is expressed in the lytic infection [15].

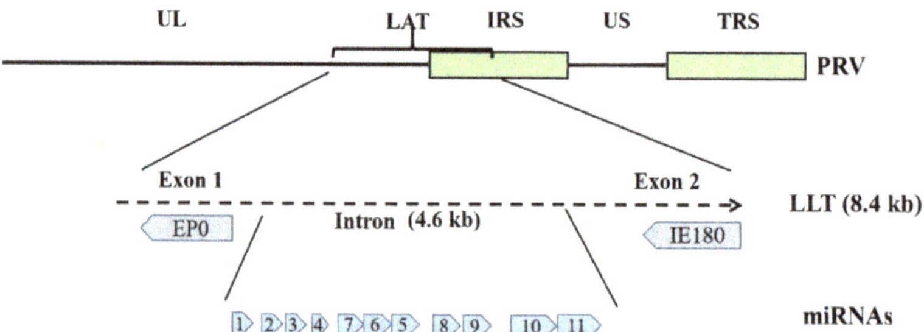

Figure 1. Map of the PRV genome containing unique long (UL), unique short (US), LAT and inverted repeat (IRS and TRS) sequences. The positions of LLT, *EP0*, *IE180*, and miRNAs are annotated, respectively.

In HSV-1, its LAT is also expressed during acute and latent infection, and viruses lacking LAT do not establish latency as efficiently as wild-type HSV-1 and exhibit delayed reactivation [16–18], suggesting that the LAT is involved in HSV-1 latency establishment and facilitates HSV-1 reactivation from latency. In VZV, although a latency-associated transcript that lies antisense to the viral transactivator gene 61 (*ORF61*) predominates during virus latency [19], transcripts from 12 VZV genes (e.g., *ORF63*) are identified in human ganglia removed at autopsy. But it is difficult to ascribe these as transcripts present during latency as virus reactivation may occur in the post-mortem time period in the ganglia [20]. Although the transcription of several LATs and relevant genes in alphaherpesviruses latently infected neurons is evident, their functions in virus latency and reactivation are still not clear.

3. Viral Non-Coding RNAs in PRV and Other Alphaherpesviruses Latency

Most herpesviruses produce various non-coding RNAs such as LAT-encoded miRNAs during virus latency. Several HSV-1 miRNAs have been found to reduce the expression of viral lytic regulatory proteins ICP0 and ICP4, suggesting that the viral miRNAs may facilitate establishment and maintenance of viral latency by post-transcriptional regulation of viral lytic gene expression [21]. As for PRV, previous studies found that the intron region of LLT functioned as a primary miRNA precursor, which encodes a cluster of 11 miRNA genes [13,14]. To assess the importance of this miRNA cluster in establishment of PRV latency in swine, Mahjoub et al. generated a PRV mutant with a 2.5-kb portion deletion in the LLT intron harboring 9 miRNA genes. This mutant displayed almost identical acute infection properties to those of parental PRV in vitro and in vivo. It also successfully establishes latency in vivo, demonstrating that these miRNAs are nonessential for PRV infection and latency establishment in TG. However, massive host gene upregulation is found in TG during miRNA-deleted PRV latency, and several biofunctions including those related to cellular immune response and dendritic cells migration are impaired during this process [12]. These findings supported a function of PRV LAT-encoded miRNAs in the modulation of host response for maintaining a latent state. However, the role of these miRNAs in PRV reactivation from latency has not been explored, hence further studies are needed to determine whether these miRNAs play a role in the process of PRV reactivation.

Similarly, initial study on HSV-1 has found that LAT-encoded miRNAs can be detected in latently infected TGs, and the LAT promoter mutants lacking miRNA expression can still replicate normally and establish wild-type levels of latency, suggesting that these viral miRNAs might be dispensable for HSV latency and reactivation [22]. However, further study focusing on the individual miRNA function has found that single miRNAs might play important roles in HSV-1 latent infection. The miR-H2, one HSV-1 LAT-encoded miRNA molecule, is found antisense to the *ICP0* gene and appears to reduce ICP0 expression [21]. To further examine the importance of miR-H2 in HSV-1 latency-reactivation cycle, researchers used a codon redundancy strategy to construct a miR-H2 disrupted mutant without any changes in the overlapping *ICP0* gene [23]. Compared to its parental virus, ICP0 protein was much more expressed during infection, confirming that miR-H2 downregulated HSV-1 *ICP0* gene expression. Ocular infection in mice showed that miR-H2 mutant increased neurovirulence as judged by mouse survival experiment. Intriguingly, this miR-H2 KO mutant is reactivated significantly earlier than its parental strain in the mouse explant TG reactivation model, indicating that the miR-H2 is involved in reducing HSV-1 neuroinvasion and maintaining HSV-1 latency [23,24].

Besides encoding multiple miRNAs in LAT, two small noncoding RNAs (sncRNAs) are also encoded in HSV-1 LAT region. The two sncRNAs (sncRNA1 and sncRNA2) are 62 and 36 bp in length, respectively, presenting antiapoptotic activity in vitro [25,26]. Furthermore, expression of these sncRNAs by transient transfection assays inhibits cell apoptosis and virus productive infection in vitro [27]. Additionally, these two sncRNAs were able to induce beta interferon promoter activity and cell survival in vitro [28]. All of these results suggest that sncRNAs play important roles in HSV-1 productive infection in vitro, while its role in virus latency and pathogenesis in vivo is not clear. Recently, researchers constructed a recombinant HSV 1 lacking the 62-bp sncRNA1 region in wild-type strain McKrae using dLAT2903 (LAT-minus) virus and compared this mutant virus with its parental virus and McKrae in vitro and in vivo. The replication kinetics of sncRNA1 mutant virus were similar to that of dLAT2903 and McKrae. Meanwhile, ocular virus titers, eye disease, and levels of latency and reactivation were similar in mice infected with either McKrae or the mutant virus. However, absence of sncRNA1 significantly decreased the ocularly infected mice survival and led to virus increased virulence [29]. These suggest that sncRNA1 has a protective function during acute ocular infection, although the presence of sncRNA1 sequence has no significant effect on virus latency or reactivation. Yet it remains to be determined whether PRV encodes the similar sncRNAs in LAT region and how these potential sncRNAs function in virus latency.

4. The Viral Proteins Directly and Indirectly Participated in Alphaherpesviruses Latency

Several viral proteins can play important roles in alphaherpesviruses neuroinvasion. For years, studies upon these proteins have highlighted their potential functions in virus latency and reactivation process directly and indirectly (Table 1). For example, thymidine kinase (*TK*) is an important virulence gene of alphaherpesviruses. The relationship between TK activity and HSV-1 virulence has been characterized in a mouse model. Following corneal inoculation, the *TK* (+) strain produced high central nervous system (CNS) titers and established latency in 78% of surviving mice. In contrast, the *TK* (−) strain did not invade the CNS or establish latency [30]. In another study, the HSV-1 *TK* (−) mutant (F)A305 showed reduced neurovirulence and failed to establish latency in mice, but was able to establish latency in rabbits [31]. Contrary to the above studies, Coen et al. found that *TK* (−) mutants established latent infections in mouse TG, but were severely impaired for acute replication in TG and failed to reactivate from ganglia upon cocultivation with permissive cells [32]. It is indicated that *TK* appears to be necessary for HSV-1 reactivation and unnecessary for latency establishment.

Table 1. The viral proteins participated in alphaherpesviruses latency as described in this study.

Viral Genes	Viral Proteins	Virus	Proposed Function
UL23	Thymidine kinase (TK)	HSV-1	Necessary for HSV-1 reactivation and unnecessary for latency establishment
		HSV-2	Important but nonessential for latency or reactivation
		BHV-1	Probably unnecessary for latency-reactivation cycle
		PRV	Crucial for the latent PRV reactivation
US8	gE	HSV-1/PRV/BHV-1	Important for efficient virus latency establishment and reactivation in neurons due to its capacity for promoting virus neuroinvasion
US9	11K	HSV-1/PRV	Affect virus reactivation due to its critical role in promoting virus anterograde transport in neurons
EP0	EP0	HSV-1/PRV	Crucial for the latency and reactivation
IE180	IE180	PRV	Has important roles in PRV reactivation; Important for swiching from latency to reactivation state
ORF1	ORF1	BHV-1	Might be important for the virus latency and reactivation
ORF2	ORF2	BHV-1	Inhibits apoptosis; interferes with Notch1-mediated transactivation of ICP0; probably inhibit virus productive infection and promote virus latency

To determine the ability of HSV-2 *TK* (−) strain to establish latency and reactivate in vivo, an intranasal or ocular infection model in rabbits was used [33]. The results showed that the *TK* (−) strain can replicate and shed virus in the eye after intranasal or ocular infection, and can cause acute keratitis and establish TG latency. Moreover, 30% of rabbits initially infected in the eyes reactivated following drug-induced immunosuppression with latently infected control rabbits uniformly reactivated. The study shows that the *TK* gene is important but not essential for latency or reactivation in this model [33]. In the relevant studies of BHV-1, a *TK* (−) strain was initially found to be incapable of establishing latency in rabbits following intranasal infection [34]. However, later it was demonstrated that after intranasal infection, latent infection established by a BHV-1 *TK* (−) strain was reactivated by dexamethasone [35]. Similarly, an experiment was conducted to examine the neuroinvasiveness of PRV *TK* (−) mutant in pigs. Results showed that mutations in

TK gene were associated with reduced virus virulence and replication in peripheral target tissues, and also reduced migration to the olfactory and trigeminal pathways [36]. In addition, *TK* defective vaccine strain of PRV established latency in swine TG, however, dexamethasone treatment failed to reactivate the latent *TK*-defected PRV in all ten pigs. In contrast, the PRV *TK* (+) strain can be reactivated from 54 of 55 latently infected pigs following dexamethasone treatments [37]. All above suggests the effectiveness of the dexamethasone treatments experiment for PRV reactivation, and the crucial role of *TK* gene in the latent PRV reactivation process.

Homologous proteins of PRV gE are found in all members of the *alphaherpesvirinae*, suggesting a conserved biological function of this protein. Studies have revealed that homologous proteins of gE among alphaherpesviruses including PRV, HSV-1, and BHV-1 are not required for replication of these viruses in cell culture. In addition, the gE proteins of PRV and HSV-1 can induce cell fusion and promote the spread of virus from cell-to-cell [38]. VZV lacking *gE* also shows significant restriction in cell-to-cell spread and reduced yield of infectious virus production [39]. More importantly, gE is a critical virulence determinant of the alphaherpesviruses. Deletion of *gE* from HSV-1, PRV, and BHV-1 drastically decreases HSV-1 neurovirulence in mice, results in PRV virulence and spread reduction in the nervous system and abolishes BHV-1 virulence in calves, respectively [38]. These findings provide strong evidence that the gE protein plays important roles in cell-to-cell spread, invasiveness, and virulence of alphaherpesviruses. Moreover, gE plays a role in alphaherpesviruses (e.g., PRV, HSV-1, and BHV-1) latency. The *gE*-deleted strain of these viruses can stay latency in TG, while the ability to replicate in all levels of host neurons were significantly reduced, so *gE*-deleted mutants establish latency in neurons less efficiently, and therefore are less likely to be reactivated than wild-type virus [38]. Further studies upon HSV-1 and PRV demonstrated that membrane proteins gE, gI, and US9 promote their anterograde transport. Anterograde transport is a process that the reactivated virus particles return to epithelial tissues by fast axonal transport moving from the neuron cell body to axon termini, which plays a key role in HSV-1 and PRV reactivation from latency. Virus transport in the anterograde direction contains at least two processes: (1) The anterograde transport in axons from cell body to axon termini, (2) virus egress from axon termini and subsequent virus spread into epithelial cell [40]. It has been shown that all *gE-*, *gI-*, and *US9-* mutants of HSV and PRV have substantial defects in anterograde transport. However, gE/gI formed protein complexes and exerted major effects in spread of virus from axon termini to epithelial cell, whereas US9 only participated in axonal events due to its lack of extracellular domain. gE/gI and US9 in both HSV and PRV act primarily in the cytoplasm to promote virus assembly and virus particles sorting into axons, which also supports the observations that HSV *gE-*/*US9-* and PRV *US9-* mutants are transported normally in axons but defected in anterograde transport [40].

In addition, the *EP0* gene of PRV is a homolog of HSV-1 *ICP0*, which has been reported to be involved in the reactivation of HSV-1. Upon deletion of *EP0* in PRV genome, the latent viral DNA level in swine TG is reduced and subsequently the reactivation of virus is also inhibited [1,41]. In a similar study, an *EP0*-deficient mutant was able to establish latency in mouse TG, but latent virus cannot be reactivated in explant reactivation assays [42]. Inactivation of the *EP0* gene inevitably results in a mutation of LLT, which is located on the complementary DNA strand of *EP0*. Therefore, it is uncertain whether the *EP0* or LLT mutation might play a role in regulating the latency and reactivation state, but at least it suggests that the *EP0* gene locus is crucial for PRV latency and reactivation in TG. The IE180 protein is a major regulator of global gene expression of PRV. Research has found that during PRV reactivation, IE180 is readily expressed and might play important roles in PRV reactivation [43]. Furthermore, IE180 forms protein–DNA complex at the transcription start site of LAT, and therefore suppresses the activity of PRV LAT promoter by interacting with a TATA box within the promoter, which might be important for swiching from latency to reactivation state [44].

In addition, a 2.0-kb RNA transcribed from PRV LLT was found to carry polyadenylation signal (AATAAA) during virus lytic infection and therefore could be translated. The RNA contains two potential exons capable of encoding a 20-kDa ORF1 and a 47-kDa ORF2. The putative ORF1 protein presents 44% homology to the apoptosis repressor ARC, 36% homology to a protein kinase C substrate, and 30% homology to a serine/threonine protein kinase. In contrast, the predicted ORF2 protein has no homology among any known proteins [15]. It would be interesting to determine if either putative protein is actually expressed, and plays crucial roles during a productive or latent infection. Unfortunately, in recent years no further advances have been made on the roles of LLT-encoded proteins in PRV latency. However, a series of relevant progress have been obtained in BHV-1 (Table 1). Like other alphaherpesviruses, BHV-1 latency related (LR) gene, a gene similar to PRV LAT, is abundantly expressed in latently infected neurons [45]. Two potential open reading frames (ORF 1 and 2) are located in the BHV-1 LR gene [46]. In exploring the role of ORF1 and ORF2 in virus latency, researchers have constructed a BHV-1 mutant with three stop codons at the N-terminus of ORF2, which resulted in ORF2 expression abolishment and ORF1 expression reduction [47,48]. This mutant failed to reactivate from latency following dexamethasone treatment [49], suggesting that protein expression within the LR gene is important for virus latency and reactivation. Moreover, BHV-1 ORF2 can participate in multiple cellular processes to regulate virus latency-reactivation cycle. For example, abolishment of ORF2 expression in the above BHV-1 mutant induces higher levels of apoptosis in infected neurons, which is considered an important cellular process in regulating the latency-reactivation cycle [50,51]. In addition, Notch family members are membrane-tethered transcription factors that regulate neuronal maintenance, development, differentiation, and development of nearly all non-neuronal cells [52]. Notch1 activates the BHV-1 immediate early transcription unit 1 and *ICP0* early promoters as a cellular transcription factor and enhances BHV-1 productive infection. ORF2 is able to interact with the components (Notch 1 and 3) of the Notch signaling pathway in a yeast two-hybrid screen, and therefore consistently interferes with Notch1–3-mediated transactivation of cellular promoters [53,54]. Furthermore, distinct domains in ORF2 are crucial for interfering with Notch1-mediated transactivation of the *ICP0* early and glycoprotein C promoters. Given the important regulatory role of ICP0 (EP0) in both virus latent and lytic infection cycles, the interference effect of ORF2 might inhibit virus productive infection and promote latency establishment [55]. So far, however, no proteins similar as ORF1 and ORF2 in BHV-1 have been identified in PRV LAT. Hence, future studies could focus on roles of PRV LAT-associated proteins in the regulation of PRV latency-reactivation cycle.

5. Immune Regulation of PRV and Other Alphaherpesviruses Latency

The latency-reactivation cycle of alphaherpesviruses is considered to be tightly regulated by a subtle and an incompletely understood interplay between the virus, the neuron, and the immune system [56]. Moreover, it is becoming increasingly evident that both innate and adaptive immunity play important roles in the initial control of herpesviruses infection and latency establishment [57]. In host innate immunity, type I interferons (IFNs) containing IFN-α and -β-mediated immune responses serve as the front line of host defense against virus infection. Activation of the type I IFN pathway is transduced via Janus kinases (JAKs) and may result in the initiation of inflammatory response and expression of antiviral genes such as IFN-stimulated genes (ISGs). IFNs have been reported to play important roles in limiting alphaherpesviruses replication and spread [58–60]. Furthermore, IFN-α has been demonstrated to drive both HSV-1 and PRV into a latency-like quiescent state in in vitro cultures of porcine TG neurons [61]. IFN-α treatment results in suppression of the immediate-early protein ICP4 of HSV-1 or its counterpart IE180 in PRV, which might be a key step for type I IFNs in promoting the establishment of alphaherpesviruses latency [61]. In addition, a recent investigation highlighted the regulatory role of type I IFNs signaling in the relationship between HSV-1 genomes and the nuclear environment during latency establishment. It showed that without IFN-α treatment, HSV-1 infection led to the formation of replication compartment genome pattern (the lytic infection pattern) in the nucleus of ~91% neurons. However, IFN-α treatment favored the multiple latency-like pattern (the latent infection pattern), with 82% neurons showing this pattern. Moreover, nearly all HSV-1-infected TG neurons harvested from type I IFN receptor KO mice showed the formation of replication compartment pattern irrespective of treatment with IFN-α. Hence, the type I IFNs favors the formation of latency-like genome pattern in HSV-1-infected neurons [62]. Nevertheless, HSV-1 has evolved strategies to inhibit type I IFNs signaling. Its LAT transcript was shown to delay IFN-α and -β expression during acute infection. Further studies found that JAK-1 and -2, as well as several downstream effectors of the JAK pathway, were also downregulated in a LAT-dependent manner during HSV-1 latency [63]. Recently, studies indicated that PRV has developed multiple strategies to antagonize the type I IFNs-mediated innate immunity via several viral proteins, ranging from inhibiting pattern-recognition receptors induced type I IFNs production to antagonizing the IFN signaling pathway and neutralizing the antiviral functions of ISGs [64–68]. This feature might confer PRV the ability of reactivation from latency in neurons.

Peripheral neurons are the main sites for alphaherpesviruses latency, virus latency involves a precise interaction between the virus and neuron immunity. Research shows that sympathetic neurons can be cultured as a pure population with the treatment of nerve growth factor (NGF). Meanwhile, latency can be established in primary sympathetic neurons cultured in presence of NGF. Furthermore, the phosphatidylinositol 3-kinase (PI3-K) pathway triggered by NGF-binding to the TrkA receptor tyrosine kinase (RTK) is crucial in maintaining HSV-1 latency by using a primary neuronal culture system. The PI3-K p110α catalytic subunit is specifically required to activate 3-phosphoinositide-dependent protein kinase-1 (PDK1) and maintain virus latency. Depletion of PDK1 results in HSV-1 reactivation. Thus, the RTK/PI3-K/PDK1-signaling is a critical host element that regulates the HSV-1 latent-lytic switch [69]. In addition, PRV also establishes a reactivatable, quiescent infection in neurons cultured in modified Campenot tri-chambers. Further study shows that several host factors including protein kinase A (PKA) and c-Jun N-terminal kinase (JNK) in cell bodies prevent establishment of quiescent infection and promote productive replication of axonally delivered virus genomes. Treatment with forskolin (a potent adenylate cyclase activator) on cell bodies activates both PKA and JNK and therefore results in virus productive infection. However, virus lytic infection can completely lost when PKA and JNK activities are inhibited [70]. Therefore, PKA and JNK signals significantly affect the switch of PRV latent and lytic infection.

In host adaptive immunity, the $\gamma\delta$ T cells play a protective role in the immunosurveillance against alphaherpesviruses infection. Study upon HSV-1 infection in TCR$\gamma\delta$- or TCR$\alpha\beta$-deficient mice have shown that $\gamma\delta$ T cells limit HSV-1-induced epithelial lesions and protect mice from HSV-induced lethal encephalitis, which is resulted from the $\gamma\delta$ T cell-mediated arrest of virus replication and neurovirulence [71]. Furthermore, $\gamma\delta$ T cells and macrophages are able to infiltrate into the TG in HSV-infected mice, and then several cytokines in inhibiting HSV-1 replication, for example IFN-γ, TNF-α and IL-12, are expressed, suggesting a direct role of $\gamma\delta$ T cells in controlling virus replication [72]. Moreover, several studies indicate that CD4$^+$ and CD8$^+$ T cells might play crucial roles in restricting HSV-1 infection [73,74], and a study using two different pathogenic strains of HSV-1 showed that the high pathogenic strain induced a stronger CD4$^+$ and CD8$^+$ T-cell response in the draining lymph nodes and latently infected TG. It was proposed that greater viral gene expression by the high pathogenic strain during latency might result in a larger T-cell infiltrate in both the cornea and the TG [75]. Further studies found that CD8$^+$ T cells inhibited HSV-1 replication in peripheral ganglia and prevented HSV-1 reactivation without destroying the infected neurons [76,77]. It was also reported that CD8$^+$ T cells might prevent HSV-1 reactivation from latency at least in part through the secretion of IFN-γ [57]. First, IFN-γ mRNA and protein were consistently detected in latently infected ganglia [78]. Subsequently, treatment with IFN-γ could block HSV-1 reactivation from latency on cultures of latently infected TG in the early reactivation process [79]. Lastly, although knockout of IFN-γ and IFN-γR from mice failed to affect HSV-1 replication and latency establishment in TG, the incidence of stress-induced reactivation in mutant mice was significantly higher than in control mice, suggesting the crucial role of IFN-γ in inhibiting HSV-1 reactivation from latency [80].

As described above, ICP0 is a crucial transactivator that is required for efficient HSV-1 reactivation from latency. The cyclin-dependent kinase (cdk) is required for the post-translational modifications necessary for ICP0 transactivating activity [81]. Study has indicated that IFN-γ could induce production of the cdk inhibitors, which might block HSV-1 reactivation from latency probably by inhibiting ICP0 transactivating activity [81,82]. Additionally, expression of viral proteins is not completely silent but sporadically expressed at low levels during HSV-1 latency, which can be then recognized by ganglion-resident HSV-1-specific CD8$^+$ T cells. In a recent study, several HSV recombinants that have different viral promoters driving expression of the immunodominant gB epitope (gB$_{498-505}$) were constructed. They induced equivalent ganglionic CD8$^+$ T-cell responses but contained altered gB-CD8s (CD8$^+$ T cells that recognize gB$_{498-505}$) immunodominance during latency. The results indicate that the selection of epitope promoter could influence CD8$^+$ T-cell population hierarchies and their function. Since CD8$^+$ T cells can influence lytic/latent cycles in reactivating neurons, it is convinced that improving their ganglionic retention and function may offer a strategy in vaccine design to reduce virus reactivation [83]. In addition, CD8$^+$ T cells can be exhausted during viral chronic infection, which is characterized by reduced memory potential, sustained expression of inhibitory receptors, and poor effector function [84,85]. Moreover, successful reactivation of the latently infected HSV-1 from TG was assumed to be partly dependent on the exhausting CD8$^+$ T cells [86,87]. Furthermore, it was found that HSV-specific CD8$^+$ T cells from symptomatic patients were phenotypically and functionally exhausted, and the exhaustion receptors (PD-1, LAG-3, TIGIT, and TIM-3) were up-regulated on HSV-specific CD8$^+$ T cells. Moreover, blocking of LAG-3 and PD-1 synergistically restored anti-viral CD8$^+$ T-cell responses, reduced HSV-1 reactivation from latency, and prevented UV-B irradiation induced recurrent ocular herpetic infection and disease in latently infected mice [88].

The role of CD4$^+$ T cells in regulating HSV-1 latency is another area that needs to be explored. One recent study demonstrated that lack of CD4$^+$ T cells during the initial programming of HSV-specific CD8$^+$ T cells caused transient partial exhaustion of CD8$^+$ T cells characterized by elevated PD-1 levels and hence reduced the ability to control HSV-1 latency in sensory ganglia. This finding demonstrates that CD4$^+$ T cell is important for generation of a memory CD8$^+$ T-cell population, which provides immune surveillance for HSV-1 latency [89]. A recent study also found that HSV-1 infection triggered the activation of regulatory T cells (Treg cells). In an HSV-1 ocular infection model, researchers observed a strong correlation between the level of Treg cells and virus infectivity [90]. Furthermore, depletion of Treg cells largely limited HSV-1 latency establishment. Stress-induced HSV-1 reactivation was tightly dependent on enhanced Treg cell functions, which can suppress the immune surveillance by CD8$^+$ T cells and permit viral replication. Taken together, Treg cell may serve as a key target for controlling HSV latency and reactivation [90].

As for PRV, the relevant study upon immune modulation of its latency-reactivation cycle is obviously lagged behind and mainly accompanied by PRV vaccine research. As is known, modified live vaccines have been successfully used to control PRV infection for many years [91]. However, vaccination with various live attenuated strains did not induce complete inhibition of a latent PRV infection; meanwhile, live vaccine strains are able to establish latency themselves [91,92]. Furthermore, the target neurons cannot be superinfected with the wild-type virus once latently infected by a preceding live vaccine strain [93]. In addition, although the significance of CD4$^+$ T cells in protective immunity against PRV infection was demonstrated in a murine model [94], the potential role of T cell and its secreted cytokines in PRV latency-reactivation cycle still needs to be determined. Taken together, it is suggested that the immune response induced by modified live vaccines might play important roles in controlling wild-type PRV latency.

6. Chromatin Regulation of Alphaherpesviruses Latency

Chromatin epigenetics, the regulation of nucleosomes and chromatin structure on the genome, is an important determinant of cellular transcription, replication, and differentiation. Recently, several studies have found that chromatin plays an important role in the regulation of alphaherpesviruses latent and lytic infection [95] (Figure 2). During HSV-1 latency, the viral DNA persists as an episome assembled in nucleosomal chromatin structure in sensory neurons of peripheral ganglia [96,97]. Additionally, chromatin immunoprecipitation (CHIP) assay shows that HSV lytic gene promoters are complexed with modified histones associated with heterochromatin during latency establishment, and transcription of the LATs promotes the formation of heterochromatin on viral lytic gene promoters [98,99]. Furthermore, the viral genome is enriched for several chromatin modifications, including the histone H3 lysine 9 trimethyl (H3K9me3), H3K9 dimethyl (H3K9me2), and H3 lysine 27 trimethyl (H3K27me3) modifications [98–100], and H3K27me3 is a major form of heterochromatin on the viral genome [99] (Figure 2). During lytic infection, there are no histones present in the virion before infection [101,102]. Once the viral DNA enters the cell nucleus, host cell could assemble chromatin on the naked viral DNA to silence incoming genes, as is observed for transfected DNA [103]. However, there are few nucleosomes on viral genome during virus replication period as demonstrated by nuclease-digestion assays. It is found that viral DNA replicates and accumulates in replication compartments that involves disruption of the host chromatin [104,105]. Moreover, several viral proteins are involved in reducing viral DNA chromatin. VP16, the virion transactivator protein, has been found to recruit the chromatin-modifying co-activators and underrepresentation of histones at immediate-early gene promoters in addition to recruiting transcription factors to immediate-early gene promoters [106]. *VP16* defective mutant also showed increased heterochromatin association with viral lytic promoters [106]. Therefore, VP16 plays an important role in reducing total chromatin levels on immediate-early genes during lytic infection. Meanwhile, ICP0 expression during HSV lytic infection results in an increase in euchromatin formation at the lytic genes and a decrease in total histone in association with immediate-early and early gene promoters [107]. *ICP0* defective mutant shows increased heterochromatin association with viral lytic promoters [108], indicating that ICP0 contributes to both histone removal at HSV lytic genes and modifications on histones assembled on viral DNA during lytic infection.

For that of PRV, Zhang et al. establishes a detection method combined CHIP with qPCR to determine the chromatin status of PRV in host cell during lytic infection. The results show that PRV genome is associated with histone H3 and several segments of the genome are presented as chromation state [109]. In addition, IE180 is able to potentiate the activity of the major late promoter in a reconstituted chromatin assembly system. Function of IE180 requires the simultaneous action of transcription factor IID (TFIID) and results in the formation of stable preinitiation complexes within nucleosome-assembled DNA. Meanwhile, in order to stimulate subsequent transcription, IE180 is required from the onset of nucleosome assembly and it is uncapable of reversing nucleosome-mediated repression once this repression established [110]. These results indicate that IE180 stimulates TFIID binding to promoters and competes with nucleosomes during chromatin reconstitution. However, the potential regulation mechanisms of chromatin in PRV latency are still unclear, which need to be further explored in the future.

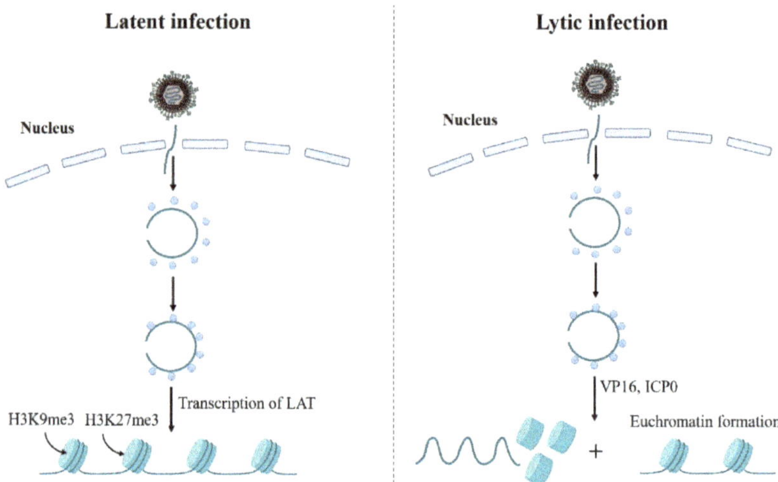

Figure 2. Chromatin can play important roles in regulation of alphaherpesvirus latent and lytic infection. During latent infection of neuronal cells, the virus capsid is transported to the nuclear pore where viral genome is released into the nucleus. When the genome enters the nucleus, it can be rapidly circularized and complexed with histones. Then the LAT is expressed and it promotes the association of heterochromatin marks on the viral genome (**left panel**). During lytic infection of epithelial cells, the virus capsid is also transported to the nuclear pore where the viral genome is released into the nucleus. In the nucleus, viral genome rapidly circularizes and becomes associated with histones. Then VP16 can decrease histone association with viral IE genes and increase euchromatin marks on the remaining histones. ICP0 can be expressed and it promotes the same processes as VP16 on the viral genome (**right panel**).

7. Conclusions and Prospects

Modified live vaccines against PRV have been successfully used for many years in pig industry. Consistent with this, PRV in pigs has been well controlled and eradicated by the use of vaccination in the United States and many European countries. However, the genetically different PRV variants have emerged in China since late 2011, which makes PRV control more difficult [3–5]. Additionally, PRV establishes a lifelong latent infection in peripheral nervous system in pigs, which is similar as other alphaherpesviruses. Vaccination with various live attenuated strains always fail to induce complete inhibition of a latent PRV infection. After stress stimulus, the latently infected PRV is able to reactivate, shed, and spread in susceptible pigs. Therefore, understanding the mechanisms behind PRV latency-reactivation cycle is important for the control and eradication of PRV infection.

Nevertheless, the relevant advances in PRV latency are largely lagged behind compared to other alphaherpesviruses such as HSV-1. An important reason is that PRV latency cannot be stably established in the mouse model due to the severe lethality (~100%) of PRV in mice. Therefore, studies upon PRV latency in vivo have to be conducted in pigs, which make the relevant study more expensive and unmanageable. Hence, to establish more convenient and stable animal model for PRV latency research, specific gene-edited transgenic mice suitable for PRV latency and reactivation might be an important direction to promote the related research in future. Although most studies on PRV latency and reactivation involve animal models, in vitro models have been developed to provide a more convenient approach for understanding the molecular biology of PRV latency. Most of the in vitro models use dissociated neuron cultures to establish a quiescent infection, and the isolated neurons usually need to be pretreated with IFN or replication inhibitors to block the initial productive infection. Recently, Koyuncu et al. used a modified Campenot tri-chambers to physically separate axons from their cell bodies. By culturing primary neurons derived from the superior cervical ganglia of rat embryos in these chambers, they successively established reactivatable quiescent infections by PRV without treatment of any replication inhibitor or cytokine [70,111]. This technique provides an important reference tool for the relevant study of PRV latent infection in vitro.

In addition, referring to the related research progress of HSV-1 and other alphaherpesviruses, researchers can explore the mechanisms of PRV latency and reactivation from aspects of PRV-encoded regulatory miRNA, LAT-encoded unknown regulatory proteins (e.g., the analogues of BHV-1 ORF2), other viral proteins involved in virus latency or reactivation, both innate and adaptive host immune regulation and chromatin epigenetics on viral genome. It is convinced that the continuous discovery of PRV latency mechanisms will contribute to proposing novel therapeutic methods for PRV control in future.

Author Contributions: Conceptualization, C.Y., Y.P. and R.F.; writing—original draft preparation, J.C., G.L. and C.W.; writing—review and editing, Y.L. and L.P. All authors have read and agreed to the published version of the manuscript.

Funding: This research was funded by the National Natural Science Foundation of China (31902256, 32172850), the Fundamental Research Funds for the Central Universities (SWU-KT22016), the Chongqing Science & Technology Commission (cstc2021jscx-lyjsAX0001), China Agriculture Research System of MOF and MARA (CARS-37), Chongqing Pig Industry Technology System (20211105), and the Foundation for Innovation Research Group in Chongqing Universities (CXQT20004).

Institutional Review Board Statement: Not applicable.

Informed Consent Statement: Not applicable.

Data Availability Statement: Not applicable.

Acknowledgments: The authors would like to thank Jinchao Guo for assistance in language editing and helpful suggestions.

References

1. Pomeranz, L.E.; Reynolds, A.E.; Hengartner, C.J. Molecular biology of pseudorabies virus: Impact on neurovirology and veterinary medicine. *Microbiol. Mol. Biol. Rev.* **2005**, *69*, 462–500. [CrossRef] [PubMed]
2. Liu, Q.Y.; Wang, X.J.; Xie, C.H.; Ding, S.F.; Yang, H.N.; Guo, S.B.; Li, J.X.; Qin, L.Z.; Ban, F.G.; Wang, D.F.; et al. A novel human acute encephalitis caused by pseudorabies virus variant strain. *Clin. Infect. Dis.* **2021**, *73*, e3690–e3700. [CrossRef] [PubMed]
3. An, T.Q.; Peng, J.M.; Tian, Z.J.; Zhao, H.Y.; Li, N.; Liu, Y.M.; Chen, J.Z.; Leng, C.L.; Sun, Y.; Chang, D.; et al. Pseudorabies virus variant in Bartha-K61–vaccinated pigs, China, 2012. *Emerg. Infect. Dis.* **2013**, *19*, 1749–1755. [CrossRef] [PubMed]
4. Yu, X.; Zhou, Z.; Hu, D.; Zhang, Q.; Han, T.; Li, X.; Gu, X.; Yuan, L.; Zhang, S.; Wang, B.; et al. Pathogenic pseudorabies virus, China, 2012. *Emerg. Infect. Dis.* **2014**, *20*, 102–104. [CrossRef] [PubMed]
5. Sun, Y.; Luo, Y.Z.; Wang, C.H.; Yuan, J.; Li, N.; Song, K.; Qiu, H.J. Control of swine pseudorabies in China: Opportunities and limitations. *Vet. Microbiol.* **2016**, *183*, 119–124. [CrossRef] [PubMed]
6. Cheung, A.K. Detection of pseudorabies virus transcripts in trigeminal ganglia of latently infected swine. *J. Virol.* **1989**, *63*, 2908–2913. [CrossRef]
7. Jin, L.; Schnitzlein, W.M.; Scherba, G. Identification of the pseudorabies virus promoter required for latency-associated transcript gene expression in the natural host. *J. Virol.* **2000**, *74*, 6333–6338. [CrossRef]
8. Cheung, A.K. DNA nucleotide sequence analysis of the immediate-early gene of pseudorabies virus. *Nucleic. Acids Res.* **1989**, *17*, 4637–4646. [CrossRef]
9. Cheung, A.K. Cloning of the latency gene and the early protein 0 gene of pseudorabies virus. *J. Virol.* **1991**, *65*, 5260–5271. [CrossRef]
10. Maes, R.K.; Sussman, M.D.; Vilnis, A.; Thacker, B.J. Recent developments in latency and recombination of Aujeszky's disease (pseudorabies) virus. *Vet. Microbiol.* **1997**, *55*, 13–27. [CrossRef]
11. Priola, S.A.; Stevens, J.G. The 5′ and 3′ limits of transcription in the pseudorabies virus latency associated transcription unit. *Virology* **1991**, *182*, 852–856. [CrossRef]
12. Mahjoub, N.; Dhorne-Pollet, S.; Fuchs, W.; Endale Ahanda, M.L.; Lange, E.; Klupp, B.; Arya, A.; Loveland, J.E.; Lefevre, F.; Mettenleiter, T.C.; et al. A 2.5-kilobase deletion containing a cluster of nine microRNAs in the latency-associated-transcript locus of the pseudorabies virus affects the host response of porcine trigeminal ganglia during established latency. *J. Virol.* **2015**, *89*, 428–442. [CrossRef] [PubMed]

13. Anselmo, A.; Flori, L.; Jaffrezic, F.; Rutigliano, T.; Cecere, M.; Cortes-Perez, N.; Lefèvre, F.; Rogel-Gaillard, C.; Giuffra, E. Co-expression of host and viral microRNAs in porcine dendritic cells infected by the pseudorabies virus. *PLoS ONE* **2011**, *6*, e17374. [CrossRef]
14. Wu, Y.Q.; Chen, D.J.; He, H.B.; Chen, D.S.; Chen, L.L.; Chen, H.C.; Liu, Z.F. Pseudorabies virus infected porcine epithelial cell line generates a diverse set of host microRNAs and a special cluster of viral microRNAs. *PLoS ONE* **2012**, *7*, e30988. [CrossRef] [PubMed]
15. Jin, L.; Scherba, G. Expression of the pseudorabies virus latency associated transcript gene during productive infection of cultured cells. *J. Virol.* **1999**, *73*, 9781–9788. [CrossRef]
16. Perng, G.C.; Dunkel, E.C.; Geary, P.A.; Slanina, S.M.; Ghiasi, H.; Kaiwar, R.; Nesburn, A.B.; Wechsler, S.L. The latency-associated transcript gene of herpes simplex virus type 1 (HSV-1) is required for efficient in vivo spontaneous reactivation of HSV-1 from latency. *J. Virol.* **1994**, *68*, 8045–8055. [CrossRef]
17. Perng, G.C.; Chokephaibulkit, K.; Thompson, R.L.; Sawtell, N.M.; Slanina, S.M.; Ghiasi, H.; Nesburn, A.B.; Wechsler, S.L. The region of the herpes simplex virus type 1 LAT gene that is colinear with the ICP34.5 gene is not involved in spontaneous reactivation. *J. Virol.* **1996**, *70*, 282–291. [CrossRef]
18. Leib, D.A.; Bogard, C.L.; Kosz-Vnenchak, M.; Hicks, K.A.; Coen, D.M.; Knipe, D.M.; Schaffer, P.A. A deletion mutant of the latency-associated transcript of herpes simplex virus type 1 reactivates from the latent state with reduced frequency. *J. Virol.* **1989**, *63*, 2893–2900. [CrossRef]
19. Depledge, D.P.; Ouwendijk, W.; Sadaoka, T.; Braspenning, S.E.; Mori, Y.; Cohrs, R.J.; Verjans, G.; Breuer, J. A spliced latency-associated VZV transcript maps antisense to the viral transactivator gene 61. *Nat. Commun.* **2018**, *9*, 1167. [CrossRef]
20. Kennedy, P.G.; Rovnak, J.; Badani, H.; Cohrs, R.J. A comparison of herpes simplex virus type 1 and varicella-zoster virus latency and reactivation. *J. Gen. Virol.* **2015**, *96*, 1581–1602. [CrossRef]
21. Umbach, J.L.; Kramer, M.F.; Jurak, I.; Karnowski, H.W.; Coen, D.M.; Cullen, B.R. MicroRNAs expressed by herpes simplex virus 1 during latent infection regulate viral mRNAs. *Nature* **2008**, *454*, 780–783. [CrossRef] [PubMed]
22. Kramer, M.F.; Jurak, I.; Pesola, J.M.; Boissel, S.; Knipe, D.M.; Coen, D.M. Herpes simplex virus 1 microRNAs expressed abundantly during latent infection are not essential for latency in mouse trigeminal ganglia. *Virology* **2011**, *417*, 239–247. [CrossRef] [PubMed]
23. Jiang, X.; Brown, D.; Osorio, N.; Hsiang, C.; Li, L.; Chan, L.; BenMohamed, L.; Wechsler, S.L. A herpes simplex virus type 1 mutant disrupted for microRNA H2 with increased neurovirulence and rate of reactivation. *J. Neurovirol.* **2015**, *21*, 199–209. [CrossRef]
24. Feldman, E.R.; Tibbetts, S.A. Emerging roles of herpesvirus microRNAs during in vivo infection and pathogenesis. *Curr. Pathobiol. Rep.* **2015**, *3*, 209–217. [CrossRef]
25. Peng, W.; Vitvitskaia, O.; Carpenter, D.; Wechsler, S.L.; Jones, C. Identification of two small RNAs within the first 1.5-kb of the herpes simplex virus type 1-encoded latency-associated transcript. *J. Neurovirol.* **2008**, *14*, 41–52. [CrossRef] [PubMed]
26. Tormanen, K.; Wang, S.; Matundan, H.H.; Yu, J.; Jaggi, U.; Ghiasi, H. Herpes simplex virus 1 small noncoding RNAs 1 and 2 activate the herpesvirus entry mediator promoter. *J. Virol.* **2022**, *96*, e0198521. [CrossRef] [PubMed]
27. Shen, W.; Silva, M.; Jaber, T.; Vitvitskaia, O.; Li, S.; Henderson, G.; Jones, C. Two small RNAs encoded within the first 1.5 kilobases of the herpes simplex virus type 1 latency-associated transcript can inhibit productive infection and cooperate to inhibit apoptosis. *J. Virol.* **2009**, *83*, 9131–9139. [CrossRef]
28. da Silva, L.F.; Jones, C. Small non-coding RNAs encoded within the herpes simplex virus type 1 latency associated transcript (LAT) cooperate with the retinoic acid inducible gene I (RIG-I) to induce beta-interferon promoter activity and promote cell survival. *Virus Res.* **2013**, *175*, 101–109. [CrossRef]
29. Tormanen, K.; Matundan, H.H.; Wang, S.; Jaggi, U.; Mott, K.R.; Ghiasi, H. Small noncoding RNA (sncRNA1) within the latency-associated transcript modulates herpes simplex virus 1 virulence and the host immune response during acute but not latent infection. *J. Virol.* **2022**, *96*, e0005422. [CrossRef]
30. Gordon, Y.J.; Gilden, D.M.; Becker, Y. HSV-1 thymidine kinase promotes virulence and latency in the mouse. *Investig. Ophthalmol. Vis. Sci.* **1983**, *24*, 599–602.
31. Meignier, B.; Longnecker, R.; Mavromara-Nazos, P.; Sears, A.E.; Roizman, B. Virulence of and establishment of latency by genetically engineered deletion mutants of herpes simplex virus 1. *Virology* **1988**, *162*, 251–254. [CrossRef]
32. Coen, D.M.; Kosz-Vnenchak, M.; Jacobson, J.G.; Leib, D.A.; Bogard, C.L.; Schaffer, P.A.; Tyler, K.L.; Knipe, D.M. Thymidine kinase-negative herpes simplex virus mutants establish latency in mouse trigeminal ganglia but do not reactivate. *Proc. Natl. Acad. Sci. USA* **1989**, *86*, 4736–4740. [CrossRef] [PubMed]
33. Stroop, W.G.; Banks, M.C.; Qavi, H.; Chodosh, J.; Brown, S.M. A thymidine kinase deficient HSV-2 strain causes acute keratitis and establishes trigeminal ganglionic latency, but poorly reactivates in vivo. *J. Med. Virol.* **1994**, *43*, 297–309. [CrossRef] [PubMed]
34. Brown, G.A.; Field, H.J. Experimental reactivation of bovine herpesvirus 1 (BHV-1) by means of corticosteroids in an intranasal rabbit model. *Arch. Virol.* **1990**, *112*, 81–101. [CrossRef] [PubMed]
35. Whetstone, C.A.; Miller, J.M.; Seal, B.S.; Bello, L.J.; Lawrence, W.C. Latency and reactivation of a thymidine kinase-negative bovine herpesvirus 1 deletion mutant. *Arch. Virol.* **1992**, *122*, 207–214. [CrossRef]
36. Ferrari, M.; Mettenleiter, T.C.; Romanelli, M.G.; Cabassi, E.; Corradi, A.; Dal Mas, N.; Silini, R. A comparative study of pseudorabies virus (PRV) strains with defects in thymidine kinase and glycoprotein genes. *J. Comp. Pathol.* **2000**, *123*, 152–163. [CrossRef]

37. Volz, D.M.; Lager, K.M.; Mengeling, W.L. Latency of a thymidine kinase-negative pseudorabies vaccine-virus detected by the polymerase chain reaction. *Arch. Virol.* **1992**, *122*, 341–348. [CrossRef]
38. Jacobs, L. Glycoprotein E of pseudorabies virus and homologous proteins in other alphaherpesvirinae. *Arch. Virol.* **1994**, *137*, 209–228. [CrossRef]
39. Yang, M.; Card, J.P.; Tirabassi, R.S.; Miselis, R.R.; Enquist, L.W. Retrograde, transneuronal spread of pseudorabies virus in defined neuronal circuitry of the rat brain is facilitated by gE mutations that reduce virulence. *J. Virol.* **1999**, *73*, 4350–4359. [CrossRef]
40. DuRaine, G.; Johnson, D.C. Anterograde transport of α-herpesviruses in neuronal axons. *Virology* **2021**, *559*, 65–73. [CrossRef]
41. Cheung, A.K. Latency characteristics of an EPO and LLT mutant of pseudorabies virus. *J. Vet. Diagn. Investig.* **1996**, *8*, 112–115. [CrossRef] [PubMed]
42. Boldogköi, Z.; Braun, A.; Fodor, I. Replication and virulence of early protein 0 and long latency transcript deficient mutants of the Aujeszky's disease (pseudorabies) virus. *Microbes Infect.* **2000**, *2*, 1321–1328. [CrossRef]
43. Wang, H.H.; Liu, J.; Li, L.T.; Chen, H.C.; Zhang, W.P.; Liu, Z.F. Typical gene expression profile of pseudorabies virus reactivation from latency in swine trigeminal ganglion. *J. Neurovirol.* **2020**, *26*, 687–695. [CrossRef] [PubMed]
44. Ou, C.J.; Wong, M.L.; Huang, C.; Chang, T.J. Suppression of promoter activity of the LAT gene by IE180 of pseudorabies virus. *Virus Genes* **2002**, *25*, 227–239. [CrossRef] [PubMed]
45. Sinani, D.; Liu, Y.; Jones, C. Analysis of a bovine herpesvirus 1 protein encoded by an alternatively spliced latency related (LR) RNA that is abundantly expressed in latently infected neurons. *Virology* **2014**, *464*, 244–252. [CrossRef]
46. Kutish, G.; Mainprize, T.; Rock, D. Characterization of the latency-related transcriptionally active region of the bovine herpesvirus 1 genome. *J. Virol.* **1990**, *64*, 5730–5737. [CrossRef]
47. Meyer, F.; Perez, S.; Jiang, Y.; Zhou, Y.; Henderson, G.; Jones, C. Identification of a novel protein encoded by the latency-related gene of bovine herpesvirus 1. *J. Neurovirol.* **2007**, *13*, 569–578. [CrossRef]
48. Jiang, Y.; Inman, M.; Zhang, Y.; Posadas, N.A.; Jones, C. A mutation in the latency-related gene of bovine herpesvirus 1 inhibits protein expression of a protein from open reading frame 2 and an adjacent reading frame during productive infection. *J. Virol.* **2004**, *78*, 3184–3189. [CrossRef]
49. Inman, M.; Lovato, L.; Doster, A.; Jones, C. A mutation in the latency related gene of bovine herpesvirus 1 interferes with the latency-reactivation cycle of latency in calves. *J. Virol.* **2002**, *76*, 6771–6779. [CrossRef]
50. Lovato, L.; Inman, M.; Henderson, G.; Doster, A.; Jones, C. Infection of cattle with a bovine herpesvirus 1 (BHV-1) strain that contains a mutation in the latency related gene leads to increased apoptosis in trigeminal ganglia during the transition from acute infection to latency. *J. Virol.* **2003**, *77*, 4848–4857. [CrossRef]
51. Shen, W.; Jones, C. Open reading frame 2, encoded by the latency-related gene of bovine herpesvirus 1, has antiapoptotic activity in transiently transfected neuroblastoma cells. *J. Virol.* **2008**, *82*, 10940–10945. [CrossRef] [PubMed]
52. Cornell, R.A.; Eisen, J.S. Notch in the pathway: The roles of Notch signaling in neural crest development. *Semin. Cell Dev. Biol.* **2005**, *16*, 663–672. [CrossRef] [PubMed]
53. Workman, A.; Sinani, D.; Pittayakhajonwut, D.; Jones, C. A protein (ORF2) encoded by the latency related gene of bovine herpesvirus 1 interacts with Notch1 and Notch3. *J. Virol.* **2011**, *85*, 2536–2546. [CrossRef] [PubMed]
54. Liu, Y.; Jones, C. Regulation of Notch-mediated transcription by a bovine herpesvirus 1 encoded protein (ORF2) that is expressed in latently infected sensory neurons. *J. Neurovirol.* **2016**, *22*, 518–528. [CrossRef] [PubMed]
55. Sinani, D.; Jones, C. Localization of sequences in a protein (ORF2) encoded by the latency-related gene of bovine herpesvirus 1 that inhibits apoptosis and interferes with Notch1-mediated trans-activation of the bICP0 promoter. *J. Virol.* **2011**, *85*, 12124–12133. [CrossRef]
56. Decman, V.; Freeman, M.L.; Kinchington, P.R.; Hendricks, R.L. Immune control of HSV-1 latency. *Viral. Immunol.* **2005**, *18*, 466–473. [CrossRef]
57. Khanna, K.M.; Lepisto, A.J.; Decman, V.; Hendricks, R.L. Immune control of herpes simplex virus during latency. *Curr. Opin. Immunol.* **2004**, *16*, 463–469. [CrossRef]
58. Hendricks, R.L.; Weber, P.C.; Taylor, J.L.; Koumbis, A.; Tumpey, T.M.; Glorioso, J.C. Endogenously produced interferon alpha protects mice from herpes simplex virus type 1 corneal disease. *J. Gen. Virol.* **1991**, *72*, 1601–1610. [CrossRef]
59. Mikloska, Z.; Cunningham, A.L. Alpha and gamma interferons inhibit herpes simplex virus type 1 infection and spread in epidermal cells after axonal transmission. *J. Virol.* **2001**, *75*, 11821–11826. [CrossRef]
60. Sainz, B., Jr.; Halford, W.P. Alpha/Beta interferon and gamma interferon synergize to inhibit the replication of herpes simplex virus type 1. *J. Virol.* **2002**, *76*, 11541–11550. [CrossRef]
61. De Regge, N.; Van Opdenbosch, N.; Nauwynck, H.J.; Efstathiou, S.; Favoreel, H.W. Interferon alpha induces establishment of alphaherpesvirus latency in sensory neurons in vitro. *PLoS ONE* **2010**, *5*, e13076. [CrossRef] [PubMed]
62. Maroui, M.A.; Callé, A.; Cohen, C.; Streichenberger, N.; Texier, P.; Takissian, J.; Rousseau, A.; Poccardi, N.; Welsch, J.; Corpet, A.; et al. Latency entry of herpes simplex virus 1 is determined by the interaction of its genome with the nuclear environment. *PLoS Pathog.* **2016**, *12*, e1005834. [CrossRef] [PubMed]
63. Tormanen, K.; Allen, S.; Mott, K.R.; Ghiasi, H. The latency-associated transcript inhibits apoptosis via downregulation of components of the type I interferon pathway during latent herpes simplex virus 1 ocular infection. *J. Virol.* **2019**, *93*, e00103-19. [CrossRef] [PubMed]

64. Zhang, R.; Tang, J. Evasion of I interferon-mediated innate immunity by pseudorabies virus. *Front. Microbiol.* **2021**, *12*, 801257. [CrossRef] [PubMed]
65. Lv, L.; Cao, M.; Bai, J.; Jin, L.; Wang, X.; Gao, Y.; Liu, X.; Jiang, P. PRV-encoded UL13 protein kinase acts as an antagonist of innate immunity by targeting IRF3-signaling pathways. *Vet. Microbiol.* **2020**, *250*, 108860. [CrossRef]
66. Zhang, R.; Chen, S.; Zhang, Y.; Wang, M.; Qin, C.; Yu, C.; Zhang, Y.; Li, Y.; Chen, L.; Zhang, X.; et al. Pseudorabies virus DNA polymerase processivity factor UL42 inhibits type I IFN response by preventing ISGF3-ISRE interaction. *J. Immunol.* **2021**, *207*, 613–625. [CrossRef]
67. Zhang, R.; Xu, A.; Qin, C.; Zhang, Q.; Chen, S.; Lang, Y.; Wang, M.; Li, C.; Feng, W.; Zhang, R.; et al. Pseudorabies virus dUTPase UL50 induces lysosomal degradation of type I interferon receptor 1 and antagonizes the alpha interferon response. *J. Virol.* **2017**, *91*, e01148-17. [CrossRef]
68. Chen, X.; Kong, N.; Xu, J.; Wang, J.; Zhang, M.; Ruan, K.; Li, L.; Zhang, Y.; Zheng, H.; Tong, W.; et al. Pseudorabies virus UL24 antagonizes OASL-mediated antiviral effect. *Virus Res.* **2021**, *295*, 198276. [CrossRef]
69. Camarena, V.; Kobayashi, M.; Kim, J.Y.; Roehm, P.; Perez, R.; Gardner, J.; Wilson, A.C.; Mohr, I.; Chao, M.V. Nature and duration of growth factor signaling through receptor tyrosine kinases regulates HSV-1 latency in neurons. *Cell Host Microbe* **2010**, *8*, 320–330. [CrossRef]
70. Koyuncu, O.O.; MacGibeny, M.A.; Hogue, I.B.; Enquist, L.W. Compartmented neuronal cultures reveal two distinct mechanisms for alpha herpesvirus escape from genome silencing. *PLoS Pathog.* **2017**, *13*, e1006608. [CrossRef]
71. Sciammas, R.; Kodukula, P.; Tang, Q.; Hendricks, R.L.; Bluestone, J.A. T cell receptor-gamma/delta cells protect mice from herpes simplex virus type 1-induced lethal encephalitis. *J. Exp. Med.* **1997**, *185*, 1969–1975. [CrossRef] [PubMed]
72. Kodukula, P.; Liu, T.; Rooijen, N.V.; Jager, M.J.; Hendricks, R.L. Macrophage control of herpes simplex virus type 1 replication in the peripheral nervous system. *J. Immunol.* **1999**, *162*, 2895–2905. [PubMed]
73. Marrack, P.; Kappler, J. The T cell receptor. *Science* **1987**, *238*, 1073–1079. [CrossRef] [PubMed]
74. Simmons, A. H-2-linked genes influence the severity of herpes simplex virus infection of the peripheral nervous system. *J. Exp. Med.* **1989**, *169*, 1503–1507. [CrossRef] [PubMed]
75. Sullivan, B.; Rowe, A.; Hendricks, R. Increased pathology in herpes stromal keratitis results from differences in local, and not systemic, immune responses (VIR5P. 1028). *J. Immunol.* **2014**, *192* (Suppl. 1), 144.11.
76. Liu, T.; Khanna, K.M.; Chen, X.; Fink, D.J.; Hendricks, R.L. CD8+ T cells can block herpes simplex virus type 1 (HSV-1) reactivation from latency in sensory neurons. *J. Exp. Med.* **2000**, *191*, 1459–1466. [CrossRef]
77. Simmons, A.; Tscharke, D.; Speck, P. The role of immune mechanisms in control of herpes simplex virus infection of the peripheral nervous system. *Curr. Top. Microbiol. Immunol.* **1992**, *179*, 31–56.
78. Shimeld, C.; Whiteland, J.L.; Williams, N.A.; Easty, D.L.; Hill, T.J. Cytokine production in the nervous system of mice during acute and latent infection with herpes simplex virus type 1. *J. Gen. Virol.* **1997**, *78*, 3317–3325. [CrossRef]
79. Liu, T.; Khanna, K.M.; Carriere, B.N.; Hendricks, R.L. Gamma interferon can prevent herpes simplex virus type 1 reactivation from latency in sensory neurons. *J. Virol.* **2001**, *75*, 11178–11184. [CrossRef]
80. Cantin, E.; Tanamachi, B.; Openshaw, H. Role for gamma interferon in control of herpes simplex virus type 1 reactivation. *J. Virol.* **1999**, *73*, 3418–3423. [CrossRef]
81. Davido, D.J.; Leib, D.A.; Schaffer, P.A. The cyclin-dependent kinase inhibitor roscovitine inhibits the transactivating activity and alters the posttranslational modification of herpes simplex virus type 1 ICP0. *J. Virol.* **2002**, *76*, 1077–1088. [CrossRef] [PubMed]
82. Mandal, M.; Bandyopadhyay, D.; Goepfert, T.M.; Kumar, R. Interferon induces expression of cyclin-dependent kinase-inhibitors p21WAF1 and p27Kip1 that prevent activation of cyclindependent kinase by CDK-activating kinase (CAK). *Oncogene* **1998**, *16*, 217–225. [CrossRef] [PubMed]
83. Treat, B.R.; Bidula, S.M.; St Leger, A.J.; Hendricks, R.L.; Kinchington, P.R. Herpes simplex virus 1-specific CD8+ T cell priming and latent ganglionic retention are shaped by viral epitope promoter kinetics. *J. Virol.* **2020**, *94*, e01193-19. [CrossRef]
84. Zajac, A.J.; Blattman, J.N.; Murali-Krishna, K.; Sourdive, D.J.; Suresh, M.; Altman, J.D.; Ahmed, R. Viral immune evasion due to persistence of activated T cells without effector function. *J. Exp. Med.* **1998**, *188*, 2205–2213. [CrossRef]
85. Wherry, E.J. T cell exhaustion. *Nat. Immunol.* **2011**, *12*, 492–499. [CrossRef] [PubMed]
86. Hoshino, Y.; Pesnicak, L.; Cohen, J.I.; Straus, S.E. Rates of reactivation of latent herpes simplex virus from mouse trigeminal ganglia ex vivo correlate directly with viral load and inversely with number of infiltrating CD8+ T cells. *J. Virol.* **2007**, *81*, 8157–8164. [CrossRef]
87. Chentoufi, A.A.; Dasgupta, G.; Christensen, N.D.; Hu, J.; Choudhury, Z.S.; Azeem, A.; Jester, J.V.; Nesburn, A.B.; Wechsler, S.L.; BenMohamed, L. A novel HLA (HLA-A*0201) transgenic rabbit model for preclinical evaluation of human CD8+ T cell epitopebased vaccines against ocular herpes. *J. Immunol.* **2010**, *184*, 2561–2571. [CrossRef]
88. Coulon, P.G.; Roy, S.; Prakash, S.; Srivastava, R.; Dhanushkodi, N.; Salazar, S.; Amezquita, C.; Nguyen, L.; Vahed, H.; Nguyen, A.M.; et al. Upregulation of multiple CD8+ T cell exhaustion pathways is associated with recurrent ocular herpes simplex virus type 1 infection. *J. Immunol.* **2020**, *205*, 454–468. [CrossRef]
89. Frank, G.M.; Lepisto, A.J.; Freeman, M.L.; Sheridan, B.S.; Cherpes, T.L.; Hendricks, R.L. Early CD4+ T cell help prevents partial CD8+ T cell exhaustion and promotes maintenance of herpes simplex virus 1 latency. *J. Immunol.* **2010**, *184*, 277–286. [CrossRef]
90. Yu, W.; Geng, S.; Suo, Y.; Wei, X.; Cai, Q.; Wu, B.; Zhou, X.; Shi, Y.; Wang, B. Critical role of regulatory T cells in the latency and stress-induced reactivation of HSV-1. *Cell Rep.* **2018**, *25*, 2379–2389.e3. [CrossRef]

91. Mengeling, W.L.; Lager, K.M.; Volz, D.M.; Brockmeier, S.L. Effect of various vaccination procedures on shedding, latency, and reactivation of attenuated and virulent pseudorabies virus in swine. *Am. J. Vet. Res.* **1992**, *53*, 2164–2173. [PubMed]

92. Vilnis, A.; Sussman, M.D.; Thacker, B.L.; Sennm, M.; Maes, R.K. Vaccine genotype and route of administration affect pseudorabies field virus latency load after challenge. *Vet. Microbiol.* **1998**, *62*, 81–96. [CrossRef]

93. Schang, L.M.; Kutish, G.F.; Osorio, F.A. Correlation between precolonization of trigeminal ganglia by attenuated strains of pseudorabies virus and resistance to wild-type virus latency. *J. Virol.* **1994**, *68*, 8470–8476. [CrossRef] [PubMed]

94. Bianchi, A.T.; Moonen-Leusen, H.W.; van Milligen, F.J.; Savelkoul, H.F.; Zwart, R.J.; Kimman, T.G. A mouse model to study immunity against pseudorabies virus infection: Significance of CD4$^+$ and CD8$^+$ cells in protective immunity. *Vaccine* **1998**, *16*, 1550–1558. [CrossRef]

95. Knipe, D.M.; Lieberman, P.M.; Jung, J.U.; McBride, A.A.; Morris, K.V.; Ott, M.; Margolis, D.; Nieto, A.; Nevels, M.; Parks, R.J.; et al. Snapshots: Chromatin control of viral infection. *Virology* **2013**, *435*, 141–156. [CrossRef] [PubMed]

96. Rock, D.L.; Fraser, N.W. Detection of HSV-1 genome in central nervous system of latently infected mice. *Nature* **1983**, *302*, 523–525. [CrossRef] [PubMed]

97. Deshmane, S.L.; Fraser, N.W. During latency, herpes simplex virus type 1 DNA is associated with nucleosomes in a chromatin structure. *J. Virol.* **1989**, *63*, 943–947. [CrossRef]

98. Wang, O.Y.; Zhou, C.; Johnson, K.E.; Colgrove, R.C.; Coen, D.M.; Knipe, D.M. Herpesviral latency-associated transcript gene promotes assembly of heterochromatin on viral lytic-gene promoters in latent infection. *Proc. Natl. Acad. Sci. USA* **2005**, *102*, 16055–16059. [CrossRef]

99. Cliffe, A.R.; Garber, D.A.; Knipe, D.M. Transcription of the herpes simplex virus latency-associated transcript promotes the formation of facultative heterochromatin on lytic promoters. *J. Virol.* **2009**, *83*, 8182–8190. [CrossRef]

100. Kwiatkowski, D.L.; Thompson, H.W.; Bloom, D.C. The polycomb group protein Bmi1 binds to the herpes simplex virus 1 latent genome and maintains repressive histone marks during latency. *J. Virol.* **2009**, *83*, 8173–8181. [CrossRef]

101. Cohen, G.H.; Ponce de Leon, M.; Diggelmann, H.; Lawrence, W.C.; Vernon, S.K.; Eisenberg, R.J. Structural analysis of the capsid polypeptides of herpes simplex virus types 1 and 2. *J. Virol.* **1980**, *34*, 521–531. [CrossRef] [PubMed]

102. Pignatti, P.F.; Cassai, E. Analysis of herpes simplex virus nucleoprotein complexes extracted from infected cells. *J. Virol.* **1980**, *36*, 816–828. [CrossRef]

103. Cereghini, S.; Yaniv, M. Assembly of transfected DNA into chromatin: Structural changes in the origin-promoter-enhancer region upon replication. *EMBO J.* **1984**, *3*, 1243–1253. [CrossRef] [PubMed]

104. Monier, K.; Armas, J.C.; Etteldorf, S.; Ghazal, P.; Sullivan, K.F. Annexation of the interchromosomal space during viral infection. *Nature Cell Biol.* **2000**, *2*, 661–665. [CrossRef] [PubMed]

105. Simpson-Holley, M.; Baines, J.; Roller, R.; Knipe, D.M. Herpes simplex virus 1 UL31 and UL34 gene products promote the late maturation of viral replication compartments to the nuclear periphery. *J. Virol.* **2004**, *78*, 5591–5600. [CrossRef] [PubMed]

106. Herrera, F.J.; Triezenberg, S.J. VP16-dependent association of chromatin-modifying coactivators and underrepresentation of histones at immediate-early gene promoters during herpes simplex virus infection. *J. Virol.* **2004**, *78*, 9689–9696. [CrossRef] [PubMed]

107. Knipe, D.M.; Cliffe, A. Chromatin control of herpes simplex virus lytic and latent infection. *Nat. Rev. Microbiol.* **2008**, *6*, 211–221. [CrossRef]

108. Cliffe, A.R.; Knipe, D.M. Herpes simplex virus ICP0 promotes both histone removal and acetylation on viral DNA during lytic infection. *J. Virol.* **2008**, *82*, 12030–12038. [CrossRef]

109. Zhang, Y.J.; Han, S.S.; Shi, Q.H.; Shi, A.; Yang, L.; Chang, H.T.; Zhao, J.; Wang, C.Q.; Chen, L. Chromatin state of some genome segment of pseudorabies virus during lytic infection. *Chin. J. Vet. Sci.* **2017**, *37*, 585–591. (In Chinese)

110. Workman, J.L.; Abmayr, S.M.; Cromlish, W.A.; Roeder, R.G. Transcriptional regulation by the immediate early protein of pseudorabies virus during in vitro nucleosome assembly. *Cell* **1988**, *55*, 211–219. [CrossRef]

111. Koyuncu, O.O.; Song, R.; Greco, T.M.; Cristea, I.M.; Enquist, L.W. The number of alphaherpesvirus particles infecting axons and the axonal protein repertoire determines the outcome of neuronal infection. *MBio* **2015**, *6*, e00276-15. [CrossRef] [PubMed]

Review

The Epidemiology and Variation in Pseudorabies Virus: A Continuing Challenge to Pigs and Humans

Qingyun Liu [1,2], Yan Kuang [1,2], Yafei Li [1,2], Huihui Guo [1,2], Chuyue Zhou [1,2], Shibang Guo [1,2], Chen Tan [1,2,3,4], Bin Wu [1,2,3,4], Huanchun Chen [1,2,3,4,*] and Xiangru Wang [1,2,3,4,*]

1 State Key Laboratory of Agricultural Microbiology, College of Veterinary Medicine, Huazhong Agricultural University, Wuhan 430070, China; lqy948987886@163.com (Q.L.); ky19981459286@163.com (Y.K.); lyfovlyf@163.com (Y.L.); guohuihui0216@163.com (H.G.); dyp053224@163.com (C.Z.); shibang_guo@163.com (S.G.); tanchen@mail.hzau.edu.cn (C.T.); wub@mail.hzau.edu.cn (B.W.)
2 Key Laboratory of Preventive Veterinary Medicine in Hubei Province, The Cooperative Innovation Center for Sustainable Pig Production, Wuhan 430070, China
3 Key Laboratory of Development of Veterinary Diagnostic Products, Ministry of Agriculture of the People's Republic of China, Wuhan 430070, China
4 International Research Center for Animal Disease, Ministry of Science and Technology of the People's Republic of China, Wuhan 430070, China
* Correspondence: chenhch@mail.hzau.edu.cn (H.C.); wangxr228@mail.hzau.edu.cn (X.W.)

Abstract: Pseudorabies virus (PRV) can infect most mammals and is well known for causing substantial economic losses in the pig industry. In addition to pigs, PRV infection usually leads to severe itching, central nervous system dysfunction, and 100% mortality in its non-natural hosts. It should be noted that increasing human cases of PRV infection have been reported in China since 2017, and these patients have generally suffered from nervous system damage and even death. Here, we reviewed the current prevalence and variation in PRV worldwide as well as the PRV-caused infections in animals and humans, and briefly summarized the vaccines and diagnostic methods used for pseudorabies control. Most countries, including China, have control programs in place for pseudorabies in domestic pigs, and thus, the disease is on the decline; however, PRV is still globally epizootic and an important pathogen for pigs. In countries where pseudorabies in domestic pigs have already been eliminated, the risk of PRV transmission by infected wild animals should be estimated and prevented. As a member of the alphaherpesviruses, PRV showed protein-coding variation that was relatively higher than that of herpes simplex virus-1 (HSV-1) and varicella-zoster virus (VZV), and its evolution was mainly contributed to by the frequent recombination observed between different genotypes or within the clade. Recombination events have promoted the generation of new variants, such as the variant strains resulting in the outbreak of pseudorabies in pigs in China, 2011. There have been 25 cases of PRV infections in humans reported in China since 2017, and they were considered to be infected by PRV variant strains. Although PRV infections have been sporadically reported in humans, their causal association remains to be determined. This review provided the latest epidemiological information on PRV for the better understanding, prevention, and treatment of pseudorabies.

Keywords: pseudorabies virus; epidemiology; variation; pig; human pseudorabies encephalitis

Citation: Liu, Q.; Kuang, Y.; Li, Y.; Guo, H.; Zhou, C.; Guo, S.; Tan, C.; Wu, B.; Chen, H.; Wang, X. The Epidemiology and Variation in Pseudorabies Virus: A Continuing Challenge to Pigs and Humans. *Viruses* **2022**, *14*, 1463. https://doi.org/10.3390/v14071463

Academic Editors: Yan-Dong Tang and Xiangdong Li

Received: 25 May 2022
Accepted: 29 June 2022
Published: 1 July 2022

Publisher's Note: MDPI stays neutral with regard to jurisdictional claims in published maps and institutional affiliations.

1. Introduction

Pseudorabies virus (PRV), the causative agent for Aujeszky's disease, belongs to the family Herpesviridae, subfamily Alphaherpesvirinae, and genus Varicellovirus [1]. Similar to other members of the Varicellovirus, PRV is neurotropic and can establish latent infection in the peripheral nervous system [2,3]. Pigs are the natural hosts of PRV, showing neurological disorders in newborn piglets and reproductive failure in sows after infection [4]. Worldwide attempts to control PRV infection in pigs have been ongoing for decades by attenuated marker vaccines with virulence-associated gene deletion and respective serological diagnostic tests [5]. However, long-term immune pressure could promote PRV variation for immune escape, creating new challenges for the future prevention and control of pseudorabies. Moreover, PRV infection in humans has been reported recently, and the number of cases has been increasing since 2017, but the causative association and the pathogenic mechanism remain unclear. In addition to its pathogenicity, PRV has been widely studied as an ideal model for investigating herpesviruses' molecular biology and pathogenic mechanism [6]. It has also been utilized as a living tracer in neural circuits and a promising oncolytic virus [7]. Therefore, it is of significant importance to understand the current clinical prevalence and variation in PRV for the better understanding, prevention, and control of pseudorabies and the appropriate application of PRV.

2. Epidemiology of PRV

2.1. The Prevalence of PRV in the World

PRV infection was first defined as "mad itch" in bovines in America in 1813 [8], and PRV was successfully isolated about 100 years later [9]. With the global development of the pig industry, pseudorabies caused by PRV firstly broke out in pigs worldwide during the 1970s–1980s and was a pandemic for decades (Figure 1A). Currently, PRV is mainly circulating in domestic pigs in Argentina, Bosnia and Herzegovina, China, Croatia, Cuba, France, Hungary, Italy, Mexico, Papua New Guinea, Poland, Portugal, Spain, and the United States of America, according to OIE reports from 2019 to 2021 [10]. Due to efficient vaccination and eradication measures, pseudorabies in domestic pigs has been eliminated in Germany, the United Kingdom, Ireland, South Korea, Sweden, Colombia, Denmark, New Zealand, and many other countries. However, it is difficult to maintain the elimination status, as indicated by second outbreaks of pseudorabies in Argentina in 2019 and France and Mexico in 2020 [10].

In the countries or districts in which pseudorabies has eliminated in domestic pigs, virus transmission from infected wild boars is a critical threat for these domestic pigs. Therefore, serological investigations in wild boars have been conducted in many countries to monitor the transmission risk (Figure 1B). In Italy, the PRV prevalence in wild boars varied from 4% to 30% because of the different densities of wild boar populations, and 30.39% of 1425 sera samples collected from wild boars between 2011 and 2015 in northwest Italy were positive for PRV antibodies [11]. An overall nationwide PRV seroprevalence of 12.09% was detected from 108,748 sera samples from wild boars in Germany from 2010 to 2015 [12]. The PRV seroprevalence rate of wild boars in Switzerland is the lowest among those recorded in Europe, with samples collected between 2008 and 2013 having a seroprevalence of 0.57% [13]. In the United States, 8498 sera samples were collected from wild boars in 35 states from 2009 to 2012, among which the samples from 25 states had a total positivity rate of 18% [14]. The above data indicate a high prevalence of PRV in wild boars and a risk of transmission to domestic pigs. Therefore, routine measures, including fencing and disinfection, should be taken in the epizootic areas with pseudorabies to prevent direct transmission from contact between wild boars and domestic pigs or indirect transmission mediated by people and hunting tools. Moreover, it has been proposed to reduce PRV prevalence in wild boars by controlling the density of wild boars [11] and the reactivation and spillover of latent PRV. In summary, for the pig farms in most countries, it is essential to ensure a sufficient biosafety distance between domestic pigs and wild boars and to ensure appropriate control of pseudorabies prevalence in wild boars.

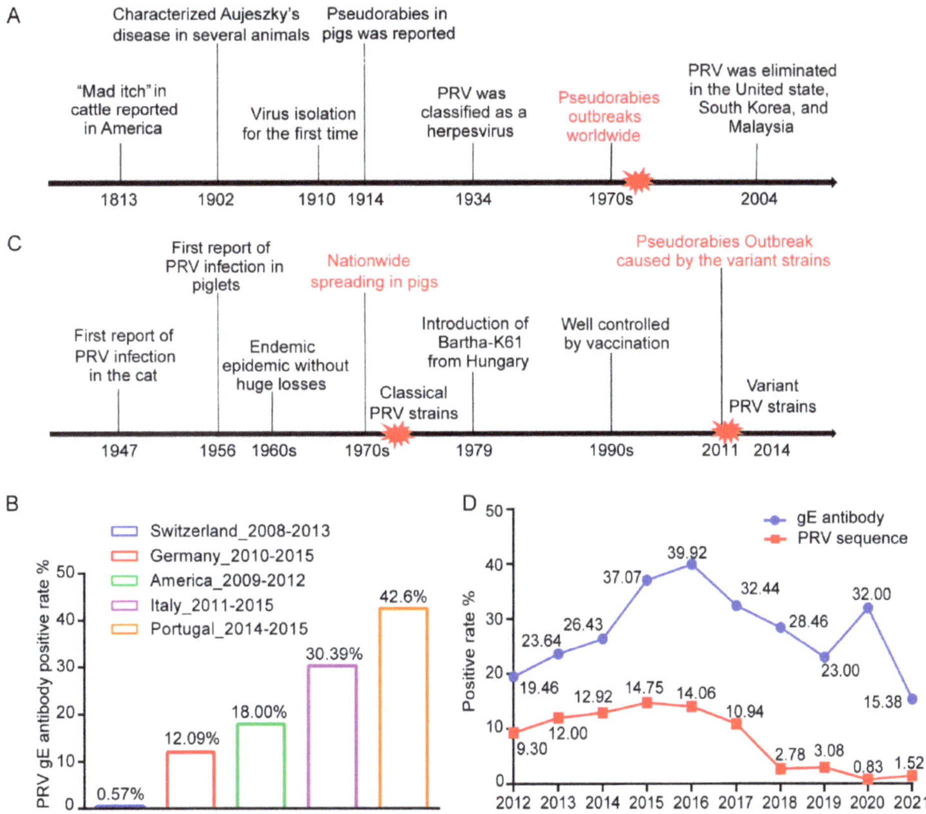

Figure 1. The prevalence of pseudorabies worldwide: (**A**) epidemic history of PRV worldwide. The red explosion shape represents outbreaks of pseudorabies. (**B**) The reported surveillance of PRV infection in wild boars, as illustrated by PRV gE antibody positive rate. (**C**) Epidemic history of PRV in China. The red explosion shape represents outbreaks of pseudorabies. (**D**) The positivity rate of PRV gE antibody and PRV *gE* nucleotide sequences detected in nationwide samples in China from 2012 to 2019.

2.2. The Prevalence of PRV in China

Pig farms in China have suffered from large-scale outbreaks of pseudorabies since the 1970s (Figure 1C). The natural attenuated vaccine strain Bartha-K61 was imported in 1979, and several attenuated strains developed from local classical PRV strains such as Ea and Fa were also utilized to control the pandemic, leading to a remarkably reduced prevalence after 1990. However, by the end of 2011, another PRV outbreak occurred from variant PRV strains, even in the pig farms with routine immunization. Since then, the PRV prevalence rate in China has raised sharply and remains high in some provinces (Table 1).

PRV *gE* sequences and antibodies in samples collected nationwide from 2012 to 2021 were detected to monitor the prevalence of pseudorabies in pigs in China. Since the commercial PRV vaccines are all strains with *gE* gene deletion, the gE antibody is considered to be an indicator of infection caused by wild strains. Additionally, the detection of the *gE* sequence indicates the presence of the virus in pigs. gE antibody prevalence has increased rapidly since the occurrence of variant strains in 2011, and it peaked at 39.92% (3733/9350) in 2016, when the positive rate of *gE* specific sequence was as high as 14.06% (399/2837). Subsequently, the PRV gE antibody and sequence positivity rate gradually decreased to 15.38% (5971/38,821) and 1.52% (53/3503) in 2021, respectively, probably attributed to by the development and application of vaccines based on the variant strains (Figure 1D). The updated variant vaccine strains and the decreased prevalence supported the importance of the high genomic identity between vaccine strains and field strains. However, several provinces in China still show a serious epidemic situation of pseudorabies, with gE antibody prevalence varying from 7.50% to 62.74% (Table 1). Although pseudorabies in domestic pigs in China is currently under control, it is necessary to monitor the variation in PRV strains and to accelerate the current elimination programs.

Table 1. The gE antibody positivity rate in different provinces in China.

Region	gE Positive Rate (gE Positive Samples/Total Samples)				Reference
	2016	2017	2018	2019	
Beijing	33.66% (662/1966)	/	20% (4/20)	/	[15,16]
Chongqing	1.6% (11/702)	9.4% (60/637)	7.5% (60/798)	11.5% (53/460)	[17]
Fujian	37.37% (111/297)	26.11% (53/203)	27.32% (50/183)	/	[18]
Guizhou	1.89% (27/1480)	16.85% (538/3192)	16.85% (538/3192)	8.5% (92/1078)	[19]
Guangdong	/	/	33.60% (1084/3226)	/	[16]
Guangxi	22.87% (854/3734)	23.71% (996/4200)	20.60% (766/3718)	/	[16,20]
Henan	26.21% (3513/13,404)	28.82% (4755/16,497)	25.31% (3000/11,854)	26.69% (3460/12,963)	[21]
Hebei	/	/	62.74% (367/585)	50.05% (5245/10,479)	[16,22]
Heilongjiang	15.36% (474/3086)	15.50% (539/3478)	11.64% (318/2731)	/	[23]
Hubei	/	/	13.21% (123/931)	/	[16]
Hunan	24.4% (344/1410)	23.2% (349/1504)	44.64% (1011/2265)	/	[24]
Jiangxi	40.1% (362/902)	34.6% (318/919)	27.41% (1769/6455)	/	[16,25]
Qinghai	28.17% (131/465)	19.75% (157/794)	/	/	[26]
Shandong	57.8% (2909/5033)	50.4% (2476/4915)	55.2% (2072/3753)	/	[27]
Sichuan	/	/	32.49% (952/2930)	/	[16]
Yunnan	/	/	17.07% (306/1793)	/	[16]
Tianjin	40.43% (970/2399)	37.02% (2219/3793)	51.59% (1957/3793)	/	[28]

/ Data not provided in the reference.

3. Genotyping and Variation in PRV

3.1. Genotyping of PRV

Different PRV strains differ in biological characteristics even though they are in one serotype. The restriction fragment length pattern (RFLP) was used in PRV genotyping [29], especially in RFLP based on *BamHI*. *BamHI*-RFLP divides the PRV strains into genotypes I–IV [30–32]. *BamHI*-mPCR is a method that combines *BamHI*-RFLP with the highly sensitive multiplex PCR. It can be applied to PRV genotyping in samples with a low DNA content without virus isolation [33]. Genotyping based on the *gC* gene and genomes has been increasingly applied in the development of sequencing technology. The *gC* gene is one of the most variable regions in the genome [34]. Based on the phylogenetic analysis of 729 global *gC* sequences, PRV can be divided into two genotypes with Chinese isolates in genotype II and with isolates from other places in genotype I. The most recent common ancestor of the two genotypes was divided into two genotypes and evolved separately around A.D. 1013 [35]. PRV strains in genotype I can be divided into six subtypes, and subtype 1.6 includes Chinese isolates that are closely related to Bartha-k61 [35]. Genotype II can be divided into two subtypes. Subtype 2.1 contains Chinese classical strains isolated in the 1990s, and subtype 2.2 mainly consists of the variant strains isolated after 2011 [36]. In addition, tandem short sequence repeats (SSRs), a class of nucleic acids motifs, might be another molecular basis for PRV genotyping in future studies. SSRs exist in almost 20% of the PRV genome. The changes in length in the SSRs have been associated with DNA binding site efficiency, transcription regulation, and protein interactions [37]. Therefore, the differences in SSR length between strains might explain the differences in the biological characteristics of different PRV strains in the same serotype.

3.2. The Evolution of PRV Based on Natural Mutation-Selection

Alphaherpesvirinae genomes are relatively stable with minor variation in the sequences among strains. The average rate of protein-coding variation in PRV was 1.6%, which is higher than the 1.3% of herpes simplex virus-1 (HSV-1) and the 0.2% of varicella-zoster virus (VZV) [37]. The mean substitution rate of the PRV genome is 4.82×10^{-5} substitutions per site annually [35]. Furthermore, Bayesian skyline coalescent reconstruction illustrated that the relative genetic diversity of genotype I remained unchanged, while in genotype II, the diversity decreased from 2004 to 2010 and increased sharply from 2010 to mid-2012 and was maintained at a high level in 2016 [36]. The time points of the diversity changes in genotype II are consistent with those of pseudorabies control and epidemic in China.

Natural mutation-selection could contribute to the diversity changes in the PRV strains in genotype II. Positive selection has been detected in the amino acid residues at site 43/75/505/834/848/908/922 of gB, site 348/575/578 of gE, and site 59/75/194 of gC but not in gD [36,38,39]. In addition, site 929/934 of gB, site 495/540 of gE, and site 59/75/76/191 of gC are involved in the adaptive evolution after cross-species transmission. The amino acid residue at site 59 of gC participates in positive selection and adaptive evolution, the function of which is related to the viral adsorption process [36]. The variation in the gB, gE, and gC proteins in Chinese variants of PRV may facilitate escaping from the host immune response and adapting to the new host after cross-species transmission.

The genetic diversity supported by SSRs might also promote PRV evolution. SSRs have been found in all herpesviruses. Their length varies in different strains of PRV and HSV-1, and a few SSRs diversity can be detected, even during the PRV plaque purification. In the SSR analysis of Kaplan, Becker, Bartha, and other strains, it was observed that SSRs existed in both coding and non-coding sequences, promoters, and open intergenic sequences, mainly in the IR-US-TR region. Furthermore, 62% of the SSRs in PRV, including most of the SSRs in the coding region, contain triplet-based repeats, such as 3-mer, 9-mer, 27-mer, etc. These triplet-based SSRs not only contribute to genetic diversity but also remain the original frame of the coding sequence [37]. These subtle changes, such as changes in SNPs and SSR length, support genetic diversity and promote PRV evolution.

3.3. Frequent Recombination between PRV Strains Significantly Contributes to Virus Evolution

The frequent inter- and intra-genotype recombination of PRV has been reported (Table 2). Recombination between the field strains is important for PRV evolution since alphaherpesviruses have DNA polymerases with high proof-reading activity and exonuclease activity [40]. There was a high recombination rate in vivo after co-inoculating different PRV strains in sheep and pigs [41,42]. In another report, a South Korean isolate (Yangsan) was located between genotype I and genotype II in the phylogenetic tree base on UL21 when located in genotype II in the phylogenetic trees based on US2, gD, and US9, which suggested recombination between genotypes I and II in UL21 [35]. Similarly, inter-clade recombination between genotypes I and II was detected in gB of PRV FJ-W2, FJ-ZXF, and FJ62 [38,43]. There was a recombination analysis of 29 full-length genomes, and more than four of the seven methods showed that almost all of the PRV strains demonstrated recombination. It was suggested that intra-clade recombination was more frequent than inter-clade recombination. Moreover, Chinese variant strains such as HeN1 and Qihe547 may have originated from the recombination between the isolates in genotype I and the vaccine isolates in genotype II (such as Ea and Fa) [36].

In addition, recombination between the field isolates and vaccine strain Bartha-K61 has been frequently detected. JSY13, which was isolated in Jiangsu in 2018, has been found to be a natural recombinant strain between Bartha-K61 in genotype I and JSY7 in genotype II. The recombination involves the genes *UL42, UL19, UL18,* and *UL10* [44]. Moreover, the earlier isolates in genotype II such as SC and LA may have originated from recombination between the foreign isolates (such as Bartha-K61) and the early epidemic PRV in China [36]. Consistently, it has been reported that SC is a recombinant strain between the Chinese early local PRV isolate and the vaccine strain Bartha-K61 [45]. In our recombination analysis based on 55 PRV genomes, a total of 23 recombination events were identified, with 16 events observed between Bartha-K61 and the Chinese strains [39]. The vaccine strain Bartha-K61 has been widely used to control porcine pseudorabies in China for decades. These recombination events, especially those between vaccine strains and field strains, suggest that long-term immunity has dramatically contributed to the variation and evolution of PRV, which may explain the pseudorabies variant outbreak in China in late 2011.

Table 2. The reported recombination events of PRV.

Strain	Isolation Country	Recombination Pattern	Recombination Site	Reference
Yangsan	South Korean	genotype I and genotype II	*UL21*	[35]
FJ-W2, FJ-ZXF	Fujian, China	genotype I and genotype II	*gB*	[38]
FJ62	Sichuan, China	genotype I (Wild boar) and genotype II	*gB*	[43]
JSY13	Jiangsu, China	genotype I (Bartha) and genotype II (JSY7)	*UL42, UL19, UL18, UL10*	[44]
SC	China	genotype I (Bartha) and genotype II	*gC*	[45]
HeN1, Qihe547	China	genotype I and genotype II (vaccine strains)	/	
SC, LA	China	genotype I and genotype II (early strains)	/	[36]
ZJ01	China	genotype I and genotype II	/	

/ Data not provided in the reference. The gene names were shown in italics.

4. PRV Infections in Animals and Potentially in Humans

4.1. PRV Infections in Pigs and Other Animals

Pigs are the natural host and reservoir of PRV. Infected newborn piglets can show neurological symptoms on the second day after birth, including screaming, ataxia, opisthotonos, and padding, and mortality can be as high as 100%. In contrast, infected fattening pigs generally show temporary temperature elevation, respiratory symptoms, and low mortality with occasional neurological symptoms. Moreover, PRV infection causes severe reproductive disorders, including orchiditis and epididymitis in boars and pregnancy failure in sows [46,47]. Additionally, PRV can establish latency in the peripheral nervous system of the tolerated pigs after infection. Latent infection is characterized without virus replication and clinical symptoms. After latency, virus reactivation can be triggered by certain factors that interfere with host immunity [6], resulting in virus spillover and disease outbreaks. Thus, latency in pigs is a major risk and an obstacle in the late stage of PRV elimination. Future research on the virus latency is critical for establishing PRV-free domestic pig herds.

PRV is also infectious to many other mammals, including ruminants, carnivores, and rodents, and is characterized by severe itching and central nervous system (CNS) dysfunction with 100% mortality [48]. PRV infections in non-natural hosts are generally experimental infections or natural infections likely associated with pigs. Natural infections in farmed cattle have been reported worldwide and are related to contact with infected pigs [49,50]. Infected cattle show mad itch, epilepsy, and paralysis [51]. In 2018, nine cattle were infected by the Chinese variant PRV strain SDLY-China-2018. The infected cattle were raised very close to the pigs positive for the gE antibody, suggesting possible virus transmission from pigs to cattle [49]. In addition, an outbreak of pseudorabies was reported in a flock of 160 ewes housed next to PRV-infected pigs under virus spillover, and 5 cats on this farm were also infected by PRV [52]. Moreover, it has been observed that PRV cannot be horizontally transmitted between infected sheep and healthy sheep [53].

In companion animals, cats and dogs can be infected by PRV through contact with infected pigs, and hunting dogs are more susceptible due to frequent contact with wild animals [54,55]. These dogs died shortly after showing neurological symptoms [56].

PRV infections in wild animals have also been widely reported, including in wild boars [57], foxes [58], wolves [59], brown bears [60], black bears [61], Florida cheetahs [62], lynx [63], and raccoons [64]. In the experimental infection of raccoons and pigs, PRV transmission did not occur between raccoons but did occur between raccoons and swine via contact or predation [64], which was similar to the transmission pattern between sheep and pigs. In 2014, a mink farm in northern China suffered from PRV infection due to feeding raw pork contaminated PRV, resulting in diarrhea, neurologic signs, and 80–90% mortality [65].

According to the above virus transmission patterns of pig–cattle, pig–sheep, pig–cat, pig-dog, and wild boar–hunting dog, it is likely that pigs are the core reservoir for the PRV cross-species transmission. However, PRV infection in non-reservoir animals is different from that in pigs. Under natural infection conditions, these non-natural hosts develop itching, severe neurological symptoms, and even death, while there is no latent infection.

4.2. Potiential PRV Infections in Humans

It is controversial whether PRV can infect humans for the past one hundred years. No PRV-specific neutralizing antibody has been detected in 455 individuals with suspicious symptoms or occupations that put them at risk for infection, and no symptoms have been observed in volunteers injected with PRV at doses of $10^{3.4}$ TCID$_{50}$ (intradermal) or $10^{6.1}$ TCID$_{50}$ (subcutaneous) [66]. These results indicate that humans are not susceptible to PRV infection, or at least not to the PRV strain used. However, in 1914, two laboratory workers had their hands injured during contact with a PRV-infected cat and developed itching and swelling of the wound. As a result, PRV infection was suspected [66]. Moreover, three human cases have been reported in Europe, showing positive responses to PRV-specific neutralizing antibodies and neurological symptoms such as dysphagia, paresthesia, and tinnitus [67]. From 1914 to 1992, there were 17 reported cases of suspected PRV infection, and these patients developed pruritus, weakness, and pain (Table 3).

Table 3. Suspected case reports of human infection with PRV between 1914 and 1992.

Case	Year	Occupation	Contact History	Clinical Symptoms	Antibody Detection	Pathogen Detection	Outcome	Reference
1	1914	Lab technician	A laboratory cat with pseudorabies	Swelling, reddening, and intense itching of the wound and the surrounding area	/	/	Survived	[66]
2	1914	Lab technician			/	/	Survived	
3	1940	Lab technician	Got injured during contacting with a dog infected with PRV	Pruritus, erythema, pain, and aphthous stomatitis	/	/	Survived	[68]
4	1940	Lab technician			/	/	Survived	
5	1963	Animal handler	A dog infected with PRV following an outbreak of pseudorabies on a pig farm	Severe throat pain and weakness in the legs	/	/	Survived	[66]
6	1963	Animal handler			/	/	Survived	
7	1963	Veterinary			/	/	Survived	
8	1963	Nightwatchman			/	/	Survived	
9	1983	Tourist in Denmark	Indirect contact with a sick cat	Anorexia, weight loss, headache, arthralgia	Neutralizing antibody Titer: 1:8–1:16	/	Survived	[67]
10	1986	Tourist in France	Close contact with cats and other domestic animals	Dysphagia, experienced strange smells and taste		/		
11	1986	Tourist in France				/		
12–17	1992	Six workers on a cattle farm	Direct contact with PRV infected cattle	Pruritus of the palms that spread onto the arms and shoulders and lasted for several days	/	/	Survived	[69]

/ Data not provided in the reference.

Since 2017, 25 more human cases of PRV infection have been reported. These cases were diagnosed by detecting PRV-specific antibodies with enzyme-linked immunosorbent assay (ELISA) and PRV nucleotides with PCR or metagenomic next-generation sequencing (Table 4). Notably, the PRV strain hSD-1/2019 was isolated from the cerebrospinal fluid sample of one patient, providing direct etiological evidence for PRV infection in humans [70]. Among the 25 cases, 100% of patients showed high fever and neurological symptoms; 56% showed severe visual impairment, including acute retinal necrosis, vitreous opacity, and blindness; and 16% of the patients died. In addition, 95% of the survivors suffered from severe sequelae, including visual impairment, vegetativeness, cognitive impairment, and memory loss. The CNS dysfunction related to PRV infection in these human cases has been defined as pseudorabies encephalitis (PRE) [71].

All 25 of these patients had a contact history with pigs or pork, indicating the importance of the infected pigs in human infection with PRV. However, more evidence is needed to support viral transmission from pigs to humans. Currently, it is believed that there are no reported cases of human–human transmission since the contacts of the patients have remained healthy. According to our investigation, the gB antibody-positive rates were 40.91% and 45.95% in the contacts of the two patients, while the gE antibodies were all negative [72]. Moreover, a retrospective investigation of 1335 serum samples from patients with encephalitis in 2012, 2013, and 2017 showed gB antibody positivity rates of 12.16%, 14.25%, and 6.52%, respectively [73]. Therefore, the positivity rates of the gB antibody in the associated populations were unable to be ignored, and gE antibody seroconversion could be an essential basis for diagnosis.

Table 4. Case reports of human infection with PRV between 2017 and 2021.

Case	Year	Occupation	Contact History	Clinical Symptoms	Antibody Detection	Nucleotide	Outcome	Reference
1	2017	Swineherder	Sewage spilled into eyes	Fever, headache, visually impaired, endophthalmitis	gB antibody	+	Survived	[74]
2	2017	Pork dealer	Cut hand by a meat cleaver	Fever, headache, consciousness disorders, seizures, retinitis, encephalitis	PRV antibody-positive in three patients	+	Survived	[71]
3	2017	Cook	/	Fever, headache, seizures, consciousness disorders		+	Died	
4	2017	Pig butcher	/	Fever, headache, seizures, consciousness disorders		+	Survived	
5	2018	Pig butcher	/	Fever, seizures, consciousness disorders, retinitis		+	Survived	
6	2018	Veterinary	Hands were punctured by a knife used for the autopsy of dead swine	Fever, headache, seizures, respiratory failure, disturbance of consciousness, encephalitis	gB antibody gE antibody	+	Survived	[75]
7	2018	Swineherder	Needlestick injury	Fever, seizures, consciousness disorders, encephalitis	neutralizing antibody	+	Survived	[76]
8	2018	Pig butcher	Finger hurt by a pig	Fever, headache, visual disturbances, convulsions		+	Survived	[77]
9	2018	Pig butcher	Hand injury before hospitalization	Fever, memory loss, consciousness disorders, convulsions, respiratory failure	/	+	Survived	
10	2018	Swineherder	Hand injury before hospitalization	Fever, extremity tremors, respiratory failure, vision loss		+	Survived	
11	2018	Porker cutter	Hand injury at work	Fever, convulsions, respiratory failure		+	Survived	
12	2018	Porker cutter	No injury	Fever, extremity tremors, respiratory failure, vision loss		+	Survived	
13	2011	Pork dealer	/	Fever, psychotic behavior, seizures			Died	[78]
14	2018	Pig butcher	/	Fever, seizures, consciousness loss, retinal necrosis		+	Died	
15	2018	Swineherder	/	Fever, seizures, cognitive decline, respiratory failure, blindness		+	Survived	
16	2018	Driver	/	Fever, seizures, consciousness loss		+	Survived	
17	2019	Pork dealer	Contact with pork with injured fingers	Fever, seizures, consciousness disorder, encephalitis	PRV antibody positive	+	Survived	[79]

Table 4. *Cont.*

Case	Year	Occupation	Contact History	Clinical Symptoms	Antibody Detection	Nucleotide	Outcome	Reference
18	2018	Veterinary	/	Fever, headache, memory loss, seizures, consciousness disorders	gB antibody gE antibody Neutralizing antibody	+	Survived	
19	2019	Pig butcher	Hand injury	Fever, headache, respiratory failure, memory loss, seizures, consciousness disorders		+	Survived	[70]
20	2019	Pig butcher	Finger injury	Fever, headache, respiratory failure, memory loss, seizures, consciousness disorders		+	Survived	
21	2019	Pig butcher	/	Fever, headache, consciousness loss, seizures, bilateral retinal detachment, encephalitis	/	+	Survived	[80]
22	2020	Swineherder	/	Fever, coma, endophthalmitis	/	+	Survived	[81]
23	2021	Housewife	/	Fever, headache, seizures, coma, respiratory failure	/	+	Survived	[82]
24	2021	Swineherder	/		/	+	Died	
25	2021	Pig butcher	Hand injury at work	Fever, consciousness loss, seizures, respiratory failure	/	+	discharged with ventilator support	[83]

/ Data not provided in the reference. + Nucleotide sequences were detected positive in the cases.

Comparing the cases listed in Tables 3 and 4, it seemed that the infectivity and infection characteristics of PRV in humans have significantly changed. The cases reported between 1914 and 1992 were diagnosed by clinical symptoms and contact history. The patients had contact with infected cats, dogs, or cattle and showed cold-like symptoms such as fever, sore throat, limb weakness, and itching in most cases (Table 3). With the development of detection technology, the diagnostic basis has become more detailed. The cases reported after 2017 were diagnosed by PRV-specific antibodies and nucleic acid. Infected pigs and contaminated pork were the common contact history. The patients generally started with influenza-like symptoms that quickly developed into neurological symptoms within five days, with some even dying or experiencing disability at the end of the disease (Table 4). Thence, the possible virus source in these cases and the virulence of the PRV strains might have changed. The PRV strains resulting in infection in patients have been reported to be phylogenetically closer to the PRV variant strains currently circulating in Chinese pig populations [70,74,76]. The variant strains isolated in China after 2012 have been sequenced and found to be quite different from foreign strains and Chinese classical strains such as Ea, Fa, LA, and SC. Based on genome sequencing of the variant stains TJ, HNX, and ZJ01 and a comparative analysis with the classical strains, VP1/2 (UL36), ICP22 (US1), and ICP4 (IE180) are the most variable proteins, and gE (US8), gB (UL27), gC (UL44), and gD (US6) are the main variable glycoproteins [84–86]. PRV *gE* is a crucial virulence factor related to the anterograde transport of viral particles in neurons [87] and is one of the genes commonly deleted in live attenuated vaccine strains. The experimentally constructed rLA-ZJ01/gEI developed by replacing the gE and gI of LA with the gE and gI of ZJ01 was more pathogenic to piglets than LA, implying that the changes in the gE and gI proteins partially contribute to the enhanced virulence of ZJ01 [86]. gB and gC are the core proteins required for the invasion of all herpesviruses and also the major immunogenic proteins [88]. PRV BJB that was reconstructed by replacing the gB of Bartha-K61 with the gB of JS-2012 showed increased protective efficacy against JS-2012 than Bartha-K61 [89]. Therefore, changes in these proteins are associated with the different biological characteristics of the PRV variant strains. It would be interesting to investigate the transmission and infection of PRV variant strains to humans based on these variations.

Additionally, to assess the risk of PRV infection in humans, it is vital to analyze whether all of the PRV variant strains or only specific PRV strains can infect humans. However, it is difficult to identify the general characteristics of PRV strains that infectious to humans since only one human-originated PRV strain has been isolated. In the phylogenetic analyses based on the *gE* and *gC* sequences of 54 PRV strains isolated from domestic pigs (44), dogs (9), and bovine (1) in Italy, most of the PRV strains from pigs, three of the strains from dogs working on pig farms, and PRV from bovine were closely related in the same clade, while five strains isolated from hunting dogs were highly close to the PRV strains from wild boars [90]. Therefore, it is presumed that the contact degree between different susceptible hosts was one of the critical points accounting for PRV cross-species transmission.

In fact, Nectin-1, Nectin-2, and HveD have been confirmed to mediate PRV infection in human and mouse cells [91]. Nectin-1 is highly conserved in mammals. Swine nectin-1 and human nectin-1 share 96% identity in amino acids, and they can both mediate the entry of HSV-1, herpes simplex virus-2 (HSV-2), PRV, and bovine herpesvirus 1 (BHV-1) [92,93]. PRV gD has been shown to bind swine nectin-1 and human nectin-1 with similar affinity, and the key residues of the interaction interface are conservative, providing structural evidence for PRV infection in humans [94]. However, there may be more ligands and receptors since it has been found that PRV can still infect Chinese hamster ovary cells, even without gD receptors [95]. PRV mutants without gD can be cultured and passaged to reach a high virus titer through cell-to-cell transmission [96]. Through porcine genome-wide CRISPR/Cas9 library screening, sphingomyelin synthase 1 (SMS1) was identified to be critical for PRV mutants without the *gD* gene to infect porcine kidney cells. When SMS1 was knocked out in the cells, the infection efficiency of PRV mutants without the *gD* gene decreased by 90%. This indicates that SMS1 plays a crucial role in PRV infection when the gD-mediated invasion pathway is blocked [97].

Moreover, HVEM mRNA and membrane-bound proteins have been shown to be expressed in the human adult retinal pigment epithelial cell line-19 (ARPE-19) [98], corneal fibroblasts cells [99], trabecular meshwork cells [100], conjunctival epithelial cells [101], and corneal epithelial cells [102]. Neutralizing antibodies or interfering RNA against HVEM could significantly reduce the entry of HSV-1 to these cells. Furthermore, previous studies have shown that HVEM can promote HSV-1 replication in mouse eyes [103,104]. Obviously, HVEM is associated with HSV-1 infection and pathogenicity in the eyes, so the common visual impairment in patients infected with PRV might be correlated to HVEM.

5. Vaccines and Diagnosis Methods for Pseudorabies

PRV infection in domestic pigs has been well controlled and even eliminated in many countries using vaccines and diagnostic tests, supporting the effectiveness of the DIVA concept. DIVA means the differentiation infected from vaccinated animals through the use of marker vaccines and respective serological diagnostic tests. After classically attenuated live vaccines developed by passaging, such as Bartha-K61, live virus vaccines lacking the major virulence-determining genes were developed by genetic engineering. The deletion of one or more genes targeting the *gE, gI, TK,* and *gG* genes are the typical choices [105–108]. Currently, based on homologous recombination, CRISPR/Cas9, bacterial artificial chromosome (BAC), and other genetic engineering technologies, gene-deletion strains can be rapidly constructed and assessed [108].

It should be noted that vaccine strains ought to be constructed based on the epizootic strains in the field to ensure the highest protection efficacy and reduce virus variation caused by recombination. Before 2011, pseudorabies in China had been well controlled by vaccination with Bartha-K61 and other vaccines constructed based on local classical strains. However, since late 2011, PRV variant strains have caused pseudorabies outbreaks in China. It has been reported that the variant strains are more virulent than the classical strains and that the classical vaccines can no longer provide sufficient protection against the variant strains [4,84]. The live attenuated vaccines based on variant strains such as SMX, TJ, ZJ01, and HN1201 were developed and showed adequate protection against the variant strains [107,109–111]. Therefore, it is necessary to monitor the changes of field strains continuously and to periodically construct new vaccine candidates. Field strains should be isolated from the wild boars in the countries in which pseudorabies has been eliminated in domestic pigs. Additionally, more detailed PRV typing methods are required to distinguish the differences among PRV strains and to select strains for vaccine development.

Meanwhile, diagnostic tests together with PRV gene-deletion vaccines are essential for applying DIVA. The indirect ELISA targeting of the gB and gE antibodies is one of the most widely applied serological approaches for differential diagnosis [5]. In addition, diverse molecular biological approaches targeting PRV genes have been established, such as PCR, real-time PCR, nano PCR, loop-mediated isothermal amplification (LAMP), and droplet digital PCR [108]. The sensitivity and specificity of these diagnostic approaches will undoubtedly be further improved. For areas with a low prevalence of pseudorabies, sensitivity is the primary concern of the diagnosis so that sporadically infected pigs can be diagnosed and eliminated. Timely diagnosis is vital for reducing the losses caused by the virus spreading among the pig population. Therefore, easy-to-operate, accurate, and on-site testing are required to develop new diagnostic methods. One of the difficulties among the current pseudorabies diagnostic technologies is that they cannot detect PRV during latency. In the final stage of pseudorabies eradication programs, infected pigs should be culled, while latently infected pigs cannot be detected using existing methods. They will be excluded through herd updating on the farms if there is no viral activation. However, many risk factors are associated with viral reactivation on pig farms. Therefore, developing specific methods to detect latently infected pigs is particularly important for the future prevention, elimination, and eradication of pseudorabies.

6. Conclusions

PRV is an important pathogen for pigs and other animals. As an alphaherpesvirus showing a relatively high rate of protein-coding variation, it is necessary to monitor the epidemiology and variations of this virus. In this review, we summarized PRV prevalence in China and worldwide, how PRV evolution was contributed to by natural selection and recombination, and PRV infections in animals and humans. All of this information facilitates future research and the control of pseudorabies. PRV elimination in the swine population should be further accelerated with better vaccines and diagnostic approaches. PRV can potentially infect humans, and further investigation is warranted.

Author Contributions: Data collection, Q.L., Y.K., Y.L., H.G., C.Z. and S.G.; writing—original draft, Q.L.; writing—review and editing, X.W., C.T., B.W. and H.C. All authors have read and agreed to the published version of the manuscript.

Funding: This work was supported by grants from the National Natural Science Foundation of China (32122086), the National Key Research and Development Program of China (2021YFD1800800), the China Agriculture Research System of MOF and MARA, and the Walmart Foundation as well as the Walmart Food Safety Collaboration Center (Project # 61626817). The funder had no role in the study design, data collection, data analysis, data interpretation, or in the writing of the manuscript.

Institutional Review Board Statement: Not applicable.

Informed Consent Statement: Not applicable.

Data Availability Statement: Not applicable.

Acknowledgments: We thank the Animal Disease Diagnostic Center of Wuhan Keqian Biology Co., Ltd. for sharing the PRV prevalence data from 2012 to 2021.

Conflicts of Interest: The authors declare no conflict of interest.

References

1. Kolb, A.W.; Lewin, A.C.; Moeller Trane, R.; McLellan, G.J.; Brandt, C.R. Phylogenetic and recombination analysis of the herpesvirus genus varicellovirus. *BMC Genom.* **2017**, *18*, 887. [CrossRef]
2. Szpara, M.L.; Kobiler, O.; Enquist, L.W. A common neuronal response to alphaherpesvirus infection. *J. Neuroimmune Pharmacol.* **2010**, *5*, 418–427. [CrossRef]
3. Lu, J.J.; Yuan, W.Z.; Zhu, Y.P.; Hou, S.H.; Wang, X.J. Latent pseudorabies virus infection in medulla oblongata from quarantined pigs. *Transbound. Emerg. Dis.* **2021**, *68*, 543–551. [CrossRef]
4. An, T.Q.; Peng, J.M.; Tian, Z.J.; Zhao, H.Y.; Li, N.; Liu, Y.M.; Chen, J.Z.; Leng, C.L.; Sun, Y.; Chang, D.; et al. Pseudorabies virus variant in Bartha-K61-vaccinated pigs, China, 2012. *Emerg. Infect. Dis.* **2013**, *19*, 1749–1755. [CrossRef]

5. Freuling, C.M.; Muller, T.F.; Mettenleiter, T.C. Vaccines against pseudorabies virus (PrV). *Vet. Microbiol.* **2017**, *206*, 3–9. [CrossRef]
6. Pomeranz, L.E.; Reynolds, A.E.; Hengartner, C.J. Molecular biology of pseudorabies virus: Impact on neurovirology and veterinary medicine. *Microbiol. Mol. Biol. Rev. MMBR* **2005**, *69*, 462–500. [CrossRef]
7. Csabai, Z.; Tombacz, D.; Deim, Z.; Snyder, M.; Boldogkoi, Z. Analysis of the Complete Genome Sequence of a Novel, Pseudorabies Virus Strain Isolated in Southeast Europe. *Can. J. Infect. Dis. Med. Microbiol.* **2019**, *2019*, 1806842. [CrossRef]
8. Hanson, R.P. The history of pseudorabies in the United States. *J. Am. Vet. Med. Assoc.* **1954**, *124*, 259–261.
9. Petrovskis, E.A.; Timmins, J.G.; Gierman, T.M.; Post, L.E. Deletions in vaccine strains of pseudorabies virus and their effect on synthesis of glycoprotein gp63. *J. Virol.* **1986**, *60*, 1166–1169. [CrossRef]
10. OIE. OIE World Animal Health Information System. Available online: https://wahis.oie.int/#/dashboards/country-or-disease-dashboard (accessed on 26 April 2022).
11. Caruso, C.; Vitale, N.; Prato, R.; Radaelli, M.C.; Zoppi, S.; Possidente, R.; Dondo, A.; Chiavacci, L.; Moreno Martin, A.M.; Masoero, L. Pseudorabies virus in North-West Italian wild boar (*Sus scrofa*) populations: Prevalence and risk factors to support a territorial risk-based surveillance. *Vet. Ital.* **2018**, *54*, 337–341. [CrossRef]
12. Denzin, N.; Conraths, F.J.; Mettenleiter, T.C.; Freuling, C.M.; Muller, T. Monitoring of Pseudorabies in Wild Boar of Germany-A Spatiotemporal Analysis. *Pathogens* **2020**, *9*, 276. [CrossRef] [PubMed]
13. Meier, R.K.; Ruiz-Fons, F.; Ryser-Degiorgis, M.P. A picture of trends in Aujeszky's disease virus exposure in wild boar in the Swiss and European contexts. *BMC Vet. Res.* **2015**, *11*, 277. [CrossRef] [PubMed]
14. Pedersen, K.; Bevins, S.N.; Baroch, J.A.; Cumbee, J.C., Jr.; Chandler, S.C.; Woodruff, B.S.; Bigelow, T.T.; DeLiberto, T.J. Pseudorabies in feral swine in the United States, 2009–2012. *J. Wildl. Dis.* **2013**, *49*, 709–713. [CrossRef] [PubMed]
15. Quan, Y.; Jia, Z.; Zhang, D.; Liu, X.; Liu, D.; Guo, Y.; Liu, L.; Sun, Y.; Liu, Y.; Chen, H. Serological investigation of pseudorabies in large-scale pig farms in Beijing. *Gansu Anim. Husb. Vet.* **2017**, *47*, 65–68. (In Chinese) [CrossRef]
16. Song, Q.; Liu, G.; Lin, W.; Wang, K.; Liu, Y.; Zhang, Y. Big data analysis of pseudorabies immune detection in the medium and large pig farms nationwide. *Vet. Orientat.* **2020**, *17*, 12–14. (In Chinese)
17. Chen, C. Analysis on Surveillance of Pig Pseudorabies Antibody in Chongqing from 2015 to 2019. Master's Thesis, Southwest University, Chongqing, China, 2020. (In Chinese).
18. Lin, R.; Xia, L.; Yin, H.; Fei, S.; Fan, K.; Huang, S.; Li, X.; Yang, X.; Dai, A. Epidemiological survey of the wild pseudorabies virus infection in some large scale pig farms of southwest Fujian from 2013 to 2018 in Chinese. *J. Anim. Infect. Dis.* **2021**, *29*, 85–90. (In Chinese)
19. Liu, X. Investigation and Control of Pseudorabies Infection in Large-Scale Pig Farms in Guizhou Province. Ph.D. Dissertation, Gansu Agricultural University, Lanzhou, China, 2019. (In Chinese).
20. He, H.; Hu, S.; Li, J.; Feng, S.; Zhong, S.; Liu, F.; Chen, Z.; Pan, Y. Seroepidemiological investigation of the major viral diseases on Large-scale Pig Farms in Guangxi in 2015–2017. *Heilongjiang Anim. Sci. Vet. Med.* **2019**, *14*, 83–86+175. (In Chinese) [CrossRef]
21. Yuan, S.; Zhao, S.; Ran, X.; Yan, R. Surveillance of gE antibody against pseudorabies virus in pigs in Henan Province from 2017 to 2019. *Anim. Breed. Feed.* **2020**, *19*, 15–17. (In Chinese) [CrossRef]
22. Zuo, Y.; Wang, B.; Han, L.; Wang, J.; Yuan, G.; Zhang, J.; Fan, J.; Zhong, F. Investigation of pseudorabies virus infection in Hebei province and phylogenetic analysis of its gE gene. *Chin. J. Vet. Sci.* **2021**, *41*, 224–230. (In Chinese)
23. Zhou, H.; Pan, Y.; Liu, M.; Han, Z. Prevalence of Porcine Pseudorabies Virus and Its Coinfection Rate in Heilongjiang Province in China from 2013 to 2018. *Viral Immunol.* **2020**, *33*, 550–554. [CrossRef]
24. Wang, H.; Yang, J.; Du, L.; Peng, M.; Liu, B.; Qiu, M. Investigation on the prevalence of main viral diseases in pig farms in some areas of Hunan Province in recent five years. *Hunan J. Anim. Sci. Vet. Med.* **2018**, *4*, 35–37. (In Chinese)
25. Li, H.; Kang, Z.; Tan, M.; Zeng, Y.; Ji, H. Seroepidemiological survey of swine pseudorabies on large-scale pig farms in Jiangxi Province from 2016 to 2017. *Acta Agric. Univ. Jiangxiensis* **2018**, *40*, 1037–1041. (In Chinese) [CrossRef]
26. Cui, W.; Fu, Y.; Wang, Y.; Lin, Y.; Li, X.; Zhang, Y.; Ying, L. Spotted surveillance and analysis of swine pseudorabies in Qinghai Province in 2016 and 2017. *Shandong J. Anim. Sci. Vet. Med.* **2019**, *40*, 13–14. (In Chinese)
27. Ma, Z.; Han, Z.; Liu, Z.; Meng, F.; Wang, H.; Cao, L.; Li, Y.; Jiao, Q.; Liu, S.; Liu, M. Epidemiological investigation of porcine pseudorabies virus and its coinfection rate in Shandong Province in China from 2015 to 2018. *J. Vet. Sci.* **2020**, *21*, e36. [CrossRef]
28. Zhang, L.; Ren, W.; Chi, J.; Lu, C.; Li, X.; Li, C.; Jiang, S.; Tian, X.; Li, F.; Wang, L.; et al. Epidemiology of Porcine Pseudorabies from 2010 to 2018 in Tianjin, China. *Viral Immunol.* **2021**, *34*, 714–721. [CrossRef]
29. Christensen, L.S.; Soerensen, K.J.; Lei, J.C. Restriction fragment pattern (RFP) analysis of genomes from Danish isolates of suid herpesvirus 1 (Aujezsky's disease virus). *Arch. Virol.* **1987**, *97*, 215–224. [CrossRef]
30. Herrmann, S.-C.; Heppner, B.; Ludwig, H. Pseudorabies Viruses from Clinical Outbreaks and Latent Infections Grouped into Four Major Genome Types. In Proceedings of the Latent Herpes Virus Infections in Veterinary Medicine: A Seminar in the CEC Programme of Coordination of Research on Animal Pathology, Tübingen, Germany, 21–24 September 1982; Wittmann, G., Gaskell, R.M., Rziha, H.J., Eds.; Springer: Dordrecht, The Netherlands, 1984; pp. 387–401.
31. Christensen, L.S. The population biology of suid herpesvirus 1. *Apmis Suppl.* **1995**, *48*, 1–48.
32. Muller, T.; Klupp, B.G.; Freuling, C.; Hoffmann, B.; Mojcicz, M.; Capua, I.; Palfi, V.; Toma, B.; Lutz, W.; Ruiz-Fon, F.; et al. Characterization of pseudorabies virus of wild boar origin from Europe. *Epidemiol. Infect.* **2010**, *138*, 1590–1600. [CrossRef]

33. Fonseca, A.A., Jr.; Magalhaes, C.G.; Sales, E.B.; D'Ambros, R.M.; Ciacci-Zanella, J.; Heinemann, M.B.; Leite, R.C.; Dos Reis, J.K. Genotyping of the pseudorabies virus by multiplex PCR followed by restriction enzyme analysis. *ISRN Microbiol.* **2011**, *2011*, 458294. [CrossRef]
34. Deblanc, C.; Oger, A.; Simon, G.; Le Potier, M.F. Genetic Diversity among Pseudorabies Viruses Isolated from Dogs in France from 2006 to 2018. *Pathogens* **2019**, *8*, 266. [CrossRef]
35. Ye, C.; Zhang, Q.Z.; Tian, Z.J.; Zheng, H.; Zhao, K.; Liu, F.; Guo, J.C.; Tong, W.; Jiang, C.G.; Wang, S.J.; et al. Genomic characterization of emergent pseudorabies virus in China reveals marked sequence divergence: Evidence for the existence of two major genotypes. *Virology* **2015**, *483*, 32–43. [CrossRef] [PubMed]
36. He, W.; Auclert, L.Z.; Zhai, X.; Wong, G.; Zhang, C.; Zhu, H.; Xing, G.; Wang, S.; He, W.; Li, K.; et al. Interspecies Transmission, Genetic Diversity, and Evolutionary Dynamics of Pseudorabies Virus. *J. Infect. Dis.* **2019**, *219*, 1705–1715. [CrossRef] [PubMed]
37. Szpara, M.L.; Tafuri, Y.R.; Parsons, L.; Shamim, S.R.; Verstrepen, K.J.; Legendre, M.; Enquist, L.W. A wide extent of inter-strain diversity in virulent and vaccine strains of alphaherpesviruses. *PLoS Pathog.* **2011**, *7*, e1002282. [CrossRef] [PubMed]
38. Zhai, X.; Zhao, W.; Li, K.; Zhang, C.; Wang, C.; Su, S.; Zhou, J.; Lei, J.; Xing, G.; Sun, H.; et al. Genome Characteristics and Evolution of Pseudorabies Virus Strains in Eastern China from 2017 to 2019. *Virol. Sin.* **2019**, *34*, 601–609. [CrossRef]
39. Hu, R.; Wang, L.; Liu, Q.; Hua, L.; Huang, X.; Zhang, Y.; Fan, J.; Chen, H.; Song, W.; Liang, W.; et al. Whole-Genome Sequence Analysis of Pseudorabies Virus Clinical Isolates from Pigs in China between 2012 and 2017 in China. *Viruses* **2021**, *13*, 1322. [CrossRef] [PubMed]
40. Thiry, E.; Meurens, F.; Muylkens, B.; McVoy, M.; Gogev, S.; Thiry, J.; Vanderplasschen, A.; Epstein, A.; Keil, G.; Schynts, F. Recombination in alphaherpesviruses. *Rev. Med. Virol.* **2005**, *15*, 89–103. [CrossRef] [PubMed]
41. Henderson, L.M.; Katz, J.B.; Erickson, G.A.; Mayfield, J.E. In vivo and in vitro genetic recombination between conventional and gene-deleted vaccine strains of pseudorabies virus. *Am. J. Vet. Res.* **1990**, *51*, 1656–1662. [PubMed]
42. Christensen, L.S.; Lomniczi, B. High frequency intergenomic recombination of suid herpesvirus 1 (SHV-1, Aujeszky's disease virus). *Arch. Virol.* **1993**, *132*, 37–50. [CrossRef]
43. Huang, J.; Zhu, L.; Zhao, J.; Yin, X.; Feng, Y.; Wang, X.; Sun, X.; Zhou, Y.; Xu, Z. Genetic evolution analysis of novel recombinant pseudorabies virus strain in Sichuan, China. *Transbound. Emerg. Dis.* **2020**, *67*, 1428–1432. [CrossRef] [PubMed]
44. Bo, Z.; Miao, Y.; Xi, R.; Gao, X.; Miao, D.; Chen, H.; Jung, Y.S.; Qian, Y.; Dai, J. Emergence of a novel pathogenic recombinant virus from Bartha vaccine and variant pseudorabies virus in China. *Transbound. Emerg. Dis.* **2021**, *68*, 1454–1464. [CrossRef] [PubMed]
45. Ye, C.; Guo, J.C.; Gao, J.C.; Wang, T.Y.; Zhao, K.; Chang, X.B.; Wang, Q.; Peng, J.M.; Tian, Z.J.; Cai, X.H.; et al. Genomic analyses reveal that partial sequence of an earlier pseudorabies virus in China is originated from a Bartha-vaccine-like strain. *Virology* **2016**, *491*, 56–63. [CrossRef] [PubMed]
46. Nauwynck, H.J.; Pensaert, M.B. Abortion induced by cell-associated pseudorabies virus in vaccinated sows. *Am. J. Vet. Res.* **1992**, *53*, 489–493.
47. Salogni, C.; Lazzaro, M.; Giacomini, E.; Giovannini, S.; Zanoni, M.; Giuliani, M.; Ruggeri, J.; Pozzi, P.; Pasquali, P.; Boniotti, M.B.; et al. Infectious agents identified in aborted swine fetuses in a high-density breeding area: A three-year study. *J. Vet. Diagn. Investig.* **2016**, *28*, 550–554. [CrossRef]
48. Sehl, J.; Teifke, J.P. Comparative Pathology of Pseudorabies in Different Naturally and Experimentally Infected Species—A Review. *Pathogens* **2020**, *9*, 633. [CrossRef] [PubMed]
49. Cheng, Z.; Kong, Z.; Liu, P.; Fu, Z.; Zhang, J.; Liu, M.; Shang, Y. Natural infection of a variant pseudorabies virus leads to bovine death in China. *Transbound. Emerg. Dis.* **2020**, *67*, 518–522. [CrossRef]
50. Ciarello, F.P.; Capucchio, M.T.; Ippolito, D.; Colombino, E.; Gibelli, L.R.M.; Fiasconaro, M.; Moreno Martin, A.M.; Di Marco Lo Presti, V. First Report of a Severe Outbreak of Aujeszky's Disease in Cattle in Sicily (Italy). *Pathogens* **2020**, *9*, 954. [CrossRef]
51. McFerran, J.B.; Dow, C. Virus Studies on Experimental Aujeszky's Disease In Calves. *J. Comp. Pathol.* **1964**, *74*, 173–179. [CrossRef]
52. Henderson, J.P.; Graham, D.A.; Stewart, D. An outbreak of Aujeszky's disease in sheep in Northern Ireland. *Vet. Rec.* **1995**, *136*, 555–557. [CrossRef] [PubMed]
53. Mocsari, E.; Toth, C.; Meder, M.; Saghy, E.; Glavits, R. Aujeszky's disease of sheep: Experimental studies on the excretion and horizontal transmission of the virus. *Vet. Microbiol.* **1987**, *13*, 353–359. [CrossRef]
54. Cramer, S.D.; Campbell, G.A.; Njaa, B.L.; Morgan, S.E.; Smith, S.K., 2nd; McLin, W.R.T.; Brodersen, B.W.; Wise, A.G.; Scherba, G.; Langohr, I.M.; et al. Pseudorabies virus infection in Oklahoma hunting dogs. *J. Vet. Diagn. Investig.* **2011**, *23*, 915–923. [CrossRef]
55. Pedersen, K.; Turnage, C.T.; Gaston, W.D.; Arruda, P.; Alls, S.A.; Gidlewski, T. Pseudorabies detected in hunting dogs in Alabama and Arkansas after close contact with feral swine (*Sus scrofa*). *BMC Vet. Res.* **2018**, *14*, 388. [CrossRef] [PubMed]
56. Zhang, L.; Zhong, C.; Wang, J.; Lu, Z.; Liu, L.; Yang, W.; Lyu, Y. Pathogenesis of natural and experimental Pseudorabies virus infections in dogs. *Virol. J.* **2015**, *12*, 44. [CrossRef] [PubMed]
57. Muller, T.; Hahn, E.C.; Tottewitz, F.; Kramer, M.; Klupp, B.G.; Mettenleiter, T.C.; Freuling, C. Pseudorabies virus in wild swine: A global perspective. *Arch. Virol.* **2011**, *156*, 1691–1705. [CrossRef]
58. Caruso, C.; Dondo, A.; Cerutti, F.; Masoero, L.; Rosamilia, A.; Zoppi, S.; D'Errico, V.; Grattarola, C.; Acutis, P.L.; Peletto, S. Aujeszky's disease in red fox (*Vulpes vulpes*): Phylogenetic analysis unravels an unexpected epidemiologic link. *J. Wildl. Dis.* **2014**, *50*, 707–710. [CrossRef] [PubMed]
59. Lian, K.; Zhang, M.; Zhou, L.; Song, Y.; Wang, G.; Wang, S. First report of a pseudorabies-virus-infected wolf (*Canis lupus*) in China. *Arch. Virol.* **2020**, *165*, 459–462. [CrossRef] [PubMed]

60. Zanin, E.; Capua, I.; Casaccia, C.; Zuin, A.; Moresco, A. Isolation and characterization of Aujeszky's disease virus in captive brown bears from Italy. *J. Wildl. Dis.* **1997**, *33*, 632–634. [CrossRef]

61. Schultze, A.E.; Maes, R.K.; Taylor, D.C. Pseudorabies and volvulus in a black bear. *J. Am. Vet. Med. Assoc.* **1986**, *189*, 1165–1166. [PubMed]

62. Glass, C.M.; McLean, R.G.; Katz, J.B.; Maehr, D.S.; Cropp, C.B.; Kirk, L.J.; McKeirnan, A.J.; Evermann, J.F. Isolation of pseudorabies (Aujeszky's disease) virus from a Florida panther. *J. Wildl. Dis.* **1994**, *30*, 180–184. [CrossRef]

63. Masot, A.J.; Gil, M.; Risco, D.; Jimenez, O.M.; Nunez, J.I.; Redondo, E. Pseudorabies virus infection (Aujeszky's disease) in an Iberian lynx (*Lynx pardinus*) in Spain: A case report. *BMC Vet. Res.* **2017**, *13*, 6. [CrossRef] [PubMed]

64. Kirkpatrick, C.M.; Kanitz, C.L.; McCrocklin, S.M. Possible role of wild mammals in transmission of pseudorabies to swine. *J. Wildl. Dis.* **1980**, *16*, 601–614. [CrossRef]

65. Liu, H.; Li, X.T.; Hu, B.; Deng, X.Y.; Zhang, L.; Lian, S.Z.; Zhang, H.L.; Lv, S.; Xue, X.H.; Lu, R.G.; et al. Outbreak of severe pseudorabies virus infection in pig-offal-fed farmed mink in Liaoning Province, China. *Arch. Virol.* **2017**, *162*, 863–866. [CrossRef] [PubMed]

66. Skinner, G.R.; Ahmad, A.; Davies, J.A. The infrequency of transmission of herpesviruses between humans and animals; postulation of an unrecognised protective host mechanism. *Comp. Immunol. Microbiol. Infect. Dis.* **2001**, *24*, 255–269. [CrossRef]

67. Mravak, S.; Bienzle, U.; Feldmeier, H.; Hampl, H.; Habermehl, K.O. Pseudorabies in man. *Lancet* **1987**, *1*, 501–502. [CrossRef] [PubMed]

68. Schükrü-Aksel, I.; Tunman, Z. Aujeskysche Erkrankung in der Türkei bei Mensch und Tier. *Z. Gesamte Neurol. Psychiatr.* **1940**, *169*, 598–606. [CrossRef]

69. Anusz, Z.; Szweda, W.; Popko, J.; Trybala, E. Is Aujeszky's disease a zoonosis? *Prz. Epidemiol.* **1992**, *46*, 181–186.

70. Liu, Q.; Wang, X.; Xie, C.; Ding, S.; Yang, H.; Guo, S.; Li, J.; Qin, L.; Ban, F.; Wang, D.; et al. A Novel Human Acute Encephalitis Caused by Pseudorabies Virus Variant Strain. *Clin. Infect. Dis.* **2021**, *73*, e3690–e3700. [CrossRef] [PubMed]

71. Zhao, W.L.; Wu, Y.H.; Li, H.F.; Li, S.Y.; Fan, S.Y.; Wu, H.L.; Li, Y.J.; Lu, Y.L.; Han, J.; Zhang, W.C.; et al. Clinical experience and next-generation sequencing analysis of encephalitis caused by pseudorabies virus. *Zhonghua Yi Xue Za Zhi* **2018**, *98*, 1152–1157. [CrossRef] [PubMed]

72. Liu, Q.; Wang, X.; Chen, H.; Yan, R.; Li, W.; Wang, X. Reply to Kitaura and Okamoto. *Clin. Infect. Dis.* **2021**, *72*, e693–e694. [CrossRef] [PubMed]

73. Li, X.D.; Fu, S.H.; Chen, L.Y.; Li, F.; Deng, J.H.; Lu, X.C.; Wang, H.Y.; Tian, K.G. Detection of Pseudorabies Virus Antibodies in Human Encephalitis Cases. *Biomed. Environ. Sci. BES* **2020**, *33*, 444–447. [CrossRef]

74. Ai, J.W.; Weng, S.S.; Cheng, Q.; Cui, P.; Li, Y.J.; Wu, H.L.; Zhu, Y.M.; Xu, B.; Zhang, W.H. Human Endophthalmitis Caused by Pseudorabies Virus Infection, China, 2017. *Emerg. Infect. Dis.* **2018**, *24*, 1087–1090. [CrossRef] [PubMed]

75. Yang, H.; Han, H.; Wang, H.; Cui, Y.; Liu, H.; Ding, S. A Case of Human Viral Encephalitis Caused by Pseudorabies Virus Infection in China. *Front. Neurol.* **2019**, *10*, 534. [CrossRef] [PubMed]

76. Zheng, L.; Liu, X.; Yuan, D.; Li, R.; Lu, J.; Li, X.; Tian, K.; Dai, E. Dynamic cerebrospinal fluid analyses of severe pseudorabies encephalitis. *Transbound. Emerg. Dis.* **2019**, *66*, 2562–2565. [CrossRef] [PubMed]

77. Yang, X.; Guan, H.; Li, C.; Li, Y.; Wang, S.; Zhao, X.; Zhao, Y.; Liu, Y. Characteristics of human encephalitis caused by pseudorabies virus: A case series study. *Int. J. Infect. Dis. IJID* **2019**, *87*, 92–99. [CrossRef] [PubMed]

78. Fan, S.; Yuan, H.; Liu, L.; Li, H.; Wang, S.; Zhao, W.; Wu, Y.; Wang, P.; Hu, Y.; Han, J.; et al. Pseudorabies virus encephalitis in humans: A case series study. *J. Neurovirol.* **2020**, *26*, 556–564. [CrossRef]

79. Wang, D.; Tao, X.; Fei, M.; Chen, J.; Guo, W.; Li, P.; Wang, J. Human encephalitis caused by pseudorabies virus infection: A case report. *J. Neurovirol.* **2020**, *26*, 442–448. [CrossRef]

80. Hu, F.; Wang, J.; Peng, X.Y. Bilateral Necrotizing Retinitis following Encephalitis Caused by the Pseudorabies Virus Confirmed by Next-Generation Sequencing. *Ocul. Immunol. Inflamm.* **2021**, *29*, 922–925. [CrossRef]

81. Ying, M.; Hu, X.; Wang, M.; Cheng, X.; Zhao, B.; Tao, Y. Vitritis and retinal vasculitis caused by pseudorabies virus. *J. Int. Med. Res.* **2021**, *49*, 3000605211058990. [CrossRef]

82. Zhou, Y.; Nie, C.; Wen, H.; Long, Y.; Zhou, M.; Xie, Z.; Hong, D. Human viral encephalitis associated with suid herpesvirus 1. *Neurol. Sci.* **2021**, *43*, 2681–2692. [CrossRef]

83. Yan, W.; Hu, Z.; Zhang, Y.; Wu, X.; Zhang, H. Case Report: Metagenomic Next-Generation Sequencing for Diagnosis of Human Encephalitis and Endophthalmitis Caused by Pseudorabies Virus. *Front. Med.* **2021**, *8*, 753988. [CrossRef]

84. Luo, Y.; Li, N.; Cong, X.; Wang, C.H.; Du, M.; Li, L.; Zhao, B.; Yuan, J.; Liu, D.D.; Li, S.; et al. Pathogenicity and genomic characterization of a pseudorabies virus variant isolated from Bartha-K61-vaccinated swine population in China. *Vet. Microbiol.* **2014**, *174*, 107–115. [CrossRef]

85. Yu, T.; Chen, F.; Ku, X.; Fan, J.; Zhu, Y.; Ma, H.; Li, S.; Wu, B.; He, Q. Growth characteristics and complete genomic sequence analysis of a novel pseudorabies virus in China. *Virus Genes* **2016**, *52*, 474–483. [CrossRef] [PubMed]

86. Dong, J.; Gu, Z.; Jin, L.; Lv, L.; Wang, J.; Sun, T.; Bai, J.; Sun, H.; Wang, X.; Jiang, P. Polymorphisms affecting the gE and gI proteins partly contribute to the virulence of a newly-emergent highly virulent Chinese pseudorabies virus. *Virology* **2018**, *519*, 42–52. [CrossRef] [PubMed]

87. Husak, P.J.; Kuo, T.; Enquist, L.W. Pseudorabies virus membrane proteins gI and gE facilitate anterograde spread of infection in projection-specific neurons in the rat. *J. Virol.* **2000**, *74*, 10975–10983. [CrossRef] [PubMed]

88. Heldwein, E.E.; Krummenacher, C. Entry of herpesviruses into mammalian cells. *Cell. Mol. Life Sci. CMLS* **2008**, *65*, 1653–1668. [CrossRef] [PubMed]
89. Yu, Z.Q.; Tong, W.; Zheng, H.; Li, L.W.; Li, G.X.; Gao, F.; Wang, T.; Liang, C.; Ye, C.; Wu, J.Q.; et al. Variations in glycoprotein B contribute to immunogenic difference between PRV variant JS-2012 and Bartha-K61. *Vet. Microbiol.* **2017**, *208*, 97–105. [CrossRef]
90. Sozzi, E.; Moreno, A.; Lelli, D.; Cinotti, S.; Alborali, G.L.; Nigrelli, A.; Luppi, A.; Bresaola, M.; Catella, A.; Cordioli, P. Genomic characterization of pseudorabies virus strains isolated in Italy. *Transbound. Emerg. Dis.* **2014**, *61*, 334–340. [CrossRef]
91. Spear, P.G.; Eisenberg, R.J.; Cohen, G.H. Three classes of cell surface receptors for alphaherpesvirus entry. *Virology* **2000**, *275*, 1–8. [CrossRef]
92. Menotti, L.; Lopez, M.; Avitabile, E.; Stefan, A.; Cocchi, F.; Adelaide, J.; Lecocq, E.; Dubreuil, P.; Campadelli-Fiume, G. The murine homolog of human Nectin1delta serves as a species nonspecific mediator for entry of human and animal alpha herpesviruses in a pathway independent of a detectable binding to gD. *Proc. Natl. Acad. Sci. USA* **2000**, *97*, 4867–4872. [CrossRef]
93. Milne, R.S.; Connolly, S.A.; Krummenacher, C.; Eisenberg, R.J.; Cohen, G.H. Porcine HveC, a member of the highly conserved HveC/nectin 1 family, is a functional alphaherpesvirus receptor. *Virology* **2001**, *281*, 315–328. [CrossRef]
94. Li, A.; Lu, G.; Qi, J.; Wu, L.; Tian, K.; Luo, T.; Shi, Y.; Yan, J.; Gao, G.F. Structural basis of nectin-1 recognition by pseudorabies virus glycoprotein D. *PLoS Pathog.* **2017**, *13*, e1006314. [CrossRef]
95. Nixdorf, R.; Schmidt, J.; Karger, A.; Mettenleiter, T.C. Infection of Chinese hamster ovary cells by pseudorabies virus. *J. Virol.* **1999**, *73*, 8019–8026. [CrossRef] [PubMed]
96. Schmidt, J.; Klupp, B.G.; Karger, A.; Mettenleiter, T.C. Adaptability in herpesviruses: Glycoprotein D-independent infectivity of pseudorabies virus. *J. Virol.* **1997**, *71*, 17–24. [CrossRef] [PubMed]
97. Holper, J.E.; Grey, F.; Baillie, J.K.; Regan, T.; Parkinson, N.J.; Hoper, D.; Thamamongood, T.; Schwemmle, M.; Pannhorst, K.; Wendt, L.; et al. A Genome-Wide CRISPR/Cas9 Screen Reveals the Requirement of Host Sphingomyelin Synthase 1 for Infection with Pseudorabies Virus Mutant gD(-)Pass. *Viruses* **2021**, *13*, 1574. [CrossRef]
98. Tiwari, V.; Oh, M.J.; Kovacs, M.; Shukla, S.Y.; Valyi-Nagy, T.; Shukla, D. Role for nectin-1 in herpes simplex virus 1 entry and spread in human retinal pigment epithelial cells. *FEBS J.* **2008**, *275*, 5272–5285. [CrossRef] [PubMed]
99. Tiwari, V.; Shukla, S.Y.; Yue, B.; Shukla, D. Herpes simplex virus type 2 entry into cultured human corneal fibroblasts is mediated by herpesvirus entry mediator. *J. Gen. Virol.* **2007**, *88*, 2106–2110. [CrossRef] [PubMed]
100. Tiwari, V.; Clement, C.; Scanlan, P.M.; Kowlessur, D.; Yue, B.Y.; Shukla, D. A role for herpesvirus entry mediator as the receptor for herpes simplex virus 1 entry into primary human trabecular meshwork cells. *J. Virol.* **2005**, *79*, 13173–13179. [CrossRef]
101. Akhtar, J.; Tiwari, V.; Oh, M.J.; Kovacs, M.; Jani, A.; Kovacs, S.K.; Valyi-Nagy, T.; Shukla, D. HVEM and nectin-1 are the major mediators of herpes simplex virus 1 (HSV-1) entry into human conjunctival epithelium. *Investig. Ophthalmol. Vis. Sci.* **2008**, *49*, 4026–4035. [CrossRef]
102. Shah, A.; Farooq, A.V.; Tiwari, V.; Kim, M.J.; Shukla, D. HSV-1 infection of human corneal epithelial cells: Receptor-mediated entry and trends of re-infection. *Mol. Vis.* **2010**, *16*, 2476–2486.
103. Karaba, A.H.; Kopp, S.J.; Longnecker, R. Herpesvirus entry mediator and nectin-1 mediate herpes simplex virus 1 infection of the murine cornea. *J. Virol.* **2011**, *85*, 10041–10047. [CrossRef]
104. Edwards, R.G.; Kopp, S.J.; Karaba, A.H.; Wilcox, D.R.; Longnecker, R. Herpesvirus entry mediator on radiation-resistant cell lineages promotes ocular herpes simplex virus 1 pathogenesis in an entry-independent manner. *mBio* **2015**, *6*, e01532-15. [CrossRef]
105. Tang, Y.D.; Liu, J.T.; Wang, T.Y.; An, T.Q.; Sun, M.X.; Wang, S.J.; Fang, Q.Q.; Hou, L.L.; Tian, Z.J.; Cai, X.H. Live attenuated pseudorabies virus developed using the CRISPR/Cas9 system. *Virus Res.* **2016**, *225*, 33–39. [CrossRef] [PubMed]
106. Tong, W.; Li, G.; Liang, C.; Liu, F.; Tian, Q.; Cao, Y.; Li, L.; Zheng, X.; Zheng, H.; Tong, G. A live, attenuated pseudorabies virus strain JS-2012 deleted for gE/gI protects against both classical and emerging strains. *Antivir. Res.* **2016**, *130*, 110–117. [CrossRef] [PubMed]
107. Hu, R.M.; Zhou, Q.; Song, W.B.; Sun, E.C.; Zhang, M.M.; He, Q.G.; Chen, H.C.; Wu, B.; Liu, Z.F. Novel pseudorabies virus variant with defects in TK, gE and gI protects growing pigs against lethal challenge. *Vaccine* **2015**, *33*, 5733–5740. [CrossRef] [PubMed]
108. Tan, L.; Yao, J.; Yang, Y.; Luo, W.; Yuan, X.; Yang, L.; Wang, A. Current Status and Challenge of Pseudorabies Virus Infection in China. *Virol. Sin.* **2021**, *36*, 588–607. [CrossRef]
109. Cong, X.; Lei, J.L.; Xia, S.L.; Wang, Y.M.; Li, Y.; Li, S.; Luo, Y.; Sun, Y.; Qiu, H.J. Pathogenicity and immunogenicity of a gE/gI/TK gene-deleted pseudorabies virus variant in susceptible animals. *Vet. Microbiol.* **2016**, *182*, 170–177. [CrossRef]
110. Lv, L.; Liu, X.; Jiang, C.; Wang, X.; Cao, M.; Bai, J.; Jiang, P. Pathogenicity and immunogenicity of a gI/gE/TK/UL13-gene-deleted variant pseudorabies virus strain in swine. *Vet. Microbiol.* **2021**, *258*, 109104. [CrossRef]
111. Yan, S.; Huang, B.; Bai, X.; Zhou, Y.; Guo, L.; Wang, T.; Shan, Y.; Wang, Y.; Tan, F.; Tian, K. Construction and Immunogenicity of a Recombinant Pseudorabies Virus Variant With TK/gI/gE/11k/28k Deletion. *Front. Vet. Sci.* **2021**, *8*, 797611. [CrossRef]

Review

Pseudorabies Virus: From Pathogenesis to Prevention Strategies

Hui-Hua Zheng [1,†], Peng-Fei Fu [1,2,†], Hong-Ying Chen [1,*] and Zhen-Ya Wang [3,*]

1   Zhengzhou Major Pig Disease Prevention and Control Laboratory, College of Veterinary Medicine, Henan Agricultural University, Zhengdong New District Longzi Lake 15#, Zhengzhou 450046, China; zhenghh112@163.com (H.-H.Z.); fpfwdm@126.com (P.-F.F.)

2   College of Life Science and Engineering, Henan University of Urban Construction, Pingdingshan 467044, China

3   Key Laboratory of "Runliang" Antiviral Medicines Research and Development, Institute of Drug Discovery & Development, Zhengzhou University, Zhengzhou 450001, China

*   Correspondence: chhy927@163.com (H.-Y.C.); zhenyawang@zzu.edu.cn (Z.-Y.W.); Tel.: +86-371-55369208 (H.-Y.C.); Fax: +86-371-55369208 (H.-Y.C.)

†   These authors contributed equally to this work.

Abstract: Pseudorabies (PR), also called Aujeszky's disease (AD), is a highly infectious viral disease which is caused by pseudorabies virus (PRV). It has been nearly 200 years since the first PR case occurred. Currently, the virus can infect human beings and various mammals, including pigs, sheep, dogs, rabbits, rodents, cattle and cats, and among them, pigs are the only natural host of PRV infection. PRV is characterized by reproductive failure in pregnant sows, nervous disorders in newborn piglets, and respiratory distress in growing pigs, resulting in serious economic losses to the pig industry worldwide. Due to the extensive application of the attenuated vaccine containing the Bartha-K61 strain, PR was well controlled. With the variation of PRV strain, PR re-emerged and rapidly spread in some countries, especially China. Although researchers have been committed to the design of diagnostic methods and the development of vaccines in recent years, PR is still an important infectious disease and is widely prevalent in the global pig industry. In this review, we introduce the structural composition and life cycle of PRV virions and then discuss the latest findings on PRV pathogenesis, following the molecular characteristic of PRV and the summary of existing diagnosis methods. Subsequently, we also focus on the latest clinical progress in the prevention and control of PRV infection via the development of vaccines, traditional herbal medicines and novel small RNAs. Lastly, we provide an outlook on PRV eradication.

Keywords: pseudorabies virus; pathogenesis; infection; prevention and control

Citation: Zheng, H.-H.; Fu, P.-F.; Chen, H.-Y.; Wang, Z.-Y. Pseudorabies Virus: From Pathogenesis to Prevention Strategies. *Viruses* **2022**, *14*, 1638. https://doi.org/10.3390/v14081638

Academic Editors: Yan-Dong Tang and Xiangdong Li

Received: 31 May 2022
Accepted: 25 July 2022
Published: 27 July 2022

Publisher's Note: MDPI stays neutral with regard to jurisdictional claims in published maps and institutional affiliations.

1. Introduction

Pseudorabies (PR), as known as Aujeszky's disease, was first described in America as early as 1813 and has spread nearly globally since the early 1980s [1]. Its etiological agent is pseudorabies virus (PRV), which has a wide range of hosts; among them, pigs are the natural host and reservoir of the virus. It displays different symptoms at distinct growth phases after being infected with PRV, including the reproductive failure of sows, fatal encephalitis and 100% mortality of newborn pigs, and respiratory distress and growth block of young pigs [2]. For other susceptible animals (ruminants, carnivores and rodents), PRV generally ends with death [3]. In addition, PRV infection might cause endophthalmitis and encephalitis in human beings. Some studies determined the specific sequences of PRV in the patients' tissues using metagenomic next-generation sequencing, and a human-originated PRV strain hSD-1/2019 was isolated from the cerebrospinal fluid of a patient with acute encephalitis [4–9]. Over the years, PRV infection has brought a huge economic loss to the pig industry worldwide and is a serious threat to the health of humans. Although the disease was transiently controlled globally as the result of the use of the glycoprotein E (gE)-negative vaccine Bartha-K61 from Hungary in 1961, PR re-emerged and rapidly spread with a variation of PRV, and the traditional vaccine only offers partial protection against the variant stains [10,11]. Several studies suggest that the PRV variant strains were more virulent to animals and humans than the classical strains [3,5,12]. Moreover, PRV can build a lifelong latent infection in the host's peripheral nervous system, and infected pigs can potentially be a source of reinfections once the latent viral genome is reactivated. The infection characteristics of PRV variant strains have led to the fact that PR is once again circulating in almost the whole world. It raised scientists' awareness of the serious threat posed by PRV and encouraged researchers to develop effective interventions.

PRV is an enveloped, linear double-stranded DNA herpes virus belonging to the *Varicellovirus* genus of subfamily *Alphaherpesvirinae* in the family *Herpesviridae* [13]. Its infection generally starts by viral replication in the epithelial cells of the nasal and oropharyngeal mucosa and then spreads to the peripheral nervous system neurons innervating the infected epithelium. Viral particles travel via retrograde transport to the sensory and autonomic peripheral ganglia, where a latent lifelong infection is established [14–16]. Upon reactivation, viral replication occurs, and particles spread in the anterograde direction along the sensory nerves back to the mucosal surfaces where the infection initiated. This makes adult pigs and piglets typically exhibit symptoms of respiratory disease and acute neurological disease, respectively [17]. Additionally, PRV infection can also spread via a cell-associated viremia in peripheral blood mononuclear cells from the primary replication site to target organs such as the pregnant uterus, and then secondary replication ensues in the endothelial cells of the pregnant uterus, which can result in vasculitis and multifocal thrombosis, usually leading to abortion [18,19].

Currently, there are no effective means for eliminating PR in the pig population, so the diagnosis, prevention and control of PR are particularly important for the pig industry. In this review, we briefly describe the structure of PRV virions and the viral life cycle. Then, we discuss recent advances in understanding the pathogenesis of PRV infection and its molecular characteristics. Subsequently, the current PRV diagnosis methods are summed up, and we also highlight the latest progress in the prevention and control of PRV infection, including the development of vaccines, Chinese herbal medicines and novel small RNAs. Finally, we look forward to the prospects of PRV eradication in the future.

2. The Virion Structure, Genome Structure and the Life Cycle

2.1. The Virion Structure

PRV mainly contains two subtypes (I and II). Similarly to other Herpes virions, PRV virions (Figure 1) are approximately 225 nm in diameter and consist of four morphologically distinct structural components, including a linear double-stranded DNA genome, an icosahedral protein capsid, a protein tegument layer, and a lipid envelope containing viral glycoproteins [2,20–22]. The double-stranded DNA genome of ~145 kb in length, which can encode more than 70 proteins, is encapsulated in an icosahedral capsid. The tegument is a collection of approximately 12 proteins organized into at least two layers, one of which interacts with envelope proteins, while the other is closely associated with the capsid. The envelope is a lipid bilayer infused with transmembrane proteins, many of which are modified by glycosylation.

2.2. Genome, Gene Content and Role in Viral Replication

Alphaherpesvirus genomes have a partial colinear arrangement of genes encoding similar functions. Based on the overall arrangement of repeat sequences and unique regions, the herpesvirus genomes can be divided into six classes, designated by the letters A to F [23]. The PRV genome belongs to the D class and is a linear, double-stranded and sense viral DNA genome of approximately 145 kb, with a 74% content of G + C. It is also characterized by two unique regions, which are the unique long region (UL) and unique short region (US), and the US region is flanked by the internal and terminal repeat sequences (IRS and TRS, respectively) of 15 kb in length (Figure 2) [2]. The sequence and gene arrangement of the entire PRV genome are known, and a map of the likely transcript organization, well supported by experimental data, has been established. Recombination between the inverted repeats can produce two possible isomers of the genome, with the US region in opposite orientation, and two isomers are both infectious. PRV has three origins of replication, with one of *OriL* located in the UL region and the others (*OriS*) located in the inverted repeats [24,25]. In addition, the full-length genome of PRV contains 73 different genes encoding a total of 70~100 proteins, which are mainly the capsid proteins, envelope proteins, tegument proteins and various enzymes. Half of them are nonessential proteins for PRV replication [2]. The PRV genes can be divided into three types based on their different functions: structural genes, virulence genes and regulatory genes. They can also be divided into immediate-early genes, early genes and late genes according to the transcription sequence of PRV invading host cells [26].

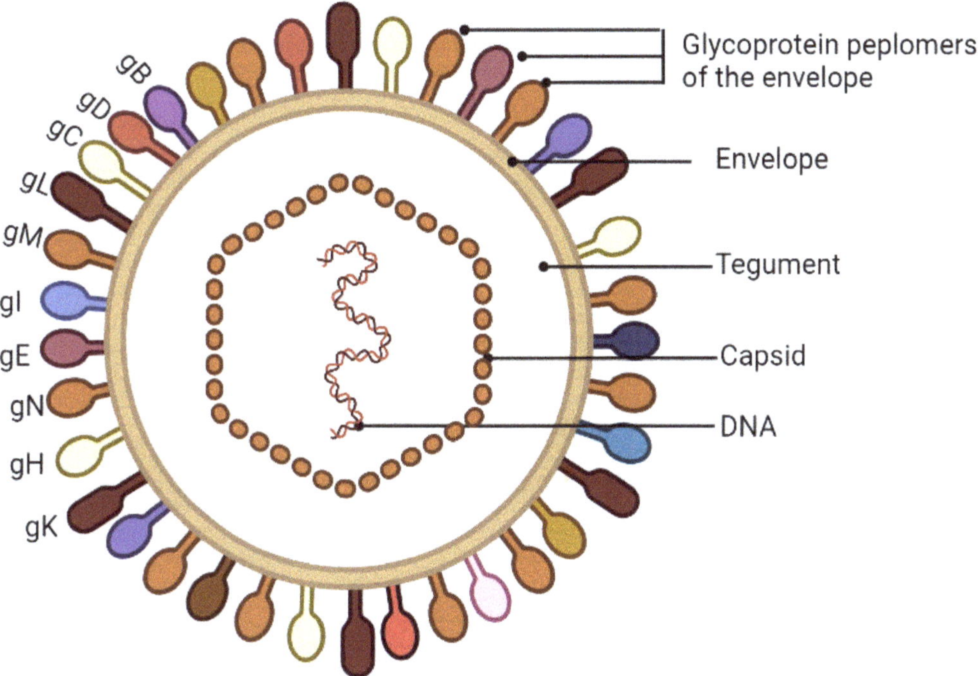

Figure 1. Schematic diagram of the PRV virion. PRV virions are composed of four structural elements, including a linear double-stranded DNA genome, an icosahedral protein capsid, a protein tegument layer, and a lipid envelope containing viral glycoproteins.

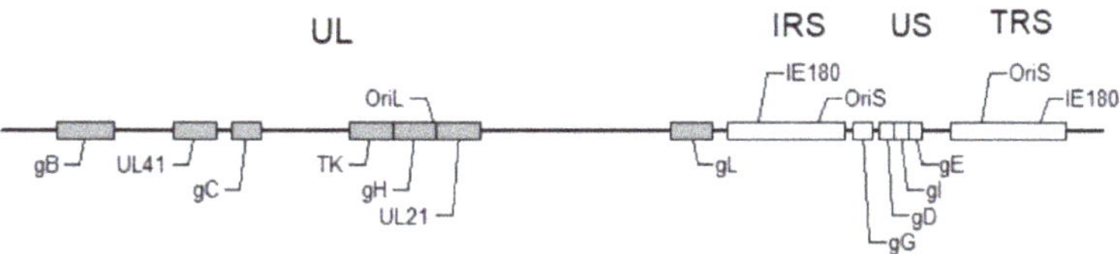

Figure 2. A map of the PRV genome showing the location of PRV genes discussed in this review. The PRV genome consists of the unique long region (UL) and unique short region (US), which is flanked by the internal (IRS) and terminal (TRS) repeat sequences. The genes represented by gray boxes locate in the UL, including gB, UL41, gC, TK, gH, UL21 and gL; the genes represented by white boxes locate in the US, including gG, gD, gI and gE.

Some major genes are associated with the process of PRV infection. PRV thymidine kinase (TK), namely the UL23 gene, plays a decisive role in the virulence of the virus. It was primarily described as being related to the replication and neuro-invasiveness of PRV in the central nervous system and is also involved in re-activating the virus during the latent infection period [27]. The lack of the TK gene significantly reduces the ability of replication and transmission in nerve cells without affecting its immunogenicity [28]. As one of the virulence genes, gE, located in the US region, is not essential for viral replication and has no effect on viral immunogenicity. gE glycoprotein can promote the fusion of PRV and cells and mediate the spread of the virus between cells. PRV without the gE gene can only infect the primary trigeminal nerve and sympathetic nerve regulating nasal mucosa, but not the secondary ganglion and sympathetic neurons [29]. In the processes of PRV invading and spreading into the nervous system, gE and gI proteins generally exist in a complex, which is often distributed in the cell membrane of infected cells and in the virus envelope [30]. gI protein is a membrane protein that can not only promote the secretion of gE protein in the endoplasmic reticulum to ensure its correct glycosylation, but also facilitate the transmission of the virus between cells [31]. Furthermore, gI glycoprotein can also cooperate with gC protein in the release process of the virus. The gC protein plays a role in the first step of PRV replication, that is, the adhesion process to cells. It can initiate virus attachment to cells by binding to heparan sulfate (HS) proteoglycans and also participate in the process of virus release from cells [32,33]. The sequence of the gB gene is more conservative than other genes, and gB glycoprotein is an important structural protein of the virus envelope. Both gB glycoprotein and the gH/gL complex can jointly promote the fusion of cell membrane and virus envelope when the virus invades cells [34]. gB protein is a typical class III post-fusion trimer that binds membranes via its fusion loops (FLs) in a cholesterol-dependent manner [35]. gD glycoprotein accelerates the rapid fusion between the PRV capsule membrane and the cytoplasmic membrane of target cells and boosts the fusion of gB/gH/gI complex [36]. Yet, it cannot participate in virus spread between cells, which is not essential for viral replication. Compared with herpes simplex virus (HSV), the gD protein of PRV can only bind to the cell surface receptor Nectin-1 at sites N77, I80, M85 and F129, with a binding ratio of 1:1, and the binding site F129 plays an important role in invading cells [37]. The gG gene locates in the US region. The gG protein belongs to a larger complex, which is synthesized, secreted and released by infected cells, but virus particles do not contain this glycoprotein [36]. Furthermore, gG protein has very good immunogenicity and can effectively stimulate the organism to produce antibodies [38]. At the same time, gG protein can bind to chemokines produced by the organism, leading to the immune escape of the virus. IE180 gene encodes 1460 amino acids with a protein of ~153 kDa in molecular mass, and it is the only immediate early gene in PRV that can be transcribed independently. The accumulation of IE180 protein can start the transcription of other genes, and it is highly similar to some regions of herpes simplex virus type I ICP4 protein, which is complementary to both proteins in some of their functions [39]. The IE180 3′UTR end sequence can form a G-quadruplex structure and inhibit gene transcription [39]. The G-quadruplex ligand small molecule TmPyP4 (meso-Tetrakis (N-methyl-4-pyridiniumyl) porphyrin) can stabilize this structure and further inhibit the early proliferation of PRV [40]. Additionally, IE180 gene transcription locates at the beginning of the PRV replication cycle, which is of great significance for PRV replication, suggesting that the IE180 gene may also become an important target for the development of anti-PRV drugs. As one of the non-essential genes for virus replication, the EP0 gene, located in the UL region, can encode 1230 amino acids and has an early protein of about 45 ku. It can transactivate the immediate early gene IE180. The replication ability of PRV without the EP0 gene in cells is weakened, but it does not affect the virulence. EP0 protein can interact with IE180 protein to activate the transcription of the TK and gG genes; besides, it is located in the nucleus, and the nuclear entry of EP0 protein is co-regulated by Ran protein and input proteins α1, α3 and β1 [41]. The UL21 gene, located in the UL region, is a non-essential gene for PRV replication, and the protein encoded by UL21 belongs to the tegument proteins. The proliferation ability of PRV without the UL21 gene is reduced, but can be restored on pUL21 compensatory cells [42]. Exogenous pUL21 can inhibit the NF-kB pathway, with a positive correlation, and its carboxyl end can interact with cytoplasmic dynamic protein Roadblock-1, which further influences the nerve infectivity of PRV [42]. The UL41 gene is an essential gene for virus replication and encodes host closure protein (named as vhs), with a molecular weight of about 40 ku. The vhs protein has ribonuclease activity both in vivo and in vitro, and can degrade the mRNA of host cells and inhibit gene expression. The vhs protein can also cleave the downstream region of internal ribosome entry site (IRES), and the translation initiation factors eIF4H and eIF4B can significantly increase the RNase activity of recombinant PRV vhs against capped RNA [43,44].

2.3. The Life Cycle of PRV

The process by which PRV virions enter host cells is primarily initiated by the binding of virions to the surface molecules of host cells and the fusion of the virus and host cell membranes (Figure 3). PRV virions first attach to cells by the interaction of gC with heparan sulfate proteoglycans in the extracellular matrix. PRV gD then binds to specific cellular receptors to stabilize the virion-cell interaction. Finally, PRV gB, gH and gL mediate the fusion of the viral envelope and the cellular plasma membrane to allow penetration of the viral capsid and tegument into the cell cytoplasm, and tegument proteins in the outer layer quickly dissociate from the capsid following their fusion [45,46]. Then, the capsid interacts with dynein for transport along microtubules from the cell periphery to the nuclear pore [45,47]. After capsid docking at the nuclear pore, the PRV DNA is released into the nucleus from intact capsids [45,47].

For the viral transcription and cascade, the immediate-early IE180 transcript is detected within 40 min of infection, and its protein is synthesized up until 2.5 h post-infection (hpi). The IE180 protein regulates the early gene expression related to replication. The early EP0 transcript is detected at 2 hpi, and its expression in vivo activates gene expression from PRV promoters, such as IE180, TK and gG, which have the characteristics of transcription activators. Other regulators of gene expression also participate in viral transcription, including UL54, UL41, and UL48, and among them, UL54 and UL41 are likely to encode potent regulators of both viral and cellular gene expression. UL41 encodes the vhs protein, which is responsible for the virion host shut-off of cellular protein synthesis. UL48 encodes the tegument protein VP16, which enhances the expression of viral immediate-early genes in newly infected host cells [48].

Upon entry into the host nucleus, the linear viral DNA genomes assume a circular form and are quickly repaired of nicks and misincorporated deoxyribonucleotides [2]. The circular genomes serve as the template for DNA synthesis, and the initial theta replication mechanism quickly switches towards a rolling-circle mechanism of DNA replication. The latter process produces replicated DNA in the form of long linear concatemeric genomes that serve as the substrate for genome encapsidation [49,50]. In this process, many of the enzymes encoded by PRV also participate in viral DNA replication, such as UL52, UL42, UL30, UL29, UL9, UL8, and UL5. Several enzymes encoded by PRV genomes can be involved in nucleotide metabolism, for example, dUTPase (UL50), thymidine kinase (UL23) and two-subunit ribonucleotide reductase (UL39/UL40). The viral genome also encodes a uracil DNA glycosylase (UL2) as well as an alkaline nuclease (UL12), which both serve in viral DNA repair, recombination and DNA concatemer resolution [27,51,52]. When the viral genome completes transcription, translation and DNA replication, the capsid protein automatically enters the nucleus of the cell to form the basic assembly unit of the capsid and assembles into the nucleocapsid in the nucleus of the cell. After assembly, PRV nucleocapsids cross the inner and outer nuclear membrane by budding and fusion, respectively, following release into the cytoplasm, and then are formed into the mature enveloped virions and released to the outer cell by budding for the next round of infection [53–55].

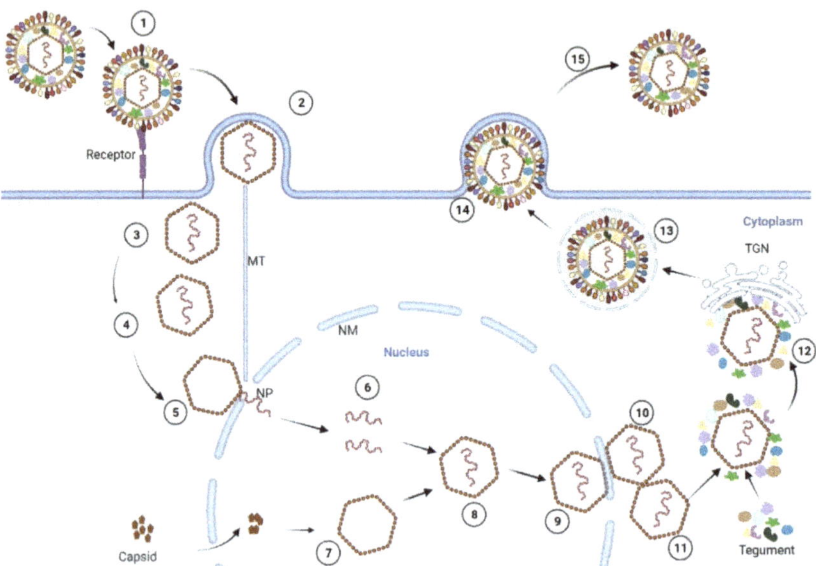

Figure 3. The replication cycle of PRV. After adsorption (1) and penetration (2), capsids are transported to the nucleus (3) via interaction with microtubules (MT) (4), docking at the nuclear pore (NP) (5) where the viral genome is released into the nucleus. In the nucleus, DNA replications occur (6). The capsid proteins are transported to the nucleus and are assembled around a scaffold (7), and then are assembled into a nucleocapsid with the insertion of the genomic DNA (8). The nucleocapsid leaves the nucleus by budding at the inner nuclear membrane (INM) (9), followed by fusion of the envelope of these primary virions located in the perinuclear space (10) with the outer nuclear membrane (11). Final maturation then occurs in the cytoplasm by the secondary envelopment of intra-cytosolic capsids via budding into vesicles of the trans-Golgi network (TGN) (12) containing viral glycoproteins, resulting in an enveloped virion within a cellular vesicle (13). After transport to the cell surface (14), vesicle and plasma membranes fuse, releasing a mature, enveloped virion from the cell (15).

3. Occurrence and Development of PRV Infection

PRV can cause respiratory disease, neurological disorders and abortion in pigs. Its transmission mainly occurs through direct contact between oral and nasal secretions but can also occur by aerosols, transplacental contact and blood [56,57]. In this work, we reviewed the main steps in the pathogenesis of PRV in pigs (Figure 4).

3.1. PRV Primary Replication in the Upper Respiratory Tract

After PRV enters a natural host, it firstly replicates by an infection foci manner in the epithelial cells lining the upper respiratory tract (URT), including nasal septa, tonsils, nasopharynxes, trachea and lungs [14,17,58,59]. In vivo, viral DNA was detected in the nasal mucosa, tonsils and lungs of 2-week-old piglets starting 24 hpi, and PRV-induced plaques can be observed in the epithelium of porcine nasal mucosa explants after 24 hpi in ex vivo experiments [17,60,61]. Primary PRV infection in multiple tissues of porcine URT causes the destruction and erosion of epithelium, with slight respiratory symptoms consisting of sneezing, coughing, dyspnea and nasal discharge after 3 to 6 days post-inoculation (dpi), which normally disappear quickly [59,62]. Besides, viral shedding can be detected in nasal secretions starting from 1 to 14 dpi [63].

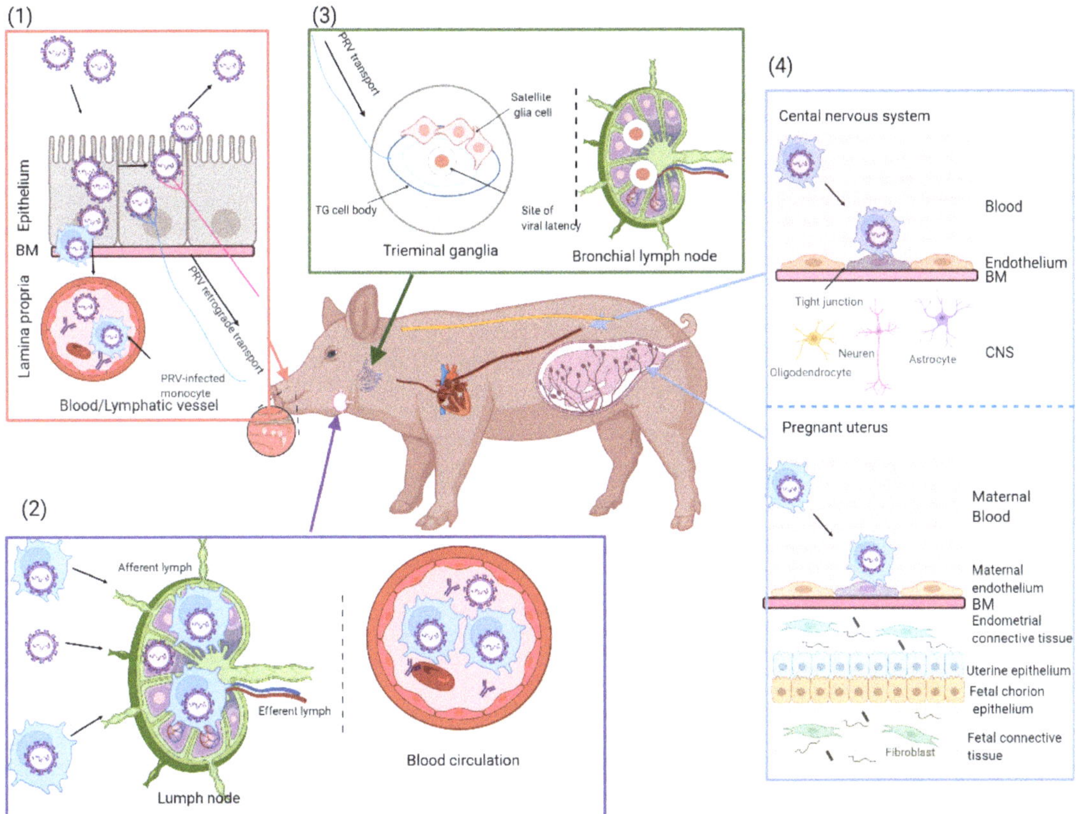

Figure 4. Schematic representation of the pathogenesis of PRV in pigs in different stages of growth. (1) Primary viral replication in the epithelial cells (ECs) of the upper respiratory tract: PRV first infects epithelial cells, with a viral spread and shedding, and then crosses the basement membrane (BM) and lamina propria by using single infected leukocytes to reach the blood circulation and draining lymph nodes. Lastly, PRV entry occurs at nerve endings of the peripheral nervous system and diffuses retrogradely to trigeminal ganglia (TG). (2) PRV replication in the draining lymph nodes and cell-associated viremia. (3) Establishment of PRV latency in the trigeminal ganglia (TG) neurons. (4) Secondary replication in target organs (the pregnant uterus and the central nervous system (CNS)): the secondary replication in the ECs of the pregnant uterus can lead to vasculitis and multifocal thrombosis, with an abortion of sows, and in newborn piglets, sudden death usually occurs in the absence of clinical signs.

3.2. PRV Replication in the Draining Lymph Nodes and Viremia

After respiratory epithelium infection, PRV can pass the basement membrane (BM) by infected leukocytes to penetrate the connective tissues and further reach the bloodstream and the draining lymph nodes [61,64]. The process of viral invasion into lamina propria through BM is mediated by the activity of trypsin-like serine protease excreted by the virus [65]. PRV antigens or DNA are both detected in the inguinal lymph nodes of pigs starting from 24 hpi to 48 hpi, and the virus can persist for 35 days in pharyngeal lymph nodes [66–69]. PRV infection is also amplified in the draining lymph nodes, and infected leukocytes are discharged into the blood circulation through the efferent lymph. Hence, PRV induces cell-associated viremia in peripheral blood monocytes and promotes its transmission in pigs [18,70]. Meanwhile, viremia also occurs in cell-free form after PRV infection and can be regularly detected between 1 dpi and 14 dpi [18]. Cell-associated viremia is considered a prerequisite for the dissemination of PRV to the pregnant uterus in pigs.

3.3. PRV Entry into the Peripheral Nervous System (PNS) Neurons and Spread to the Central Nervous System (CNS)

Upon primary replication of respiratory epithelium in adult pigs after PRV primary infection, it enters PNS' nerve endings, which contain those coming from the sensory trigeminal ganglia (TG) and olfactory bulb, and other facial, parasympathetic, sympathetic nerve neurons that innervate the epithelium [71,72]. PRV particles can be transported retrogradely to sensory and autonomic peripheral ganglia. Generally, the herpesviruses are able to establish a reactivable, latent infection in their hosts. PRV can also set up a lifelong potential infection in PNS neurons of pigs, whereas the infected pigs cannot display any clinical symptoms after recovering from the respiratory disease [15,16,63,73,74]. After stress-induced reactivation, viral replication happens in the PNS ganglia, and virions spread in the anterograde along the nerves to the mucosal surfaces where the infection initiated, further causing mild respiratory signs in adult pigs upon viral reactivation [75]. In addition, PRV barely propagates to the CNS in the retrograde direction to result in the encephalitis of adult pigs, but its latent period and reactive cycles in pigs cause the infectious virus to fall off and spread to uninfected pigs, which is conducive to viral accumulation in pig populations. Intriguingly, the herpesviruses of human beings and other animals, such as varicella-zoster virus (VZV) and bovine herpesvirus type 1 (BHV-1) have a similar way of spreading to invade PNS neurons [76,77].

3.4. Secondary Replication in the Swine Pregnant Uterus

Once in the blood circulation, PRV-infected monocytes can cross the endothelial cells (EC) barrier of the maternal blood vessels to reach the pregnant uterus of sows via adhesion and fusion of these monocytes with EC, further transmitting PRV [78]. The adhesion molecules on the surface of EC and leukocytes play a significant role in the infection of the vascular endothelium, and the secondary replication in the EC of the pregnant uterus can lead to vasculitis and multifocal thrombosis, with an abortion of a sow [18,79,80]. The EC infection in the vasculature of the pregnant uterus is usually mediated by intercellular contact between infected monocytes and EC [78]. The occurrence of abortion may depend on the hormonal activity and immune status of sows during pregnancy. In fact, it has been proved that the expression of adhesion molecules on EC is induced by cytokines and hormones in the local environment during pregnancy [81–83]. These cytokines may accelerate the adhesion of infected monocytes to the endothelial cells.

Upon the intranasal, intra-uterine, and intra-fetal inoculations of vaccinated pregnant sows, PRV antigen can be detected in vaginal and sacral ganglia [84]. An extensive EC infection can lead to the fetal membranes shedding in early pregnancy, resulting in a virus-negative fetal abortion or fetal reabsorption in sows. Little uterine vascular pathology may cause transplacental infection and the abortion of virus-positive fetuses in the second and third trimester of gestation or a stillborn pig [59,85]. A fetus with viral abortion generally displays several lesions, such as necrosis of the liver, spleen and lungs, and PRV strains can be isolated from the above organs [19,86].

3.5. PRV Infection in Suckling and Weaned Piglets

PRV infection usually brings more severe lethality to piglets than to adult swine and sows [59]. In newborn piglets, sudden death usually occurs in the absence of clinical signs. Instead, before the death of suckling pigs, some signs are found in infected pigs, including fever, vomiting and CNS symptoms, which consist of coordination problems, hindquarter weakness, convulsions and paralysis. It is worth noting that the mortality rate of newborn and suckling pigs is close to 100%. In weaner pigs, clinical signs are similar to those of suckling pigs, with a mortality rate of 5% to 10%. Nevertheless, pigs of any age cannot have itching. Infectious viruses can be isolated from brain tissue samples of piglets naturally infected with PRV [64]. The severity of symptoms diminishes with age, as adult pigs have a more effective immunity than piglets.

3.6. PRV Infection in Humans

In the past century, the viewpoint of PRV infecting humans has been controversial due to the absence of unequivocal etiology or serological diagnosis [87]. The first case of humans with suspected PRV infections was reported in 1914, and they were previously in contact with PRV-infected cats. Subsequently, several patients with PRV infection were reported after long-term exposure to PRV-susceptible animals (pigs, cats, dogs). These patients mainly displayed pruritus, headache, fever, swelling, sweating, dysphagia, aphthous ulcer, altered mental status, seizure and coma [4–8]. Notably, Chinese patients mainly showed encephalitis and endophthalmitis, and most of them were working on pig farms and had injuries to their fingers or other places on their bodies. In addition, the first human-originated PRV strain, designated hSD-1/2019, was isolated and identified from the cerebrospinal fluid (CSF) of PRV-infected patients worldwide, which provided direct evidence of PRV infection in human beings [5]. In terms of the PRV infectious pathway in humans, it has yet been unclear. However, from the perspective of animal research, it was supposed that PRV might not only affect the brain, but also affect other human organ systems, further causing serious consequences. Although these patients all survived until hospital discharge, and clinical symptoms disappeared, it was unclear whether the recovered patients still carry PRV owing to the long-term latent period of PRV in pigs. Therefore, it is necessary to ensure thorough skin protection in humans, particularly those who have been in close contact with pigs.

4. Genetic Evolution of PRV

Currently, PRV strains in the world can be divided into two genotypes, I and II. The genotype I strains are mainly prevalent in Europe, America and parts of China, while the majority of genotype II strains are isolated from Asia, particularly China [88]. The latter has undergone mutations caused by host immune pressure over a long period of time and evolved into novel PRV variant strains, which then led to the co-prevalence of both variant and classical strains, which again poses great threats to the swine industry in China [89]. Because gB, gC and gD glycoproteins are major immune-related proteins and gE is a significant virulence protein, these four gene sequences are often used to analyse PRV's genetic evolution [90,91]. To investigate the genetic characteristics of PRV strains from various countries and species, phylogenetic trees based on the complete length of gE, gC, gB and gD gene sequences were constructed using MEGA software (version 7.0) by the neighbor-joining method with 1000 bootstrap replicates [92]. The results showed that these PRV strains were grouped into two genotypes as expected (Figure 5). The genotype I group is composed of European–American PRV strains, but the genotype II group is formed by Asiatic strains (China and Japan), and among them, Chinese PRV variant strains were clustered in one subgroup. Findings revealed that PRV genotype II strains had become dominant prevalent strains instead of genotype I in China, which is in agreement with previous studies [88,90,93]. This finding again explains why Bartha strains (genotype I) did not provide full protection against variant PRV strains (genotype II). In addition, PRV strains isolated from other animals are randomly distributed in two genotypes (Figure 5). Combined with previous research [94], this suggests that PRV strains isolated from animals and human beings may have a similar ancestor to those of pigs.

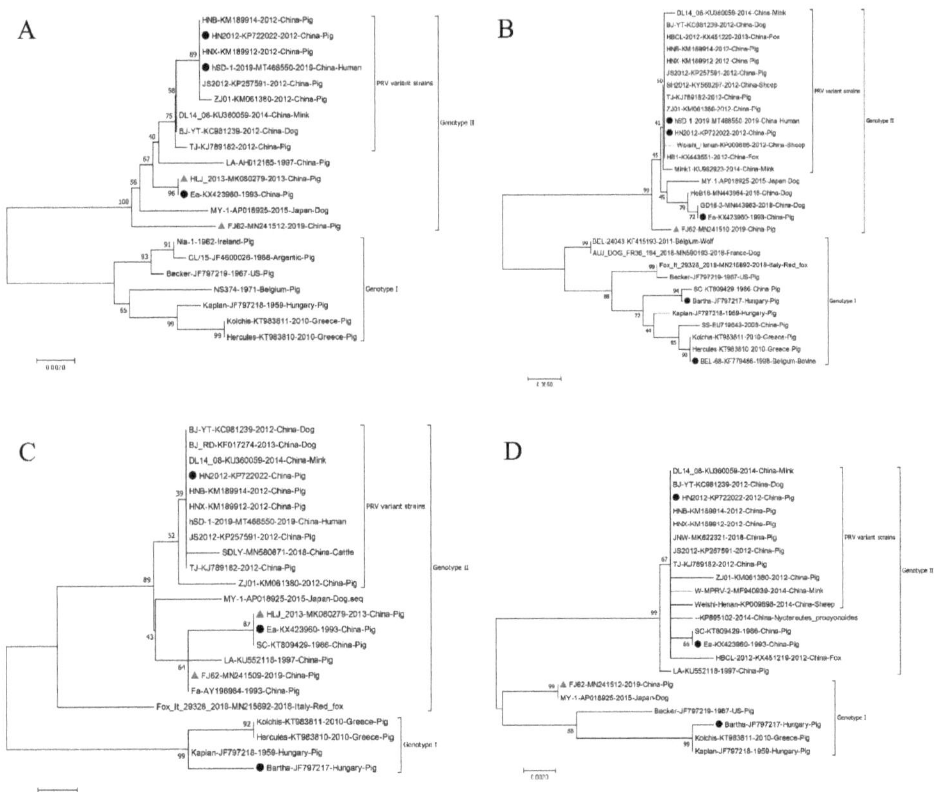

Figure 5. Phylogenetic trees based on the nucleotide sequences of the gE (**A**), gC (**B**), gD (**C**) and gB (**D**) genes of PRV strains from different hosts/regions using the neighbour-joining method with a bootstrap test of 1000 replicates using MEGA 7.0 software (www.megasoftware.net; Access date: 15 April 2022). The hosts, countries, years, names and GenBank accession numbers of the reference strains employed in this phylogenetic tree are labeled. Black circles and red triangles represent vaccine strains and recombinant strains, respectively.

In mutation analysis of PRV variant strains, He et al. found that the average amino acid (aa) differences of the Qihe547 variant strain are 4.94%, 1.16% and 0.46% compared with PRV strains of genotype I and the classical and variant strains of genotype II, respectively [88]. Compared with genotype I strains, PRV genotype II strains present high genetic mutations in internal and terminal repeat regions, and nucleotide insertions, deletions and mutations are commonly observed in the different PRV genes [95]. For example, there are two insertions of discontinuous six nucleotides at sites 142–144 (GAC) and 1488–1490 (CGA) in the gE gene of PRV strains in genotype II [96,97]. In Sun's study, PRV strains in genotype II are found to have three aa deletions (75VPG77) in the gB protein and seven aa insertions (63AASTPAA69) in the gC protein [98]. Interestingly, two aa deletions (288SP289) have been identified in the gD protein of Chinese PRV variant strains by comparison with classical Chinese strains [99]. Furthermore, inter-clade and intra-clade recombinant events can accelerate the evolution of the PRV genome and alter viral virulence, which has been demonstrated in several studies [88,90,100]. In Zhai's study, two inter-clade recombinant PRV strains (FJ-W2 and FJ-ZXF) were reported, and their gE, gC and gD genes were assigned to genotype II, whereas gB genes belong to genotype I [90]. Subsequently, Huang et al. analyzed the genetic evolution of the primary immune-related gene sequences of PRV variants, and the results showed that the gB gene of the PRV variant strain FJ62 isolated from piglets in Sichuan, China, is identical (100%) to the MY-1 strain (No. AP018925) from a wild boar in Japan, with low sequence homologies (98.4–98.5%) of Chinese PRV strains. However, its gC, gD and gE genes have high sequence similarities of 99.5%, 99.9%, and 99.9%, respectively, demonstrating that PRV variant strain FJ62 may appear from a recombinant event of PRV strains of genotypes I (Japan) and II (China) spanning different countries [91]. In another report, PRV HLJ-2013 was isolated from pigs in Heilongjiang province of China belonging to genotype II, and its genome sequences are derived from three viruses (including a yet unknown parental virus, the European viruses and the same ancestor of all Chinese strains) based on the phylogenetic trees of both protein-coding genes and non-coding regions [100]. The recombination analysis showed that there are six recombinant events in SC strains (No. KT809429) belonging to genotype II, and among these events, HLJ-2013 is predicted to be the major parent of the SC strain, with a minor parent of Bartha [100]. Additionally, Tan et al. also found that a naturally recombinant event might occur in the genome of the HN-2019 strain isolated from a sick piglet in Hunan of China, between the PRV classical strain and the HB-98 vaccine strain, which again confirmed the presence of a recombinant event in PRV [101].

5. Diagnostic Methods

At present, serological technologies and molecular biology methods have become the common diagnostic approaches for PRV detection (Figure 6) because traditional clinical and pathological diagnostic methods cannot accurately diagnose PR. Two common methods are used to verify the PRV infection based on PRV-specific antibodies and nucleic acids, respectively, which have their own features (Table 1).

5.1. Serological Approaches for the Detection of PRV Antibodies

Due to the extensive use of PRV gE-deleted vaccines worldwide, gE as the marker antigen has become widely used in serological approaches, and many PRV gE antibodies have been developed to quickly and effectively differentiate infected from vaccinated animals (DIVA). gB antibodies based on serological methods can also be used for monitoring the immune level induced by vaccine immunization. Up to date, various serological methods can be used for the detection of PRV antibodies, such as the direct-immunofluorescence method (DFM), indirect immunofluorescence assay (IFA), serum neutralization test (SNT), enzyme-linked immunosorbent assays (ELISA), blocking immunoperoxidase monolayer assay (b-IPMA), latex agglutination test, agar diffusion test, particle concentration fluorescence immunoassay (PCFIA) and immunochromatographic strip [1,93,101–109]. Among them, ELISA remains the most common method in the clinical detection of PRV antibodies because it has high specificity and sensitivity compared with other screening assays. There are three single commercial serum antibody ELISAs, including the gB blocking ELISA (gB bELISA), gI blocking ELISA (gI bELISA) and gE indirect ELISA (gE iELISA) [110]. To make detection more convenient, competitive ELISAs (cELISA) targeting the gB or gE antibody have been developed and extensively applied in China [111–114]. Interestingly, a novel serological technology based on the blocking fluorescent lateral flow immunoassay takes less time for PRV detection and is sensitive to differentiate wild PRV-infected and vaccinated pigs, whereas a commercial gE-ELISA kit is not [102]. Subsequently, another new detection method, dual fluorescent microsphere immunological assay (FMIA), was developed for detecting PRV gE and gB IgG antibodies simultaneously, and it also has accuracy for gE detection with high sensitivity (92.3%) and specificity (99.26%) compared to a commercial gE/gB ELISA kit, with less time and cost expenses [115]. Furthermore, both the immunochromatographic assay and liquid chip technology methods also exhibit higher sensitivity than that of cELISAs [116]. Hence, blocking fluorescent lateral flow immunoassay, FMIA, immunochromatogragphic assay and liquid chip technology are expected to become new clinical laboratory diagnostic methods for detecting PRV antibodies, although these methods are not universal thus far.

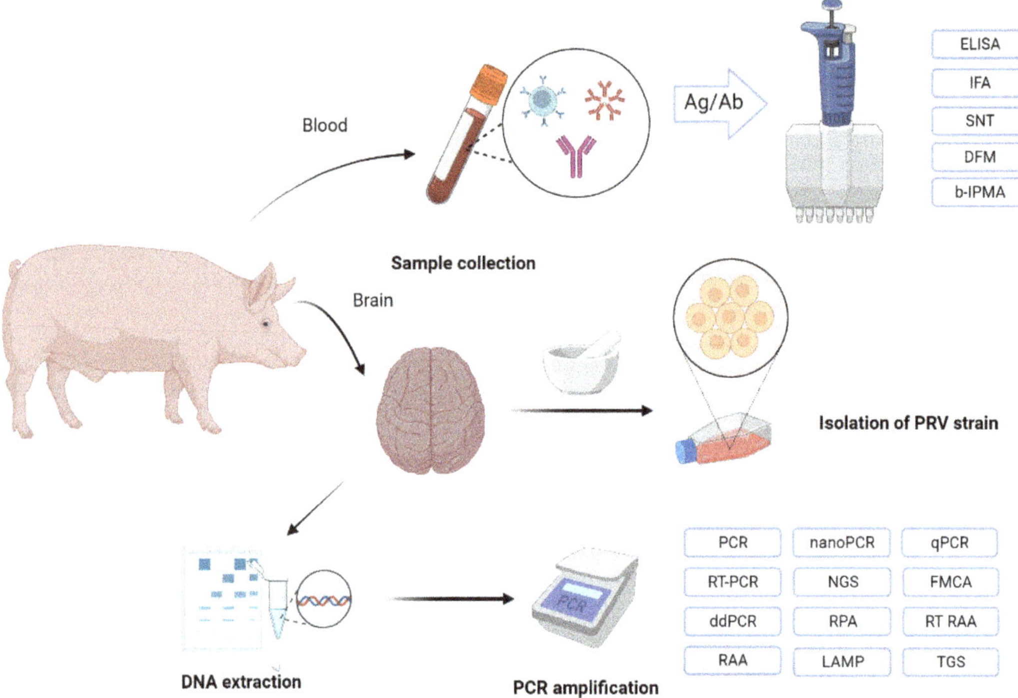

Figure 6. Flowchart of common diagnostic methods for PRV infection. Two types of common methods are used to verify the PRV infection based on the PRV-specific antibodies and nucleic acids. Among them, serological approaches for the detection of PRV infection include the enzyme-linked immunosorbent assays (ELISA), indirect immunofluorescence assay (IFA), serum neutralization test (SNT), direct-immunofluorescence method (DFM) and blocking immunoperoxidase monolayer assay (b-IPMA). Molecular biology approaches include polymerase chain reaction (PCR), real-time PCR (RT-PCR), TaqMan real-time PCR (qPCR), nano PCR, droplet digital PCR (ddPCR), real-time recombinase-aided amplification (RT RAA), loop-mediated isothermal amplification (LAMP), real-time fluorescent detection (real-time RPA assay), duplex fluorescence melting curve analysis (FMCA), next-generation sequencing (NGS), probe-based fluorescence melting curve analysis (FMCA), real-time recombinase-aided amplification assay (RT RAA) and third-generation sequencing (TGS).

Table 1. Comparison of diagnostic methods of PRV infection.

	Molecular	Serology
Test type	Viral	Antibody
Description	Nucleic acid amplification test to detect viral DNA	Detects the presence of IgA, IgM/IgG antibodies against PRV
Platform technology	PCR, RT-PCR, LAMP, qPCR, ddPCR, FMCA	ELISA, SNT, IFA, IPMA, DFM
Sample type	Brains, Hearts, livers, spleens, lungs, kidneys and lymph nodes	Plasma, serum, whole blood
Result turnaround time	<5 h	15–30 min

5.2. Molecular Biology Approaches for the Detection of PRV Infection

To further improve the sensitivity and specificity of PRV detection, molecular biology methods targeting the specific sequences of PRV genes, including the gE, gI, gC, gD, gB, and gG genes, have been established (Table 2), such as polymerase chain reaction (PCR), real-time PCR (RT-PCR), TaqMan real-time PCR (qPCR), nano PCR, droplet digital PCR (ddPCR), real-time recombinase-aided amplification (RT RAA), loop-mediated isothermal amplification (LAMP), and duplex fluorescence melting curve analysis (FMCA) [117–131]. Generally, PCR and RT-PCR are the most frequently used approaches to quickly detect PRV infection or distinguish between the PRV wild-type and vaccine strains [124]. In particular, multiplex PCR assays have also been developed for simultaneously detecting PRV and other pathogens, for example, porcine circovirus 3 (PCV3), porcine circovirus 2 (PCV2), porcine parvovirus (PPV) and porcine cytomegalovirus (PCMV), and Torque teno sus virus 1 and 2 [118,132–136]. In addition, the novel high-throughput sequencing, next- and third-generation sequencing (NGS and TGS) methods are used to survey the transcriptome of PRV, and the existence of PRV can be detected in patients by NGS, which is the most powerful and supersensitive assay [6,137]. Nevertheless, high-throughput sequencing is not suitable for wide-range clinical detection owing to its high cost [6]. Compared with conventional PCR, the LAMP assay is more sensitive, specific, rapid and cost-effective, so it is more suitable for PR diagnosis in the field, with a huge potential application in the prevention and control of PRV [122].

Table 2. List of molecular diagnostic methods of PRV infection.

Name of Diagnostic Assay	Sensitivity	Target Gene	Turnaround Time	Samples Used	References
Conventional polymerase chain reaction (PCR)	——	gE gene	Result in <5 h	Various tissue	[119,121]
Duplex droplet digital PCR (ddPCR) assary	4.75 copies/μL	Both gE and gB genes	Result in <2 h	Lung, brain, liver and spleen	[124]
SYBR green I-based duplex real-time PCR assay	37.8 copies/μL	gE gene	Result within 50 min	Hearts, livers, spleens, lungs, kidneys, brains and lymph nodes	[118]
Real-time recombinase-aided amplification assay (RAA)	Three 50% TCID$_{50}$	gE gene	Result in 75 min	Lung, lymph node, tonsil and spleen	[125]
Triplex real-time PCR	0.5 TCID$_{50}$ for classical strains, 0.2 TCID$_{50}$ for variant strains and 0.05 TCID$_{50}$ for vaccine strains	gE and gI genes	Result within 1 h	PRV strains	[127]
Probe-based fluorescence melting curve analysis (FMCA)	1×10^0 copies per reaction	gC and gE genes	Result in <2 h	PRV strains	[120]
Loop-mediated isothermal amplification (LAMP) assay	10 copies per sample	gE and gG genes	Result in <2 h	PRV strains and clinical tissue samples	[122]
Duplex nanoparticle-assisted polymerase chain reaction (nanoPCR)	6 copies/μL	gE gene	Result in 80 min	The recombinant plasmids pET30a-PRV-gE and pUC57-PBoVNS1	[128]
Real-time quantitative PCR (RT-qPCR)	Oral fluid of 53% and nasal swab of 70%	gB gene	Result in <1 h	Oral fluid and nasal swab	[129]
Metagenomic next-and third-generation sequencing (mNGS/TGS)	——	Short- and long-read sequencing	——	Brains	[137]
Real-time fluorescent detection (real-time RPA assay)	100 copies per reaction	gD gene	Result within 20 min	Tissue	[130]
Lateral flow dipstick (RPA LFD assay)	160 copies per reaction	gD gene	Result within 20 min	Tissue	[130]
Magnetic beads-based chemiluminescent assay	100 μmol/5 pM	——	Result in 20 min	Serum samples	[131]

5.3. Other Approaches for the Detection of PRV Infection

In addition to the above methods, direct detection of pathogens is another effective candidate, which also greatly facilitates the vaccine design to prevent PR. The methods mainly include virus isolation and subsequent laboratory diagnosis, involving serology, molecular biology and electron microscopy. Virus isolation (VI) is considered the "gold standard" for pathogen diagnosis, and a large number of PRV strains from different hosts have been successfully isolated, though this method is only applicable to professional laboratories [138,139]. Because the traditional diagnostic methods (VI identification and animal experiments), serological diagnostic approaches and molecular biological methods have some limitations, such as complicated operation and high technical requirements, and are time-consuming, a paper biosensor doped with $Fe_3O_4@SiO_2$–NH_2 and multi-walled carbon nanotubes (MWCNTs) for the rapid detection of PRV has been developed [140]. In this assay, $Fe_3O_4@SiO_2$–NH_2 can provide magnetic response characteristics, and MWCNTs are able to increase the electrical conductivity [140].

6. The Prevention of PR

To better prevent and control PR, numerous efforts have been made for the development of effective means to control PRV infection, mainly including vaccines and other novel viral inhibitors.

6.1. Main Vaccines against PRV Infection

As a great challenge, PRV has been prevalent in pig farms worldwide for nearly two hundred years. Vaccination is one of the most effective ways to prevent disease and minimize the economic losses caused by PR [10]. Most PRV vaccines are live gene-modified virus vaccines (Table 3). The initial live gene-modified vaccines (attenuated Bartha-K61 strain and PRV Bucharest strain) are usually obtained from extensive passaging of virulent field isolates in cell cultures from 1961, which, following the wide application in pig herds, and effectively controlled PR worldwide [141]. In China, a live gene-modified vaccine (attenuated Bartha-K61 strain) was imported from Hungary in the 1970s and was widely inoculated in pig farms, with effective control of PR [12,142]. With further understanding of PRV genetics and molecular biology, gE protein has become regarded as the important swine neurovirulence factor [143]. TK is also essential for virus replication in nonmitotic tissues (neurons), and other proteins are considered the factors of viral replication, such as gI, Us9, Us2, gC, gG and PK proteins [144]. Therefore, the genes encoding these proteins can be deleted to mediate viral attenuation, especially the gE gene. In fact, TK-negative stains (ΔTK) is the first genetically modified live vaccine and was licensed for application in 1985 [27]. Subsequently, other gene deletion vaccines also generated by various technologies, such as gE gene deletion and double-gene gE/TK deletion of the NIA3 strain, double-gene gD/gI deletion of the NIV-3 strain, double-gene TK/gG deletion of the virus strain, triple-gene TK/gE/gI deletion of the BUK strain and Fa strains, four-gene gD/gG/gI/gE deletion of the PrV(376) strain, and a triple-gene-deleted (gE/gI/TK) vaccine generated based on the PRV Fa strain licensed in 2003, which was regarded as the first genetically modified vaccine against PR in China [145–150]. However, since 2011, PR outbreaks caused by emerging PRV variants have occurred in Chinese pig herds which were immunized with the Bartha-K61 strain, indicating that the classical attenuated PRV vaccine cannot provide complete protection for swine [111,151–153]. With the rapid development of biotechnologies and in-depth understanding of the biological functions of PRV encoding genes, some gene-modified vaccines and other types of vaccines have also been generated based on emerging virulent PRV strains. As ideal target genes, the gE, gI, TK, Us9, Us2, gC, gG and PK genes can be deleted to develop genetically engineered vaccines against re-emerging PR, particularly gE, gI and TK genes. Therefore, similar to single- or multiple-genes deletion of classical PRV strains, some novel gene-modified vaccines based on the variant PRV strains have been generated in recent years, but only two types of vaccines have been licensed thus far, including the gE-gene-deleted inactivated vaccine on the basis of the PRV HeN1201 strain isolated from 2019 and another natural four-gene-deleted (gI/gE/Us9/Us2) vaccine based on the PRV C strain in 2017 [29,154,155]. There are other candidates, including killed and live attenuated vaccines based on the variant PRV strains, and the latter can be generated by homologous DNA recombination, clustered regularly interspaced palindromic repeats (CRISPR)/associated (Cas9) system and bacterial artificial chromosome (BAC) [155–158]. Thus far, the gene-modified vaccine of PRV based on the variant PRV strains contains a double-gene deletion of the ZJ01 strain (vZJ01-ΔgE/gI), the PRV-XJ strain (rPRVXJ-delgI/gE-EGFP), and the AH02LA strain (PRV B-gD&gC^S), three-gene deletion based on the HN1201 strain (vPRV

HN1201 TK⁻/gE⁻/gI⁻), the TJ strain (rPRVTJ-delgE/gI/TK), the NY strains (rPRV NY-gE⁻/gI⁻/TK⁻), the GX strain (rGXΔTK/gE/gI), the XJ5 strain (rPRV/XJ5-gE⁻/gI⁻/TK⁻) and the ZJ01 strain (rZJ01-ΔTK/gE/gI), four-gene deletion of the PRV-GDFS (PRV GDFS-delgI/gE/US9/US2) and ZJ01 strains (rZJ01-ΔgI/gE/TK/UL13), and five-gene deletion of HN1201 (rHN1201$^{TK-/gE-/gI-/11k-/28k-}$), which have been proved to be effective in preventing PR caused by mutant strains [29,155–165]. Considering the safety of these candidate strains in field applications, further clinical trials must be conducted.

Table 3. Overview of genetically modified live PRV strains for vaccination in pigs.

Gene-Deleted Vaccines	Vaccine Strains	Progenitor Strains	Deleted Gene	Technology Used	Authorization	References
Single gene-deleted vaccine	Omnivac	BUK	TK gene	Natural losses	Licensed	[27]
	2.4N3A	NIA-3 (field strain)	gE gene	HR	Licensed	[166]
	PRV(LA-A^B)	AH02LA (field strain)	gE gene	BCA	Not available	[167]
	HN1201ΔgE (inactivated)	HN1201 (field strain)	gE gene	HR	Licensed	[154]
	rPRVTJ-delgE	TJ (field strain)	gE gene	HR	Not available	[11]
Double gene-deleted vaccine	Omnimark	Omnivac (BUK)	TK and gIII genes	Natural losses	Licensed	[145]
	Begonia	2.4N3A (NIA-3)	TK and gE genes	Natural losses	Licensed	[168]
	NIA3-783	2.4N3A (NIA-3)	TK and gE genes	HR	Licensed	[146]
	Tolvi	field strain	TK and gpX genes	HR	Licensed	[169]
	D1200/D560	NIA-3	gD and gI genes	HR	Not available	[148]
	AD-YS400	Yangsan (field strain)	TK and gE genes	HR	Not available	[170]
	JS-2012-ΔgE/gI	JS-2012 (field strain)	gE and gI genes	HR	Not available	[171]
	gE-TK-PRV	TNL (field strain)	TK and gE genes	HR	Not available	[172]
	vZJ01ΔgE/gI (inactivated)	ZJ01 (field strain)	gE and gI genes	BCA	Not available	[157]
	PRV (PRV$^{ΔTK\&gE-AH02}$)	AH02LA (field strain)	TK and gE genes	HR	Not available	[173]
Triple gene-deleted vaccine	6C2	Field strain	TK, gE and gI genes	HR	Not available	[174]
	SA215	Fa (classical strain)	gE, gI and TK genes	HR	Licensed	[175]
	rSMXΔgI/gEΔTK	Field strain	TK, gE and gI genes	HR	Not available	[176]
	rPRVTJ-delgE/gI/TK-	rPRVTJ-delgE (TJ strain)	TK, gE and gI genes	HR	Not available	[29]
	vPRV HN1201	HN1201 (field strain)	TK, gE and gI genes	HR	Not available	[158]
	gE⁻/gI⁻/TK⁻ PRV	HeN1 (field strain)	TK, gE and gI genes	CRISPR/Cas9	Not available	[177]
	rPRV NY-gE⁻/gI⁻/TK⁻	NY (field strain)	TK, gE and gI genes	HR and CRISPR/Cas9	Not available	[155]
		201715 (field strain)	gE, gC and TK genes	CRISPR/Cas9	Not available	[178]
	rPRV/XJ5-gE⁻/gI⁻/TK⁻	XJ5 (field strain)	gE, gI and TK genes	HR	Not available	[159]
	rGXΔTK/gE/gI	GX (field strain)	TK, gE and gI genes		Not available	[165]
Four gene-deleted vaccine	PrV (376)	PrV (376)	gD, gG, gI and gE genes		Not available	[147]
	———	C (field strain)	gI, gE, Us9 and Us2 genes	Natura losses	Licensed	[179]
	PRV GDFS-delgI/gE/US9/US2	GDFS (field strain)	gI, gE, Us9 and Us2 genes	CRISPR/Cas9	Not available	[161]
	rZJ01-ΔgI/gE/TK/UL13	ZJ01	gI, gE, TK and UL13 genes	CRISPR/Cas9	Not available	[29]
Five gene-deleted vaccine	PRV rHN1201$^{TK-/gE-/gI-/11k-/28k-}$	HN1201 (field strain))	TK, gI, gE, 11k and 28k genes	BCA	Not available	[164]

Note: HR is the homologous DNA recombination; BCA is the bacterial artificial chromosome; CRISPR/Cas9 is the clustered regularly interspaced short palindromic repeats/Cas9.

Additionally, the large genome of PRV can serve as vaccine vectors for expressing exogenous antigens without affecting its infectivity and immunogenicity [180]. Several live PRV-based vector vaccines encoding significant antigens of other animal pathogens have been generated and represented in the previous overview [181]. Similar to the gene-modified vaccine, gG, gI, gE, and TK genes generally are used to insert exogenous sequences [182–184]. At present, varied foreign genes encoding protective antigens of other pathogens have been successfully inserted into the large genome of PRV, including the PPV VP2 gene, classical swine fever virus (CSFV) E2 gene, the Brucella melitensis Bp26 gene, and some genes of the African swine fever virus which include CP204L (p30), CP530R (pp62), E183L (p54), B646L (p72) and EP402R (CD2v) [116,183–188]. Generally speaking, most PRV recombinant vaccines have not yet been licensed so far, although these vaccines can effectively prevent multiple infectious diseases simultaneously. Surprisingly, recombinant targeted *Bacillus subtilis* vaccine expressing PRV gC and gD proteins can effectively induce a mucosal immune response against this disease in recent study [189], and efficient protection provided by a gD-based subunit vaccine against PRV variant infection in pig models was also confirmed [190].

For China, many researchers have developed diverse types of anti-PRV vaccines, mainly including inactivated vaccines, live attenuated vaccines and live PRV-based vector vaccines, and most of them can produce high levels of neutralizing and gB antibodies, with effective protection against PRV. Furthermore, these candidate vaccines also follow the distinction between the infected and vaccinated animals (DIVA), which is beneficial to PR eradication in China. Different types of vaccines have different advantages. The inactivated vaccine is highly safe for vaccinated animals without viral virulence reversion, but it is generally less effective than both live vaccines [10]. Notably, two live vaccines display disadvantages, for example, their safety in pigs and non-target animals. Some studies reported that vaccination with a gene-modified PRV strain could lead to PR in sheep with severe clinical signs, as well as adult red foxes, and also pose a potential threat to the health of dogs [191–193]. In general, lengthy testing should be conducted on genetically modified live vaccines [167].

6.2. Chinese Herbal Medicines as Potential Anti-PRV Drugs

Chinese herbal medicine has a history spanning thousands of years and has been extensively applied in the treatment of various diseases of human beings, such as atherosclerosis, sepsis, diabetes, cancers, chronic kidney disease and anti-SARS-CoV-2 (COVID-19) infection [194–200]. Some researchers have also found that Chinese herbal medicine can improve some animal diseases [201–204]. Sinomenine, a comment agent in Chinese herbal medicines, can decrease the incidence and severity of certain LPS-induced toxicities, for example, cell adhesion, systemic inflammation and multiple organ dysfunction [201]. Ginger extracts can relax and vasoprotect porcine coronary arteries [203]. With the international recognition of traditional Chinese herbal medicine in disease treatment, this medicine has gradually gained comprehensive attention. In light of the tremendous impact of PR on the swine industry, some scientists have also committed to looking for inhibitors against PRV infection (Table 4) [205–209]. As a natural phenolic compound, resveratrol (trans-3, 4, 5-trihydroxystilbene; Res) has a variety of properties, such as immunomodulatory, anti-inflammatory and antiviral activities, and especially, its antiviral activities against PRV infection have been fully recognized in numerous examinations [209,210]. Based on these bioactivities, it appears that Res can inhibit the proliferation in PRV-infected piglets and protect rotavirus-infected piglets by reducing the inflammatory response and enhancing immune function [211]. In vitro, Res effectively inhibits PRV replication in a dose-dependent manner, with a 50% inhibition concentration of 17.17 μM, and its inhibition of PRV-induced cell death and gene expression may be related to IκB kinase degradation [211]. The ability of Res anti-PRV and immune-adjuvant was also corroborated in both mice and pigs, and Res could inhibit the replication of ASFV in vitro [142,186,212]. Taken together, these traditional Chinese herbal medicines have great potential value on the inhibitive ability of PRV infection, and they need to be further studied to be promoted to be an effective choice for animals or human beings against PRV challenge.

In addition to Res, other types of compounds with anti-PRV activity have been identified to be effective in vitro, such as kaempferol, panax notoginseng polysaccharides, germacrone, plantago, quercetin, isatis indigotica, radix isatidis, marine *Bacillus* S-12–86 lysozyme, diammonium glycyrrhizin, vanadium-substituted Heteropolytungstate, graphene Oxide, ivermectin and phosphonoformate sodium, whereas some of them cannot be verified for the inhibition of viral replication in vivo (Table 4) [99,213–221]. As a novel anti-PRV drug, quercetin is a natural product that has anti-oxidant, anti-bacterial, anti-cancer and anti-viral activities [216,222]. In the latest research, it was found that quercetin can indeed reduce the extent of PRV infection in virus-infected cells in a concentration-dependent manner, suggesting that quercetin mainly holds back the entry of PRV into the host cell by preventing its adsorption to the cell surface [216]. In addition, it is important for a viral infection that quercetin is able to insert into the substrate-binding pocket of PRV gD protein on the PRV surface and connect the N-ring and spiral alpha3 by hydrogen bonding [216]. The anti-PRV activity of quercetin has been demonstrated in mice, indicating that mice inoculated with quercetin can resist the lethal challenge of PRV and decrease the viral loads in the brain [216]. Although quercetin has a powerful therapeutic property against PRV infection both in vitro and vivo, it still needs long-term validation before it can be widely applied in veterinary clinics.

Table 4. Overview of compounds with anti-PRV infection activity.

Source	Mechanism	50% Effective Concentration	50% Cytotoxic Concentration	PRV Strain	In Vitro	In Vivo	References
Resveratrol (Res)	The inhibition of viral proliferation, IκB kinase activation	17.17 ± 0.35 μM	Above 262.87 μM	Rong A strain	Yes	Yes	[142,209,211]
Kaempferol	The inhibition of viral proliferation	25.57 μM of 50% inhibited concentration	No mention	Ra strain	Yes	Yes	[213]
Panax notoginseng polysaccharides	The inhibition of viral adsorption and replication	No mention	No mention	PRV XJ5 strain	Yes	No	[214]
Germacrone	The inhibition of viral proliferation	54.51 μM for Vero cells and 88.78 μM for LLC-PK-1 cells	233.5 μM for Vero cells and 184.1 μM for LLC-PK-1 cells	Variant PRV and PRV vaccine strain Barth K61	Yes	No	[206]
Plantago	The inhibition of viral attachment and penetration; decreasing ROS (reactive oxygen species) production	No mention	No mention	PRV XJ5	Yes	No	[215]
Quercetin	The inhibition of viral adsorption	2.618 ± 0.673 μM of 50% inhibited concentration	Above 599 μM	HNX strain	Yes	Yes	[216]
Isatis indigotica	The inhibition of viral proliferation	11 μg/mL	299 μg/mL	TNL strain	Yes	No	[217]
Radix isatidis	The inhibition of viral proliferation; killing virus directly	The inhibition rate of viral replication by 14.674–30.84%	No mention	Min A strain	Yes	No	[207]
Marine *Bacillus* S-12–86 lysozyme	The inhibition of viral proliferation; killing virus directly	0.46 mg/L	100 mg/L	Min A strain	Yes	No	[218]
Diammonium glycyrrhizin	Killing virus directly	No mention	Above 1250 μg/mL	Bartha K-61	Yes	No	[219]
Vanadium-substituted Heteropolytungstate	Killing virus directly	3.5–5.0 mg/L	400–420 mg/L	Bartha strain	Yes	No	[220]
Graphene Oxide	Killing virus directly	No mention	No mention	HNX strain	Yes	No	[223]
Ivermectin	The inhibition of viral DNA polymerase UL42 in entering the nucleus	No mention	No mention	No mention	Yes	Yes	[208]
Phosphonoformate sodium	Inhibition of viral DNA polymerase	Nearly 60 μg/mL of 50% inhibited concentration	No mention	Kaplan	Yes	No	[221]

6.3. Novel Small RNAs

Owing to the characteristic of targeting mRNA degradation, small RNAs are extensively used for searching gene functions and are also considered a novel therapeutic approach that effectively inhibits viral replication and interferes in protein synthesis, including small interfering RNAs (siRNAs) and microRNAs (miRNAs) [224–226]. In a previous study, it was proved that the PRV processivity factor UL42 is critical for viral replication and can improve the catalytic activity of the DNA polymerase, suggesting that it may be a latent drug target for antiviral treatment against PRV infection [2]. To verify this guess, three siRNAs (siR-386, siR-517, and siR-849) directed against UL42 were synthesized and defined their anti-PRV activities in cell culture, and the results showed that these three siRNAs induce great inhibitory effects on UL42 expression after PRV infection and impair viral replication [227]. miR-21 is one of the earliest miRNAs to be discovered, and it is associated with the immune response, viral replication, cell apoptosis and cancer [228–230]. In Huang's study, it demonstrated that miR-21 plays a crucial role in the immune response to PRV infection and can directly target interferon-γ inducible protein-10 (IP-10) to inhibit PRV replication in PK15 cells [230]. Subsequently, the detailed function of a large latency transcript (LLT) miRNA cluster was further studied, and PRV-encoded prv-miR-LLT11a appeared during initial downregulation and following upregulation in PK15 cells with PRV-infection, suggesting that it may have obviously repressed viral replication [231]. However, their potential mechanism remains unclear so far, which needs deeper research.

7. Conclusions and Future Perspectives

It has been nearly 200 years since PR was first discovered in 1813 in America, and PRV infection has become one of the most important pathogens leading to reproductive failure of pregnant sows, nervous disorders in newborn piglets, and respiratory distress in growing pigs [2]. Although few countries have eradicated PR from their swine populations due to the application of gE-deleted PRV vaccines and the DIVA strategy, PR is still prevalent in most countries, especially in China, which has resulted in huge economic losses to the swine industry during the past several decades [157]. With continuous exploration, the roles of the various components, pathogenesis, diagnosis method and prevention of PRV are being developed in depth. In the past decades, the infected host of PRV has ranged from various animals to human beings, including pigs, dogs, cats, cattle, sheep, goats, captive mink, wild foxes, captive foxes, wolves and lynxes [232]. PRV infection invades the peripheral nervous system in pigs with the occasional invasion of the central nervous system after primary infection at mucosal surfaces [15]. For its diagnosis, different types of detection technology have been generated, such as serological approaches, molecular biology approaches, and other approaches (VI and the paper biosensor doped with $Fe_3O_4@SiO_2–NH_2$ and MWCNTs). As one of the serological assays, ELISA is most widely used in detecting the infection of PRV wild strains due to its sensitive and rapid features [1]. It is worth nothing that PRV infection in humans is confirmed by high-through sequencing technology, which is time-consuming and expensive [6,138]. Thus, these problems should be the focus of future efforts.

In terms of the prevention and control of PR, vaccination with the Bartha-K61 strain is the most common method, and PR is well controlled worldwide. However, PR re-emerged at the end of 2011 in Chinese pig herds which were vaccinated with the Bartha-K61 strain, and it was demonstrated that re-emerging PR is caused by variant PRV strains, suggesting that the commercial vaccine containing the Bartha-K61 strain cannot provide full protection against the variant PRV challenge [111]. With continuous research, various types of vaccine candidates have been created, such as inactivated vaccines, live gene-deleted vaccines, live attenuated recombinant vaccines, DNA vaccines and subunit vaccines, though the last two types of vaccines are rarely used in the clinical prevention of PRV. At present, PRV vaccines that are widely used in pig farms are mainly the licensed live attenuated gene-deleted vaccines, but they may appear to cause viral virulence reversion and influence the safety of vaccinated pigs. Another disadvantage is that these commercial attenuated vaccines can lead to various susceptibility and immune responses in some species (goat, dog and mink) [99]. Therefore, it is very necessary to develop safer vaccines with efficient protection against PRV infection in the future.

In conclusion, the current study and clinical progress data on PR prevention and control are optimistic, and we believe that we can achieve the goal of eradicating PR worldwide in the near future.

Author Contributions: Conceptualization, H.-H.Z. and H.-Y.C.; software, P.-F.F.; writing—original draft preparation, H.-H.Z.; writing—review and editing, H.-H.Z.; funding acquisition, H.-Y.C. and Z.-Y.W. All authors have read and agreed to the published version of the manuscript.

Funding: This work was supported by the National Key Research and Development Program (No. 2021YFD1801105), the Zhongyuan High Level Talents Special Support Plan (No. 204200510015), Henan open competition mechanism to select the best candidates Program (No. 211110111000), and Fang's family (Hong Kong) foundation.

Institutional Review Board Statement: This article does not contain any studies with human participants or animals performed by any of the authors.

Informed Consent Statement: Not applicable.

Data Availability Statement: Not applicable.

Conflicts of Interest: The authors declare that they have no conflict of interest.

References

1. Zheng, H.H.; Jin, Y.; Hou, C.Y.; Li, X.S.; Zhao, L.; Wang, Z.Y.; Chen, H.Y. Seroprevalence investigation and genetic analysis of pseudorabies virus within pig populations in Henan province of China during 2018–2019. *Infect. Genet. Evol.* **2021**, *92*, 104835. [CrossRef]
2. Pomeranz, L.E.; Reynolds, A.E.; Hengartner, C.J. Molecular biology of pseudorabies virus: Impact on neurovirology and veterinary medicine. *Microbiol. Mol. Biol. Rev. MMBR* **2005**, *69*, 462–500. [CrossRef] [PubMed]
3. Laval, K.; Vernejoul, J.; Van Cleemput, J.; Koyuncu, O.; Enquist, L. Virulent Pseudorabies Virus Infection Induces a Specific and Lethal Systemic Inflammatory Response in Mice. *J. Virol.* **2018**, *92*, e01614-18. [CrossRef]
4. Yang, H.; Han, H.; Wang, H.; Cui, Y.; Liu, H.; Ding, S. A Case of Human Viral Encephalitis Caused by Pseudorabies Virus Infection in China. *Front. Neurol.* **2019**, *10*, 534. [CrossRef] [PubMed]
5. Liu, Q.; Wang, X.; Xie, C.; Ding, S.; Yang, H.; Guo, S.; Li, J.; Qin, L.; Ban, F.; Wang, D.; et al. A Novel Human Acute Encephalitis Caused by Pseudorabies Virus Variant Strain. *Clin. Infect. Dis.* **2021**, *73*, e3690–e3700. [CrossRef] [PubMed]
6. Ai, J.; Weng, S.; Cheng, Q.; Cui, P.; Li, Y.; Wu, H.; Zhu, Y.; Xu, B.; Zhang, W. Human Endophthalmitis Caused By Pseudorabies Virus Infection, China, 2017. *Emerg. Infect. Dis.* **2018**, *24*, 1087–1090. [CrossRef]
7. Li, X.; Fu, S.; Chen, L.; Li, F.; Deng, J.; Lu, X.; Wang, H.; Tian, K. Detection of Pseudorabies Virus Antibodies in Human Encephalitis Cases. *Biomed. Environ. Sci. BES* **2020**, *33*, 444–447. [CrossRef]
8. Wang, Y.; Nian, H.; Li, Z.; Wang, W.; Wang, X.; Cui, Y. Human encephalitis complicated with bilateral acute retinal necrosis associated with pseudorabies virus infection: A case report. *Int. J. Infect. Dis. IJID* **2019**, *89*, 51–54. [CrossRef]
9. Ying, M.; Hu, X.; Wang, M.; Cheng, X.; Zhao, B.; Tao, Y. Vitritis and retinal vasculitis caused by pseudorabies virus. *J. Int. Med. Res.* **2021**, *49*, 3000605211058990. [CrossRef]
10. Freuling, C.M.; Muller, T.F.; Mettenleiter, T.C. Vaccines against pseudorabies virus (PrV). *Vet. Microbiol.* **2017**, *206*, 3–9. [CrossRef] [PubMed]
11. Wang, C.H.; Yuan, J.; Qin, H.Y.; Luo, Y.; Cong, X.; Li, Y.; Chen, J.; Li, S.; Sun, Y.; Qiu, H.J. A novel gE-deleted pseudorabies virus (PRV) provides rapid and complete protection from lethal challenge with the PRV variant emerging in Bartha-K61-vaccinated swine population in China. *Vaccine* **2014**, *32*, 3379–3385. [CrossRef] [PubMed]
12. Zhou, J.; Li, S.; Wang, X.; Zou, M.; Gao, S. Bartha-k61 vaccine protects growing pigs against challenge with an emerging variant pseudorabies virus. *Vaccine* **2017**, *35*, 1161–1166. [CrossRef]
13. Mettenleiter, T. Aujeszky's disease (pseudorabies) virus: The virus and molecular pathogenesis–state of the art, June 1999. *Vet. Res.* **2000**, *31*, 99–115. [CrossRef] [PubMed]
14. Masić, M.; Ercegan, M.; Petrović, M. The significance of the tonsils in the pathogenesis and diagnosis of Aujeszyk's disease in pigs. *Zent. Fur Veterinarmedizin. Reihe B J. Vet. Medicine. Ser. B* **1965**, *12*, 398–405.
15. Tirabassi, R.; Townley, R.; Eldridge, M.; Enquist, L. Molecular mechanisms of neurotropic herpesvirus invasion and spread in the CNS. *Neurosci. Biobehav. Rev.* **1998**, *22*, 709–720. [CrossRef]
16. Gutekunst, D.; Pirtle, E.; Miller, L.; Stewart, W. Isolation of pseudorabies virus from trigeminal ganglia of a latently infected sow. *Am. J. Vet. Res.* **1980**, *41*, 1315–1316.
17. Verpoest, S.; Cay, B.; Favoreel, H.; De Regge, N. Age-Dependent Differences in Pseudorabies Virus Neuropathogenesis and Associated Cytokine Expression. *J. Virol.* **2017**, *91*, e02058-16. [CrossRef] [PubMed]
18. Nawynck, H.; Pensaert, M. Cell-free and cell-associated viremia in pigs after oronasal infection with Aujeszky's disease virus. *Vet. Microbiol.* **1995**, *43*, 307–314. [CrossRef]
19. Nauwynck, H.; Pensaert, M. Abortion induced by cell-associated pseudorabies virus in vaccinated sows. *Am. J. Vet. Res.* **1992**, *53*, 489–493. [PubMed]
20. Mettenleiter, T.C.; Klupp, B.G.; Granzow, H. Herpesvirus assembly: An update. *Virus Res.* **2009**, *143*, 222–234. [CrossRef] [PubMed]
21. Daniel, G.R.; Pegg, C.E.; Smith, G.A.; Sandri-Goldin, R.M. Dissecting the Herpesvirus Architecture by Targeted Proteolysis. *J. Virol.* **2018**, *92*, e00738-18. [CrossRef] [PubMed]
22. Kramer, T.; Greco, T.M.; Enquist, L.W.; Cristea, I.M. Proteomic characterization of pseudorabies virus extracellular virions. *J. Virol.* **2011**, *85*, 6427–6441. [CrossRef] [PubMed]
23. Romero, C.H.; Meade, P.; Santagata, J.; Gillis, K.; Lollis, G.; Hahn, E.C.; Gibbs, E.P. Genital infection and transmission of pseudorabies virus in feral swine in Florida, USA. *Vet. Microbiol.* **1997**, *55*, 131–139. [CrossRef]
24. Fuchs, W.; Ehrlich, C.; Klupp, B.; Mettenleiter, T. Characterization of the replication origin (Ori(S)) and adjoining parts of the inverted repeat sequences of the pseudorabies virus genome. *J. Gen. Virol.* **2000**, *81*, 1539–1543. [CrossRef]
25. Wu, C.; Harper, L.; Ben-Porat, T. Molecular basis for interference of defective interfering particles of pseudorabies virus with replication of standard virus. *J. Virol.* **1986**, *59*, 308–317. [CrossRef]
26. Wesley, R.D.; Cheung, A.K. A Pseudorabies Virus Mutant with Deletions in the Latency and Early Protein O Genes: Replication, Virulence, and Immunity in Neonatal Piglets. *J. Vet. Diagn. Investig.* **1996**, *8*, 21–24. [CrossRef]

27. Kit, S.; Kit, M.; Pirtle, E. Attenuated properties of thymidine kinase-negative deletion mutant of pseudorabies virus. *Am. J. Vet. Res.* **1985**, *46*, 1359–1367. [PubMed]

28. Xu, L.; Wei, J.; Zhao, J.; Xu, S.; Lee, F.; Nie, M.; Xu, Z.; Zhou, Y.; Zhu, L. The Immunity Protection of Central Nervous System Induced by Pseudorabies Virus DelgI/gE/TK in Mice. *Front. Microbiol.* **2022**, *13*, 862907. [CrossRef] [PubMed]

29. Cong, X.; Lei, J.L.; Xia, S.L.; Wang, Y.M.; Li, Y.; Li, S.; Luo, Y.; Sun, Y.; Qiu, H.J. Pathogenicity and immunogenicity of a gE/gI/TK gene-deleted pseudorabies virus variant in susceptible animals. *Vet. Microbiol.* **2016**, *182*, 170–177. [CrossRef]

30. Rauh, I.; Mettenleiter, T.C. Pseudorabies virus glycoproteins gII and gp50 are essential for virus penetration. *J. Virol.* **1991**, *65*, 5348–5356. [CrossRef] [PubMed]

31. Peeters, B.; de Wind, N.; Hooisma, M.; Wagenaar, F.; Gielkens, A.; Moormann, R. Pseudorabies virus envelope glycoproteins gp50 and gII are essential for virus penetration, but only gII is involved in membrane fusion. *J. Virol.* **1992**, *66*, 894–905. [CrossRef] [PubMed]

32. Maeda, K.; Hayashi, S.; Tanioka, Y.; Matsumoto, Y.; Otsuka, H. Pseudorabies virus (PRV) is protected from complement attack by cellular factors and glycoprotein C (gC). *Virus Res.* **2002**, *84*, 79–87. [CrossRef]

33. Rue, C.A.; Ryan, P. Pseudorabies virus glycoprotein C attachment-proficient revertants isolated through a simple, targeted mutagenesis scheme. *J. Virol. Methods* **2008**, *151*, 101–106. [CrossRef]

34. Sapir, A.; Avinoam, O.; Podbilewicz, B.; Chernomordik, L.V. Viral and developmental cell fusion mechanisms: Conservation and divergence. *Dev. Cell* **2008**, *14*, 11–21. [CrossRef]

35. Vallbracht, M.; Brun, D.; Tassinari, M.; Vaney, M.C.; Pehau-Arnaudet, G.; Guardado-Calvo, P.; Haouz, A.; Klupp, B.G.; Mettenleiter, T.C.; Rey, F.A.; et al. Structure-Function Dissection of Pseudorabies Virus Glycoprotein B Fusion Loops. *J. Virol.* **2018**, *92*, e01203-17. [CrossRef]

36. Mettenleiter, T.C.; Lukàcs, N.; Rziha, H.J. Pseudorabies virus avirulent strains fail to express a major glycoprotein. *J. Virol.* **1985**, *56*, 307–311. [CrossRef] [PubMed]

37. Li, A.; Lu, G.; Qi, J.; Wu, L.; Tian, K.; Luo, T.; Shi, Y.; Yan, J.; Gao, G.F. Structural basis of nectin-1 recognition by pseudorabies virus glycoprotein D. *PLoS Pathog.* **2017**, *13*, e1006314. [CrossRef]

38. Thomsen, D.R.; Marchioli, C.C.; Yancey, R.J., Jr.; Post, L.E. Replication and virulence of pseudorabies virus mutants lacking glycoprotein gX. *J. Virol.* **1987**, *61*, 229–232. [CrossRef]

39. Lerma, L.; Muñoz, A.L.; García Utrilla, R.; Sainz, B., Jr.; Lim, F.; Tabarés, E.; Gómez-Sebastián, S. Partial complementation between the immediate early proteins ICP4 of herpes simplex virus type 1 and IE180 of pseudorabies virus. *Virus Res.* **2020**, *279*, 197896. [CrossRef] [PubMed]

40. Zhang, Y.; El Omari, K.; Duman, R.; Liu, S.; Haider, S.; Wagner, A.; Parkinson, G.N.; Wei, D. Native de novo structural determinations of non-canonical nucleic acid motifs by X-ray crystallography at long wavelengths. *Nucleic Acids Res.* **2020**, *48*, 9886–9898. [CrossRef] [PubMed]

41. Cai, M.; Wang, P.; Wang, Y.; Chen, T.; Xu, Z.; Zou, X.; Ou, X.; Li, Y.; Chen, D.; Peng, T.; et al. Identification of the molecular determinants for nuclear import of PRV EP0. *Biol. Chem.* **2019**, *400*, 1385–1394. [CrossRef] [PubMed]

42. Yan, K.; Liu, J.; Guan, X.; Yin, Y.X.; Peng, H.; Chen, H.C.; Liu, Z.F. The Carboxyl Terminus of Tegument Protein pUL21 Contributes to Pseudorabies Virus Neuroinvasion. *J. Virol.* **2019**, *93*, e02052-18. [CrossRef]

43. Liu, Y.F.; Tsai, P.Y.; Chulakasian, S.; Lin, F.Y.; Hsu, W.L. The pseudorabies virus vhs protein cleaves RNA containing an IRES sequence. *FEBS J.* **2016**, *283*, 899–911. [CrossRef]

44. Liu, Y.F.; Tsai, P.Y.; Lin, F.Y.; Lin, K.H.; Chang, T.J.; Lin, H.W.; Chulakasian, S.; Hsu, W.L. Roles of nucleic acid substrates and cofactors in the vhs protein activity of pseudorabies virus. *Vet. Res.* **2015**, *46*, 141. [CrossRef]

45. Granzow, H.; Klupp, B.G.; Mettenleiter, T.C. Entry of pseudorabies virus: An immunogold-labeling study. *J. Virol.* **2005**, *79*, 3200–3205. [CrossRef] [PubMed]

46. Luxton, G.W.; Haverlock, S.; Coller, K.E.; Antinone, S.E.; Pincetic, A.; Smith, G.A. Targeting of herpesvirus capsid transport in axons is coupled to association with specific sets of tegument proteins. *Proc. Natl. Acad. Sci. USA* **2005**, *102*, 5832–5837. [CrossRef]

47. Granzow, H.; Weiland, F.; Jöns, A.; Klupp, B.G.; Karger, A.; Mettenleiter, T.C. Ultrastructural analysis of the replication cycle of pseudorabies virus in cell culture: A reassessment. *J. Virol.* **1997**, *71*, 2072–2082. [CrossRef] [PubMed]

48. Fuchs, W.; Granzow, H.; Klupp, B.G.; Kopp, M.; Mettenleiter, T.C. The UL48 tegument protein of pseudorabies virus is critical for intracytoplasmic assembly of infectious virions. *J. Virol.* **2002**, *76*, 6729–6742. [CrossRef] [PubMed]

49. Ben-Porat, T.; Veach, R.A.; Ihara, S. Localization of the regions of homology between the genomes of herpes simplex virus, type 1, and pseudorabies virus. *Virology* **1983**, *127*, 194–204. [CrossRef]

50. Lehman, I.; Boehmer, P. Replication of herpes simplex virus DNA. *J. Biol. Chem.* **1999**, *274*, 28059–28062. [CrossRef]

51. De Wind, N.; Peeters, B.P.; Zuderveld, A.; Gielkens, A.L.; Berns, A.J.; Kimman, T.G. Mutagenesis and characterization of a 41-kilobase-pair region of the pseudorabies virus genome: Transcription map, search for virulence genes, and comparison with homologs of herpes simplex virus type 1. *Virology* **1994**, *200*, 784–790. [CrossRef]

52. Jöns, A.; Mettenleiter, T.C. Identification and characterization of pseudorabies virus dUTPase. *J. Virol.* **1996**, *70*, 1242–1245. [CrossRef] [PubMed]

53. Fuchs, W.; Klupp, B.G.; Granzow, H.; Rziha, H.J.; Mettenleiter, T.C. Identification and characterization of the pseudorabies virus UL3.5 protein, which is involved in virus egress. *J. Virol.* **1996**, *70*, 3517–3527. [CrossRef] [PubMed]

54. Fuchs, W.; Klupp, B.G.; Granzow, H.; Osterrieder, N.; Mettenleiter, T.C. The interacting UL31 and UL34 gene products of pseudorabies virus are involved in egress from the host-cell nucleus and represent components of primary enveloped but not mature virions. *J. Virol.* **2002**, *76*, 364–378. [CrossRef] [PubMed]
55. Morrison, E.E.; Wang, Y.F.; Meredith, D.M. Phosphorylation of structural components promotes dissociation of the herpes simplex virus type 1 tegument. *J. Virol.* **1998**, *72*, 7108–7114. [CrossRef]
56. Hahn, E.C.; Page, G.R.; Hahn, P.S.; Gillis, K.D.; Romero, C.; Annelli, J.A.; Gibbs, E.P. Mechanisms of transmission of Aujeszky's disease virus originating from feral swine in the USA. *Vet. Microbiol.* **1997**, *55*, 123–130. [CrossRef]
57. Romero, C.H.; Meade, P.N.; Shultz, J.E.; Chung, H.Y.; Gibbs, E.P.; Hahn, E.C.; Lollis, G. Venereal transmission of pseudorabies viruses indigenous to feral swine. *J. Wildl. Dis.* **2001**, *37*, 289–296. [CrossRef]
58. Nauwynck, H.; Glorieux, S.; Favoreel, H.; Pensaert, M. Cell biological and molecular characteristics of pseudorabies virus infections in cell cultures and in pigs with emphasis on the respiratory tract. *Vet. Res.* **2007**, *38*, 229–241. [CrossRef]
59. Crandell, R.A. Pseudorabies (Aujeszky's disease). *Vet. Clin. N. Am. Large Anim. Pract.* **2014**, *4*, 321. [CrossRef]
60. Glorieux, S.; Van den Broeck, W.; van der Meulen, K.M.; Van Reeth, K.; Favoreel, H.W.; Nauwynck, H.J. In vitro culture of porcine respiratory nasal mucosa explants for studying the interaction of porcine viruses with the respiratory tract. *J. Virol. Methods* **2007**, *142*, 105–112. [CrossRef]
61. Glorieux, S.; Favoreel, H.W.; Meesen, G.; de Vos, W.; Van den Broeck, W.; Nauwynck, H.J. Different replication characteristics of historical pseudorabies virus strains in porcine respiratory nasal mucosa explants. *Vet. Microbiol.* **2009**, *136*, 341–346. [CrossRef] [PubMed]
62. Narita, M.; Kawashima, K.; Matsuura, S.; Uchimura, A.; Miura, Y. Pneumonia in pigs infected with pseudorabies virus and Haemophilus parasuis serovar 4. *J. Comp. Pathol.* **1994**, *110*, 329–339. [CrossRef]
63. Maes, R.K.; Kanitz, C.L.; Gustafson, D.P. Shedding patterns in swine of virulent and attenuated pseudorabies virus. *Am. J. Vet. Res.* **1983**, *44*, 2083–2086. [PubMed]
64. Lamote, J.A.S.; Glorieux, S.; Nauwynck, H.J.; Favoreel, H.W. The US3 Protein of Pseudorabies Virus Drives Viral Passage across the Basement Membrane in Porcine Respiratory Mucosa Explants. *J. Virol.* **2016**, *90*, 10945–10950. [CrossRef]
65. Glorieux, S.; Favoreel, H.W.; Steukers, L.; Vandekerckhove, A.P.; Nauwynck, H.J. A trypsin-like serine protease is involved in pseudorabies virus invasion through the basement membrane barrier of porcine nasal respiratory mucosa. *Vet. Res.* **2011**, *42*, 58. [CrossRef]
66. Jamrichová, O.; Skoda, R. Multiplication of pseudorabies virus in the inguinal lymph nodes of pigs. *Acta Virol.* **1968**, *12*, 555.
67. Wittmann, G.; Ohlinger, V.; Rziha, H.J. Occurrence and reactivation of latent Aujeszky's disease virus following challenge in previously vaccinated pigs. *Arch. Virol.* **1983**, *75*, 29–41. [CrossRef]
68. Mulder, W.A.; Jacobs, L.; Priem, J.; Kok, G.L.; Wagenaar, F.; Kimman, T.G.; Pol, J.M. Glycoprotein gE-negative pseudorabies virus has a reduced capability to infect second- and third-order neurons of the olfactory and trigeminal routes in the porcine central nervous system. *J. Gen. Virol.* **1994**, *75 Pt 11*, 3095–3106. [CrossRef]
69. Sabó, A.; Rajcáni, J.; Blaskovic, D. Studies on the pathogenesis of Aujeszky's disease. 3. The distribution of virulent virus in piglets after intranasal infection. *Acta Virol.* **1969**, *13*, 407–414.
70. Nauwynck, H.J.; Pensaert, M.B. Interactions of Aujeszky's disease virus and porcine blood mononuclear cells in vivo and in vitro. *Acta Vet. Hung.* **1994**, *42*, 301–308.
71. Babic, N.; Mettenleiter, T.C.; Ugolini, G.; Flamand, A.; Coulon, P. Propagation of pseudorabies virus in the nervous system of the mouse after intranasal inoculation. *Virology* **1994**, *204*, 616–625. [CrossRef] [PubMed]
72. Kramer, T.; Greco, T.M.; Taylor, M.P.; Ambrosini, A.E.; Cristea, I.M.; Enquist, L.W. Kinesin-3 mediates axonal sorting and directional transport of alphaherpesvirus particles in neurons. *Cell Host Microbe* **2012**, *12*, 806–814. [CrossRef] [PubMed]
73. Grinde, B. Herpesviruses: Latency and reactivation—Viral strategies and host response. *J. Oral Microbiol.* **2013**, *5*, 22766. [CrossRef] [PubMed]
74. Wheeler, J.G.; Osorio, F.A. Investigation of sites of pseudorabies virus latency, using polymerase chain reaction. *Am. J. Vet. Res.* **1991**, *52*, 1799–1803. [PubMed]
75. van Oirschot, J.T.; Gielkens, A.L. In vivo and in vitro reactivation of latent pseudorabies virus in pigs born to vaccinated sows. *Am. J. Vet. Res.* **1984**, *45*, 567–571.
76. Wigdahl, B.; Rong, B.L.; Kinney-Thomas, E. Varicella-zoster virus infection of human sensory neurons. *Virology* **1986**, *152*, 384–399. [CrossRef]
77. Jones, C. Bovine Herpes Virus 1 (BHV-1) and Herpes Simplex Virus Type 1 (HSV-1) Promote Survival of Latently Infected Sensory Neurons, in Part by Inhibiting Apoptosis. *J. Cell Death* **2013**, *6*, 1–16. [CrossRef]
78. Van de Walle, G.R.; Favoreel, H.W.; Nauwynck, H.J.; Mettenleiter, T.C.; Pensaert, M.B. Transmission of pseudorabies virus from immune-masked blood monocytes to endothelial cells. *J. Gen. Virol.* **2003**, *84 Pt 3*, 629–637. [CrossRef]
79. Kluge, J.P.; Maré, C.J. Swine pseudorabies: Abortion, clinical disease, and lesions in pregnant gilts infected with pseudorabies virus (Aujeszky's disease). *Am. J. Vet. Res.* **1974**, *35*, 991–995.
80. I Isu, F.S.; Chu, R.M.; Lee, R.C.; Chu, S.H. Placental lesions caused by pseudorabies virus in pregnant sows. *J. Am. Vet. Med. Assoc.* **1980**, *177*, 636–641.
81. Ka, H.; Seo, H.; Choi, Y.; Yoo, I.; Han, J. Endometrial response to conceptus-derived estrogen and interleukin-1β at the time of implantation in pigs. *J. Anim. Sci. Biotechnol.* **2018**, *9*, 44. [CrossRef] [PubMed]

82. Vélez, C.; Barbeito, C.; Koncurat, M. αvβ3 Integrin and fibronectin expressions and their relation to estrogen and progesterone during placentation in swine. *Biotech. Histochem.* **2018**, *93*, 15–24. [CrossRef]
83. Bidarimath, M.; Tayade, C. Pregnancy and spontaneous fetal loss: A pig perspective. *Mol. Reprod. Dev.* **2017**, *84*, 856–869. [CrossRef] [PubMed]
84. Dieuzy, I.; Vannier, P.; Jestin, A. Effects of experimental pseudorabies virus infection on vaccinated pregnant sows. *Ann. De Rech. Vet. Ann. Vet. Res.* **1987**, *18*, 233–240.
85. Iglesias, J.G.; Harkness, J.W. Studies of transplacental and perinatal infection with two clones of a single Aujeszky's disease (pseudorabies) virus isolate. *Vet. Microbiol.* **1988**, *16*, 243–254. [CrossRef]
86. Ceriatti, F.S.; Sabini, L.I.; Bettera, S.G.; Zanón, S.M.; Ramos, B.A. Experimental infection of pregnant gilts with Aujeszky's disease virus strain RC/79. *Rev. Argent. Microbiol.* **1992**, *24*, 102–112. [PubMed]
87. Wong, G.; Lu, J.; Zhang, W.; Gao, G.F. Pseudorabies virus: A neglected zoonotic pathogen in humans? *Emerg. Microbes Infect.* **2019**, *8*, 150–154. [CrossRef]
88. He, W.; Auclert, L.Z.; Zhai, X.; Wong, G.; Zhang, C.; Zhu, H.; Xing, G.; Wang, S.; He, W.; Li, K.; et al. Interspecies Transmission, Genetic Diversity, and Evolutionary Dynamics of Pseudorabies Virus. *J. Infect. Dis.* **2019**, *219*, 1705–1715. [CrossRef]
89. Wang, X.; Wu, C.X.; Song, X.R.; Chen, H.C.; Liu, Z.F. Comparison of pseudorabies virus China reference strain with emerging variants reveals independent virus evolution within specific geographic regions. *Virology* **2017**, *506*, 92–98. [CrossRef]
90. Zhai, X.; Zhao, W.; Li, K.; Zhang, C.; Wang, C.; Su, S.; Zhou, J.; Lei, J.; Xing, G.; Sun, H.; et al. Genome Characteristics and Evolution of Pseudorabies Virus Strains in Eastern China from 2017 to 2019. *Virol. Sin.* **2019**, *34*, 601–609. [CrossRef] [PubMed]
91. Huang, J.; Zhu, L.; Zhao, J.; Yin, X.; Feng, Y.; Wang, X.; Sun, X.; Zhou, Y.; Xu, Z. Genetic evolution analysis of novel recombinant pseudorabies virus strain in Sichuan, China. *Transbound. Emerg. Dis.* **2020**, *67*, 1428–1432. [CrossRef] [PubMed]
92. Kumar, S.; Stecher, G.; Tamura, K. MEGA7: Molecular Evolutionary Genetics Analysis Version 7.0 for Bigger Datasets. *Mol. Biol. Evol.* **2016**, *33*, 1870–1874. [CrossRef] [PubMed]
93. Lin, Y.; Tan, L.; Wang, C.; He, S.; Fang, L.; Wang, Z.; Zhong, Y.; Zhang, K.; Liu, D.; Yang, Q.; et al. Serological Investigation and Genetic Characteristics of Pseudorabies Virus in Hunan Province of China From 2016 to 2020. *Front. Vet. Sci.* **2021**, *8*, 762326. [CrossRef] [PubMed]
94. Liu, J.; Chen, C.; Li, X. Novel Chinese pseudorabies virus variants undergo extensive recombination and rapid interspecies transmission. *Transbound. Emerg. Dis.* **2020**, *67*, 2274–2276. [CrossRef]
95. Ye, C.; Zhang, Q.; Tian, Z.; Zheng, H.; Zhao, K.; Liu, F.; Guo, J.; Tong, W.; Jiang, C.; Wang, S.; et al. Genomic characterization of emergent pseudorabies virus in China reveals marked sequence divergence: Evidence for the existence of two major genotypes. *Virology* **2015**, *483*, 32–43. [CrossRef]
96. Fan, J.; Zeng, X.; Zhang, G.; Wu, Q.; Niu, J.; Sun, B.; Xie, Q.; Ma, J. Molecular characterization and phylogenetic analysis of pseudorabies virus variants isolated from Guangdong province of southern China during 2013-2014. *J. Vet. Sci.* **2016**, *17*, 369–375. [CrossRef] [PubMed]
97. Ren, Q.; Ren, H.; Gu, J.; Wang, J.; Jiang, L.; Gao, S. The Epidemiological Analysis of Pseudorabies Virus and Pathogenicity of the Variant Strain in Shandong Province. *Front. Vet. Sci.* **2022**, *9*, 806824. [CrossRef]
98. Sun, Y.; Liang, W.; Liu, Q.; Zhao, T.; Zhu, H.; Hua, L.; Peng, Z.; Tang, X.; Stratton, C.; Zhou, D.; et al. Epidemiological and genetic characteristics of swine pseudorabies virus in mainland China between 2012 and 2017. *PeerJ* **2018**, *6*, e5785. [CrossRef]
99. Tan, L.; Yao, J.; Yang, Y.; Luo, W.; Yuan, X.; Yang, L.; Wang, A. Current Status and Challenge of Pseudorabies Virus Infection in China. *Virol. Sin.* **2021**, *36*, 588–607. [CrossRef]
100. Liu, H.; Shi, Z.; Liu, C.; Wang, P.; Wang, M.; Wang, S.; Liu, Z.; Wei, L.; Sun, Z.; He, X.; et al. Implication of the Identification of an Earlier Pseudorabies Virus (PRV) Strain HLJ-2013 to the Evolution of Chinese PRVs. *Front. Microbiol.* **2020**, *11*, 612474. [CrossRef]
101. Tan, L.; Yao, J.; Lei, L.; Xu, K.; Liao, F.; Yang, S.; Yang, L.; Shu, X.; Duan, D.; Wang, A. Emergence of a Novel Recombinant Pseudorabies Virus Derived From the Field Virus and Its Attenuated Vaccine in China. *Front. Vet. Sci.* **2022**, *9*, 872002. [CrossRef]
102. Ma, Z.; Han, Z.; Liu, Z.; Meng, F.; Liu, M. Epidemiological investigation of porcine pseudorabies virus and its coinfection rate in Shandong Province in China from 2015 to 2018. *J. Vet. Sci.* **2020**, *21*, e36. [CrossRef] [PubMed]
103. Kou, X.; Gao, F.; Guo, H.; Liu, J. Establishment of Indirect ELISA Antibody Detection Kit for gE Protein of Porcine Pseudorabies Virus. *China Anim. Health Insp.* **2018**, *35*, 91–94.
104. Morenkov, O.S.; Sobko, Y.A.; Panchenko, O.A. Glycoprotein gE blocking ELISAs to differentiate between Aujeszky's disease-vaccinated and infected animals. *J. Virol. Methods* **1997**, *65*, 83–94. [CrossRef]
105. Xu, L.; Peng, Z.; Zhao, T.T.; Xu, S.; Chen, H.C.; Bin, W.U. Development and preliminary application of a direct immunofluorescence method for the detection of pseudorabies virus. *Chin. J. Prev. Vet. Med.* **2017**, *39*, 993–997.
106. Zhang, P.; Lv, L.; Sun, H.; Li, S.; Fan, H.; Wang, X.; Bai, J.; Jiang, P. Identification of linear B cell epitope on gB, gC, and gE proteins of porcine pseudorabies virus using monoclonal antibodies. *Vet. Microbiol.* **2019**, *234*, 83–91. [CrossRef] [PubMed]
107. Li, X.; Sun, Y.; Yang, S.; Wang, Y.; Yang, J.; Liu, Y.; Jin, Q.; Li, X.; Guo, C.; Zhang, G. Development of an immunochromatographic strip for antibody detection of pseudorabies virus in swine. *J. Vet. Diagn. Investig.* **2015**, *27*, 739–742. [CrossRef]
108. Wang, Y.B.; Li, Y.H.; Li, Q.M.; Xie, W.T.; Guo, C.L.; Guo, J.Q.; Deng, R.G.; Zhang, G.P. Development of a blocking immunoperoxidase monolayer assay for differentiation between pseudorabies virus-infected and vaccinated animalss. *Pol. J. Vet. Sci.* **2019**, *22*, 717–723. [CrossRef] [PubMed]

109. Kinker, D.R.; Swenson, S.L.; Wu, L.L.; Zimmerman, J.J. Evaluation of serological tests for the detection of pseudorabies gE antibodies during early infection. *Vet. Microbiol.* **1997**, *55*, 99–106. [CrossRef]
110. Panyasing, Y.; Kedkovid, R.; Kittawornrat, A.; Ji, J.; Zimmerman, J.; Thanawongnuwech, R. Detection of Aujeszky's disease virus DNA and antibody in swine oral fluid specimens. *Transbound. Emerg. Dis.* **2018**, *65*, 1828–1835. [CrossRef]
111. Hu, D.; Zhang, Z.; Lv, L.; Xiao, Y.; Qu, Y.; Ma, H.; Niu, Y.; Wang, G.; Liu, S. Outbreak of variant pseudorabies virus in Bartha-K61-vaccinated piglets in central Shandong Province, China. *J. Vet. Diagn. Investig.* **2015**, *27*, 600–605. [CrossRef] [PubMed]
112. Liu, Y.; Zhang, S.; Xu, Q.; Wu, J.; Zhai, X.; Li, S.; Wang, J.; Ni, J.; Yuan, L.; Song, X.; et al. Investigation on pseudorabies prevalence in Chinese swine breeding farms in 2013–2016. *Trop. Anim. Health Prod.* **2018**, *50*, 1279–1285. [CrossRef]
113. Wu, Q.; Zhang, H.; Dong, H.; Mehmood, K.; Chang, Z.; Li, K.; Liu, S.; Rehman, M.U.; Nabi, F.; Javed, M.T.; et al. Seroprevalence and risk factors associated with Pseudorabies virus infection in Tibetan pigs in Tibet. *BMC Vet. Res.* **2018**, *14*, 25. [CrossRef]
114. Xia, L.; Sun, Q.; Wang, J.; Chen, Q.; Liu, P.; Shen, C.; Sun, J.; Tu, Y.; Shen, S.; Zhu, J.; et al. Epidemiology of pseudorabies in intensive pig farms in Shanghai, China: Herd-level prevalence and risk factors. *Prev. Vet. Med.* **2018**, *159*, 51–56. [CrossRef]
115. Ji, C.; Wei, Y.; Wang, J.; Zeng, Y.; Pan, H.; Liang, G.; Ma, J.; Gong, L.; Zhang, W.; Zhang, G.; et al. Development of a Dual Fluorescent Microsphere Immunological Assay for Detection of Pseudorabies Virus gE and gB IgG Antibodies. *Viruses* **2020**, *12*, 912. [CrossRef] [PubMed]
116. Lei, J.L.; Xia, S.L.; Wang, Y.; Du, M.; Xiang, G.T.; Cong, X.; Luo, Y.; Li, L.F.; Zhang, L.; Yu, J.; et al. Safety and immunogenicity of a gE/gI/TK gene-deleted pseudorabies virus variant expressing the E2 protein of classical swine fever virus in pigs. *Immunol. Lett.* **2016**, *174*, 63–71. [CrossRef]
117. Aytogu, G.; Toker, E.B.; Yavas, O.; Kadiroglu, B.; Ates, O.; Ozyigit, M.O.; Yesilbag, K. First isolation and molecular characterization of pseudorabies virus detected in Turkey. *Mol. Biol. Rep.* **2022**, *49*, 1679–1686. [CrossRef]
118. Tian, R.B.; Jin, Y.; Xu, T.; Zhao, Y.; Wang, Z.Y.; Chen, H.Y. Development of a SYBR green I-based duplex real-time PCR assay for detection of pseudorabies virus and porcine circovirus 3. *Mol. Cell. Probes* **2020**, *53*, 101593. [CrossRef]
119. Song, C.; Gao, L.; Bai, W.; Zha, X.; Yin, G.; Shu, X. Molecular epidemiology of pseudorabies virus in Yunnan and the sequence analysis of its gD gene. *Virus Genes* **2017**, *53*, 392–399. [CrossRef]
120. Liu, Z.; Zhang, C.; Shen, H.; Sun, J.; Zhang, J. Duplex fluorescence melting curve analysis as a new tool for rapid detection and differentiation of genotype I, II and Bartha-K61 vaccine strains of pseudorabies virus. *BMC Vet. Res.* **2018**, *14*, 372. [CrossRef] [PubMed]
121. Zheng, H.H.; Bai, Y.L.; Xu, T.; Zheng, L.L.; Li, X.S.; Chen, H.Y.; Wang, Z.Y. Isolation and Phylogenetic Analysis of Reemerging Pseudorabies Virus Within Pig Populations in Central China During 2012 to 2019. *Front. Vet. Sci.* **2021**, *8*, 764982. [CrossRef] [PubMed]
122. Zhang, C.F.; Cui, S.J.; Zhu, C. Loop-mediated isothermal amplification for rapid detection and differentiation of wild-type pseudorabies and gene-deleted virus vaccines. *J. Virol. Methods* **2010**, *169*, 239–243. [CrossRef] [PubMed]
123. Zhang, Y.; Nan, W.; Qin, L.; Gong, M.; Wu, F.; Hu, S.; Chen, Y. A NanoPCR Assay for Detection of Pseudorabies Virus. *China Anim. Health Insp.* **2017**, *34*, 102–105.
124. Ren, M.; Lin, H.; Chen, S.; Yang, M.; An, W.; Wang, Y.; Xue, C.; Sun, Y.; Yan, Y.; Hu, J. Detection of pseudorabies virus by duplex droplet digital PCR assay. *J. Vet. Diagn. Investig.* **2018**, *30*, 105–112. [CrossRef] [PubMed]
125. Tu, F.; Zhang, Y.; Xu, S.; Yang, X.; Zhou, L.; Ge, X.; Han, J.; Guo, X.; Yang, H. Detection of pseudorabies virus with a real-time recombinase-aided amplification assay. *Transbound. Emerg. Dis.* **2022**, *69*, 2266–2274. [CrossRef] [PubMed]
126. Wang, J.; Han, H.; Liu, W.; Li, S.; Guo, D. Diagnosis and gI antibody dynamics of pseudorabies virus in an intensive pig farm in Hei Longjiang Province. *J. Vet. Sci.* **2021**, *22*, e23. [CrossRef] [PubMed]
127. Meng, X.Y.; Luo, Y.; Liu, Y.; Shao, L.; Sun, Y.; Li, Y.; Li, S.; Ji, S.; Qiu, H.J. A triplex real-time PCR for differential detection of classical, variant and Bartha-K61 vaccine strains of pseudorabies virus. *Arch. Virol.* **2016**, *161*, 2425–2430. [CrossRef] [PubMed]
128. Luo, Y.; Liang, L.; Zhou, L.; Zhao, K.; Cui, S. Concurrent infections of pseudorabies virus and porcine bocavirus in China detected by duplex nanoPCR. *J. Virol. Methods* **2015**, *219*, 46–50. [CrossRef] [PubMed]
129. Cheng, T.Y.; Henao-Diaz, A.; Poonsuk, K.; Buckley, A.; van Geelen, A.; Lager, K.; Harmon, K.; Gauger, P.; Wang, C.; Ambagala, A.; et al. Pseudorabies (Aujeszky's disease) virus DNA detection in swine nasal swab and oral fluid specimens using a gB-based real-time quantitative PCR. *Prev. Vet. Med.* **2021**, *189*, 105308. [CrossRef]
130. Yang, Y.; Qin, X.; Zhang, W.; Li, Z.; Zhang, S.; Li, Y.; Zhang, Z. Development of an isothermal recombinase polymerase amplification assay for rapid detection of pseudorabies virus. *Mol. Cell. Probes* **2017**, *33*, 32–35. [CrossRef]
131. Yang, H.; Guo, Y.; Li, S.; Lan, G.; Jiang, Q.; Yang, X.; Fan, J.; Ali, Z.; Tang, Y.; Mou, X.; et al. Magnetic beads-based chemiluminescent assay for ultrasensitive detection of pseudorabies virus. *J. Nanosci. Nanotechnol.* **2014**, *14*, 3337–3342. [CrossRef] [PubMed]
132. Li, H.; Wei, X.; Zhang, X.; Xu, H.; Zhao, X.; Zhou, S.; Huang, S.; Liu, X. Establishment of a multiplex RT-PCR assay for identification of atmospheric virus contamination in pig farms. *Environ. Pollut. (Barking Essex 1987)* **2019**, *253*, 358–364. [CrossRef] [PubMed]
133. Sunaga, F.; Tsuchiaka, S.; Kishimoto, M.; Aoki, H.; Kakinoki, M.; Kure, K.; Okumura, H.; Okumura, M.; Okumura, A.; Nagai, M.; et al. Development of a one-run real-time PCR detection system for pathogens associated with porcine respiratory diseases. *J. Vet. Med. Sci.* **2020**, *82*, 217–223. [CrossRef]
134. Lee, C.-S.; Moon, H.-J.; Yang, J.-S.; Park, S.-J.; Song, D.-S.; Kang, B.-K.; Park, B.-K. Multiplex PCR for the simultaneous detection of pseudorabies virus, porcine cytomegalovirus, and porcine circovirus in pigs. *J. Virol. Methods* **2007**, *139*, 39–43. [CrossRef] [PubMed]

135. Pérez, L.J.; Perera, C.L.; Frías, M.T.; Núñez, J.I.; Ganges, L.; de Arce, H.D. A multiple SYBR Green I-based real-time PCR system for the simultaneous detection of porcine circovirus type 2, porcine parvovirus, pseudorabies virus and Torque teno sus virus 1 and 2 in pigs. *J. Virol. Methods* **2012**, *179*, 233–241. [CrossRef]
136. Huang, C.; Hung, J.J.; Wu, C.Y.; Chien, M.S. Multiplex PCR for rapid detection of pseudorabies virus, porcine parvovirus and porcine circoviruses. *Vet. Microbiol.* **2004**, *101*, 209–214. [CrossRef] [PubMed]
137. Tombácz, D.; Sharon, D.; Szűcs, A.; Moldován, N.; Snyder, M.; Boldogkői, Z. Transcriptome-wide survey of pseudorabies virus using next- and third-generation sequencing platforms. *Sci. Data* **2018**, *5*, 180119. [CrossRef] [PubMed]
138. Liu, H.; Li, X.T.; Hu, B.; Deng, X.Y.; Zhang, L.; Lian, S.Z.; Zhang, H.L.; Lv, S.; Xue, X.H.; Lu, R.G.; et al. Outbreak of severe pseudorabies virus infection in pig-offal-fed farmed mink in Liaoning Province, China. *Arch. Virol.* **2017**, *162*, 863–866. [CrossRef] [PubMed]
139. Lian, K.; Zhang, M.; Zhou, L.; Song, Y.; Wang, G.; Wang, S. First report of a pseudorabies-virus-infected wolf (Canis lupus) in China. *Arch. Virol.* **2020**, *165*, 459–462. [CrossRef]
140. Guo, X.; Hou, J.; Yuan, Z.; Li, H.; Sang, S. A novel paper biosensor based on Fe(3)O(4)@SiO(2)-NH(2)and MWCNTs for rapid detection of pseudorabies virus. *Nanotechnology* **2021**, *32*, 355102. [CrossRef] [PubMed]
141. McFerran, J.B.; Dow, C. Experimental Aujeszky's disease (pseudorabies) in rats. *Br. Vet. J.* **1970**, *126*, 173–179. [CrossRef]
142. Su, D.; Wu, S.; Guo, J.; Wu, X.; Yang, Q.; Xiong, X. Protective effect of resveratrol against pseudorabies virus-induced reproductive failure in a mouse model. *Food Sci. Biotechnol.* **2016**, *25*, 103–106. [CrossRef]
143. Lomniczi, B.; Watanabe, S.; Ben-Porat, T.; Kaplan, A.S. Genetic basis of the neurovirulence of pseudorabies virus. *J. Virol.* **1984**, *52*, 198–205. [CrossRef]
144. Yin, H.; Li, Z.; Zhang, J.; Huang, J.; Kang, H.; Tian, J.; Qu, L. Construction of a US7/US8/UL23/US3-deleted recombinant pseudorabies virus and evaluation of its pathogenicity in dogs. *Vet. Microbiol.* **2020**, *240*, 108543. [CrossRef]
145. Kit, S.; Sheppard, M.; Ichimura, H.; Kit, M. Second-generation pseudorabies virus vaccine with deletions in thymidine kinase and glycoprotein genes. *Am. J. Vet. Res.* **1987**, *48*, 780–793.
146. Moormann, R.J.; de Rover, T.; Briaire, J.; Peeters, B.P.; Gielkens, A.L.; van Oirschot, J.T. Inactivation of the thymidine kinase gene of a gI deletion mutant of pseudorabies virus generates a safe but still highly immunogenic vaccine strain. *J. Gen. Virol.* **1990**, *71 Pt 7*, 1591–1595. [CrossRef]
147. Mettenleiter, T.C.; Klupp, B.G.; Weiland, F.; Visser, N. Characterization of a quadruple glycoprotein-deleted pseudorabies virus mutant for use as a biologically safe live virus vaccine. *J. Gen. Virol.* **1994**, *75 Pt 7*, 1723–1733. [CrossRef]
148. Peeters, B.; Bouma, A.; de Bruin, T.; Moormann, R.; Gielkens, A.; Kimman, T. Non-transmissible pseudorabies virus gp50 mutants: A new generation of safe live vaccines. *Vaccine* **1994**, *12*, 375–380. [CrossRef]
149. Ling, Z.; Wan-Zhu, G.; Zhi-Wen, X.U. Fluctuant Rule of Colostral Antibodies and the Date of Initial Immunization for the Piglet from Sows Inoculated with Pseudorabies Virus Gene-deleted Vaccine SA215. *Chin. J. Vet.* **2004**, *24*, 320–322.
150. Xu, X.J.; Xu, G.Y.; Chen, H.C.; Liu, Z.F.; He, Q.G. Construction and characterization of a pseudorabies virus TK-/gG- mutant. *China J. Biotechnol.* **2004**, *20*, 532–535.
151. Ye, C.; Guo, J.C.; Gao, J.C.; Wang, T.Y.; Zhao, K.; Chang, X.B.; Wang, Q.; Peng, J.M.; Tian, Z.J.; Cai, X.H.; et al. Genomic analyses reveal that partial sequence of an earlier pseudorabies virus in China is originated from a Bartha-vaccine-like strain. *Virology* **2016**, *491*, 56–63. [CrossRef]
152. Tong, W.; Liu, F.; Zheng, H.; Liang, C.; Zhou, Y.J.; Jiang, Y.F.; Shan, T.L.; Gao, F.; Li, G.X.; Tong, G.Z. Emergence of a Pseudorabies virus variant with increased virulence to piglets. *Vet. Microbiol.* **2015**, *181*, 236–240. [CrossRef]
153. Sun, Y.; Luo, Y.; Wang, C.H.; Yuan, J.; Li, N.; Song, K.; Qiu, H.J. Control of swine pseudorabies in China: Opportunities and limitations. *Vet. Microbiol.* **2016**, *183*, 119–124. [CrossRef]
154. Wang, T.; Xiao, Y.; Yang, Q.; Wang, Y.; Sun, Z.; Zhang, C.; Yan, S.; Wang, J.; Guo, L.; Yan, H.; et al. Construction of a gE-Deleted Pseudorabies Virus and Its Efficacy to the New-Emerging Variant PRV Challenge in the Form of Killed Vaccine. *BioMed Res. Int.* **2015**, *2015*, 684945. [CrossRef]
155. Zhao, Y.; Wang, L.Q.; Zheng, H.H.; Yang, Y.R.; Liu, F.; Zheng, L.L.; Jin, Y.; Chen, H.Y. Construction and immunogenicity of a gE/gI/TK-deleted PRV based on porcine pseudorabies virus variant. *Mol. Cell. Probes* **2020**, *53*, 101605. [CrossRef]
156. Lv, L.; Liu, X.; Jiang, C.; Wang, X.; Cao, M.; Bai, J.; Jiang, P. Pathogenicity and immunogenicity of a gI/gE/TK/UL13-gene-deleted variant pseudorabies virus strain in swine. *Vet. Microbiol.* **2021**, *258*, 109104. [CrossRef]
157. Gu, Z.; Dong, J.; Wang, J.; Hou, C.; Sun, H.; Yang, W.; Bai, J.; Jiang, P. A novel inactivated gE/gI deleted pseudorabies virus (PRV) vaccine completely protects pigs from an emerged variant PRV challenge. *Virus Res.* **2015**, *195*, 57–63. [CrossRef]
158. Zhang, C.; Guo, L.; Jia, X.; Wang, T.; Wang, J.; Sun, Z.; Wang, L.; Li, X.; Tan, F.; Tian, K. Construction of a triple gene-deleted Chinese Pseudorabies virus variant and its efficacy study as a vaccine candidate on suckling piglets. *Vaccine* **2015**, *33*, 2432–2437. [CrossRef]
159. Wang, J.; Cui, X.; Wang, X.; Wang, W.; Gao, S.; Liu, X.; Kai, Y.; Chen, C. Efficacy of the Bartha-K61 vaccine and a gE(-)/gI(-)/TK(-) prototype vaccine against variant porcine pseudorabies virus (vPRV) in piglets with sublethal challenge of vPRV. *Res. Vet. Sci.* **2020**, *128*, 16–23. [CrossRef]
160. Dong, J.; Bai, J.; Sun, T.; Gu, Z.; Wang, J.; Sun, H.; Jiang, P. Comparative pathogenicity and immunogenicity of triple and double gene-deletion pseudorabies virus vaccine candidates. *Res. Vet. Sci.* **2017**, *115*, 17–23. [CrossRef]

161. Sun, L.; Tang, Y.; Yan, K.; Zhang, H. Construction of a quadruple gene-deleted vaccine confers complete protective immunity against emerging PRV variant challenge in piglets. *Virol. J.* **2022**, *19*, 19. [CrossRef]

162. Zhang, C.; Liu, Y.; Chen, S.; Qiao, Y.; Guo, M.; Zheng, Y.; Xu, M.; Wang, Z.; Hou, J.; Wang, J. A gD&gC-substituted pseudorabies virus vaccine strain provides complete clinical protection and is helpful to prevent virus shedding against challenge by a Chinese pseudorabies variant. *BMC Vet. Res.* **2019**, *15*, 2. [CrossRef]

163. Yin, Y.; Xu, Z.; Liu, X.; Li, P.; Yang, F.; Zhao, J.; Fan, Y.; Sun, X.; Zhu, L. A live gI/gE-deleted pseudorabies virus (PRV) protects weaned piglets against lethal variant PRV challenge. *Virus Genes* **2017**, *53*, 565–572. [CrossRef] [PubMed]

164. Yan, S.; Huang, B.; Bai, X.; Zhou, Y.; Guo, L.; Wang, T.; Shan, Y.; Wang, Y.; Tan, F.; Tian, K. Construction and Immunogenicity of a Recombinant Pseudorabies Virus Variant With TK/gI/gE/11k/28k Deletion. *Front. Vet. Sci.* **2021**, *8*, 797611. [CrossRef] [PubMed]

165. Li, J.; Fang, K.; Rong, Z.; Li, X.; Ren, X.; Ma, H.; Chen, H.; Li, X.; Qian, P. Comparison of gE/gI- and TK/gE/gI-Gene-Deleted Pseudorabies Virus Vaccines Mediated by CRISPR/Cas9 and Cre/Lox Systems. *Viruses* **2020**, *12*, 369. [CrossRef]

166. Quint, W.; Gielkens, A.; Van Oirschot, J.; Berns, A.; Cuypers, H.T. Construction and characterization of deletion mutants of pseudorabies virus: A new generation of 'live' vaccines. *J. Gen. Virol.* **1987**, *68 Pt 2*, 523–534. [CrossRef]

167. Wang, J.; Guo, R.; Qiao, Y.; Xu, M.; Wang, Z.; Liu, Y.; Gu, Y.; Liu, C.; Hou, J. An inactivated gE-deleted pseudorabies vaccine provides complete clinical protection and reduces virus shedding against challenge by a Chinese pseudorabies variant. *BMC Vet. Res.* **2016**, *12*, 277. [CrossRef]

168. Visser, N.; Luetticken, D. Experiences with a gI-/TK- modified live pseudorabies virus vaccine: Strain Begonia. *Curr. Top. Vet. Med.* **1989**, *48*, 37–44.

169. Wardley, R.C.; Post, L.E. The use of the gX deleted vaccine PRV.TK.gX-1 in the control of Aujeszky's disease. In Proceedings of the CEC Seminar on Vaccination and Control of Aujeszky's Disease, Brussels, Belgium, 5–6 July 1988; Van Oirschot, J.T., Ed.; Kluwer Academic: Dordrecht, The Netherlands, 1989; pp. 13–25.

170. Yang, D.K.; Kim, H.H.; Choi, S.S.; Hyun, B.H.; Song, J.Y. An oral Aujeszky's disease vaccine (YS-400) induces neutralizing antibody in pigs. *Clin. Exp. Vaccine Res.* **2016**, *5*, 132–137. [CrossRef] [PubMed]

171. Tong, W.; Li, G.; Liang, C.; Liu, F.; Tian, Q.; Cao, Y.; Li, L.; Zheng, X.; Zheng, H.; Tong, G. A live, attenuated pseudorabies virus strain JS-2012 deleted for gE/gI protects against both classical and emerging strains. *Antivir. Res.* **2016**, *130*, 110–117. [CrossRef]

172. Wu, C.Y.; Liao, C.M.; Chi, J.N.; Chien, M.S.; Huang, C. Growth properties and vaccine efficacy of recombinant pseudorabies virus defective in glycoprotein E and thymidine kinase genes. *J. Biotechnol.* **2016**, *229*, 58–64. [CrossRef] [PubMed]

173. Wang, J.; Song, Z.; Ge, A.; Guo, R.; Qiao, Y.; Xu, M.; Wang, Z.; Liu, Y.; Zheng, Y.; Fan, H.; et al. Safety and immunogenicity of an attenuated Chinese pseudorabies variant by dual deletion of TK&gE genes. *BMC Vet. Res.* **2018**, *14*, 287. [CrossRef]

174. Ferrari, M.; Brack, A.; Romanelli, M.G.; Mettenleiter, T.C.; Corradi, A.; Dal Mas, N.; Losio, M.N.; Silini, R.; Pinoni, C.; Pratelli, A. A study of the ability of a TK-negative and gI/gE-negative pseudorabies virus (PRV) mutant inoculated by different routes to protect pigs against PRV infection. *J. Vet. Med. B* **2000**, *47*, 753–762. [CrossRef] [PubMed]

175. Zhu, L.; Yi, Y.; Xu, Z.; Cheng, L.; Tang, S.; Guo, W. Growth, physicochemical properties, and morphogenesis of Chinese wild-type PRV Fa and its gene-deleted mutant strain PRV SA215. *Virol. J.* **2011**, *8*, 272. [CrossRef] [PubMed]

176. Hu, R.M.; Zhou, Q.; Song, W.B.; Sun, E.C.; Zhang, M.M.; He, Q.G.; Chen, H.C.; Wu, B.; Liu, Z.F. Novel pseudorabies virus variant with defects in TK, gE and gI protects growing pigs against lethal challenge. *Vaccine* **2015**, *33*, 5733–5740. [CrossRef]

177. Tang, Y.D.; Liu, J.T.; Wang, T.Y.; An, T.Q.; Sun, M.X.; Wang, S.J.; Fang, Q.Q.; Hou, L.L.; Tian, Z.J.; Cai, X.H. Live attenuated pseudorabies virus developed using the CRISPR/Cas9 system. *Virus Res.* **2016**, *225*, 33–39. [CrossRef] [PubMed]

178. Lin, J.; Li, Z.; Feng, Z.; Fang, Z.; Chen, J.; Chen, W.; Liang, W.; Chen, Q. Pseudorabies virus (PRV) strain with defects in gE, gC, and TK genes protects piglets against an emerging PRV variant. *J. Vet. Med. Sci.* **2020**, *82*, 846–855. [CrossRef]

179. Gao, J.F.; Lai, Z.; Shu, Y.H.; Qi, S.; Ma, J.; Wu, B.; Gong, J.P. Isolation and identification of porcine pseudorabies virus(PRV)C strain. *Acta Agric. Shanghai* **2015**, *31*, 32–36.

180. Kimman, T.G.; de Wind, N.; Oei-Lie, N.; Pol, J.M.; Berns, A.J.; Gielkens, A.L. Contribution of single genes within the unique short region of Aujeszky's disease virus (suid herpesvirus type 1) to virulence, pathogenesis and immunogenicity. *J. Gen. Virol.* **1992**, *73 Pt 2*, 243–251. [CrossRef]

181. Dong, B.; Zarlenga, D.S.; Ren, X. An overview of live attenuated recombinant pseudorabies viruses for use as novel vaccines. *J. Immunol. Res.* **2014**, *2014*, 824630. [CrossRef]

182. Chen, Y.; Guo, W.; Xu, Z.; Yan, Q.; Luo, Y.; Shi, Q.; Chen, D.; Zhu, L.; Wang, X. A novel recombinant pseudorabies virus expressing parvovirus VP2 gene: Immunogenicity and protective efficacy in swine. *Virol. J.* **2011**, *8*, 307. [CrossRef] [PubMed]

183. Tong, W.; Zheng, H.; Li, G.X.; Gao, F.; Shan, T.L.; Zhou, Y.J.; Yu, H.; Jiang, Y.F.; Yu, L.X.; Li, L.W.; et al. Recombinant pseudorabies virus expressing E2 of classical swine fever virus (CSFV) protects against both virulent pseudorabies virus and CSFV. *Antivir. Res.* **2020**, *173*, 104652. [CrossRef]

184. Zheng, H.H.; Wang, L.Q.; Fu, P.F.; Zheng, L.L.; Chen, H.Y.; Liu, F. Characterization of a recombinant pseudorabies virus expressing porcine parvovirus VP2 protein and porcine IL-6. *Virol. J.* **2020**, *17*, 19. [CrossRef]

185. Wang, Y.; Yuan, J.; Cong, X.; Qin, H.Y.; Wang, C.H.; Li, Y.; Li, S.; Luo, Y.; Sun, Y.; Qiu, H.J. Generation and Efficacy Evaluation of a Recombinant Pseudorabies Virus Variant Expressing the E2 Protein of Classical Swine Fever Virus in Pigs. *Clin. Vaccine Immunol.* **2015**, *22*, 1121–1129. [CrossRef]

186. Yao, L.; Wu, C.X.; Zheng, K.; Xu, X.J.; Zhang, H.; Chen, C.F.; Liu, Z.F. Immunogenic response to a recombinant pseudorabies virus carrying bp26 gene of Brucella melitensis in mice. *Res. Vet. Sci.* **2015**, *100*, 61–67. [CrossRef] [PubMed]

187. Feng, Z.; Chen, J.; Liang, W.; Chen, W.; Li, Z.; Chen, Q.; Cai, S. The recombinant pseudorabies virus expressing African swine fever virus CD2v protein is safe and effective in mice. *Virol. J.* **2020**, *17*, 180. [CrossRef] [PubMed]

188. Chen, L.; Zhang, X.; Shao, G.; Shao, Y.; Hu, Z.; Feng, K.; Xie, Z.; Li, H.; Chen, W.; Lin, W.; et al. Construction and Evaluation of Recombinant Pseudorabies Virus Expressing African Swine Fever Virus Antigen Genes. *Front. Vet. Sci.* **2022**, *9*, 832255. [CrossRef] [PubMed]

189. Wang, J.; Wang, Y.; Zhang, E.; Zhou, M.; Lin, J.; Yang, Q. Intranasal administration with recombinant Bacillus subtilis induces strong mucosal immune responses against pseudorabies. *Microb. Cell Factories* **2019**, *18*, 103. [CrossRef]

190. Zhang, T.; Liu, Y.; Chen, Y.; Wang, A.; Feng, H.; Wei, Q.; Zhou, E.; Zhang, G. A single dose glycoprotein D-based subunit vaccine against pseudorabies virus infection. *Vaccine* **2020**, *38*, 6153–6161. [CrossRef]

191. Kong, H.; Zhang, K.; Liu, Y.; Shang, Y.; Wu, B.; Liu, X. Attenuated live vaccine (Bartha-K16) caused pseudorabies (Aujeszky's disease) in sheep. *Vet. Res. Commun.* **2013**, *37*, 329–332. [CrossRef]

192. Lin, W.; Shao, Y.; Tan, C.; Shen, Y.; Zhang, X.; Xiao, J.; Wu, Y.; He, L.; Shao, G.; Han, M.; et al. Commercial vaccine against pseudorabies virus: A hidden health risk for dogs. *Vet. Microbiol.* **2019**, *233*, 102–112. [CrossRef]

193. Moreno, A.; Chiapponi, C.; Sozzi, E.; Morelli, A.; Silenzi, V.; Gobbi, M.; Lavazza, A.; Paniccià, M. Detection of a gE-deleted Pseudorabies virus strain in an Italian red fox. *Vet. Microbiol.* **2020**, *244*, 108666. [CrossRef]

194. Cheng, C.; Yu, X. Research Progress in Chinese Herbal Medicines for Treatment of Sepsis: Pharmacological Action, Phytochemistry, and Pharmacokinetics. *Int. J. Mol. Sci.* **2021**, *22*, 11078. [CrossRef] [PubMed]

195. Song, L.; Zhang, J.; Lai, R.; Li, Q.; Ju, J.; Xu, H. Chinese Herbal Medicines and Active Metabolites: Potential Antioxidant Treatments for Atherosclerosis. *Front. Pharmacol.* **2021**, *12*, 675999. [CrossRef]

196. Sun, J.; Ren, J.; Hu, X.; Hou, Y.; Yang, Y. Therapeutic effects of Chinese herbal medicines and their extracts on diabetes. *Biomed. Pharmacother.* **2021**, *142*, 111977. [CrossRef]

197. Hsiao, Y.H.; Lin, C.W.; Wang, P.H.; Hsin, M.C.; Yang, S.F. The Potential of Chinese Herbal Medicines in the Treatment of Cervical Cancer. *Integr. Cancer Ther.* **2019**, *18*, 1534735419861693. [CrossRef]

198. Wang, Z.; Qi, F.; Cui, Y.; Zhao, L.; Sun, X.; Tang, W.; Cai, P. An update on Chinese herbal medicines as adjuvant treatment of anticancer therapeutics. *Biosci. Trends* **2018**, *12*, 220–239. [CrossRef]

199. Zhao, M.; Yu, Y.; Wang, R.; Chang, M.; Ma, S.; Qu, H.; Zhang, Y. Mechanisms and Efficacy of Chinese Herbal Medicines in Chronic Kidney Disease. *Front. Pharmacol.* **2020**, *11*, 619201. [CrossRef]

200. Wang, Z.; Yang, L. Chinese herbal medicine: Fighting SARS-CoV-2 infection on all fronts. *J. Ethnopharmacol.* **2021**, *270*, 113869. [CrossRef]

201. Yang, H.; Jiang, C.; Chen, X.; He, K.; Hu, Y. Protective effects of sinomenine against LPS-induced inflammation in piglets. *Microb. Pathog.* **2017**, *110*, 573–577. [CrossRef]

202. Zhang, J.Z.; Zhang, X.L.; Liang, X.Q.; Gu, H.G.; Zhu, P.T. Effects of different Chinese herbal medicines on biochemical parameters in guinea-pig with pigment gallstones. *J. Chin. Integr. Med.* **2008**, *6*, 856–859. [CrossRef] [PubMed]

203. Wu, H.C.; Horng, C.T.; Tsai, S.C.; Lee, Y.L.; Hsu, S.C.; Tsai, Y.J.; Tsai, F.J.; Chiang, J.H.; Kuo, D.H.; Yang, J.S. Relaxant and vasoprotective effects of ginger extracts on porcine coronary arteries. *Int. J. Mol. Med.* **2018**, *41*, 2420–2428. [CrossRef] [PubMed]

204. Lin, R.; Duan, J.; Mu, F.; Bian, H.; Zhao, M.; Zhou, M.; Li, Y.; Wen, A.; Yang, Y.; Xi, M. Cardioprotective effects and underlying mechanism of Radix Salvia miltiorrhiza and Lignum Dalbergia odorifera in a pig chronic myocardial ischemia model. *Int. J. Mol. Med.* **2018**, *42*, 2628–2640. [CrossRef] [PubMed]

205. Chen, M.; Chen, X.; Song, X.; Muhammad, A.; Jia, R.; Zou, Y.; Yin, L.; Li, L.; He, C.; Ye, G.; et al. The immune-adjuvant activity and the mechanism of resveratrol on pseudorabies virus vaccine in a mouse model. *Int. Immunopharmacol.* **2019**, *76*, 105876. [CrossRef]

206. He, W.; Zhai, X.; Su, J.; Ye, R.; Zheng, Y.; Su, S. Antiviral Activity of Germacrone against Pseudorabies Virus in Vitro. *Pathogens* **2019**, *8*, 258. [CrossRef] [PubMed]

207. Tong, C.; Chen, Z.; Liu, F.; Qiao, Y.; Chen, T.; Wang, X. Antiviral activities of Radix isatidis polysaccharide against pseudorabies virus in swine testicle cells. *BMC Complementary Med. Ther.* **2020**, *20*, 48. [CrossRef]

208. Lv, C.; Liu, W.; Wang, B.; Dang, R.; Qiu, L.; Ren, J.; Yan, C.; Yang, Z.; Wang, X. Ivermectin inhibits DNA polymerase UL42 of pseudorabies virus entrance into the nucleus and proliferation of the virus in vitro and vivo. *Antivir. Res.* **2018**, *159*, 55–62. [CrossRef]

209. Chen, X.; Song, X.; Li, L.; Chen, Y.; Jia, R.; Zou, Y.; Wan, H.; Zhao, L.; Tang, H.; Lv, C.; et al. Resveratrol Inhibits Pseudorabies Virus Replication by Targeting IE180 Protein. *Front. Microbiol.* **2022**, *13*, 891978. [CrossRef]

210. Pezzuto, J.M. Resveratrol: Twenty Years of Growth, Development and Controversy. *Biomol. Ther.* **2019**, *27*, 1–14. [CrossRef]

211. Zhao, X.; Cui, Q.; Fu, Q.; Song, X.; Jia, R.; Yang, Y.; Zou, Y.; Li, L.; He, C.; Liang, X.; et al. Antiviral properties of resveratrol against pseudorabies virus are associated with the inhibition of IκB kinase activation. *Sci. Rep.* **2017**, *7*, 8782. [CrossRef]

212. Galindo, I.; Hernáez, B.; Berná, J.; Fenoll, J.; Cenis, J.L.; Escribano, J.M.; Alonso, C. Comparative inhibitory activity of the stilbenes resveratrol and oxyresveratrol on African swine fever virus replication. *Antivir. Res.* **2011**, *91*, 57–63. [CrossRef] [PubMed]

213. Li, L.; Wang, R.; Hu, H.; Chen, X.; Yin, Z.; Liang, X.; He, C.; Yin, L.; Ye, G.; Zou, Y.; et al. The antiviral activity of kaempferol against pseudorabies virus in mice. *BMC Vet. Res.* **2021**, *17*, 247. [CrossRef]

214. Huan, C.; Zhou, Z.; Yao, J.; Ni, B.; Gao, S. The Antiviral Effect of Panax Notoginseng Polysaccharides by Inhibiting PRV Adsorption and Replication In Vitro. *Molecules* **2022**, *27*, 1254. [CrossRef]

215. Huan, C.; Zhang, W.; Xu, Y.; Ni, B.; Gao, S. Plantago asiaticaAntiviral Activity of Polysaccharide against Pseudorabies Virus. *Oxidative Med. Cell. Longev.* **2022**, *2022*, 3570475. [CrossRef]
216. Sun, Y.; Li, C.; Li, Z.; Shangguan, A.; Jiang, J.; Zeng, W.; Zhang, S.; He, Q. Quercetin as an antiviral agent inhibits the Pseudorabies virus in vitro and in vivo. *Virus Res.* **2021**, *305*, 198556. [CrossRef]
217. Hsuan, S.L.; Chang, S.C.; Wang, S.Y.; Liao, T.L.; Jong, T.T.; Chien, M.S.; Lee, W.C.; Chen, S.S.; Liao, J.W. The cytotoxicity to leukemia cells and antiviral effects of Isatis indigotica extracts on pseudorabies virus. *J. Ethnopharmacol.* **2009**, *123*, 61–67. [CrossRef]
218. Zhu, W.; Zou, M.; Sun, M.; Liu, J.; Wang, Y. The anti-PRV and anti-PRRSV effects of Marine Bacillus S-12 low temperature lysozyme in vitro. *China J. Mar. Drugs* **2013**, *32*, 26–32. [CrossRef]
219. Sui, X.; Yin, J.; Ren, X. Antiviral effect of diammonium glycyrrhizinate and lithium chloride on cell infection by pseudorabies herpesvirus. *Antivir. Res.* **2010**, *85*, 346–353. [CrossRef]
220. Liu, J.; Tu, C.; Zhang, M.; Hou, S.; Wang, E. Activity of vanadium-substituted heteropolytungstate to antipseudorabies virus. *China J. Appl. Chem.* **1998**, *15*, 41–44.
221. Ren, X.; Li, G.; Sui, X. Antiviral activities of phosphonoformate sodium to pseudorabies herpesvirus infection in vitro. *Pharm. Biol.* **2011**, *49*, 608–613. [CrossRef]
222. Wang, Q.; Xie, X.; Chen, Q.; Yi, S.; Chen, J.; Xiao, Q.; Yu, M.; Wei, Y.; Hu, T. Effects of Quercitrin on PRV-Induced Secretion of Reactive Oxygen Species and Prediction of lncRNA Regulatory Targets in 3D4/2 Cells. *Antioxidants* **2022**, *11*, 631. [CrossRef] [PubMed]
223. Ye, S.; Shao, K.; Li, Z.; Guo, N.; Zuo, Y.; Li, Q.; Lu, Z.; Chen, L.; He, Q.; Han, H. Antiviral Activity of Graphene Oxide: How Sharp Edged Structure and Charge Matter. *ACS Appl. Mater. Interfaces* **2015**, *7*, 21571–21579. [CrossRef] [PubMed]
224. Du, L.; He, Y.; Zhou, Y.; Liu, S.; Zheng, B.J.; Jiang, S. The spike protein of SARS-CoV—A target for vaccine and therapeutic development. *Nat. Reviews. Microbiol.* **2009**, *7*, 226–236. [CrossRef] [PubMed]
225. Wong, R.R.; Abd-Aziz, N.; Affendi, S.; Poh, C.L. Role of microRNAs in antiviral responses to dengue infection. *J. Biomed. Sci.* **2020**, *27*, 4. [CrossRef]
226. Tan, L.; Yuan, X.; Liu, Y.; Cai, X.; Guo, S.; Wang, A. Non-muscle Myosin II: Role in Microbial Infection and Its Potential as a Therapeutic Target. *Front. Microbiol.* **2019**, *10*, 401. [CrossRef]
227. Wang, Y.P.; Huang, L.P.; Du, W.J.; Wei, Y.W.; Wu, H.L.; Feng, L.; Liu, C.M. Targeting the pseudorabies virus DNA polymerase processivity factor UL42 by RNA interference efficiently inhibits viral replication. *Antivir. Res.* **2016**, *132*, 219–224. [CrossRef]
228. Lagos-Quintana, M.; Rauhut, R.; Lendeckel, W.; Tuschl, T. Identification of novel genes coding for small expressed RNAs. *Science (New York, NY, USA)* **2001**, *294*, 853–858. [CrossRef]
229. Stagakis, E.; Bertsias, G.; Verginis, P.; Nakou, M.; Hatziapostolou, M.; Kritikos, H.; Iliopoulos, D.; Boumpas, D.T. Identification of novel microRNA signatures linked to human lupus disease activity and pathogenesis: miR-21 regulates aberrant T cell responses through regulation of PDCD4 expression. *Ann. Rheum. Dis.* **2011**, *70*, 1496–1506. [CrossRef]
230. Huang, J.; Ma, G.; Fu, L.; Jia, H.; Zhu, M.; Li, X.; Zhao, S. Pseudorabies viral replication is inhibited by a novel target of miR-21. *Virology* **2014**, *456*, 319–328. [CrossRef]
231. Liu, H.; Yang, L.; Shi, Z.; Lv, R.; Yang, X.; Wang, C.; Chen, L.; Chang, H. Functional analysis of prv-miR-LLT11a encoded by pseudorabies virus. *J. Vet. Sci.* **2019**, *20*, e68. [CrossRef]
232. Laval, K.; Enquist, L. The Neuropathic Itch Caused by Pseudorabies Virus. *Pathogens* **2020**, *9*, 254. [CrossRef] [PubMed]

Review

The Role of Latency-Associated Transcripts in the Latent Infection of Pseudorabies Virus

Jiahuan Deng, Zhuoyun Wu [†], Jiaqi Liu, Qiuyun Ji and Chunmei Ju *,[†]

Key Laboratory of Zoonosis Prevention and Control of Guangdong Province, College of Veterinary Medicine, South China Agricultural University, Guangzhou 510642, China; 20213073021@stu.scau.edu.cn (J.D.); wzzy0615@stu.scau.edu.cn (Z.W.); 15013279249@stu.scau.edu.cn (J.L.); 20193073034@stu.scau.edu.cn (Q.J.)
* Correspondence: juchunmei@scau.edu.cn
† These authors contributed equally to this work.

Abstract: Pseudorabies virus (PRV) can cause neurological, respiratory, and reproductive diseases in pigs and establish lifelong latent infection in the peripheral nervous system (PNS). Latent infection is a typical feature of PRV, which brings great difficulties to the prevention, control, and eradication of pseudorabies. The integral mechanism of latent infection is still unclear. Latency-associated transcripts (LAT) gene is the only transcriptional region during latent infection of PRV which plays the key role in regulating viral latent infection and inhibiting apoptosis. Here, we review the characteristics of PRV latent infection and the transcriptional characteristics of the LAT gene. We also analyzed the function of non-coding RNA (ncRNA) produced by the LAT gene and its importance in latent infection. Furthermore, we provided possible strategies to solve the problem of latent infection of virulent PRV strains in the host. In short, the detailed mechanism of PRV latent infection needs to be further studied and elucidated.

Keywords: pseudorabies virus; latent infection; latency-associated transcripts; non-coding RNA

Citation: Deng, J.; Wu, Z.; Liu, J.; Ji, Q.; Ju, C. The Role of Latency-Associated Transcripts in the Latent Infection of Pseudorabies Virus. *Viruses* **2022**, *14*, 1379. https://doi.org/10.3390/v14071379

Academic Editors: Yan-Dong Tang and Xiangdong Li

Received: 30 May 2022
Accepted: 22 June 2022
Published: 24 June 2022

Publisher's Note: MDPI stays neutral with regard to jurisdictional claims in published maps and institutional affiliations.

1. Introduction

Pseudorabies virus (PRV) belongs to the family *Herpesviridae*, subfamily *Alphaherpesvirinae*, and the *Varicellovirus* genus [1]. The PRV genome is 142 kb of linear double-stranded DNA with 70 different coding genes and one latency-associated transcript (LAT) site. It consists of a unique long region (U_L), a unique short region (U_S), internal repetitive sequences (IRS), and terminal repetitive sequences (TRS) [2]. The natural host of PRV is pigs, but it can infect most mammals, including cattle, sheep, cats, dogs, mink, and rodents [3–9]. There are significant differences in PRV infection between natural and non-natural hosts [10,11]. In natural hosts, PRV can cause neurological, respiratory, and reproductive diseases and establish latent infection in the peripheral nervous system (PNS) of surviving pigs, but death in adult pigs is uncommon [11–14]. In non-natural hosts, PRV infection is characterized by severe pruritus, a short duration of disease, and rapid death [10,11]. The mortality rate of non-natural hosts is up to 100%, and consequently latent infection rarely occurs [10,11,15]. However, under laboratory conditions, PRV can establish activatable latent infection in non-natural hosts, which is of great significance in the study of herpesvirus latent infection [16–19].

PRV has been eradicated or controlled through the use of gene-deficient vaccines and differentiating infected from vaccinated animals (DIVA) strategy in many countries. However, since 2011, the emergence of mutant strains of PRV has made pseudorabies come back in China, one of the world's largest pig breeding countries [20–24]. Tong et al. found that PRV mutant strain JS-2012 caused earlier clinical symptoms and higher mortality compared to PRV classic strain SC in 15, 30, and 60-day-old pigs [25]. In protection assays, the Bartha-K61 vaccine provided 100% protection against classic strain, but only partial protection against JS-2012 strain or HeN1 strain [25–27].The mutant strains have been prevalent in pig farms immunized with PRV vaccine, which shows that the existing PRV vaccine cannot prevent infection caused by the new PRV mutant strains [20,22,25–27]. In recent years, human infections of PRV have been increasingly reported. From 2018 to 2022, more than 20 cases of human PRV infection have been found [28–39]. Liu et al. isolated and identified a human PRV strain hSD-1/2019 which had high pathogenicity to mice and pigs [28]. Most of the patients infected with PRV are workers related to the pig industry, and they are directly or indirectly infected with PRV through conjunctiva, skin wounds, and syringe stab wounds. PRV is not only costly to the pig industry but also a serious threat to humans. Therefore, the eradication of PRV should be accelerated all over the world.

Latent infection is the major impediment to eradication of PRV. PRV can establish latent infection in the PNS of pigs. During the latent infection of PRV, no clinical symptoms and infectious virions exist in pigs. When stimulated by stressors, the latent virus can be reactivated, and then productive infection occurs. According to the investigation of pig farms, PRV was in a state of latent infection most of the time and latent virions were prone to reactivation in winter and spring [40]. Interestingly, although reactivated virions were detected in pigs, no clinical symptoms were observed. These virions were excreted into the environment, resulting in the spread of PRV and the infection of other susceptible animals. Therefore, controlling latent infection plays an important role in the eradication of PRV.

The integral mechanism of latent infection is still unclear. The LAT gene is the only active gene during the latent infection. It can transcribe a variety of non-coding RNAs (ncRNAs) which are involved in the establishment, maintenance, and reactivation of viral latent infection, as well as the inhibition of productive infection and anti-apoptosis [41]. Here, we mainly summarized the characteristics of PRV latent infection, the transcription and function of the LAT gene, so as to provide new perspectives for future research on PRV latent infection.

2. The Characteristics of PRV Latent Infection

During the natural infection of pigs, PRV replicates in the epithelial cells of the nasal mucosa and invades sensory nerve endings by membrane fusion. The virus particles entering the axon terminals are retrogradely transported to the neuron nucleus [42,43]. Then, the capsid docks near the nuclear foramen, PRV genome is released, and the tegument protein VP16 activates the immediate early gene IE180 of PRV to form a productive infection [1]. During latent infection, the expression of immediate early genes (IE gene) is affected by many factors, such as VP16, Oct-1 (a member of the Oct protein family), HCF (a cellular protein), and LAT [44]. In sensory neurons, Oct proteins (except Oct-1) can prevent the formation of VP16/Oct-1/HCF complexes, thus inhibiting the transcription of IE genes [44]. Moreover, in the nucleus, the viral genome binds to the nucleosome and further inhibits the expression of IE180 during latent infection [41]. IE180 is a potent transcriptional activator which is required for efficient transcription of early (E) gene and late (L) gene of PRV, so it is essential for viral replication [1]. The expression product of IE180 can bind to the promoter of LAT and inhibit the transcription of LAT genes [45]. Therefore, when the IE180 gene is silenced, IE180-mediated transcription of E gene and L gene is restricted, while LAT gene transcriptional activity is enhanced. Thus, LAT gene is the only transcriptional region during latent infection [46].

The PRV genome mainly exists in the nerve tissue, especially in the trigeminal ganglion, which is the most reliable tissue for detecting latent PRV during latent infection [47]. The olfactory bulb and medulla oblongata can also contain the latent genome. The PRV latent genome is mainly confined to the nucleus in the form of linear and unintegrated, and a small number of latent genomes exist in the form of ring [47]. The positive cells of latent infection are distributed in different regions of the neural tissue in the form of aggregation [47]. In latently infected neurons, the number of PRV genome is stable and is not related to the length of the latency period [47]. Latent infections are generally stable but can be reactivated under stress, such as restraint, exposure to cold, or transport [47,48].

Latent infection requires the co-regulation of viruses, neurons, and the host immune system [49,50]. When the three are balanced, the virus can establish a latent infection in the host. The conditions for the establishment of latent infection are as follows: firstly, the viral genome enters the nucleus of neurons, and the vast majority of genes are restricted for transcription and translation. Secondly, in order to avoid latent genome loss, the virus takes certain measures to promote the survival of infected cells and evade host immunity [51]. For example, LAT has an anti-apoptotic effect and can prolong the survival time of neurons [41]. Finally, latent viruses can monitor and manipulate the environment of host cells for reactivation.

3. Transcriptional Characteristics of LAT Gene

3.1. Transcriptional Region and Sizes of the LAT Gene

The study of latent viral gene expression is restricted because less than 1% of ganglion neurons contain latent viral genomes [47]. Among the latently infected neurons, Rock et al. detected DNA and mRNA of the virus by in situ hybridization, which proved that the PRV genome has transcriptional activity during the latent infection [46]. Researchers soon discovered that mRNAs produced during latent infection were transcribed from a region between 0.69 and 0.77 map units of the PRV genome [52]. This region is about 11 kb, which covers the early protein 0 (EP0) gene and IE180 gene, and the transcriptional direction is opposite to that of the IE180 and EP0 genes [53–55]. These mRNAs are collectively referred to as LATs, including various sizes of mRNA, in which 0.95 kb, 1.0 kb, 2.0 kb, 8.0 kb, and 8.4 kb mRNA are generally detected [52,56,57]. The mRNA of 8.4 kb (some refer to 8.5 kb) is called large latency transcript (LLT) [52,55,57,58]. Through in situ hybridization analysis of LATs, it was found that LATs were mainly confined to the nucleus of neurons and a small part of them existed in the cytoplasm [44,46].

3.2. The Structure and Function of LAT Promoter

The structure of the LAT promoter (LAP) overlaps with the promoter of UL1-3.5 gene cluster in the opposite direction [57]. LAP contains 2 TATA boxes, 3 CAAT boxes, and 2 GC boxes, which is a dual regulatory promoter. The first latent activation promoter (LAP1) contains the first TATA box and three CAAT boxes, and the second latent activation promoter (LAP2) contains the second TATA box and two GC boxes. LAP1 is the basic promoter of LAT gene expression during PRV latent infection. It initiates transcription of LAT in nerve tissue and produces LLT with 4.6 kb intron [58]. LLT starts at 34 nucleotides downstream of the first TATA box in LAP1 and can be spliced into different sizes of RNAs. The whole nucleotide sequence of 2.0 kb mRNA is contained in the LLT, and it lacks the intron of 4.6 kb. However, the LAT of 2.0 kb is regulated by LAP2, which starts at about 243 bp downstream of the LLT transcription initiation site and ends at the junction of *Bam*HI fragments 8′ and 8 [52]. Whether in latent or lytic infection, nerve or non-nerve cells, in vivo or in vitro, LAP2 has no specific activity, and it is responsible for regulating the transcription of 1.0 kb, 2.0 kb, and 8.0 kb LATs [52,58]. When LAP2 and LAP1 coexist, LAP2 can enhance the activity of LAP1 [57].

LAP1 mediates the transcription of LLT [58]. The nerve cells were infected by the recombinant PRV strain with a deletion of the LAP1 region in vitro. The mRNA of 2.0 kb and 8.0 kb could be obtained from the infected nerve cells, but the LLT of 8.4 kb could not be detected. The recombinant strain could establish latent infection in the pigs, but the LLT of 8.4 kb could not be detected in trigeminal nerve. Therefore, LAP1 is the key promoter of LLT, and LLT is not required for the establishment of PRV latent infection [58]. The role of LLT needs to be further understood in PRV latent infection.

LAP is neuron-specific in vivo [59,60]. Taharaguchi et al. established a transgenic mouse line containing LAP linked with the chloramphenicol acetyltransferase (CAT) gene [60]. The expression level of the CAT gene in different tissues of transgenic mice was evaluated by enzyme-linked immunosorbent assay (ELISA). It was found that CAT was almost exclusively expressed in nerve tissue, and the expression level was the highest in the trigeminal nerve [60]. The expression of CAT in trigeminal ganglion neurons was further verified by in situ hybridization. In the absence of viral proteins, LAP is not only active but also neuron-specific, indicating that LAP may be regulated by neuronal transcription factors, and is independent of viral proteins. However, in the studies by Ou and Taharaguchi et al., it has been shown that the inhibition of LAP by IE180 is caused by the formation of a stable complex of IE180, cellular protein(s), and the IE180 binding site located on LAP, suggesting that LAP can be regulated by viral and host protein(s) [45,60,61]. In order to further understand the molecular regulatory mechanism of LAP, the host protein that regulates the neuron-specificity of LAP needs to be discovered. In the study of Taharaguchi et al., there were significant differences in the expression level of the CAT gene in different neural tissues, suggesting that LAP activity varies in different neuronal environments, which may be related to the differences of neuronal transcription factors and neuronal morphology in various nerve tissues [60,62]. It has been proved that dexamethasone can activate latent infection of PRV [48,63]. When transgenic mice were treated with dexamethasone, it was found that dexamethasone did not affect LAP-mediated CAT transcription and translation. We speculate that dexamethasone induced viral reactivation may be irrelevant to LAP [60]. Therefore, it is important to analyze the interaction between LAP, viral protein, and host protein for the study of PRV latent infection.

4. The Role of LAT Gene in Latent Infection

Non-coding RNA (ncRNA) molecules are small and have various regulatory functions in virus replication, virus persistence, immune escape, and cellular transformation [64]. Compared with proteins, the regulation of latent infection by ncRNAs is more desirable: first, the LAT gene is the only gene with transcriptional activity during latent infection, with can transcribe a variety of ncRNAs, but does not produce bioactive proteins [65–70]; second, ncRNAs lack antigenicity and are more likely to evade host cellular immunity [64]; third, the rich functions of ncRNAs are suitable for the regulation of latent infection [64]; finally, compared with proteins, the regulatory function of ncRNAs is mild, and the regulatory mode is ideal in the setting of latent infection [71]. Therefore, the ncRNA is of great significance for the research of PRV latent infection. The LAT gene can transcribe several different types of ncRNAs, such as microRNA (miRNA), small RNA (sRNA), long non-coding RNA (lncRNA), and short non-coding RNA (sncRNA), etc. [65–70]. At present, studies on PRV latent infection are mainly focused on miRNA, and few reports on other ncRNAs. PRV LAT can transcribe many kinds of ncRNAs just like that of Herpes simplex virus type 1 (HSV-1). PRV EP0 and IE180 are homologues of HSV-1 ICP0 (infected cell polypeptide 0) and ICP4 (infected cell polypeptide 4), respectively [72]. The ncRNAs produced by HSV-1 LAT, such as miR-H2 and miR-H6, can inhibit the expression of ICP0 and ICP4 [73,74]. The ncRNAs transcribed by PRV LAT can target EP0 and IE180 mRNAs. Based on the structural similarity of homologous proteins, we speculate that some ncRNAs transcribed by PRV may have functions analogous to those of HSV. HSV's research on LAT ncRNA is more abundant than PRV's. Therefore, we will analyze the function of PRV LAT ncRNAs combined with HSV to comprehensively elucidate their role in latent infection.

4.1. MicroRNAs Transcribed by the LAT Gene and Host Cell

MicroRNA (miRNA) is about 20–24 nucleotides (nt) in size and has a variety of important regulatory functions [75]. It can regulate target mRNA by altering its stability or inhibiting its transcription. With the development of sequencing technology, the miRNAs of most herpesviruses have been identified, including PRV. However, the specific function of PRV miRNAs in the process of PRV infection is still unclear.

Alphaherpesvirinae has been shown to transcribe a variety of miRNAs which are usually clustered in the viral genome. The viral miRNAs are limited to LAT sites or adjacent regions and can be transcribed by each strand of the genome [72]. Anselmo et al. identified five viral miRNAs (prv-miR-LLT 1 to prv-miR-LLT 5) in PRV-infected porcine dendritic cells (DCs) by deep sequencing [76]. These miRNAs are all transcribed by the intron of

LLT [76]. The sizes of prv-miR-LLT 1, prv-miR-LLT 2, prv-miR-LLT 3, and prv-miR-LLT 5 are between 21 and 23nt. Prv-miR-LLT 4 is a mature miRNA with a size of 18nt. Using gene target analysis of miRNAs (prv-miR-1,2,3,4,5), it was found that the possible targets of Prv-miR-LLT 1-5 located in LLT, EP0, and IE180. Based on Anselmo's study, Wu et al. identified 11 viral miRNAs (prv-miR-LLT 1 to prv-miR-LLT 11) in porcine epithelial cell line (PK-15) infected with PRV by the same method [77] (Figure 1). Gene target analysis showed that prv-miR-LLT 1 and prv-miR-LLT 9 could target IE180 and LLT, and prv-miR-LLT 2 could target EP0 and LLT. It was also found that 11 viral miRNAs could target 235 host genes. GO enrichment analysis showed that these 235 host genes are involved in apoptosis, host immune response, cell metabolism, and virus replication [77]. These results suggest that viral miRNAs can play an important role in regulating the interaction between virus and host.

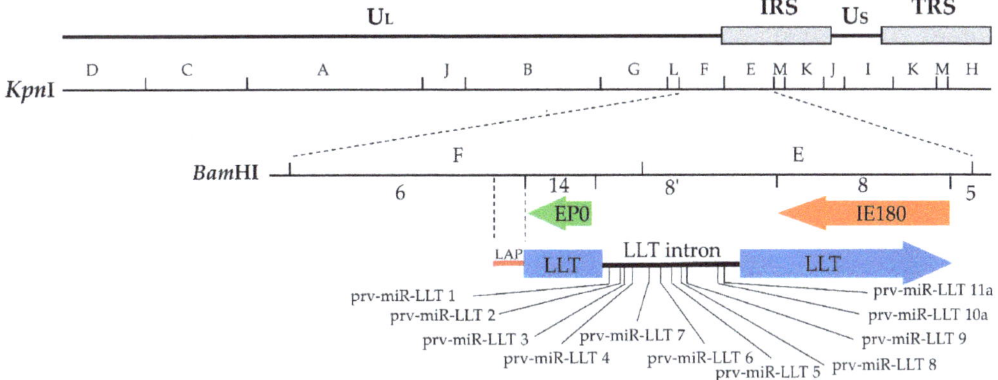

Figure 1. Schematic of the PRV genome and location of the LAT gene. The images from top to bottom show the structure of the PRV genome, the *Kpn*I restriction enzyme map, the *Bam*HI restriction enzyme map, the transcription location and direction of LAT, IE180, and EP0 genes, the LAP location, and the location of prv-miR-LLT 1-11 in the LLT intron.

Although these viral miRNAs were detected during productive infection of PRV, they were produced by LAT and could regulate the latent infection of the virus. As a member of the *Alphaherpesvirinae* subfamily, the viral miRNA of PRV may have similar functions to that of HSV. MiRNAs of HSV can regulate viral latent infection by down-regulating the expression of IE genes or E genes [73,78]. In HSV-1 infected cells, LAT, as the primary miRNA precursor, can transcribe various miRNAs. Among them, HSV-1 miR-H2 can inhibit the expression of ICP0 through targeting its mRNA [73]. ICP0 is an effective activator of virus reactivation, so miR-H2 inhibits the expression of ICP0, thereby hindering virus reactivation [73,79]. When interfering with the transcription of HSV-1 miR-H2, reverse consequences will occur, including the increase of ICP0 expression, viral reactivation, and the neurovirulence of HSV [74]. Therefore, miR-H2 can regulate the conversion between latency and reactivation of the virus. In addition, HSV-1 miR-H6 can inhibit the expression of ICP4. The LAT gene of herpes simplex virus type 2 (HSV-2) can also produce many different miRNAs, among which miR-I and miR-II can reduce the expression of the neurovirulence factor ICP34.5, and miR-III can block the expression of the ICP0 gene [78,80]. MiRNAs not only regulate the latent infection, but also the productive infection. Timoneda et al. detected 8 miRNAs transcribed by LLT intron in PRV infected pigs The expression of these viral miRNAs was obviously changed at different times of acute infection and significantly increased only in the early stage of virus infection. Thus, the authors speculated that the 8 miRNAs could be involved in the establishment of productive infection of the virus [65]. Another interesting phenomenon was also observed.

The 20 viral miRNAs detected from PK15 cell line by using Illumina deep sequencing were derived from the open reading frame (ORF), IRS, and TRS regions of the PRV genome [81].

Host miRNAs, like viral miRNAs, can participate in the regulation of virus infection. Many studies have shown that PRV infection can affect the expression of miRNAs in host cells [76,77,81,82]. Members of the miR-146 family can regulate host inflammatory and immune responses [83,84]. In PRV infected mouse neuroblastoma cells, miR-146b-5p was significantly up-regulated after PRV infection, which could promote PRV replication and negatively regulate type I interferon response [82]. Furthermore, other host miRNAs can also target multiple viral genes such as miR-1249-3p, miR-6538, miR-466k, and miR-714 to regulate the PRV infection [82].

4.2. Other Non-Coding RNAs Transcribed by LAT Gene

Long non-coding RNA (lncRNA) has the function of regulating gene expression [67]. In latently infected ganglia, the LAT gene of HSV can express different sizes of lncRNAs. They can accumulate in the nucleus of latently infected neurons to induce the formation of facultative heterochromatin, promote lytic gene silencing, and evade the host immune response [66,70]. PRV can up-regulate the expression of host lncRNAs in infected cells, thus promoting the replication of itself [85].

Peng et al. identified two ncRNAs in the first 1.5 kb LAT region of HSV which were different from the typical miRNA structure [86]. They possessed the feature of sRNA, so the two ncRNAs were called LAT sRNA1 and sRNA2.The first 1.5 kb LAT region of HSV plays an important role in suppressing productive infection, resisting apoptosis, maintaining latent infection, and ensuring a high rate of viral reactivation [87,88]. According to the complementary pairing of the two sRNAs with ICP4 mRNA and the position of their corresponding DNA sequence, it is speculated that sRNA1 and sRNA2 could inhibit the translation of ICP4 mRNA and apoptosis. Shen et al. confirmed this speculation [69]. ICP4 is essential for productive infection and viral reactivation. The LAT sRNA2 can inhibit the translation of ICP4 mRNA, so the functions of the latter can be affected in regulating the virus infection. LAT sRNA1 has a stronger ability to suppress productive infection than sRNA2, but it has no significant effect on the expression of ICP0 or ICP4 protein. It is predicted that LAT sRNA1 may target mRNA of VP16 and UL8 to inhibit the production of regulatory proteins necessary for viral replication. Single point mutations in LAT sRNA1 and sRNA2 can reduce the ability of LAT to inhibit apoptosis [69]. Therefore, sRNA1 and sRNA2 take an important role in the anti-apoptotic effect of LAT.

In some studies, LAT sRNA1 and sRNA2 are named LAT sncRNA1 and LAT sncRNA2, respectively. LAT sncRNA1 and sncRNA2 can regulate the signal pathway of retinoic acid-induced gene I (RIG-I) to improve cell survival [68]. Herpesvirus entry medium (HVEM) is a cell surface protein that mediates the attachment and entry of HSV into cells. HVEM can regulate the cellular immune response and inhibit apoptosis. In latently infected neurons, LAT can up-regulate the expression of HVEM through the interaction of LAT sncRNA1 and sncRNA2 with the HVEM promoters, thereby promoting cell survival and helping the virus escape host immunity [89].

4.3. LAT Encoding Protein in Latent Infection

Open reading frames (ORFs) are DNA sequences that are capable of encoding proteins. Eight potential ORFs exist in the first 1.5 kb LAT of HSV-1. Introducing point mutations into the ATG of ORFs can reduce the activity of LAT in inhibiting apoptosis [90]. Although no bioactive LAT encoding protein was detected in latently infected neurons, its biological function was confirmed in vitro [91]. HSV-1 LAT encoding protein can restore the replication level of the ICP0 gene-deficient strain in vitro and improve the growth level of the virus in vivo. Therefore, it was implied that the LAT encoding protein might have functions similar to ICP0 in the reactivation of virus latent infection [91,92].

5. Effect of Pseudorabies Vaccine on Virus Latent Infection

The PRV vaccine could not prevent the establishment of latent infection of wild-type strains. As early as 1981, it was reported that latent infection of PRV still existed in pig farms vaccinated with live attenuated PRV vaccine [93]. Since then, a large number of studies have confirmed that the clinical symptoms and mortality of infected pigs can be reduced by active immunization of the inactivated vaccine, live attenuated vaccine, and subunit vaccine or passive immunization of maternal antibody, but latent infection of the virulent PRV strain cannot be prevented in the PNS of the host [18,26,94–98]. Inoculation with the PRV vaccine cannot prevent the virulent strain from establishing latent infection in host pigs, but it can reduce the amount and time of virus excretion after the virulent strain reactivated [99].

Live attenuated PRV vaccines can also establish latent infections in PNS of pigs like virulent strains [100]. The latent infection of the PRV gene deletion vaccine can affect that of wild-type strains, and there is a significant negative correlation between them [101]. By increasing the level of latent infection of the gene deletion vaccine, we can reduce or eliminate that of wild-type PRV in the host. However, one problem that needs to be considered is genetic recombination. Virus recombination is affected by the dose of the inoculated virus, the time interval between the two viruses, the distance between marker mutations, genetic homology, virulence, and latency [102–104]. Homologous recombination often occurs among the same *Alphaherpesvirinae* [104]. The PRV gene deletion vaccine may recombine with the wild-type virus strain to produce a highly virulent variant strain, or a variant strain that cannot be differentiated by serological method [105]. To avoid this, the biosafety of live attenuated vaccines must be carefully evaluated. The LAT gene is a nonessential factor for the establishment of viral latent infection, but it plays a pivotal role in the reactivation of the virus [58]. Mahjoub et al. obtained a PRV mutant with 9 LAT miRNAs deletions. This mutant could establish latent infection, but it could not be reactivated [106]. In HSV, the deletion of LAT fragments can affect the reactivation rate, the virulence, and the ability of anti-apoptosis [74,107,108]. The specific molecular mechanism of LAT still needs to be explored. LAT plays a key role in regulating the conversion between virus latency and reactivation, inhibiting apoptosis to prolong the survival time of neurons. Perhaps we could take some measures to modify the LAT gene to develop a non-reactivated PRV vaccine.

6. Conclusions and Perspectives

In this review, we summarize the characteristics of PRV latent infection, the transcriptional characteristics and functions of LAT gene, and the effects of PRV vaccine on the establishment of latent infection of virulent PRV strains. Studying LAT ncRNA can provide a new perspective for elucidating the molecular mechanism of PRV latent infection. We can also take some effective measures to control the latent infection of wild-type PRV in pigs, such as developing effective vaccines or drugs to inhibit the establishment or reactivation of latent infection of wild type PRV. During latent infection of PRV, no infectious virions are produced, but the genome of the virus can be detected in the host. However, latent infection of the virus can be detected by serology, in situ hybridization, tissue co-culture, polymerase chain reaction (PCR), fluorescence quantitative PCR, and real-time recombinant enzyme-assisted amplification [46,109–113]. Similar to the PRV virulent strain, the PRV gene deletion vaccine can also establish latent infection in the PNS of pigs. When pigs are inoculated with two vaccines of different gene deletions, genetic recombination may occur between them, and then new strains may be produced. Therefore, only one gene deletion vaccine is recommended in the same pig farm or the same animal individual to avoid genetic recombination between vaccine strains [102,103].

The LAT gene is closely related to latent infection. The latent infection of the PRV gene deletion vaccine was negatively correlated with that of wild-type PRV strains, and the LAT gene could affect the reactivation rate of the virus. Therefore, we can modify the LAT gene to develop a genetic engineering vaccine to restrict the latent infection of PRV wild-type strain in pigs. The ncRNAs transcribed by the LAT gene play an important role in the process of viral latent infection. Therefore, RNA interference and RNA silencing are applied to these ncRNAs to regulate the latency and reactivation of PRV. At present, there is little research on the ncRNA of PRV LAT. Therefore, in order to further understand the molecular mechanism of PRV latent infection, we can emphasize the study of ncRNAs produced by PRV LAT.

PRV can be transmitted from pigs to humans, threatening human health [28–39]. PRV infection can cause fever, headache, endophthalmitis, and acute encephalitis [28–39]. Antiherpes drugs such as valaciclovir, penciclovir, and phosphonoformate can control the symptoms of patients, but the visual and nerve damage caused by PRV is irreversible. PRV can establish reactivatable latent infection in mice [16,114,115]. Humans are non-natural hosts like mice, so it is possible for PRV to establish reactivatable latent infection in the body, resulting in irregular recurrence of the disease. Therefore, the subsequent progress of PRV-infected patients should be tracked in the long-term to evaluate the probability. If the molecular mechanism of latent infection is clear, some specific targets may be found, and then relevant drugs can be developed to block the latency and reactivation of PRV.

Author Contributions: Conceptualization of the manuscript, J.D., C.J., Z.W., J.L. and Q.J.; writing—original draft preparation, J.D. and C.J.; writing—review and editing, J.D., Z.W., J.L., Q.J. and C.J. All authors have read and agreed to the published version of the manuscript.

Funding: This research was supported by the grants for Da Hua Nong Award Fund for training young teachers of Veterinary College (No. 5500-A17003) and Natural Science Foundation of Guangdong Province (No. 2018A030313163).

Institutional Review Board Statement: Not applicable.

Informed Consent Statement: Not applicable.

Data Availability Statement: Not applicable.

Conflicts of Interest: The authors declare no conflict of interest.

References

1. Hengartner, L.E.; Reynolds, A.E.; Hengartner, C.J. Molecular Biology of Pseudorabies Virus: Impact on Neurovirology and Veterinary Medicine. *Microbiol. Mol. Biol.* **2005**, *69*, 462–500. [CrossRef]
2. Klupp, B.G.; Hengartner, C.J.; Mettenleiter, T.C.; Enquist, L.W. Complete, Annotated Sequence of the Pseudorabies Virus Genome. *J. Virol.* **2004**, *78*, 17. [CrossRef]
3. Hugoson, G.; Rockborn, G. On the Occurrence of Pseudorabies in Sweden II. An Outbreak in Dogs Caused by Feeding Abattoir Offal. *Zentralbl Vet. B* **2010**, *19*, 641–645. [CrossRef] [PubMed]
4. Capua, I.; Fico, R.; Banks, M.; Tamba, M.; Calzetta, G. Isolation and Characterisation of an Aujeszky's Disease Virus Naturally Infecting a Wild Boar (Sus Scrofa). *Vet. Microbiol.* **1997**, *55*, 141–146. [CrossRef]
5. Cheng, Z.; Kong, Z.; Liu, P.; Fu, Z.; Zhang, J.; Liu, M.; Shang, Y. Natural Infection of a Variant Pseudorabies Virus Leads to Bovine Death in China. *Transbound. Emerg. Dis.* **2020**, *67*, 518–522. [CrossRef]
6. Kaneko, C.; Kaneko, Y.; Sudaryatma, P.E.; Mekata, H.; Kirino, Y.; Yamaguchi, R.; Okabayashi, T. Pseudorabies Virus Infection in Hunting Dogs in Oita, Japan: Report from a Prefecture Free from Aujeszky's Disease in Domestic Pigs. *J. Vet. Med. Sci.* **2021**, *83*, 680–684. [CrossRef]
7. Marcaccini, A.; López Peña, M.; Quiroga, M.I.; Bermúdez, R.; Nieto, J.M.; Alemañ, N. Pseudorabies Virus Infection in Mink: A Host-Specific Pathogenesis. *Vet. Immunol. Immunopathol.* **2008**, *124*, 264–273. [CrossRef]
8. Laval, K.; Vernejoul, J.B.; Van Cleemput, J.; Koyuncu, O.O.; Enquist, L.W. Virulent Pseudorabies Virus Infection Induces a Specific and Lethal Systemic Inflammatory Response in Mice. *J. Virol.* **2018**, *92*, e01614-18. [CrossRef]
9. Di Marco Lo Presti, V.; Moreno, A.; Castelli, A.; Ippolito, D.; Aliberti, A.; Amato, B.; Vitale, M.; Fiasconaro, M.; Pruiti Ciarello, F. Retrieving Historical Cases of Aujeszky's Disease in Sicily (Italy): Report of a Natural Outbreak Affecting Sheep, Goats, Dogs, Cats and Foxes and Considerations on Critical Issues and Perspectives in Light of the Recent EU Regulation 429/2016. *Pathogens* **2021**, *10*, 1301. [CrossRef]
10. Laval, K.; Enquist, L.W. The Neuropathic Itch Caused by Pseudorabies Virus. *Pathogens* **2020**, *9*, 254. [CrossRef]
11. Sehl, J.; Teifke, J.P. Comparative Pathology of Pseudorabies in Different Naturally and Experimentally Infected Species—A Review. *Pathogens* **2020**, *9*, 633. [CrossRef] [PubMed]
12. Gutekunst, D.C.; Pirtle, E.C.; Miller, L.D.; Stewart, W.C. Isolation of Pseudorabies Virus from Trigeminal Ganglia of a Latently Infected Sow. *Am. J. Vet. Res.* **1980**, *41*, 1315–1316. [PubMed]

13. Brown, T.M.; Osorio, F.A.; Rock, D.L. Detection of Latent Pseudorabies Virus in Swine Using in Situ Hybridization. *Vet. Microbiol.* **1990**, *24*, 273–280. [CrossRef]
14. Romero, C.H.; Meade, P.N.; Homer, B.L.; Shultz, J.E.; Lollis, G. Potential Sites of Virus Latency Associated with Indigenous Pseudorabies Viruses in Feral Swine. *J. Wildl. Dis.* **2003**, *39*, 567–575. [CrossRef]
15. Brittle, E.E.; Reynolds, A.E.; Enquist, L.W. Two Modes of Pseudorabies Virus Neuroinvasion and Lethality in Mice. *J. Virol.* **2004**, *78*, 12951–12963. [CrossRef]
16. Seiichi, T.; Takashi, I.; Masashi, S.; Kazuaki, M. Acetylcholine Reactivates Latent Pseudorabies Virus in Mice. *J. Virol. Methods* **1998**, *70*, 103–106. [CrossRef]
17. Seiichi, T.; Kazuaki, M. Activation of Latent Pseudorabies Virus Infection in Mice Treated with Acetylcholine. *Exp. Anim.* **2002**, *51*, 407–709. [CrossRef]
18. Osorio, F.A.; Rock, D.L. A Murine Model of Pseudorabies Virus Latency. *Microb. Pathog.* **1992**, *12*, 39–46. [CrossRef]
19. Seiichi, T.; Kazuaki, M. Analysis of the Mechanism of Reactivation of Latently Infecting Pseudorabies Virus by Acetylcholine. *J. Vet. Med. Sci.* **2014**, *76*, 719–722. [CrossRef]
20. Sun, Y.; Liang, W.; Liu, Q.; Zhao, T.; Zhu, H.; Hua, L.; Peng, Z.; Tang, X.; Stratton, C.; Zhou, D.; et al. Epidemiological and Genetic Characteristics of Swine Pseudorabies Virus in Mainland China between 2012 and 2017. *PeerJ* **2018**, *6*, e5785. [CrossRef]
21. Zheng, H.; Jin, Y.; Hou, C.; Li, X.; Zhao, L.; Wang, Z.; Chen, H. Seroprevalence Investigation and Genetic Analysis of Pseudorabies Virus within Pig Populations in Henan Province of China during 2018–2019. *Infect. Genet. Evol.* **2021**, *92*, 104835. [CrossRef] [PubMed]
22. Ren, Q.; Ren, H.; Gu, J.; Wang, J.; Jiang, L.; Gao, S. The Epidemiological Analysis of Pseudorabies Virus and Pathogenicity of the Variant Strain in Shandong Province. *Front. Vet. Sci.* **2022**, *9*, 806824. [CrossRef] [PubMed]
23. Lin, Y.; Tan, L.; Wang, C.; He, S.; Fang, L.; Wang, Z.; Zhong, Y.; Zhang, K.; Liu, D.; Yang, Q.; et al. Serological Investigation and Genetic Characteristics of Pseudorabies Virus in Hunan Province of China from 2016 to 2020. *Front. Vet. Sci.* **2021**, *8*, 762326. [CrossRef] [PubMed]
24. Sun, Y.; Luo, Y.; Wang, C.H.; Yuan, J.; Li, N.; Song, K.; Qiu, H.J. Control of Swine Pseudorabies in China: Opportunities and Limitations. *Vet. Microbiol.* **2016**, *183*, 119–124. [CrossRef]
25. Tong, W.; Liu, F.; Zheng, H.; Liang, C.; Zhou, Y.; Jiang, Y.; Shan, T.; Gao, F.; Li, G.; Tong, G. Emergence of a Pseudorabies Virus Variant with Increased Virulence to Piglets. *Vet. Microbiol.* **2015**, *181*, 236–240. [CrossRef] [PubMed]
26. An, T.; Peng, J.; Tian, Z.; Zhao, H.; Li, N.; Liu, Y.; Chen, J.; Leng, C.; Sun, Y.; Chang, D.; et al. Pseudorabies Virus Variant in Bartha-K61-Vaccinated Pigs, China, 2012. *Emerg. Infect. Dis.* **2013**, *19*, 1749–1755. [CrossRef]
27. Yu, Z.; Tong, W.; Zheng, H.; Li, L.; Li, G.; Gao, F.; Wang, T.; Liang, C.; Ye, C.; Wu, J.; et al. Variations in Glycoprotein B Contribute to Immunogenic Difference between PRV Variant JS-2012 and Bartha-K61. *Vet. Microbiol.* **2017**, *208*, 97–105. [CrossRef]
28. Liu, Q.; Wang, X.; Xie, C.; Ding, S.; Yang, H.; Guo, S.; Li, J.; Qin, L.; Ban, F.; Wang, D.; et al. A Novel Human Acute Encephalitis Caused by Pseudorabies Virus Variant Strain. *Clin. Infect. Dis.* **2021**, *73*, e3690–e3700. [CrossRef]
29. Ai, J.; Weng, S.; Cheng, Q.; Cui, P.; Li, Y.; Wu, H.; Zhu, Y.; Xu, B.; Zhang, W. Human Endophthalmitis Caused By Pseudorabies Virus Infection, China, 2017. *Emerg. Infect. Dis.* **2018**, *24*, 1087–1090. [CrossRef]
30. Fan, S.; Yuan, H.; Liu, L.; Li, H.; Wang, S.; Zhao, W.; Wu, Y.; Wang, P.; Hu, Y.; Han, J.; et al. Pseudorabies Virus Encephalitis in Humans: A Case Series Study. *J. Neurovirol.* **2020**, *26*, 556–564. [CrossRef]
31. Wang, Y.; Nian, H.; Li, Z.; Wang, W.; Wang, X.; Cui, Y. Human Encephalitis Complicated with Bilateral Acute Retinal Necrosis Associated with Pseudorabies Virus Infection: A Case Report. *Int. J. Infect. Dis.* **2019**, *89*, 51–54. [CrossRef] [PubMed]
32. Yang, X.; Guan, H.; Li, C.; Li, Y.; Wang, S.; Zhao, X.; Zhao, Y.; Liu, Y. Characteristics of Human Encephalitis Caused by Pseudorabies Virus: A Case Series Study. *Int. J. Infect. Dis.* **2019**, *87*, 92–99. [CrossRef]
33. Yang, H.; Han, H.; Wang, H.; Cui, Y.; Liu, H.; Ding, S. A Case of Human Viral Encephalitis Caused by Pseudorabies Virus Infection in China. *Front. Neurol.* **2019**, *10*, 534. [CrossRef] [PubMed]
34. Zheng, L.; Liu, X.; Yuan, D.; Li, R.; Lu, J.; Li, X.; Tian, K.; Dai, E. Dynamic Cerebrospinal Fluid Analyses of Severe Pseudorabies Encephalitis. *Transbound. Emerg. Dis.* **2019**, *66*, 2562–2565. [CrossRef] [PubMed]
35. Hu, F.; Wang, J.; Peng, X.Y. Bilateral Necrotizing Retinitis Following Encephalitis Caused by the Pseudorabies Virus Confirmed by Next-Generation Sequencing. *Ocul. Immunol. Inflamm.* **2021**, *29*, 922–925. [CrossRef]
36. Yan, W.; Hu, Z.; Zhang, Y.; Wu, X.; Zhang, H. Case Report: Metagenomic Next-Generation Sequencing for Diagnosis of Human Encephalitis and Endophthalmitis Caused by Pseudorabies Virus. *Front. Med.* **2022**, *8*, 753988. [CrossRef] [PubMed]
37. Wang, D.; Tao, X.; Fei, M.; Chen, J.; Guo, W.; Li, P.; Wang, J. Human Encephalitis Caused by Pseudorabies Virus Infection: A Case Report. *J. Neurovirol.* **2020**, *26*, 442–448. [CrossRef]
38. Liu, Y.; Li, Y.; Tong, F.; Tian, M.; Li, M.; Wang, L.; Zou, Y.; Duan, J.; Bu, H.; He, J. Human Encephalitis Complicated With Ocular Symptoms Associated With Pseudorabies Virus Infection: A Case Report. *Front. Neurol.* **2022**, *13*, 878007.
39. Zhou, Y.; Nie, C.; Wen, H.; Long, Y.; Zhou, M.; Xie, Z.; Hong, D. Human Viral Encephalitis Associated with Suid Herpesvirus 1. *Neurol. Sci.* **2022**, *43*, 2681–2692. [CrossRef]
40. Motovski, A.; Kunev, Z.; Stoichev, P. Epizootic Process on a Farm Chronically Infected with Aujeszky's Disease. *Vet.-Meditsinski Nauki* **1977**, *14*, 16–21.

41. Zhang, Y.; Zeng, L.-S.; Wang, J.; Cai, W.-Q.; Cui, W.; Song, T.-J.; Peng, X.-C.; Ma, Z.; Xiang, Y.; Cui, S.-Z.; et al. Multifunctional Non-Coding RNAs Mediate Latent Infection and Recurrence of Herpes Simplex Viruses. *Infect. Drug Resist.* **2021**, *14*, 5335–5349. [CrossRef] [PubMed]

42. Smith, G. Herpesvirus Transport to the Nervous System and Back Again. *Annu. Rev. Microbiol.* **2012**, *66*, 153–176. [CrossRef] [PubMed]

43. Koyuncu, O.O.; Hogue, L.B.; Enquist, L.W. Virus Infections in the Nervous System. *Cell Host Microbe* **2013**, *13*, 379–393. [CrossRef] [PubMed]

44. Preston, C.M. Repression of Viral Transcription during Herpes Simplex Virus Latency. *Microbiology* **2000**, *81*, 1–19. [CrossRef] [PubMed]

45. Ou, C.J.; Wong, M.; Huang, C.; Chang, T.J. Suppression of Promoter Activity of the LAT Gene by IE180 of Pseudorabies Virus. *Virus Genes* **2002**, *13*, 227–239. [CrossRef] [PubMed]

46. Rock, D.L.; Hagemoser, W.A.; Osorio, F.A.; McAllister, H.A. Transcription from the Pseudorabies Virus Genome during Latent Infection. Brief Report. *Arch. Virol.* **1988**, *98*, 99–106. [CrossRef]

47. Rziha, H.J.; Mettenleiter, T.C.; Ohlinger, V.; Wittmann, G. Herpesvirus (Pseudorabies Virus) Latency in Swine: Occurrence and Physical State of Viral DNA in Neural Tissues. *Virology* **1986**, *155*, 600–613. [CrossRef]

48. Tanaka, S.; Mannen, K. Effect of Mild Stress in Mice Latently Infected Pseudorabies Virus. *Exp. Anim.* **2003**, *52*, 383–386. [CrossRef]

49. Szpara, M.L.; Kobiler, O.; Enquist, L.W. A Common Neuronal Response to Alphaherpesvirus Infection. *J. Neuroimmune Pharmacol.* **2010**, *5*, 418–427. [CrossRef]

50. Shu, M.; Du, T.; Zhou, G.; Roizman, B. Role of Activating Transcription Factor 3 in the Synthesis of Latency-Associated Transcript and Maintenance of Herpes Simplex Virus 1 in Latent State in Ganglia. *Proc. Natl. Acad. Sci. USA* **2015**, *112*, E5420–E5426. [CrossRef]

51. Zhang, R.; Tang, J. Evasion of I Interferon-Mediated Innate Immunity by Pseudorabies Virus. *Front. Microbiol.* **2021**, *12*, 801257. [CrossRef] [PubMed]

52. Jin, L.; Scherba, G. Expression of the Pseudorabies Virus Latency-Associated Transcript Gene during Productive Infection of Cultured Cells. *J. Virol.* **1999**, *73*, 9781–9788. [CrossRef] [PubMed]

53. Priola, S.A.; Gustafson, D.P.; Wagner, E.K.; Stevens, J.G. A Major Portion of the Latent Pseudorabies Virus Genome Is Transcribed in Trigeminal Ganglia of Pigs. *J. Virol.* **1990**, *64*, 4755–4760. [CrossRef] [PubMed]

54. Priola, S.A.; Stevens, J.G. The 5′ and 3′ Limits of Transcription in the Pseudorabies Virus Latency Associated Transcription Unit. *Virology* **1991**, *182*, 852–856. [CrossRef]

55. Cheung, A.K. Cloning of the Latency Gene and the Early Protein 0 Gene of Pseudorabies Virus. *J. Virol.* **1991**, *65*, 5260–5271. [CrossRef]

56. Cheung, A.K. Detection of Pseudorabies Virus Transcripts in Trigeminal Ganglia of Latently Infected Swine. *J. Virol.* **1989**, *63*, 2908–2913. [CrossRef]

57. Cheung, A.K.; Smith, T.A. Analysis of the Latency-Associated Transcript/UL1-3.5 Gene Cluster Promoter Complex of Pseudorabies Virus. *Arch. Virol.* **1999**, *144*, 381–391. [CrossRef]

58. Jin, L.; Schnitzlein, W.M.; Scherba, G. Identification of the Pseudorabies Virus Promoter Required for Latency-Associated Transcript Gene Expression in the Natural Host. *J. Virol.* **2000**, *74*, 6333–6338. [CrossRef]

59. Ou, C.J.; Chen, Y.; Huang, C. Cloning and Characterization of the Pseudorabies Virus Latency-Associated Transcript Promoter. *Taiwan Vet. J.* **2002**, *28*, 252–259.

60. Taharaguchi, S.; Yoshino, S.; Amagai, K.; Ono, E. The Latency-Associated Transcript Promoter of Pseudorabies Virus Directs Neuron-Specific Expression in Trigeminal Ganglia of Transgenic Mice. *J. Gen. Virol.* **2003**, *84*, 2015–2022. [CrossRef]

61. Taharaguchi, S.; Kobayashi, T.; Yoshino, S.; Ono, E. Analysis of Regulatory Functions for the Region Located Upstream from the Latency-Associated Transcript (LAT) Promoter of Pseudorabies Virus in Cultured Cells. *Vet. Microbiol.* **2002**, *85*, 197–208. [CrossRef]

62. Papageorgiou, K.; Grivas, I.; Chiotelli, M.; Theodoridis, A.; Panteris, E.; Papadopoulos, D.; Petridou, E.; Papaioannou, N.; Nauwynck, H.; Kritas, S.K. Age-Dependent Invasion of Pseudorabies Virus into Porcine Central Nervous System via Maxillary Nerve. *Pathogens* **2022**, *11*, 157. [CrossRef] [PubMed]

63. Tham, K.M.; Motha, M.X.J.; Horner, G.W.; Ralston, J.C. Polymerase Chain Reaction Amplification of Latent Aujeszky's Disease Virus in Dexamethasone Treated Pigs. *Arch. Virol.* **1994**, *136*, 197–205. [CrossRef]

64. Tycowski, K.T.; Guo, Y.E.; Lee, N.; Moss, W.N.; Vallery, T.K.; Xie, M.; Steitz, J.A. Viral Noncoding RNAs: More Surprises. *Genes Dev.* **2015**, *29*, 567–584. [CrossRef] [PubMed]

65. Timoneda, O.; Núñez-Hernández, F.; Balcells, I.; Muñoz, M.; Castelló, A.; Vera, G.; Pérez, L.J.; Egea, R.; Mir, G.; Córdoba, S.; et al. The Role of Viral and Host MicroRNAs in the Aujeszky's Disease Virus during the Infection Process. *PLoS ONE* **2014**, *9*, e86965. [CrossRef] [PubMed]

66. Ahmed, W.; Liu, Z.F. Long Non-Coding RNAs: Novel Players in Regulation of Immune Response Upon Herpesvirus Infection. *Front. Immunol.* **2018**, *9*, 761. [CrossRef] [PubMed]

67. Cheng, J.T.; Wang, L.; Wang, H.; Tang, F.R.; Cai, W.Q.; Sethi, G.; Xin, H.W.; Ma, Z. Insights into Biological Role of LncRNAs in Epithelial-Mesenchymal Transition. *Cells* **2019**, *8*, 1178. [CrossRef] [PubMed]

68. Silva, L.F.D.; Jones, C. Small Non-Coding RNAs Encoded within the Herpes Simplex Virus Type 1 Latency Associated Transcript (LAT) Cooperate with the Retinoic Acid Inducible Gene I (RIG-I) to Induce Beta-Interferon Promoter Activity and Promote Cell Survival. *Virus Res.* **2013**, *175*, 101–109. [CrossRef]
69. Shen, W.; Silva, M.S.E.; Jaber, T.; Vitvitskaia, O.; Li, S.; Henderson, G.; Jones, C. Two Small RNAs Encoded within the First 1.5 Kilobases of the Herpes Simplex Virus Type 1 Latency-Associated Transcript Can Inhibit Productive Infection and Cooperate To Inhibit Apoptosis. *J. Virol.* **2009**, *83*, 9131–9139. [CrossRef]
70. Cliffe, A.R.; Garber, D.A.; Knipe, D.M. Transcription of the Herpes Simplex Virus Latency-Associated Transcript Promotes the Formation of Facultative Heterochromatin on Lytic Promoters. *J. Virol.* **2009**, *83*, 8182–8190. [CrossRef]
71. Grey, F. Role of MicroRNAs in Herpesvirus Latency and Persistence. *J. Gen. Virol.* **2015**, *96*, 739–751. [CrossRef] [PubMed]
72. Jurak, L.; Griffiths, A.; Coen, D.M. Mammalian Alphaherpesvirus MiRNAs. *Biochim. Biophys. Acta BBA Gene Regul. Mech.* **2011**, *1809*, 641–653. [CrossRef] [PubMed]
73. Umbach, J.L.; Kramer, M.F.; Jurak, L.; Karnowski, H.W.; Coen, D.M.; Cullen, B.R. MicroRNAs Expressed by Herpes Simplex Virus 1 during Latent Infection Regulate Viral MRNAs. *Nature* **2008**, *454*, 780–783. [CrossRef] [PubMed]
74. Jiang, X.; Brown, D.; Osorio, N.; Hsiang, C.; BenMohamed, L.; Wechsler, S.L. Increased Neurovirulence and Reactivation of the Herpes Simplex Virus Type 1 Latency-Associated Transcript (LAT)-Negative Mutant DLAT2903 with a Disrupted LAT MiR-H2. *J. Neurovirol.* **2016**, *22*, 38–49. [CrossRef]
75. Kincaid, R.P.; Sullivan, C.S. Virus-Encoded MicroRNAs: An Overview and a Look to the Future. *PLoS Pathog.* **2012**, *8*, e1003018. [CrossRef]
76. Anselmo, A.; Flori, L.; Jaffrezic, F.; Rutigliano, T.; Cecere, M.; Cortes-Perez, N.; Lefevre, F.; Rogel-Gaillard, C.; Giuffra, E. Co-Expression of Host and Viral MicroRNAs in Porcine Dendritic Cells Infected by the Pseudorabies Virus. *PLoS ONE* **2011**, *6*, e17374. [CrossRef]
77. Wu, Y.Q.; Chen, Q.J.; He, H.B.; Chen, D.S.; Chen, L.L.; Chen, H.C.; Liu, Z.F. Pseudorabies Virus Infected Porcine Epithelial Cell Line Generates a Diverse Set of Host MicroRNAs and a Special Cluster of Viral MicroRNAs. *PLoS ONE* **2012**, *7*, e30988. [CrossRef]
78. Tang, S.; Patel, A.; Krause, P.R. Novel Less-Abundant Viral MicroRNAs Encoded by Herpes Simplex Virus 2 Latency-Associated Transcript and Their Roles in Regulating ICP34.5 and ICP0 MRNAs. *J. Virol.* **2009**, *83*, 1433–1442. [CrossRef]
79. Hobbs, W.E.; Brough, D.E.; Kovesdi, I.; DeLuca, N.A. Efficient Activation of Viral Genomes by Levels of Herpes Simplex Virus ICP0 Insufficient To Affect Cellular Gene Expression or Cell Survival. *J. Virol.* **2001**, *75*, 3391–3403. [CrossRef]
80. Tang, S.; Bertke, A.S.; Patel, A.; Wang, K.; Cohen, J.I.; Krause, P.R. An Acutely and Latently Expressed Herpes Simplex Virus 2 Viral MicroRNA Inhibits Expression of ICP34.5, a Viral Neurovirulence Factor. *Proc. Natl. Acad. Sci. USA* **2008**, *105*, 10931–10936. [CrossRef]
81. Liu, F.; Zheng, H.; Tong, W.; Li, G.X.; Tian, Q.; Liang, C.; Li, L.W.; Zheng, X.C.; Tong, G.Z. Identification and Analysis of Novel Viral and Host Dysregulated MicroRNAs in Variant Pseudorabies Virus-Infected PK15 Cells. *PLoS ONE* **2016**, *11*, e0151546. [CrossRef] [PubMed]
82. Li, Y.; Zheng, G.; Zhang, Y.; Yang, X.; Liu, H.; Chang, H.; Wang, X.; Zhao, J.; Wang, C.; Chen, L. MicroRNA Analysis in Mouse Neuro-2a Cells after Pseudorabies Virus Infection. *J. Neurovirol.* **2017**, *23*, 430–440. [CrossRef] [PubMed]
83. Saba, R.; Sorensen, D.L.; Booth, S.A. MicroRNA-146a: A Dominant, Negative Regulator of the Innate Immune Response. *Front. Immunol.* **2014**, *5*, 578. [CrossRef]
84. Taganov, K.D.; Boldin, M.P.; Chang, K.J.; Baltimore, D. NF-KB-Dependent Induction of MicroRNA MiR-146, an Inhibitor Targeted to Signaling Proteins of Innate Immune Responses. *Proc. Natl. Acad. Sci. USA* **2006**, *103*, 12481–12486. [CrossRef] [PubMed]
85. Fang, L.; Gao, Y.; Liu, X.; Bai, J.; Jiang, P.; Wang, X. Long Non-Coding RNA LNC_000641 Regulates Pseudorabies Virus Replication. *Vet. Res.* **2021**, *52*, 1–13. [CrossRef]
86. Peng, W.; Vitvitskaia, O.; Carpenter, D.; Wechsler, S.L.; Jones, C. Identification of Two Small RNAs within the First 1.5-Kb of the Herpes Simplex Virus Type 1–Encoded Latency-Associated Transcript. *J. Neurovirol.* **2008**, *14*, 41–52. [CrossRef]
87. Ahmed, M.; Lock, M.; Miller, C.G.; Fraser, N.W. Regions of the Herpes Simplex Virus Type 1 Latency-Associated Transcript That Protect Cells from Apoptosis In Vitro and Protect Neuronal Cells In Vivo. *J. Virol.* **2002**, *76*, 717–729. [CrossRef]
88. Nicoll, M.P.; Proenc, J.T.; Efstathiou, S. The Molecular Basis of Herpes Simplex Virus Latency. *FEMS Microbiol. Rev.* **2012**, *36*, 684–705. [CrossRef]
89. Allen, S.J.; Rhode-Kurnow, A.; Mott, K.R.; Jiang, X.; Carpenter, D.; Rodriguez-Barbosa, J.I.; Jones, C.; Wechsler, S.L.; Ware, C.F.; Ghiasi, H. Interactions between Herpesvirus Entry Mediator (TNFRSF14) and Latency-Associated Transcript during Herpes Simplex Virus 1 Latency. *J. Virol.* **2014**, *88*, 1961–1971. [CrossRef]
90. Carpenter, D.; Henderson, G.; Hsiang, C.; Osorio, N.; BenMohamed, L.; Jones, C.; Wechsler, S.L. Introducing Point Mutations into the ATGs of the Putative Open Reading Frames of the HSV-1 Gene Encoding the Latency Associated Transcript (LAT) Reduces Its Anti-Apoptosis Activity. *Microb. Pathog.* **2008**, *44*, 98–102. [CrossRef]
91. Thomas, S.K.; Lilley, C.E.; Latchman, D.S.; Coffin, R.S. A Protein Encoded by the Herpes Simplex Virus (HSV) Type 1 2-Kilobase Latency-Associated Transcript Is Phosphorylated, Localized to the Nucleus, and Overcomes the Repression of Expression from Exogenous Promoters When Inserted into the Quiescent HSV Genome. *J. Virol.* **2002**, *76*, 4056–4067. [CrossRef] [PubMed]
92. Thomas, S.K.; Gough, G.; Latchman, D.S.; Coffin, R.S. Herpes Simplex Virus Latency-Associated Transcript Encodes a Protein Which Greatly Enhances Virus Growth, Can Compensate for Deficiencies in Immediate-Early Gene Expression, and Is Likely To Function during Reactivation from Virus Latency. *J. Virol.* **1999**, *73*, 6618–6625. [CrossRef]

93. Mock, R.E.; Crandell, R.A.; Mesfin, G.M. Induced Latency in Pseudorabies Vaccinated Pigs. *Can. J. Comp. Med. Rev. Can. Med. Comp.* **1981**, *45*, 56–59.

94. Mengeling, W.L. Virus Reactivation in Pigs Latently Infected with a Thymidine Kinase Negative Vaccine Strain of Pseudorabies Virus. *Arch. Virol.* **1991**, *120*, 57–70. [CrossRef] [PubMed]

95. Volz, D.M.; Lager, K.M.; Mengeling, W.L. Latency of a Thymidine Kinase-Negative Pseudorabies Vaccine Virus Detected by the Polymerase Chain Reaction. *Arch. Virol.* **1992**, *122*, 341–348. [CrossRef] [PubMed]

96. van Oirschot, J.T.; Gielkens, A.L. In Vivo and in Vitro Reactivation of Latent Pseudorabies Virus in Pigs Born to Vaccinated Sows. *Am. J. Vet. Res.* **1984**, *45*, 567–571.

97. McCaw, M.B.; Osorio, F.A.; Wheeler, J.; Xu, J.; Erickson, G.A. Effect of Maternally Acquired Aujeszky's Disease (Pseudorabies) Virus-Specific Antibody in Pigs on Establishment of Latency and Seroconversion to Differential Glycoproteins after Low Dose Challenge. *Vet. Microbiol.* **1997**, *55*, 91–98. [CrossRef]

98. Wittmann, G.; Ohlinger, V.; Rziha, J.H. Occurrence and Reactivation of Latent Aujeszky's Disease Virus Following Challenge in Previously Vaccinated Pigs. *Arch. Virol.* **1983**, *75*, 29–41. [CrossRef]

99. Schoenbaum, M.A.; Beran, G.W.; Murphy, D.P. Pseudorabies Virus Latency and Reactivation in Vaccinated Swine. *Am. J. Vet. Res.* **1990**, *51*, 334–338.

100. Lu, J.J.; Yuan, W.Z.; Zhu, Y.P.; Hou, S.H.; Wang, X.J. Latent Pseudorabies Virus Infection in Medulla Oblongata from Quarantined Pigs. *Transbound. Emerg. Dis.* **2021**, *68*, 543–551. [CrossRef]

101. Schang, L.M.; Kutish, G.F.; Osorio, F.A. Correlation between Precolonization of Trigeminal Ganglia by Attenuated Strains of Pseudorabies Virus and Resistance to Wild-Type Virus Latency. *J. Virol.* **1994**, *68*, 8470–8476. [CrossRef] [PubMed]

102. Thiry, E.; Meurens, F.; Muylkens, B.; McVoy, M.; Gogev, S.; Thiry, J.; Vanderplasschen, A.; Epstein, A.; Keil, G.; Schynts, F. Recombination in Alphaherpesviruses. *Rev. Med. Virol.* **2005**, *15*, 89–103. [CrossRef] [PubMed]

103. Thiry, E.; Muylkens, B.; Meurens, F.; Gogev, S.; Thiry, J.; Vanderplasschen, A.; Schynts, F. Recombination in the Alphaherpesvirus Bovine Herpesvirus 1. *Vet. Microbiol.* **2006**, *113*, 171–177. [CrossRef]

104. Meurens, F.; Schynts, F.; Keil, G.M.; Muylkens, B.; Vanderplasschen, A.; Gallego, P.; Thiry, E. Superinfection Prevents Recombination of the Alphaherpesvirus Bovine Herpesvirus 1. *J. Virol.* **2004**, *78*, 3872–3879. [CrossRef] [PubMed]

105. Maes, R.K.; Sussman, M.D.; Vilnis, A.; Thacker, B.J. Recent Developments in Latency and Recombination of Aujeszky's Disease (Pseudorabies) Virus. *Vet. Microbiol.* **1997**, *55*, 13–27. [CrossRef]

106. Mahjoub, N.; Dhorne-Pollet, S.; Fuchs, W.; Ahanda, E.; Lange, E.; Klupp, B.; Arya, A.; Loveland, J.E.; Lefevre, F.; Mettenleiter, T.C.; et al. A 2.5-Kilobase Deletion Containing a Cluster of Nine MicroRNAs in the Latency-Associated-Transcript Locus of the Pseudorabies Virus Affects the Host Response of Porcine Trigeminal Ganglia during Established Latency. *J. Virol.* **2015**, *89*, 428–442. [CrossRef]

107. Mott, K.R. The Bovine Herpesvirus-1 LR ORF2 Is Critical for This Gene's Ability to Restore the High Wild-Type Reactivation Phenotype to a Herpes Simplex Virus-1 LAT Null Mutant. *J. Gen. Virol.* **2003**, *84*, 2975–2985. [CrossRef]

108. Harrison, K.S.; Zhu, L.; Thunuguntla, P.; Jones, C. Herpes Simplex Virus 1 Regulates β-Catenin Expression in TG Neurons during the Latency-Reactivation Cycle. *PLoS ONE* **2020**, *15*, e0230870. [CrossRef]

109. Thiery, R.; Boutin, P.; Arnauld, C.; Jestin, A. Pseudorabies Virus Latency: A Quantitative Approach by Polymerase Chain Reaction. *Acta Vet. Hung.* **1994**, *42*, 277–287.

110. White, A.K.; Ciacci-Zanella, J.; Galeota, J.; Ele, S.; Osorio Fernando, A. Comparison of the Abilities of Serologic Tests to Detect Pseudorabies-Infected Pigs during the Latent Phase of Infection. *Am. J. Vet. Res.* **1996**, *57*, 608–611.

111. Tu, F.; Zhang, Y.; Xu, S.; Yang, X.; Zhou, L.; Ge, X.; Han, J.; Guo, X.; Yang, H. Detection of Pseudorabies Virus with a Real-time Recombinase-aided Amplification Assay. *Transbound. Emerg. Dis.* **2021**, 1–9. [CrossRef] [PubMed]

112. Yoon, H.A.; Eo, S.K.; Aleyas, A.G.; Park, S.O.; Lee, J.H.; Chae, J.S.; Cho, J.G.; Song, H.J. Molecular Survey of Latent Pseudorabies Virus Infection in Nervous Tissues of Slaughtered Pigs by Nested and Real-Time PCR. *J. Microbiol.* **2005**, *43*, 430–436. [PubMed]

113. Cheng, T.Y.; Henao-Diaz, A.; Poonsuk, K.; Buckley, A.; van Geelen, A.; Lager, K.; Harmon, K.; Gauger, P.; Wang, C.; Ambagala, A.; et al. Pseudorabies (Aujeszky's Disease) Virus DNA Detection in Swine Nasal Swab and Oral Fluid Specimens Using a GB-Based Real-Time Quantitative PCR. *Prev. Vet. Med.* **2021**, *189*, 105308. [CrossRef]

114. Flatschart, R.B.; Maurício, R. Acute and Latent Infection in Mice with a Virulent Strain of Aujeszky's Disease Virus. *Braz. J. Microbiol.* **2000**, *31*, 308–311. [CrossRef]

115. Ren, C.Z.; Hu, W.Y.; Zhang, J.W.; Wei, Y.Y.; Yu, M.L.; Hu, T.J. Establishment of Inflammatory Model Induced by *Pseudorabies* Virus Infection in Mice. *J. Vet. Sci.* **2021**, *22*, e20. [CrossRef] [PubMed]

Review

A Tug of War: Pseudorabies Virus and Host Antiviral Innate Immunity

Guangqiang Ye [1], Hongyang Liu [1], Qiongqiong Zhou [1], Xiaohong Liu [1], Li Huang [1,2] and Changjiang Weng [1,2,*]

[1] State Key Laboratory of Veterinary Biotechnology, Division of Fundamental Immunology, Harbin Veterinary Research Institute of Chinese Academy of Agricultural Sciences, Harbin 150069, China; ygqyyds123@163.com (G.Y.); lhyyyds1234@163.com (H.L.); qqhenanly@163.com (Q.Z.); lxhyyds1234@163.com (X.L.); highlight0315@163.com (L.H.)

[2] Heilongjiang Provincial Key Laboratory of Veterinary Immunology, Harbin 150069, China

* Correspondence: wengchangjiang@caas.cn

Abstract: The non-specific innate immunity can initiate host antiviral innate immune responses within minutes to hours after the invasion of pathogenic microorganisms. Therefore, the natural immune response is the first line of defense for the host to resist the invaders, including viruses, bacteria, fungi. Host pattern recognition receptors (PRRs) in the infected cells or bystander cells recognize pathogen-associated molecular patterns (PAMPs) of invading pathogens and initiate a series of signal cascades, resulting in the expression of type I interferons (IFN-I) and inflammatory cytokines to antagonize the infection of microorganisms. In contrast, the invading pathogens take a variety of mechanisms to inhibit the induction of IFN-I production from avoiding being cleared. Pseudorabies virus (PRV) belongs to the family Herpesviridae, subfamily Alphaherpesvirinae, genus Varicellovirus. PRV is the causative agent of Aujeszky's disease (AD, pseudorabies). Although the natural host of PRV is swine, it can infect a wide variety of mammals, such as cattle, sheep, cats, and dogs. The disease is usually fatal to these hosts. PRV mainly infects the peripheral nervous system (PNS) in swine. For other species, PRV mainly invades the PNS first and then progresses to the central nervous system (CNS), which leads to acute death of the host with serious clinical and neurological symptoms. In recent years, new PRV variant strains have appeared in some areas, and sporadic cases of PRV infection in humans have also been reported, suggesting that PRV is still an important emerging and re-emerging infectious disease. This review summarizes the strategies of PRV evading host innate immunity and new targets for inhibition of PRV replication, which will provide more information for the development of effective inactivated vaccines and drugs for PRV.

Keywords: pseudorabies virus; innate immune response; type I interferons; apoptosis; autophagy

Citation: Ye, G.; Liu, H.; Zhou, Q.; Liu, X.; Huang, L.; Weng, C. A Tug of War: Pseudorabies Virus and Host Antiviral Innate Immunity. *Viruses* **2022**, *14*, 547. https://doi.org/10.3390/v14030547

Academic Editor: Xiangdong Li

Received: 16 January 2022
Accepted: 1 March 2022
Published: 6 March 2022

1. Introduction

Virus infection induces host innate immune responses, which play an important and decisive role in determining the outcome of the infected host, inducing acute infection death or establishing persistent infection in mammals. The host can establish an antiviral status based on innate immunity systems to antagonize the invasion of the virus by identifying the components of invading virus through pattern recognition receptors (PRRs), which activate signaling pathways to produce type I interferons (IFN-I) [1]. The released IFN-I binds IFN receptors (IFNAR1 and/or IFNAR2) to activate the transduction of downstream JAK-STAT signaling pathway, eventually leading to the expression of a variety of interferon-stimulated genes (ISGs). Accumulating data showed that the ISGs could achieve many cellular outcomes, including antiviral defense, antiproliferative activities, and stimulation of adaptive immunity [2–4].

In the process of host resistance to virus invasion, recognition of PAMPs by host PRRs is the first step of innate immunity [5]. For RNA virus, viral RNA is often recognized by a variety of PRRs, including endosomal Toll-like receptor 3 (TLR3), cytosolic retinoic acid-inducible gene I (RIG-I), melanoma differentiation-associated gene 5 (MDA5), and LGP2/DHX58 sense viral RNA, NOD-like receptor protein 3 (NLRP3), nucleotide-binding oligomerization domain containing 2 (NOD2) [6,7]. For DNA virus, Toll-like receptor 9 (TLR9), cyclic GMP-AMP (cGAMP) synthase (cGAS), DAI (DLM-1/ZBP1), absent in melanoma 2 (AIM2), and IFN gamma-inducible protein 16 (IFI16) act as the main PRRs that recognize viral DNA [8,9]. PRRs then recruit a series of important signal transduction molecules, such as myeloid differentiation primary response gene 88 (MyD88), mitochondrial antiviral-signaling protein (MAVS), intracellular stimulator of IFN genes (STING). These proteins then transfer the different signals to the downstream molecules in different signaling pathways, which eventually lead to activation and translocation of several transcription factors, including NF-κB, interferon regulatory factor 3 (IRF3), and IRF7 into the nucleus to induce the expression of IFN-I and proinflammatory cytokines [10–12].

PRV, a member of alpha-herpesvirus, is the pathogen of Aujeszky's disease. Domestic pigs and wild boars are considered the natural host of the disease, but the disease also threatens most mammals [13–16]. The persistence of PRV infection is recognized only in the family Suidae due to establishing the latent infection. Acute and lethal infection in Suidae occurs in piglets. For other species, PRV mainly invades the PNS first but then progresses to the CNS, which leads to acute death of the host with serious clinical and neurological symptoms [17]. Although humans infected with PRV do not cause death, it causes strong neurological symptoms [18,19]. Recently, as an important emerging and re-emerging infectious disease, PRV cases have been frequently reported. Therefore, PRV is still an important pathogen in agriculture. Previous studies have shown that PRV is an ideal model for studying virus escape from host immune responses in different species [17,20]. The virus is also used as a "live" tracer of a neuronal pathway because it has a significant tendency to infect synaptically connected neurons [21].

IFN-I and tumor necrosis factor (TNF) play a central role in host antiviral immunity. However, excessive production of IFN-I and TNF will cause harmful immune effects. Therefore, IFNs, TNF, and activation of their downstream are strictly regulated. On the other hand, PRV has evolved various mechanisms to inhibit the induction of IFN-I from avoiding being cleared by the active antiviral innate immunity. Accumulated evidence has shown that several alpha-herpesvirus-encoded proteins can antagonize host antiviral innate immune responses by inhibiting IFN-I production, blocking downstream IFN signaling, or regulating the specific ISGs [22–26].

Recently, increasing reports indicate that PRV has evolved various mechanisms to antagonize host immune responses for efficient infection. This review summarizes the immune escape strategies of PRV, including inhibition of IFN-I production and IFN signaling, modulation of inflammatory responses, regulation of apoptosis, and autophagy. These important research advances summarized in this review will help readers understand the latest PRV immune escape strategies, which will help PRV researchers design new effective vaccines and develop new antiviral drugs to prevent and control PRV.

2. PRV Virion

All herpesvirus virions have similar virus particle size (200–250 nm) and structure, containing a double-stranded DNA (dsDNA) genome. In morphology, the complete PRV virions are round or oval, with 150–180 nm diameter. PRV virions are composed of four structural components (Figure 1): the central core containing the viral genomic DNA is packaged in the nucleocapsid; the capsid is embedded by tegument composed of a protein matrix; the envelope is a lipid membrane containing several viral glycoproteins. Previous studies showed that the structural components of mature virus particles are composed of nearly half of PRV gene products [17]. Like VZV, the PRV genome has two unique regions (UL and US). The US region is flanked by the internal and terminal repeat sequences. Most PRV proteins have orthologs in other alpha-herpesvirus. Generally, each gene's name and its corresponding PRV protein can refer to the location of the homologous protein in the region of the HSV-1 genome [17]. In Figure 1, some PRV-encoded proteins are listed that partially make up viral particles. However, not all proteins are shown in the figure. It should be noted that some HSV viral proteins are multifunctional proteins involved in antiviral innate immunity. For example, Huang et al. reported that HSV-1 VP22 counteracts the cGAS/STING-mediated IFN production by inhibiting the enzymatic activity of cGAS [27]. Maruzuru et al. reported that HSV-1 VP22 interacts with AIM2 and prevents its oligomerization [28]. Deschamps et al. reported that HSV-1 UL46 blocks STING-mediated IFN production signaling pathways by eliminating STING and IFI16 [29]. However, whether the homologs of PRV are functionally equivalent to these HSV-1 tegument proteins in immune evasion is still unclear.

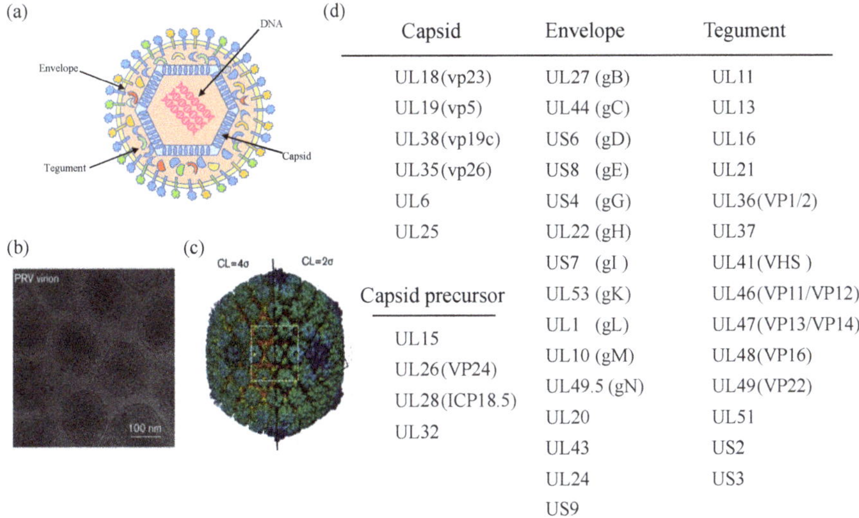

Capsid	Envelope	Tegument
UL18(vp23)	UL27 (gB)	UL11
UL19(vp5)	UL44 (gC)	UL13
UL38(vp19c)	US6 (gD)	UL16
UL35(vp26)	US8 (gE)	UL21
UL6	US4 (gG)	UL36(VP1/2)
UL25	UL22 (gH)	UL37
	US7 (gI)	UL41(VHS)
Capsid precursor	UL53 (gK)	UL46(VP11/VP12)
	UL1 (gL)	UL47(VP13/VP14)
UL15	UL10 (gM)	UL48(VP16)
UL26(VP24)	UL49.5 (gN)	UL49(VP22)
UL28(ICP18.5)	UL20	UL51
UL32	UL43	US2
	UL24	US3
	US9	

Figure 1. Structure and associated proteins of pseudorabies virus (PRV) virion. (**a**) Structure diagram of PRV virion. (**b**) CryoEM image of PRV virions, adapted from [30]. (**c**) 3D reconstruction of the PRV virion at 4.9 Å, colored radially, adapted from [30]. The left and right halves of the capsid are rendered at a contour level (CL) of four and two times standard deviations above the mean density (σ), respectively. The white box and the black dashed box demark a hexon and a vertex region containing five CATCs, respectively [30]. (**d**) Distribution of some proteins.

3. PRV Entry

The entry of PRV virions into cells is mediated by several viral glycoproteins [31]. The attachment process of PRV virions is mediated by the interaction between glycoprotein gC and heparan sulfate proteoglycans in the extracellular matrix of the permissive cells. PRV gD is also involved in viral entry by stabilizing the virus-cell interaction. Subsequently, PRV gB, gH, and gL mediate the viral membrane fusion process by fusing the envelope of virions with the plasma membrane, resulting in the penetration of virus capsid and envelope into the cytoplasm [32]. After membrane fusion, PRV capsid binds to microtubule-like structures to be transported to the nuclear pore [33]. Subsequently, the envelope proteins (UL11, UL47, UL48, and UL49) are rapidly separated from the capsid. Finally, intracellular PRV genomic DNA is released from the morphologically intact capsid into the host nucleus and replicated. In this process, DNA located in the nuclear triggers the host antiviral immune responses and transmits the signals to the cytoplasm.

4. Viral DNA Are Recognized by DNA Sensors

It is well known that sensing viral DNA and activating its downstream cascades play core roles in host antiviral innate immunity. At present, more than 20 DNA sensors have been identified. TLR9, AIM2, cGAS, and IFI16 are reported for recognizing viral DNA.

TLR9 is the first identified DNA sensor to recognize endogenous damaged or exogenous pathogenic DNA on the endoplasmic membrane, especially the CpG DNA motif of un-methylated bacteria [34]. AIM2 and IFI16 are the two most intensive studied AIM2-like receptors (ALRs) family members [35]. AIM2 recognizes double-stranded viral DNA and then is assembled into AIM2 inflammasome to activate inflammatory response [36]. AIM2 binds double-stranded DNA through its hematopoietic IFN-inducible nuclear protein (HIN) domain, then recruits downstream caspase-1 and adaptor apoptosis-associated speck-like protein (ASC) by pyrin domain (PYD) to assemble AIM2 inflammasome, resulting in the release of mature IL-1β/IL-18 and pyroptosis due to the cleavage of gasdermin D (GS-DMD) [37,38]. IFI16 can play a variety of functions in inflammatory responses and IFN-I responses. IFI16 senses the Kaposi sarcoma-associated herpesvirus (KSHV) infection and induces the activation of inflammasomes [39]. IFI16 can recognize HSV DNA and partic-ipate in IFI16-STING and NF-κB signaling pathways [40]. Recently, cGAS was defined as a dsDNA sensor [41]. cGAS recognizes cytoplasmic dsDNA to produce secondary messenger cGAMP. Subsequently, STING located on the endoplasmic reticulum is poly-merized and translocated to the Golgi, where TBK1 and IRF3 are recruited and activated to phosphorylated IRF3, the dimerized IRF3 is translocated into the nucleus, resulting in the transcriptional induction of IFN-I and the NF-κB-dependent expression [42–44]. The TLR signal transduction adaptor protein TRIF is also reported to be involved in the cGAS-STING signaling pathway [45]. For PRV, it has been reported that DNA-dependent activator of interferon (IFN)-regulatory factors (DAI), a DNA sensor of the PRV genome, triggers the production of IFN-I [46]. DEAD (Asp-Glu-Ala-Asp) box polypeptide 41 (DDX41), a member of the DEXDc helicase family, also has been identified as another intracellular DNA sensor of the genomic DNA of PRV in porcine kidney cells [47]. Subsequently, Wang et al. also demonstrated that cGAS senses cytosolic PRV DNA and promotes IFN-β expression [48]. However, why cells use different DNA sensors to defend against PRV infection and the coordinated roles of these PRRs in recognizing the PRV genome remain unknown.

5. PRV Infection Inhibits IFN-I Production

This section summarizes the PRV-encoded proteins that participate in regulating NF-κB and cGAS-STING signaling pathways (Table 1 and Figure 2).

Table 1. NF-κB, cGAS-STING, and IFN signaling pathways are negatively regulated by PRV.

PRV Factors	Immune Elements and Mechanisms	Targeting Pathway	References
UL13	Degradation of PRDX1	cGAS-STING pathway	[49]
	Degradation of IRF3	cGAS-STING pathway	[50]
	Disrupts IRF3 binding to the IRF3-responsive promoter	cGAS-STING pathway	[51]
UL24	Degradation of IRF7	cGAS-STING pathway	[52]
	Degradation of p65	NF-κB signaling pathway	[53]
	Inhibit the transcription of OASL	IFN signaling pathway	[54]
	Inhibit the transcription of ISG20	IFN signaling pathway	[55]
US3	Degradation of IRF3	NF-κB signaling pathway	[56]
	Degradation of Bclaf1	IFN signaling pathway	[57]
gE	Degradation of CBP/p300	NF-κB signaling pathway	[58]
UL50	Degradation of IFNAR1	IFN signaling pathway	[59]
UL42	Competes with ISG factor 3 (ISGF3) by binding to ISRE	IFN signaling pathway	[60]

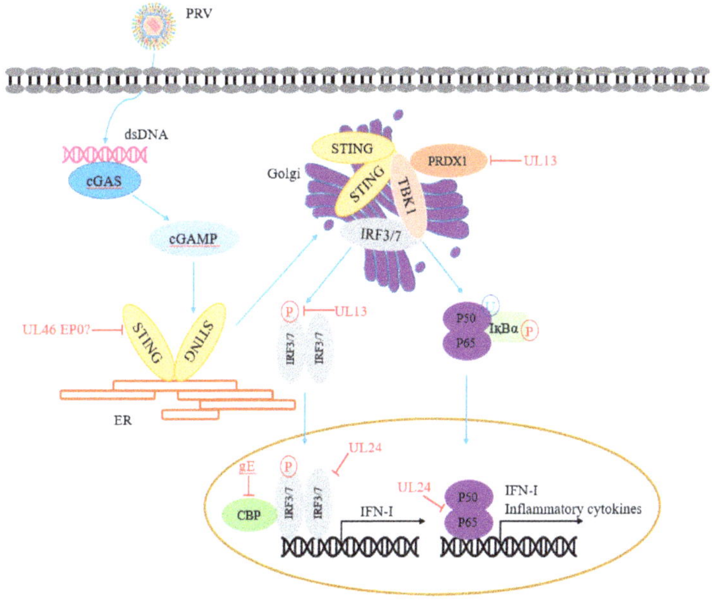

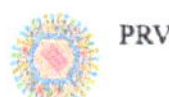

Figure 2. PRV evades the host antiviral immune responses by regulating the DNA-sensing signaling pathway. PRV invasion induces type I interferons (IFN-I) production through IRFs or NF-κB signal pathways. PRV-encoded multiple proteins can target various steps involved in this process.

5.1. cGAS

During PRV infection, cGAS senses cytosolic PRV DNA to synthesize cGAMP and then activate the STING/TBK1/IRF3 signaling pathway to produce IFN-I [48]. Histone deacetylases (HDACs) are epigenetic regulators that regulate the histones, chromatin conformation, protein-DNA interaction, and even transcription. Recently, Guo et al. found that genetic and pharmacological inhibition of HDAC1 significantly influences PRV replication in that the inhibition of HDAC1 induces DNA damage response, resulting in the release of damaged DNA into the cytosol, then activates cGAS and the downstream STING/TBK1/IRF3 signaling pathway [61]. Walters et al. found that the *US3* gene of PRV is conserved in HSV-1 and varicella-zoster virus (VZV). US3 is a serine/threonine (S/T) kinase. HDAC2 is hyperphosphorylated in cells infected with PRV and PRV lacking US3 kinase. However, specific chemical inhibition of class I HDAC activity increases the plaquing efficiency of PRV lacking US3 kinase, whereas only minimal effects are observed with wild-type viruses, indicating that PRV US3 kinase activity is required for HDACs to reduce viral genome silencing and allow efficient viral replication [62].

5.2. STING

STING, an evolutionarily conserved transmembrane protein [63], plays a central role in antiviral immunity by activating its downstream TBK1-IRF3 axis [64]. A variety of viruses has evolved many strategies to target STING, resulting in the inhibition of innate immune responses. For example, HSV ICP34.5 interacts with STING and prevents its transport from the endoplasmic reticulum to Golgi, inhibiting host innate immunity [65]. PRV *UL46* gene is a late gene that encodes viral phosphoproteins 11 and 12 (VP11/12). Recently, PRV UL46 has been shown to induce phosphorylation of ERK1/2, which is not involved in impairing the integrity of the nuclear envelope [66]. Xu et al. found that PRV UL46 is a nucleocytoplasmic shuttling tegument protein. UL46 interacts with EP0, UL48, and STING [67]. However, whether these interactions affect STING-mediated host innate immune responses remain unknown.

5.3. TBK1

Tank binding kinase 1 (TBK1) is a serine/threonine kinase involved in various biological processes, including innate immune response, autophagy, and cell growth [68]. In the long-term struggle between host and virus, as a key adaptor protein of multiple signal pathways [69–71], TBK1 has become the target of many hosts' immune regulatory factors and viral proteins.

PRV UL13 is a protein serine/threonine kinase, which can be packaged into the tegument of PRV virions. Lv et al. found that peroxiredoxin 1 (PRDX1), a member of antioxidant enzymes, binds to TBK1 and IκB kinase ε (IKKε) to regulate IFN-I production positively. Studies have shown that UL13 interacts with and promotes antiviral regulator PRDX1 degradation via the ubiquitin-proteasome pathway in a kinase-dependent manner, thereby inhibiting the host's innate immune response [49].

5.4. UL13, UL24, US3, and gE Target IRF3/IRF7

IRF3 plays a key role in the induction of IFN-I production. As the central factor of antiviral response, IRF3 is targeted by many viral proteins through ubiquitination and phosphorylation. PRV has evolved a variety of antiviral strategies to antagonize IRF3 function. Recently, LV et al. found that PRV protein kinase UL13 inhibits IFN-I production. by enhancing the ubiquitination and degradation of IRF3 [50]. Bo et al. showed that PRV UL13 phosphorylates IRF3 and disrupts the interaction between IRF3 and the IRF3-responsive promoter, thereby inhibiting cGAS-STING signaling [51].

UL24 protein is a conserved protein in the herpesvirus family, required for viral growth. HSV-1 UL24 is a nucleolar protein, and the endonuclease motif of UL24 is required for viral diffusion [72,73]. HSV-1 UL24 inhibits cGAS-STING-mediated NF-κB promoter activity, dependent on the region containing 74 to 134 aa within HSV-1 UL24. Mechanistically, HSV-1 UL24 binds to NF-κB subunits p65 and p50 via the Rel homology domains (RHDs), which reduces the tumor necrosis factor-alpha (TNF-α)-mediated nuclear translocation of p65 and p50 [74]. Liu et al. showed that PRV UL24 promotes IRF7 degradation through the ubiquitin-proteasome pathway, resulting in antagonizing cGAS-STING-mediated the production of IFN-I [52].

US3 is a conserved serine/threonine kinase in all herpes viruses. It plays an important role in virus replication. It has been reported that PRV US3 is mainly involved in the nuclear export of viral capsids, which is related to the pathogenicity of the virus in vivo [75]. On the other hand, PRV US3 induces cytoskeleton changes in PRV-infected cells [76,77] and has different effects on various host defense mechanisms. Xie et al. demonstrated that PRV US3 inhibits IFN-I production by degrading IRF3. The results showed that PRV US3 could inhibit host antiviral innate immunity by targeting various host defense mechanisms [56]. CREB-binding protein (CBP/p300) is a histone acetyltransferase, which has been proved to play an important role in transcriptional regulation. Previous studies showed that CBP/p300 could interact with IRF3, NF-κB, p53, and other transcription factors [78,79]. Lu et al. found that PRV glycoprotein gE suppresses IFN-β production by targeting IRF3 and promotes CBP/p300 degradation [58].

Recently, several studies have confirmed that PRV has also evolved a variety of immune escape strategies in plasma cells, such as dendritic cells (pDCs). pDCs play a central role by producing many IFN-I in the antiviral immune response [80]. Lamote et al. showed that the lack of gE could enhance the phosphorylation of foreign signal-regulated kinase 1/2 (ERK1/2) in PDCs, thereby inducing high levels of IFN-I in pDCs. gE/gI glycoprotein complex was identified as an inhibitor of pDCs activity [81].

5.5. NF-κB Signaling Pathway

NF-κB family members play important roles in many biological processes, such as immunity, inflammation, and cell proliferation. The NF-κB transcription factor can be activated by various cell stimulants. NF-κB interacts with key adaptor proteins, which eventually leads to induce the expression of IFN-I and proinflammatory cytokines to execute its antiviral function [82–84]. PRV proteins can inhibit the transduction of the NF-κB signaling pathway in the process of infection. Wang et al. demonstrated that PRV UL24 promotes p65 degradation through the ubiquitin-proteasome pathway to eliminate TNF-α-mediated activation of NF-κB [53]. Recently, Romero et al. demonstrated that PRV infection induces DNA damage response (DDR), which then activates NF-κB via a peculiar "inside-out" nucleus-to-cytoplasm signal. The DDR-NF-κB signaling axis requires the expression of viral proteins but is initiated before active PRV replication. However, late PRV proteins inhibit NF-κB-dependent gene expression [85].

6. PRV Infection Inhibits IFN Signaling Pathway

The secreted IFN-I binds to IFNAR1 and/or IFNAR2 to activate JAK1 and Tyk2, which then phosphorylates STAT1 and STAT2 and forms a trimer together with IRF9, called IFN stimulating gene factor 3 (ISGF3). ISGF3 then binds to IFN stimulating response element (ISRE) in the nucleus and initiates the transcription of hundreds of ISGs [86]. Some of these ISGs-encoding proteins are directly involved in antiviral response. Of course, the JAK-STAT signaling pathway plays an important role in inhibiting PRV replication and transmission. On the contrary, PRV has evolved various immune escape strategies against the JAK-STAT signaling pathway. Here, we list the key adaptor proteins in the JAK-STAT signaling pathway and summarize the immune escape strategies of PRV observed in the past five years (Table 1 and Figure 3).

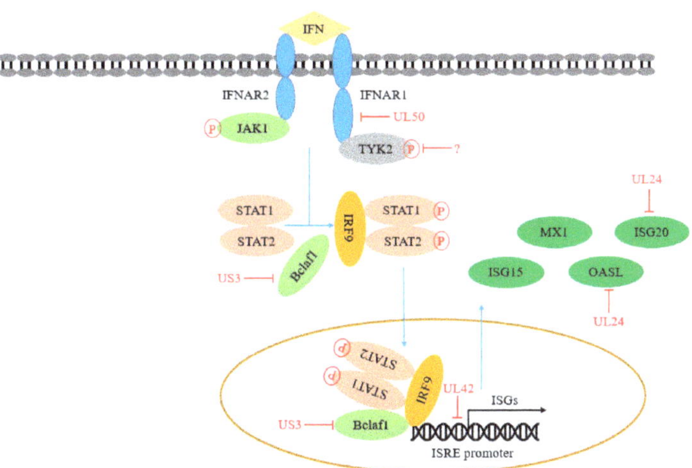

Figure 3. PRV infection inhibits IFN signaling pathway. The secreted IFN-I binds to IFNAR1 and/or IFNAR2 on the cell surface and initiates the transcription of hundreds of antiviral factors ISGs through the JAK-STAT signal pathway. PRV-encoded multiple proteins can target various steps involved in this process.

6.1. UL50 Regulates IFNARs

IFN-I binds IFNARs (IFNAR1 and/or IFNAR2) is the initial step of the JAK-STAT signaling pathway. As the important membrane receptors, IFNARs play important roles in the immune process of the host. Consistently, patients with IFNAR1 deficiency are prone to severe COVID-19 [87], indicating that IFNAR1 plays a central role in host innate immunity. Previous studies showed that IFNAR1 is necessary for IFN signaling transduction, while

IFNAR2 is necessary for STAT binding and activation. Shemesh et al. proposed that IFNAR2 is a platform for STAT activation. Interestingly, tyrosine phosphorylation of IFNAR2 may enhance JAK-STAT signal transduction by promoting the separation of activated STAT and IFNAR2 [88].

PRV has evolved an immune escape strategy by targeting IFNARs. Zhang et al. found that PRV UL50, a dUTPase, has the function of inhibiting the JAK-STAT signaling pathway. PRV UL50 degrades IFNAR1 through the lysosomal pathway to prevent STAT1 phosphorylation induced by IFN-I [59]. However, PRV UL50-mediated inhibition effect does not depend on its dUTPase activity. Its functional region for inhibiting the JAK-STAT signaling pathway corresponds to 225 to 253 aa in its C-terminal region.

6.2. PRV Infection Degrades JAK1 and Tyk2

Yin et al. found that PRV infection can lead to the degradation of JAK1 and Tyk2 through proteasome [89], inhibiting STAT1 phosphorylation and finally interfering with JAK-STAT signal transduction. However, the early viral protein EP0 does not relate to this inhibition, suggesting that other PRV proteins might recruit other E3 ligases to execute its degradation function.

6.3. US3 and UL42 Regulate ISRE

ISGF3 complex binding to the ISRE promoter is a crucial link in the JAK-STAT signaling pathway, leading to the induced expression of hundreds of ISGs. Bclaf1 (Bcl-2 related transcription factor 1) was originally found as a binding protein of adenovirus E1B 19k protein [90]. Bclaf1 exerts its functions in various biological processes, including apoptosis, cancer process, and autophagy [90–93]. Recently, Qin et al. reported that Bclaf1 is an important regulator in IFN signaling during PRV infection. Bclaf1 enhances the phosphorylation of STAT1 and STAT2 in response to IFNα. Additionally, Bclaf1 facilitates ISGF3 complex binding with ISRE, improving efficient gene transcription by directly interacting with ISRE and STAT2. The two strategies actively regulate the induction of IFN production induced by PRV. Subsequently, they found that US3 of PRV and HSV-1 can degrade Bclaf1 through proteasome pathway to resist its antiviral response [57]. In addition, Zhang et al. found that PRV UL42 competes with ISG factor 3 (ISGF3) by binding to ISRE to inhibit the transcription of the related ISGs. The four conserved DNA binding sites of PRV UL42 are necessary for its inhibitory effect on the JAK-STAT signaling pathway [60].

6.4. UL24 Regulates ISGs

ISGs can be induced and released in an IFN-dependent or independent manner during PRV infection. ISGs help the host build their antiviral status to block viral invasion. Several viruses have evolved strategies to target these ISGs to establish various immune evasion directly. For example, oligoadenylate synthase (OAS), an ISG family member, produces 2′–5′ oligoadenylate to trigger viral RNA degradation by recruiting/binding RNase L. It has been reported that oligoadenylate synthase-like (OASL) protein plays a role in regulating host innate immunity by inhibiting cGAS-induced IFN production and enhancing RIG-I-mediated IFN induction, respectively. Recently, Chen et al. showed that OASL promotes RIG-I-mediated IFN expression, resulting in the induction of ISGs to inhibit PRV proliferation. Of note, PRV UL24 reduces the transcription of OASL, thereby damaging the RIG-I signal pathway and antagonizing the antiviral effect of OASL [54]. Chen et al. recently reported that ISG20, a member of the ISGs, can upregulate the expression of IFN-β to resist PRV infection. Further studies found that PRV UL24 inhibits the transcription of ISG20, which weakens the antiviral function of ISG20 [55]. Recently, IFN-inducible transmembrane proteins (IFITM1, 2, 3) are found to work as important host self-restriction factors, possessing a broad spectrum of antiviral effects. Xie et al. IFITM2 are crucial for controlling PRV infection by interfering with cell binding and entry [94]. However, Wang et al. found that knockdown of IFITM2 and IFITM3 expression don't influence PRV infection. Knockdown of IFITM1 expression using RNA interference enhances PRV infection while overexpression of IFITM1 has the opposite effect, indicating that IFITM1 is required to inhibit PRV entry [95]. Whether PRV proteins antagonize the functions of IFITMs has not been reported.

7. Inhibition of Intrinsic Antiviral Immunity by PRV

As a specific part of innate immunity, intrinsic antiviral immunity directly restricts viral replication and assembly. Unlike PRRs indirectly inhibit viral infection by inducing interferons and other antiviral molecules, intrinsic antiviral immunity factors recognize specific viral components and directly prevent virus replication without induction of antiviral gene expression [96,97]. The nuclear domain 10 (ND10) complex consists of three key components, namely, promyelocytic leukemia antigen (PML), speckled protein of 100 kDa (Sp100), and human death domain-associated protein 6 (hDaxx) [98,99]. Yu et al. found that promyelocytic leukemia nuclear bodies (PML-NBs) inhibit PRV infection by directly engaging in the repression of viral gene transcription [100]. Everett et al. found that PRV EP0 induces a reduction in the SUMO-modified forms but not the other isoforms of PML. PRV EP0 also causes a reduction in the numbers of hDaxx foci and also in their apparent fluorescence intensity [101]. A previous study showed that HSV-1 ICP0 also induces degradation of PML and Sp100 [102,103], suggesting that herpesvirus has evolved strategies to destroy the ND10 complex to inhibit intrinsic antiviral immunity.

Recently, a study demonstrated that cholesterol 25-hydroxylase (CH25H) catalyzes the conversion of cholesterol to 25-hydroxycholesterol (25 HC), which inhibits the growth of PRV in vitro subject to its inhibitory effect on PRV attachment and entry [104]. Li et al. found that p53, a key cellular transcription factor, positively regulates viral replication and pathogenesis, providing a novel target for intrinsic host cell immunity during PRV infection [105]. However, whether PRV proteins inhibit their antiviral function is still unknown.

8. Regulation of Inflammatory Responses by PRV

Inflammatory cytokines, such as interleukin 1β (IL-1β) and IL-18, play roles in clearing various invading pathogens [106]. The host's inflammatory response is regulated by a multiprotein complex called the inflammasome. An inflammasome comprises a sensor protein, ASC, and an inflammatory pro-caspase-1 [107]. Nod-like receptors (NLRs), AIM2-like receptors (ALRs), and pyrin can be used as a sensor protein, the starting point of inflammasome assembly [107]. For example, as an NLRP3 inflammasome, NLRP3 recruits ASC and pro-caspase-1, resulting in pro-caspase-1 self-cleavage to form active caspase-1 (p10/p20 tetramer). Active caspase-1 cleaves the pro-IL-1β and pro-IL-18 to secrete mature IL-1β and IL-18 [107].

Previous studies demonstrated that PRV infection activates several inflammasomes [108]. Laval et al. found that PRV infection primes peripheral nervous system (PNS) neurons to an inflammatory state regulated by TLR2 and IFN-I signaling and induces a lethal inflammatory response in vitro and in vivo [109,110]. Ye et al. found that PRV infection can activate NLRP3-mediated inflammatory responses and significantly increase the pathogenicity of PRV in infected mice [111]. Sun et al. found that PRV infection upregulates the expression level of NLRP3, pro-caspase-1, GSDMD, pro-IL-1β, and pro-IL-18, resulting in activation of NLRP3-mediated inflammatory responses and pyroptosis by cleaving GSDMD [112].

Many studies demonstrated that some drugs could reduce the inflammatory responses induced by PRV. Liu et al. confirmed that luteolin reduces the inflammatory levels in leukemia cells and mouse macrophage (RAW264.7) during PRV infection, which is manifested in inhibiting their expression levels of proinflammatory mediators, inflammatory cytokines, and their regulatory genes, iNOS and COX-2 [113]. Vitamin A [114], Dunaliella salina alga extract [115], β-carotene [116], Resveratrol [117], and Ethyl acetate fraction of flavonoids from Polygonum hydropiper L [118] have been shown to regulate the inflammatory response induced by PRV.

9. Regulation of Apoptosis, Autophagy, ER Stress, and Stress Granules by PRV

9.1. Apoptosis

Apoptosis, also named programmed cell death, is characterized by cell shrinkage, membrane blebbing, the formation of apoptotic body, and nuclear DNA fragments, which can maintain the homeostasis of host cells [119,120]. As a cellular defense mechanism, apoptosis plays an important role in preventing virus transmission and diffusion in the early stage of virus infection [121], while apoptosis enhances virus replication and egress in the later stage of viral infection.

Several studies demonstrated that PRV infection induces apoptosis. Cheung et al. found that PRV infection undergoes apoptosis with several apoptotic characteristics, including the externalization of membrane phospholipid phosphatidylserine, the activation of caspase 3, cellular DNA degradation, and morphological changes of the nucleus [122]. Lai et al. confirmed that PRV infection induces expression of proapoptotic Bcl family proteins in PK15 cells in a dose- and time-dependent manner and trigger apoptosis. PRV infection can cause oxidative stress and free radicals, cause DNA damage, and trigger apoptosis [123]. Further studies found that caffeine, a known DNA damage inhibitor, can protect cells from PRV-induced apoptosis. Antioxidant *N*-acetyl-L-cysteine can prevent the production of reactive oxygen species (ROS) in cells and protect DNA from cutting.

Interestingly, Alemañ et al. pointed out that obvious pathological changes were observed around PRV-infected neurons, but the morphological or histochemical evidence of apoptosis was not observed. However, apoptosis was easily detected among infiltrating immune cells around PRV-infected neurons [124]. Therefore, they concluded that the apoptosis of trigeminal ganglionic neurons might be blocked during PRV acute infection, while the apoptosis of infiltrating immune cells is observed during PRV infection, indicating an important mechanism of immune evasion for the PRV. Yeh et al. found that PRV infection increases the expression levels of TNF-α and its receptor. The inhibitors of p38 and JNK/SAPK can significantly reduce the numbers of PRV infection-induced apoptosis, and the expression up-regulation of TNF-α was also inhibited in this process. So they proposed that TNF-α mediates apoptosis via the activation of p38 MAPK and JNK/SAPK signaling during PRV infection [125].

PRV has evolved many strategies to block the apoptotic signaling pathway in the long-term struggle between the virus and the hosts. For example, PRV US3 can block PRV-induced apoptosis. Chang et al. demonstrated that PRV-induced apoptosis in swine-testicle (ST) cells could be inhibited by US3 in the late stage of infection, depending on its enzyme activity [126]. Chang et al. subsequently confirmed that PRV infection could increase the expression of anti-apoptotic signaling molecules, including Akt, PDK-1, and IκBα in the trigeminal ganglion. Inhibiting the Akt and NF-κB pathways in the early stage of PRV infection can promote cell death [126]. A long isoform US3 of PRV can protect ST cells from PRV- or staurosporine-induced apoptosis. The study also pointed out that US3 is located on the mitochondrial and played an important role in inhibiting PRV-induced apoptosis [127]. Consistent with these results, a PRV-ΔUS3, a recombinant virus with deletion of *US3* gene, induced more apoptosis cells, and the virus titers were lower than wild-type PRV infection. Q-VD-OPh, a broad-spectrum caspase inhibitor, inhibited apoptosis of ST and HEp-2 cells induced by PRV-ΔUS3 or WT PRV [128].

9.2. Autophagy

Autophagy is a conserved lysosomal degradation process involved in a mechanism for cells to maintain homeostasis [129], which the host can use to antiviral and recover damaged organelles [130]. In some cases, autophagy is necessary for cells to antagonize viral replication. It has been reported that intracellular autophagosomes can be activated to target and degrade some viral proteins during HSV-1 infection [131]. Concerning HSV-1, many studies revealed its escape strategies and the potential protective effects of autophagy during infection [132,133]. However, VZV lacks these specific genes encoding proteins interfering with autophagy, suggesting that in some cases, VZV may be able to exploit autophagy for its replication [134,135].

The relationship between PRV replication and autophagy has been reported. Previous studies showed that PRV infection induces autophagy in vitro through the classical Beclin-1-ATG7-ATG5 pathway, resulting in increased PRV replication [71]. Xu et al. reported that PRV infection induces transformation of light chain 3 (LC3-I) autophagosomes in mouse neuron 2A (N2a) cells [136], indicating that PRV infection can induce autophagy in this cell line completely. It is an obvious paradox that PRV infection-induced autophagy enhances viral replication in that autophagy antagonizes viral replication as part of the host immune response [131]. Based on their results, Sun et al. put forward a new viewpoint on the relationship between autophagy and PRV replication. They found that PRV can induce autophagy in the early stage of infection. Upon PRV replication, several PRV-encoded proteins inhibit the level of autophagy, leading to an increase in the titer of PRV. Mechanistically, US3 protein reduces the PRV-infected cells' autophagy level by activating the AKT/mTOR pathway. In other words, PRV infection has a dual effect on the autophagy process of host cells [137].

Bcl2-associated athanogene 3 (BAG3) is first identified as BCL-2 binding protein, belonging to the BAG protein family [138], which has a highly conserved BAG domain at the C-terminal [138,139]. Lyu et al. found that BAG3 is involved in the autophagy process during PRV infection, negatively regulating virus replication. Recently, a study showed that overexpressed UL56 induces BAG3 degradation by using its C-terminal [140]. However, BAG3 degradation was not observed in the wild-type PRV (WT) or a PRV-ΔUL56 recombinant virus with deletion of the *UL56* gene; therefore, whether the occurrence of BAG3 degradation mediated by UL56 in the PRV infection process needs further study.

Based on the progress of PRV infection-induced autophagy, some therapeutic targets and drugs against PRV infection have been reported. Xing et al. found that PRV infection inhibits the AKT/mTOR signaling pathway while Platycodon grandiflorus polysaccharide (PGPS) upregulates the mTOR signaling pathway [141]. The findings showed that PGPS could inhibit viral replication by promoting autophagy. Recently, Ming et al. assessed some deubiquitinases (USPs) inhibitors' inhibitory effect on PRV replication [142]. Among them, USP14 inhibitor b-AP15 exhibits the most dramatic. Consistently, replenishment of USP14 in USP14 null cells restored viral replication. Inhibition of USP14 induces the K63-linked ubiquitination of PRV VP16, where USP14 directly binds to ubiquitin chains on VP16 through its UBL domain during the early stage of viral infection. Mechanistically, b-AP15 induces VP16 degradation through SQSTM1/p62-mediated selective autophagy, which is related to the EIF2AK3/PERK- and ERN1/IRE1-mediated signaling pathways [142]. Interestingly, pretreatment of mice with b-AP15 activates ER stress and autophagy, resulting in inhibition of PRV infection in vivo [142]. These results showed that USP14 might be a potential therapeutic target to treat alpha-herpesvirus-induced infectious diseases.

9.3. ER Stress

The endoplasmic reticulum (ER) is involved in protein synthesis, folding, transportation, and secretion. It has been reported that intracellular stress states will be sensed by ER, such as heat shock, viral and bacterial infection, hypoxia, and misfolded or unfolded proteins to activate the unfolded protein response (UPR) to restore ER and cell homeostasis [143]. During viral infection, the accumulation of viral proteins can cause stress in the ER and trigger the unfolded protein response (UPR) to restore ER homeostasis [144]. Yang et al. reported that the expression of glucose-regulated protein 78 (GRP78), as a marker of ER stress, is upregulated in the early stage of PRV infection, indicating that PRV infection induces ER stress and unfolded protein response (UPR). In addition, the IRE1-XBP1 and eIF2α-ATF4 pathways are activated during PRV infection [145]. Whether PRV proteins regulate ER stress is still unknown. Upon HSV-1 infection, HSV-1 UL41 protein suppresses the IRE1/XBP1 signaling pathway of the UPR via its RNase activity. Ectopic expression of HSV-1 UL41 decreases the expression of XBP1 and blocks XBP1 splicing activation induced by the ER stress inducer thapsigargin. Compared with HSV-1, the HSV-1 mutant lacking the *UL41* gene did not induce the decreased XBP1 mRNA induced by thapsigargin [144]. Only ATF6 activation is detected during early infection while the activity of the eIF2alpha/ATF4 signaling is increased at the final stage of HSV-1 replication, suggesting that HSV-1 disarms the unfolded protein response in the early stages of infection. HSV-1 may use ICP0 as a sensor to modulate the cellular stress response [146].

9.4. Stress Granules Formation

The formation of stress granules (SGs) is also involved in antiviral innate immune responses. In the case of virus infection, host cells may turn off the synthesis of intracellular protein translation by forming stress particles to resist virus replication [147]. SGs formation is related to inhibiting the host's protein translation [148,149]. SGs contain a variety of components, such as untranslated mRNA, eukaryotic translation initiation factors (such as eIF4E, eIF4G, eIF4A, eIF2), and T-cell intracellular antigen 1 (TIA-1) and two markers of SGs: TIA-1-related protein (TIAR) and Ras GTPase activating protein-binding protein 1 (G3BP1) [147]. Phosphorylation of eIF2α (eIF2α-p) is also a landmark event in the formation of SGs [150]. eIF2α-p may lead to the retention of many translation initiation complexes in the cytoplasm and eventually induce the formation of SGs [151]. Xu et al. found that SGs induced by sodium arsenate (AS) and DL-Dithiothreitol (DTT) are blocked when the phosphorylation of eIF2α kinases double-stranded RNA-activated protein kinase (PKR) is significantly inhibited during PRV infection, suggesting that PRV-encoded protein itself or by recruiting some host factors to inhibit the formation of SGs, which will benefit PRV replication [152].

10. Conclusions and Prospects

PRV had caused great loss in the livestock industry. As an emerging and re-emerging infectious disease, it still poses a great threat to animal and human health. Like other herpes viruses, PRV can establish latent infection in specific tissues, which can lurk in the host's nervous system for a long time. However, the detailed mechanisms are still not understood. Virus clearance in the body is closely related to the immune state of the host, especially innate immunity. Therefore, an in-depth understanding of the immune evade mechanism of PRV will benefit the development of effective PRV vaccines and antiviral drugs.

In this review, we summarized several strategies of PRV to escape the host's innate immunity by disrupting important adaptor proteins in different signaling pathways, which are related to IFN production, IFN signaling pathway, inflammasome activation, apoptosis, autophagy, and ER stress. Although several PRV proteins have been reported to inhibit host innate immunity, the specific mechanisms for some proteins have not been clarified. For example, PRV EP0 could inhibit IFN response in primary cells, but it does not have this function in the nonhost cells [153]. PRV EP0 executes the different antiviral functions in different hosts is still unknown. In addition, HSV-1 ICP0, a homologous of PRV EP0, is an important protein involved in resisting host innate immunity [23,154]. Whether PRV EP0 has the same function is still unknown, which needs to be further investigated.

Although we have listed many natural immune escape strategies for PRV, compared with HSV-1 and VZV, the research on the regulation of host innate immunity by PRV needs to be further studied, such as NLR signaling pathway and inflammation. For example, AIM2 is a key DNA sensor for recognizing viral DNA upon DNA virus infection. PRV can infect swine and induce a strong inflammatory response. However, there is no AIM2 and its homolog in swine. Other host proteins maybe work as viral DNA sensors. We found that vimentin interacts with viral DNA and NLRP3, involved in the NLRP3 inflammatory response. Therefore, we proposed a new model that vimentin senses viral DNA and then recruits NLRP3 inflammasome to induce inflammatory responses in PRV-infected pigs.

Author Contributions: Conceptualization, C.W.; writing—original draft preparation, G.Y.; writing and revision, G.Y., H.L., Q.Z., X.L. and L.H.; Figure preparation, G.Y.; manuscript revision and supervision, C.W. All authors have read and agreed to the published version of the manuscript.

Funding: This work was supported by the Natural Science Foundation of Heilongjiang Province of China (grants No. YQ2020C022).

Institutional Review Board Statement: Not applicable.

Informed Consent Statement: Not applicable.

Data Availability Statement: All data generated or analyzed during this study are included in this published article.

Conflicts of Interest: The authors declare no conflict of interest.

References

1. Akira, S.; Uematsu, S.; Takeuchi, O. Pathogen recognition and innate immunity. *Cell* **2006**, *124*, 783–801. [CrossRef] [PubMed]
2. Müller, U.; Steinhoff, U.; Reis, L.; Hemmi, S.; Pavlovic, J.; Zinkernagel, R.; Aguet, M. Functional role of type I and type II interferons in antiviral defense. *Science* **1994**, *264*, 1918–1921. [CrossRef] [PubMed]
3. Sadler, A.; Williams, B. Interferon-inducible antiviral effectors. *Nat. Rev. Immunol.* **2008**, *8*, 559–568. [CrossRef] [PubMed]
4. Schoggins, J. Interferon-Stimulated Genes: What Do They All Do? *Annu. Rev. Virol.* **2019**, *6*, 567–584. [CrossRef]
5. Streicher, F.; Jouvenet, N. Stimulation of Innate Immunity by Host and Viral RNAs. *Trends Immunol.* **2019**, *40*, 1134–1148. [CrossRef]
6. Yao, Q. Nucleotide-binding oligomerization domain containing 2: Structure, function, and diseases. *Semin. Arthritis Rheum.* **2013**, *43*, 125–130. [CrossRef]
7. Chen, N.; Xia, P.; Li, S.; Zhang, T.; Wang, T.; Zhu, J. RNA sensors of the innate immune system and their detection of pathogens. *IUBMB Life* **2017**, *69*, 297–304. [CrossRef]
8. Brennan, K.; Bowie, A. Activation of host pattern recognition receptors by viruses. *Curr. Opin. Microbiol.* **2010**, *13*, 503–507. [CrossRef]
9. Xia, P.; Wang, S.; Gao, P.; Gao, G.; Fan, Z. DNA sensor cGAS-mediated immune recognition. *Protein Cell* **2016**, *7*, 777–791. [CrossRef]
10. Wan, D.; Jiang, W.; Hao, J. Research Advances in How the cGAS-STING Pathway Controls the Cellular Inflammatory Response. *Front. Immunol.* **2020**, *11*, 615. [CrossRef]
11. Koyama, S.; Ishii, K.; Coban, C.; Akira, S. Innate immune response to viral infection. *Cytokine* **2008**, *43*, 336–341. [CrossRef] [PubMed]
12. Hu, M.; Shu, H. Innate Immune Response to Cytoplasmic DNA: Mechanisms and Diseases. *Annu. Rev. Immunol.* **2020**, *38*, 79–98. [CrossRef]
13. Minamiguchi, K.; Kojima, S.; Sakumoto, K.; Kirisawa, R. Isolation and molecular characterization of a variant of Chinese gC-genotype II pseudorabies virus from a hunting dog infected by biting a wild boar in Japan and its pathogenicity in a mouse model. *Virus Genes* **2019**, *55*, 322–331. [CrossRef] [PubMed]
14. Kong, H.; Zhang, K.; Liu, Y.; Shang, Y.; Wu, B.; Liu, X. Attenuated live vaccine (Bartha-K16) caused pseudorabies (Aujeszky's disease) in sheep. *Vet. Res. Commun.* **2013**, *37*, 329–332. [CrossRef] [PubMed]
15. Jin, H.; Gao, S.; Liu, Y.; Zhang, S.; Hu, R. Pseudorabies in farmed foxes fed pig offal in Shandong province, China. *Arch. Virol.* **2016**, *161*, 445–448. [CrossRef] [PubMed]
16. Cheng, Z.; Kong, Z.; Liu, P.; Fu, Z.; Zhang, J.; Liu, M.; Shang, Y. Natural infection of a variant pseudorabies virus leads to bovine death in China. *Transbound. Emerg. Dis.* **2020**, *67*, 518–522. [CrossRef] [PubMed]
17. Pomeranz, L.; Reynolds, A.; Hengartner, C. Molecular biology of pseudorabies virus: Impact on neurovirology and veterinary medicine. *Microbiol. Mol. Biol. Rev. MMBR* **2005**, *69*, 462–500. [CrossRef]
18. Guo, Z.; Chen, X.; Zhang, G. Human PRV Infection in China: An Alarm to Accelerate Eradication of PRV in Domestic Pigs. *Virol. Sin.* **2021**, *36*, 823–828. [CrossRef]
19. Liu, Q.; Wang, X.; Xie, C.; Ding, S.; Yang, H.; Guo, S.; Li, J.; Qin, L.; Ban, F.; Wang, D.; et al. A Novel Human Acute Encephalitis Caused by Pseudorabies Virus Variant Strain. *Clin. Infect. Dis. Off. Publ. Infect. Dis. Soc. Am.* **2021**, *73*, e3690–e3700. [CrossRef]
20. Brittle, E.; Reynolds, A.; Enquist, L. Two modes of pseudorabies virus neuroinvasion and lethality in mice. *J. Virol.* **2004**, *78*, 12951–12963. [CrossRef]
21. Ono, E.; Tomioka, Y.; Taharaguchi, S. Possible roles of transcription factors of pseudorabies virus in neuropathogenicity. *Fukuoka Igaku Zasshi = Hukuoka Acta Medica* **2007**, *98*, 364–372. [PubMed]
22. Ma, Y.; Jin, H.; Valyi-Nagy, T.; Cao, Y.; Yan, Z.; He, B. Inhibition of TANK binding kinase 1 by herpes simplex virus 1 facilitates productive infection. *J. Virol.* **2012**, *86*, 2188–2196. [CrossRef] [PubMed]
23. Zhang, J.; Wang, K.; Wang, S.; Zheng, C. Herpes simplex virus 1 E3 ubiquitin ligase ICP0 protein inhibits tumor necrosis factor alpha-induced NF-κB activation by interacting with p65/RelA and p50/NF-κB1. *J. Virol.* **2013**, *87*, 12935–12948. [CrossRef] [PubMed]
24. Su, C.; Zhan, G.; Zheng, C. Evasion of host antiviral innate immunity by HSV-1, an update. *Virol. J.* **2016**, *13*, 38. [CrossRef] [PubMed]

25. Gerada, C.; Campbell, T.; Kennedy, J.; McSharry, B.; Steain, M.; Slobedman, B.; Abendroth, A. Manipulation of the Innate Immune Response by Varicella Zoster Virus. *Front. Immunol.* **2020**, *11*, 1. [CrossRef] [PubMed]

26. Li, K.; Liu, Y.; Xu, Z.; Zhang, Y.; Luo, D.; Gao, Y.; Qian, Y.; Bao, C.; Liu, C.; Zhang, Y.; et al. Avian oncogenic herpesvirus antagonizes the cGAS-STING DNA-sensing pathway to mediate immune evasion. *PLoS Pathog.* **2019**, *15*, e1007999. [CrossRef]

27. Huang, J.; You, H.; Su, C.; Li, Y.; Chen, S.; Zheng, C. Herpes Simplex Virus 1 Tegument Protein VP22 Abrogates cGAS/STING-Mediated Antiviral Innate Immunity. *J. Virol.* **2018**, *92*, e00841-18. [CrossRef]

28. Maruzuru, Y.; Ichinohe, T.; Sato, R.; Miyake, K.; Okano, T.; Suzuki, T.; Koshiba, T.; Koyanagi, N.; Tsuda, S.; Watanabe, M.; et al. Herpes Simplex Virus 1 VP22 Inhibits AIM2-Dependent Inflammasome Activation to Enable Efficient Viral Replication. *Cell Host Microbe* **2018**, *23*, 254–265. [CrossRef]

29. Deschamps, T.; Kalamvoki, M. Evasion of the STING DNA-Sensing Pathway by VP11/12 of Herpes Simplex Virus 1. *J. Virol.* **2017**, *91*, e00535-17. [CrossRef]

30. Liu, Y.; Jiang, J.; Bohannon, K.; Dai, X.; Gant Luxton, G.; Hui, W.; Bi, G.; Smith, G.; Zhou, Z. A pUL25 dimer interfaces the pseudorabies virus capsid and tegument. *J. Gen. Virol.* **2017**, *98*, 2837–2849. [CrossRef]

31. Spear, P.; Longnecker, R. Herpesvirus entry: An update. *J. Virol.* **2003**, *77*, 10179–10185. [CrossRef] [PubMed]

32. Ben-Porat, T.; Rakusanova, T.; Kaplan, A. Early functions of the genome of herpesvirus. II. Inhibition of the formation of cell-specific polysomes. *Virology* **1971**, *46*, 890–899. [CrossRef]

33. Kaelin, K.; Dezélée, S.; Masse, M.; Bras, F.; Flamand, A. The UL25 protein of pseudorabies virus associates with capsids and localizes to the nucleus and to microtubules. *J. Virol.* **2000**, *74*, 474–482. [CrossRef]

34. Müller, T.; Hamm, S.; Bauer, S. TLR9-mediated recognition of DNA. *Handb. Exp. Pharmacol.* **2008**, 51–70. [CrossRef]

35. Brunette, R.; Young, J.; Whitley, D.; Brodsky, I.; Malik, H.; Stetson, D. Extensive evolutionary and functional diversity among mammalian AIM2-like receptors. *J. Exp. Med.* **2012**, *209*, 1969–1983. [CrossRef]

36. Fernandes-Alnemri, T.; Yu, J.; Datta, P.; Wu, J.; Alnemri, E. AIM2 activates the inflammasome and cell death in response to cytoplasmic DNA. *Nature* **2009**, *458*, 509–513. [CrossRef] [PubMed]

37. Hornung, V.; Ablasser, A.; Charrel-Dennis, M.; Bauernfeind, F.; Horvath, G.; Caffrey, D.; Latz, E.; Fitzgerald, K. AIM2 recognizes cytosolic dsDNA and forms a caspase-1-activating inflammasome with ASC. *Nature* **2009**, *458*, 514–518. [CrossRef]

38. Roberts, T.; Idris, A.; Dunn, J.; Kelly, G.; Burnton, C.; Hodgson, S.; Hardy, L.; Garceau, V.; Sweet, M.; Ross, I.; et al. HIN-200 proteins regulate caspase activation in response to foreign cytoplasmic DNA. *Science* **2009**, *323*, 1057–1060. [CrossRef]

39. Kerur, N.; Veettil, M.; Sharma-Walia, N.; Bottero, V.; Sadagopan, S.; Otageri, P.; Chandran, B. IFI16 acts as a nuclear pathogen sensor to induce the inflammasome in response to Kaposi Sarcoma-associated herpesvirus infection. *Cell Host Microbe* **2011**, *9*, 363–375. [CrossRef]

40. Unterholzner, L.; Keating, S.; Baran, M.; Horan, K.; Jensen, S.; Sharma, S.; Sirois, C.; Jin, T.; Latz, E.; Xiao, T.; et al. IFI16 is an innate immune sensor for intracellular DNA. *Nat. Immunol.* **2010**, *11*, 997–1004. [CrossRef]

41. Ablasser, A.; Goldeck, M.; Cavlar, T.; Deimling, T.; Witte, G.; Röhl, I.; Hopfner, K.; Ludwig, J.; Hornung, V. cGAS produces a 2′-5′-linked cyclic dinucleotide second messenger that activates STING. *Nature* **2013**, *498*, 380–384. [CrossRef] [PubMed]

42. Holm, C.; Jensen, S.; Jakobsen, M.; Cheshenko, N.; Horan, K.; Moeller, H.; Gonzalez-Dosal, R.; Rasmussen, S.; Christensen, M.; Yarovinsky, T.; et al. Virus-cell fusion as a trigger of innate immunity dependent on the adaptor STING. *Nat. Immunol.* **2012**, *13*, 737–743. [CrossRef] [PubMed]

43. Motwani, M.; Pesiridis, S.; Fitzgerald, K. DNA sensing by the cGAS-STING pathway in health and disease. *Nat. Rev. Genet.* **2019**, *20*, 657–674. [CrossRef] [PubMed]

44. Kwon, J.; Bakhoum, S. The Cytosolic DNA-Sensing cGAS-STING Pathway in Cancer. *Cancer Discov.* **2020**, *10*, 26–39. [CrossRef] [PubMed]

45. Wang, X.; Majumdar, T.; Kessler, P.; Ozhegov, E.; Zhang, Y.; Chattopadhyay, S.; Barik, S.; Sen, G. STING Requires the Adaptor TRIF to Trigger Innate Immune Responses to Microbial Infection. *Cell Host Microbe* **2017**, *21*, 788. [CrossRef] [PubMed]

46. Xie, L.; Fang, L.; Wang, D.; Luo, R.; Cai, K.; Chen, H.; Xiao, S. Molecular cloning and functional characterization of porcine DNA-dependent activator of IFN-regulatory factors (DAI). *Dev. Comp. Immunol.* **2010**, *34*, 293–299. [CrossRef] [PubMed]

47. Zhu, X.; Wang, D.; Zhang, H.; Zhou, Y.; Luo, R.; Chen, H.; Xiao, S.; Fang, L. Molecular cloning and functional characterization of porcine DEAD (Asp-Glu-Ala-Asp) box polypeptide 41 (DDX41). *Dev. Comp. Immunol.* **2014**, *47*, 191–196. [CrossRef]

48. Wang, J.; Chu, B.; Du, L.; Han, Y.; Zhang, X.; Fan, S.; Wang, Y.; Yang, G. Molecular cloning and functional characterization of porcine cyclic GMP-AMP synthase. *Mol. Immunol.* **2015**, *65*, 436–445. [CrossRef]

49. Lv, L.; Bai, J.; Gao, Y.; Jin, L.; Wang, X.; Cao, M.; Liu, X.; Jiang, P. Peroxiredoxin 1 interacts with TBK1/IKKε and negatively regulates pseudorabies virus propagation by promoting innate immunity. *J. Virol.* **2021**, *95*, e00923-21. [CrossRef]

50. Lv, L.; Cao, M.; Bai, J.; Jin, L.; Wang, X.; Gao, Y.; Liu, X.; Jiang, P. PRV-encoded UL13 protein kinase acts as an antagonist of innate immunity by targeting IRF3-signaling pathways. *Vet. Microbiol.* **2020**, *250*, 108860. [CrossRef]

51. Bo, Z.; Miao, Y.; Xi, R.; Zhong, Q.; Bao, C.; Chen, H.; Sun, L.; Qian, Y.; Jung, Y.; Dai, J. PRV UL13 inhibits cGAS-STING-mediated IFN-β production by phosphorylating IRF3. *Vet. Res.* **2020**, *51*, 118. [CrossRef]

52. Liu, X.; Zhang, M.; Ye, C.; Ruan, K.; Xu, A.; Gao, F.; Tong, G.; Zheng, H. Inhibition of the DNA-Sensing pathway by pseudorabies virus UL24 protein via degradation of interferon regulatory factor 7. *Vet. Microbiol.* **2021**, *255*, 109023. [CrossRef] [PubMed]

53. Wang, T.; Yang, Y.; Feng, C.; Sun, M.; Peng, J.; Tian, Z.; Tang, Y.; Cai, X. Pseudorabies Virus UL24 Abrogates Tumor Necrosis Factor Alpha-Induced NF-κB Activation by Degrading P65. *Viruses* **2020**, *12*, 51. [CrossRef] [PubMed]

54. Chen, X.; Kong, N.; Xu, J.; Wang, J.; Zhang, M.; Ruan, K.; Li, L.; Zhang, Y.; Zheng, H.; Tong, W.; et al. Pseudorabies virus UL24 antagonizes OASL-mediated antiviral effect. *Virus Res.* **2021**, *295*, 198276. [CrossRef] [PubMed]
55. Chen, X.; Sun, D.; Dong, S.; Zhai, H.; Kong, N.; Zheng, H.; Tong, W.; Li, G.; Shan, T.; Tong, G. Host Interferon-Stimulated Gene 20 Inhibits Pseudorabies Virus Proliferation. *Virol. Sin.* **2021**, *36*, 1027–1035. [CrossRef] [PubMed]
56. Xie, J.; Zhang, X.; Chen, L.; Bi, Y.; Idris, A.; Xu, S.; Li, X.; Zhang, Y.; Feng, R. Pseudorabies Virus US3 Protein Inhibits IFN-β Production by Interacting With IRF3 to Block Its Activation. *Front. Microbiol.* **2021**, *12*, 761282. [CrossRef] [PubMed]
57. Qin, C.; Zhang, R.; Lang, Y.; Shao, A.; Xu, A.; Feng, W.; Han, J.; Wang, M.; He, W.; Yu, C.; et al. Bclaf1 critically regulates the type I interferon response and is degraded by alphaherpesvirus US3. *PLoS Pathog.* **2019**, *15*, e1007559. [CrossRef]
58. Lu, M.; Qiu, S.; Zhang, L.; Sun, Y.; Bao, E.; Lv, Y. Pseudorabies virus glycoprotein gE suppresses interferon-β production via CREB-binding protein degradation. *Virus Res.* **2021**, *291*, 198220. [CrossRef]
59. Zhang, R.; Xu, A.; Qin, C.; Zhang, Q.; Chen, S.; Lang, Y.; Wang, M.; Li, C.; Feng, W.; Zhang, R.; et al. Pseudorabies Virus dUTPase UL50 Induces Lysosomal Degradation of Type I Interferon Receptor 1 and Antagonizes the Alpha Interferon Response. *J. Virol.* **2017**, *91*, e01148-17. [CrossRef]
60. Zhang, R.; Chen, S.; Zhang, Y.; Wang, M.; Qin, C.; Yu, C.; Zhang, Y.; Li, Y.; Chen, L.; Zhang, X.; et al. Pseudorabies Virus DNA Polymerase Processivity Factor UL42 Inhibits Type I IFN Response by Preventing ISGF3-ISRE Interaction. *J. Immunol.* **2021**, *207*, 613–625. [CrossRef]
61. Guo, Y.; Ming, S.; Zeng, L.; Chang, W.; Pan, J.; Zhang, C.; Wan, B.; Wang, J.; Su, Y.; Yang, G.; et al. Inhibition of histone deacetylase 1 suppresses pseudorabies virus infection through cGAS-STING antiviral innate immunity. *Mol. Immunol.* **2021**, *136*, 55–64. [CrossRef] [PubMed]
62. Walters, M.; Kinchington, P.; Banfield, B.; Silverstein, S. Hyperphosphorylation of histone deacetylase 2 by alphaherpesvirus US3 kinases. *J. Virol.* **2010**, *84*, 9666–9676. [CrossRef] [PubMed]
63. Ishikawa, H.; Barber, G. STING is an endoplasmic reticulum adaptor that facilitates innate immune signalling. *Nature* **2008**, *455*, 674–678. [CrossRef] [PubMed]
64. Zhong, B.; Yang, Y.; Li, S.; Wang, Y.; Li, Y.; Diao, F.; Lei, C.; He, X.; Zhang, L.; Tien, P.; et al. The adaptor protein MITA links virus-sensing receptors to IRF3 transcription factor activation. *Immunity* **2008**, *29*, 538–550. [CrossRef]
65. Pan, S.; Liu, X.; Ma, Y.; Cao, Y.; He, B. Herpes Simplex Virus 1 γ34.5 Protein Inhibits STING Activation That Restricts Viral Replication. *J. Virol.* **2018**, *92*, e01015-18. [CrossRef]
66. Heine, J.; Honess, R.; Cassai, E.; Roizman, B. Proteins specified by herpes simplex virus. XII. The virion polypeptides of type 1 strains. *J. Virol.* **1974**, *14*, 640–651. [CrossRef]
67. Xu, J.; Gao, F.; Wu, J.; Zheng, H.; Tong, W.; Cheng, X.; Liu, Y.; Zhu, H.; Fu, X.; Jiang, Y.; et al. Characterization of Nucleocytoplasmic Shuttling of Pseudorabies Virus Protein UL46. *Front. Vet. Sci.* **2020**, *7*, 484. [CrossRef]
68. Lee, S.; Shin, J.; Kim, J.; Shin, J.; Lee, S.; Park, H. Targeting TBK1 Attenuates LPS-Induced NLRP3 Inflammasome Activation by Regulating of mTORC1 Pathways in Trophoblasts. *Front. Immunol.* **2021**, *12*, 743700. [CrossRef]
69. Niederberger, E.; Möser, C.; Kynast, K.; Geisslinger, G. The non-canonical IκB kinases IKKε and TBK1 as potential targets for the development of novel therapeutic drugs. *Curr. Mol. Med.* **2013**, *13*, 1089–1097. [CrossRef]
70. Huh, J.; Saltiel, A. Roles of IκB kinases and TANK-binding kinase 1 in hepatic lipid metabolism and nonalcoholic fatty liver disease. *Exp. Mol. Med.* **2021**, *53*, 1697–1705. [CrossRef]
71. Cho, C.; Park, H.; Ho, A.; Semple, I.; Kim, B.; Jang, I.; Park, H.; Reilly, S.; Saltiel, A.; Lee, J. Lipotoxicity induces hepatic protein inclusions through TANK binding kinase 1-mediated p62/sequestosome 1 phosphorylation. *Hepatology* **2018**, *68*, 1331–1346. [CrossRef] [PubMed]
72. Lymberopoulos, M.; Pearson, A. Involvement of UL24 in herpes-simplex-virus-1-induced dispersal of nucleolin. *Virology* **2007**, *363*, 397–409. [CrossRef] [PubMed]
73. Lymberopoulos, M.; Bourget, A.; Ben Abdeljelil, N.; Pearson, A. Involvement of the UL24 protein in herpes simplex virus 1-induced dispersal of B23 and in nuclear egress. *Virology* **2011**, *412*, 341–348. [CrossRef] [PubMed]
74. Xu, H.; Su, C.; Pearson, A.; Mody, C.; Zheng, C. Herpes Simplex Virus 1 UL24 Abrogates the DNA Sensing Signal Pathway by Inhibiting NF-κB Activation. *J. Virol.* **2017**, *91*, e00025-17. [CrossRef]
75. Wagenaar, F.; Pol, J.; Peeters, B.; Gielkens, A.; de Wind, N.; Kimman, T. The US3-encoded protein kinase from pseudorabies virus affects egress of virions from the nucleus. *J. Gen. Virol.* **1995**, *76*, 1851–1859. [CrossRef]
76. Favoreel, H.; Van Minnebruggen, G.; Adriaensen, D.; Nauwynck, H. Cytoskeletal rearrangements and cell extensions induced by the US3 kinase of an alphaherpesvirus are associated with enhanced spread. *Proc. Natl. Acad. Sci. USA* **2005**, *102*, 8990–8995. [CrossRef]
77. Jacob, T.; Van den Broeke, C.; Grauwet, K.; Baert, K.; Claessen, C.; De Pelsmaeker, S.; Van Waesberghe, C.; Favoreel, H. Pseudorabies virus US3 leads to filamentous actin disassembly and contributes to viral genome delivery to the nucleus. *Vet. Microbiol.* **2015**, *177*, 379–385. [CrossRef]
78. Goodman, R.; Smolik, S. CBP/p300 in cell growth, transformation, and development. *Genes Dev.* **2000**, *14*, 1553–1577. [CrossRef]
79. Bedford, D.; Brindle, P. Is histone acetylation the most important physiological function for CBP and p300? *Aging* **2012**, *4*, 247–255. [CrossRef]
80. Summerfield, A.; Guzylack-Piriou, L.; Schaub, A.; Carrasco, C.; Tâche, V.; Charley, B.; McCullough, K. Porcine peripheral blood dendritic cells and natural interferon-producing cells. *Immunology* **2003**, *110*, 440–449. [CrossRef]

81. Lamote, J.; Kestens, M.; Van Waesberghe, C.; Delva, J.; De Pelsmaeker, S.; Devriendt, B.; Favoreel, H. The Pseudorabies Virus Glycoprotein gE/gI Complex Suppresses Type I Interferon Production by Plasmacytoid Dendritic Cells. *J. Virol.* **2017**, *91*. [CrossRef] [PubMed]

82. Zinatizadeh, M.; Schock, B.; Chalbatani, G.; Zarandi, P.; Jalali, S.; Miri, S. The Nuclear Factor Kappa B (NF-kB) signaling in cancer development and immune diseases. *Genes Dis.* **2021**, *8*, 287–297. [CrossRef] [PubMed]

83. Thoma, A.; Lightfoot, A. NF-kB and Inflammatory Cytokine Signalling: Role in Skeletal Muscle Atrophy. *Adv. Exp. Med. Biol.* **2018**, *1088*, 267–279. [CrossRef] [PubMed]

84. Iwai, K. Diverse roles of the ubiquitin system in NF-κB activation. *Biochim. Biophys. Acta* **2014**, *1843*, 129–136. [CrossRef]

85. Romero, N.; Favoreel, H. Pseudorabies Virus Infection Triggers NF-κB Activation via the DNA Damage Response but Actively Inhibits NF-κB-Dependent Gene Expression. *J. Virol.* **2021**, *95*, e0166621. [CrossRef]

86. Katze, M.; He, Y.; Gale, M. Viruses and interferon: A fight for supremacy. *Nat. Rev. Immunol.* **2002**, *2*, 675–687. [CrossRef]

87. Khanmohammadi, S.; Rezaei, N.; Khazaei, M.; Shirkani, A. A Case of Autosomal Recessive Interferon Alpha/Beta Receptor Alpha Chain (IFNAR1) Deficiency with Severe COVID-19. *J. Clin. Immunol.* **2021**, *42*, 19–24. [CrossRef]

88. Shemesh, M.; Lochte, S.; Piehler, J.; Schreiber, G. IFNAR1 and IFNAR2 play distinct roles in initiating type I interferon-induced JAK-STAT signaling and activating STATs. *Sci. Signal.* **2021**, *14*, eabe4627. [CrossRef]

89. Yin, Y.; Romero, N.; Favoreel, H. Pseudorabies virus inhibits type I and type III interferon-induced signaling via proteasomal degradation of Janus kinases. *J. Virol.* **2021**, *95*, e00793-21. [CrossRef]

90. Kasof, G.; Goyal, L.; White, E. Btf, a novel death-promoting transcriptional repressor that interacts with Bcl-2-related proteins. *Mol. Cell. Biol.* **1999**, *19*, 4390–4404. [CrossRef]

91. Lamy, L.; Ngo, V.; Emre, N.; Shaffer, A.; Yang, Y.; Tian, E.; Nair, V.; Kruhlak, M.; Zingone, A.; Landgren, O.; et al. Control of autophagic cell death by caspase-10 in multiple myeloma. *Cancer Cell* **2013**, *23*, 435–449. [CrossRef]

92. Dell'Aversana, C.; Giorgio, C.; D'Amato, L.; Lania, G.; Matarese, F.; Saeed, S.; Di Costanzo, A.; Belsito Petrizzi, V.; Ingenito, C.; Martens, J.; et al. miR-194-5p/BCLAF1 deregulation in AML tumorigenesis. *Leukemia* **2017**, *31*, 2315–2325. [CrossRef] [PubMed]

93. Zhou, X.; Li, X.; Cheng, Y.; Wu, W.; Xie, Z.; Xi, Q.; Han, J.; Wu, G.; Fang, J.; Feng, Y. BCLAF1 and its splicing regulator SRSF10 regulate the tumorigenic potential of colon cancer cells. *Nat. Commun.* **2014**, *5*, 4581. [CrossRef] [PubMed]

94. Xie, J.; Bi, Y.; Xu, S.; Han, Y.; Idris, A.; Zhang, H.; Li, X.; Bai, J.; Zhang, Y.; Feng, R. Host antiviral protein IFITM2 restricts pseudorabies virus replication. *Virus Res.* **2020**, *287*, 198105. [CrossRef] [PubMed]

95. Wang, J.; Wang, C.; Ming, S.; Li, G.; Zeng, L.; Wang, M.; Su, B.; Wang, Q.; Yang, G.; Chu, B. Porcine IFITM1 is a host restriction factor that inhibits pseudorabies virus infection. *Int. J. Biol. Macromol.* **2020**, *151*, 1181–1193. [CrossRef] [PubMed]

96. Yan, N.; Chen, Z. Intrinsic antiviral immunity. *Nat. Immunol.* **2012**, *13*, 214–222. [CrossRef] [PubMed]

97. Chen, Z.; Zhang, L.; Ying, S. SAMHD1: A novel antiviral factor in intrinsic immunity. *Future Microbiol.* **2012**, *7*, 1117–1126. [CrossRef]

98. Sanyal, A.; Wallaschek, N.; Glass, M.; Flamand, L.; Wight, D.; Kaufer, B. The ND10 Complex Represses Lytic Human Herpesvirus 6A Replication and Promotes Silencing of the Viral Genome. *Viruses* **2018**, *10*, 401. [CrossRef]

99. Tavalai, N.; Stamminger, T. New insights into the role of the subnuclear structure ND10 for viral infection. *Biochimica biophysica Acta* **2008**, *1783*, 2207–2221. [CrossRef]

100. Yu, C.; Xu, A.; Lang, Y.; Qin, C.; Wang, M.; Yuan, X.; Sun, S.; Feng, W.; Gao, C.; Chen, J.; et al. Swine Promyelocytic Leukemia Isoform II Inhibits Pseudorabies Virus Infection by Suppressing Viral Gene Transcription in Promyelocytic Leukemia Nuclear Bodies. *J. Virol.* **2020**, *94*, e01197-20. [CrossRef]

101. Everett, R.; Boutell, C.; McNair, C.; Grant, L.; Orr, A. Comparison of the biological and biochemical activities of several members of the alphaherpesvirus ICP0 family of proteins. *J. Virol.* **2010**, *84*, 3476–3487. [CrossRef] [PubMed]

102. Everett, R.; Murray, J. ND10 components relocate to sites associated with herpes simplex virus type 1 nucleoprotein complexes during virus infection. *J. Virol.* **2005**, *79*, 5078–5089. [CrossRef] [PubMed]

103. Müller, S.; Dejean, A. Viral immediate-early proteins abrogate the modification by SUMO-1 of PML and Sp100 proteins, correlating with nuclear body disruption. *J. Virol.* **1999**, *73*, 5137–5143. [CrossRef] [PubMed]

104. Wang, J.; Zeng, L.; Zhang, L.; Guo, Z.; Lu, S.; Ming, S.; Li, G.; Wan, B.; Tian, K.; Yang, G.; et al. Cholesterol 25-hydroxylase acts as a host restriction factor on pseudorabies virus replication. *J. Gen. Virol.* **2017**, *98*, 1467–1476. [CrossRef]

105. Li, X.; Zhang, W.; Liu, Y.; Xie, J.; Hu, C.; Wang, X. Role of p53 in pseudorabies virus replication, pathogenicity, and host immune responses. *Vet. Res.* **2019**, *50*, 9. [CrossRef]

106. Garlanda, C.; Dinarello, C.; Mantovani, A. The interleukin-1 family: Back to the future. *Immunity* **2013**, *39*, 1003–1018. [CrossRef]

107. Vanaja, S.; Rathinam, V.; Fitzgerald, K. Mechanisms of inflammasome activation: Recent advances and novel insights. *Trends Cell Biol.* **2015**, *25*, 308–315. [CrossRef]

108. Sharma, B.; Karki, R.; Kanneganti, T. Role of AIM2 inflammasome in inflammatory diseases, cancer and infection. *Eur. J. Immunol.* **2019**, *49*, 1998–2011. [CrossRef]

109. Laval, K.; Maturana, C.J.; Enquist, L.W. Mouse Footpad Inoculation Model to Study Viral-Induced Neuroinflammatory Responses. *J. Vis. Exp.* **2020**, *160*, e61121. [CrossRef]

110. Laval, K.; Vernejoul, J.B.; Van Cleemput, J.; Koyuncu, O.O.; Enquist, L.W. Virulent Pseudorabies Virus Infection Induces a Specific and Lethal Systemic Inflammatory Response in Mice. *J. Virol* **2018**, *92*, e01614-18. [CrossRef]

111. Ye, C.; Huang, Q.; Jiang, J.; Li, G.; Xu, D.; Zeng, Z.; Peng, L.; Peng, Y.; Fang, R. ATP-dependent activation of NLRP3 inflammasome in primary murine macrophages infected by pseudorabies virus. *Vet. Microbiol* **2021**, *259*, 109130. [CrossRef] [PubMed]

112. Sun, W.; Liu, S.; Huang, X.; Yuan, R.; Yu, J. Cytokine storms and pyroptosis are primarily responsible for the rapid death of mice infected with pseudorabies virus. *R. Soc. Open Sci.* **2021**, *8*, 210296. [CrossRef]

113. Liu, C.; Lin, H.; Yang, D.; Chen, S.; Tseng, J.; Chang, T.; Chang, Y. Luteolin inhibits viral-induced inflammatory response in RAW264.7 cells via suppression of STAT1/3 dependent NF-κB and activation of HO-1. *Free Radic. Biol. Med.* **2016**, *95*, 180–189. [CrossRef] [PubMed]

114. Fang, Z.; Xu, H.; Wu, D.; Zhuo, Y.; Lin, Y.; Luo, X. Vitamin a supplements alleviate inflammatory responses in reproductive tracts of male mice infected with pseudorabies virus. *Int. J. Vitam. Nutr. Res. Int. Z. Vitam. Ernahrungsforschung. J. Int. Vitaminol. Nutr.* **2010**, *80*, 117–130. [CrossRef] [PubMed]

115. Lin, H.; Chen, Y.; Liu, C.; Yang, D.; Chen, S.; Chang, T.; Chang, Y. Regulation of virus-induced inflammatory response by Dunaliella salina alga extract in macrophages. *Food Chem. Toxicol. Int. J. Publ. Br. Ind. Biol. Res. Assoc.* **2014**, *71*, 159–165. [CrossRef] [PubMed]

116. Lin, H.; Chang, T.; Yang, D.; Chen, Y.; Wang, M.; Chang, Y. Regulation of virus-induced inflammatory response by β-carotene in RAW264.7 cells. *Food Chem.* **2012**, *134*, 2169–2175. [CrossRef]

117. Zhao, X.; Tong, W.; Song, X.; Jia, R.; Li, L.; Zou, Y.; He, C.; Liang, X.; Lv, C.; Jing, B.; et al. Antiviral Effect of Resveratrol in Piglets Infected with Virulent Pseudorabies Virus. *Viruses* **2018**, *10*, 457. [CrossRef]

118. Ren, C.; Hu, W.; Li, J.; Xie, Y.; Jia, N.; Shi, J.; Wei, Y.; Hu, T. Ethyl acetate fraction of flavonoids from Polygonum hydropiper L. modulates pseudorabies virus-induced inflammation in RAW264.7 cells via the nuclear factor-kappa B and mitogen-activated protein kinase pathways. *J. Vet. Med. Sci.* **2020**, *82*, 1781–1792. [CrossRef]

119. Ryter, S.; Choi, A. Autophagy in lung disease pathogenesis and therapeutics. *Redox Biol.* **2015**, *4*, 215–225. [CrossRef]

120. Kennedy, P. Viruses, apoptosis, and neuroinflammation—A double-edged sword. *J. Neurovirol.* **2015**, *21*, 1–7. [CrossRef]

121. Koyama, A.; Fukumori, T.; Fujita, M.; Irie, H.; Adachi, A. Physiological significance of apoptosis in animal virus infection. *Microbes Infect.* **2000**, *2*, 1111–1117. [CrossRef]

122. Cheung, A.; Chen, Z.; Sun, Z.; McCullough, D. Pseudorabies virus induces apoptosis in tissue culture cells. *Arch. Virol.* **2000**, *145*, 2193–2200. [CrossRef] [PubMed]

123. Lai, I.; Chang, C.; Shih, W. Apoptosis Induction by Pseudorabies Virus via Oxidative Stress and Subsequent DNA Damage Signaling. *Intervirology* **2019**, *62*, 116–123. [CrossRef] [PubMed]

124. Alemañ, N.; Quiroga, M.; López-Peña, M.; Vázquez, S.; Guerrero, F.; Nieto, J. Induction and inhibition of apoptosis by pseudorabies virus in the trigeminal ganglion during acute infection of swine. *J. Virol.* **2001**, *75*, 469–479. [CrossRef]

125. Yeh, C.; Lin, P.; Liao, M.; Liu, H.; Lee, J.; Chiu, S.; Hsu, H.; Shih, W. TNF-alpha mediates pseudorabies virus-induced apoptosis via the activation of p38 MAPK and JNK/SAPK signaling. *Virology* **2008**, *381*, 55–66. [CrossRef]

126. Chang, C.; Lin, P.; Liao, M.; Chang, C.; Hsu, J.; Yu, F.; Wu, H.; Shih, W. Suppression of apoptosis by pseudorabies virus Us3 protein kinase through the activation of PI3-K/Akt and NF-κB pathways. *Res. Vet. Sci.* **2013**, *95*, 764–774. [CrossRef]

127. Deruelle, M.; Geenen, K.; Nauwynck, H.; Favoreel, H. A point mutation in the putative ATP binding site of the pseudorabies virus US3 protein kinase prevents Bad phosphorylation and cell survival following apoptosis induction. *Virus Res.* **2007**, *128*, 65–70. [CrossRef]

128. Deruelle, M.; De Corte, N.; Englebienne, J.; Nauwynck, H.; Favoreel, H. Pseudorabies virus US3-mediated inhibition of apoptosis does not affect infectious virus production. *J. Gen. Virol.* **2010**, *91*, 1127–1132. [CrossRef]

129. Mizushima, N.; Komatsu, M. Autophagy: Renovation of cells and tissues. *Cell* **2011**, *147*, 728–741. [CrossRef]

130. Klionsky, D.; Emr, S. Autophagy as a regulated pathway of cellular degradation. *Science* **2000**, *290*, 1717–1721. [CrossRef]

131. Tallóczy, Z.; Jiang, W.; Virgin, H.; Leib, D.; Scheuner, D.; Kaufman, R.; Eskelinen, E.; Levine, B. Regulation of starvation- and virus-induced autophagy by the eIF2alpha kinase signaling pathway. *Proc. Natl. Acad. Sci. USA* **2002**, *99*, 190–195. [CrossRef] [PubMed]

132. Orvedahl, A.; Alexander, D.; Tallóczy, Z.; Sun, Q.; Wei, Y.; Zhang, W.; Burns, D.; Leib, D.; Levine, B. HSV-1 ICP34.5 confers neurovirulence by targeting the Beclin 1 autophagy protein. *Cell Host Microbe* **2007**, *1*, 23–35. [CrossRef] [PubMed]

133. Wilcox, D.; Wadhwani, N.; Longnecker, R.; Muller, W. Differential reliance on autophagy for protection from HSV encephalitis between newborns and adults. *PLoS Pathog.* **2015**, *11*, e1004580. [CrossRef] [PubMed]

134. Heinz, J.; Kennedy, P.; Mogensen, T. The Role of Autophagy in Varicella Zoster Virus Infection. *Viruses* **2021**, *13*, 53. [CrossRef]

135. Lussignol, M.; Esclatine, A. Herpesvirus and Autophagy: "All Right, Everybody Be Cool, This Is a Robbery!". *Viruses* **2017**, *9*, 372. [CrossRef] [PubMed]

136. Xu, C.; Wang, M.; Song, Z.; Wang, Z.; Liu, Q.; Jiang, P.; Bai, J.; Li, Y.; Wang, X. Pseudorabies virus induces autophagy to enhance viral replication in mouse neuro-2a cells in vitro. *Virus Res.* **2018**, *248*, 44–52. [CrossRef]

137. Sun, M.; Hou, L.; Tang, Y.; Liu, Y.; Wang, S.; Wang, J.; Shen, N.; An, T.; Tian, Z.; Cai, X. Pseudorabies virus infection inhibits autophagy in permissive cells in vitro. *Sci. Rep.* **2017**, *7*, 39964. [CrossRef]

138. Antoku, K.; Maser, R.; Scully, W.; Delach, S.; Johnson, D. Isolation of Bcl-2 binding proteins that exhibit homology with BAG-1 and suppressor of death domains protein. *Biochem. Biophys. Res. Commun.* **2001**, *286*, 1003–1010. [CrossRef]

139. Doong, H.; Vrailas, A.; Kohn, E. What's in the 'BAG'?–A functional domain analysis of the BAG-family proteins. *Cancer Lett.* **2002**, *188*, 25–32. [CrossRef]

140. Lyu, C.; Li, W.; Wang, S.; Peng, J.; Yang, Y.; Tian, Z.; Cai, X. Host BAG3 Is Degraded by Pseudorabies Virus pUL56 C-Terminal L-L and Plays a Negative Regulation Role during Viral Lytic Infection. *Int. J. Mol. Sci.* **2020**, *21*, 3148. [CrossRef]

141. Xing, Y.; Wang, L.; Xu, G.; Guo, S.; Zhang, M.; Cheng, G.; Liu, Y.; Liu, J. Platycodon grandiflorus polysaccharides inhibit Pseudorabies virus replication via downregulating virus-induced autophagy. *Res. Vet. Sci.* **2021**, *140*, 18–25. [CrossRef] [PubMed]

142. Ming, S.; Zhang, S.; Wang, Q.; Zeng, L.; Zhou, L.; Wang, M.; Ma, Y.; Han, L.; Zhong, K.; Zhu, H.; et al. Inhibition of USP14 influences alphaherpesvirus proliferation by degrading viral VP16 protein via ER stress-triggered selective autophagy. *Autophagy* **2021**, 1–21. [CrossRef] [PubMed]

143. Verchot, J. How does the stressed out ER find relief during virus infection? *Curr. Opin. Virol.* **2016**, *17*, 74–79. [CrossRef] [PubMed]

144. Zhang, P.; Su, C.; Jiang, Z.; Zheng, C. Herpes Simplex Virus 1 UL41 Protein Suppresses the IRE1/XBP1 Signal Pathway of the Unfolded Protein Response via Its RNase Activity. *J. Virol.* **2017**, *91*, e02056-16. [CrossRef] [PubMed]

145. Yang, S.; Zhu, J.; Zhou, X.; Wang, H.; Li, X.; Zhao, A. Induction of the unfolded protein response (UPR) during pseudorabies virus infection. *Vet. Microbiol.* **2019**, *239*, 108485. [CrossRef]

146. Burnett, H.F.; Audas, T.E.; Liang, G.; Lu, R.R. Herpes simplex virus-1 disarms the unfolded protein response in the early stages of infection. *Cell Stress Chaperones* **2012**, *17*, 473–483. [CrossRef]

147. White, J.; Lloyd, R. Regulation of stress granules in virus systems. *Trends Microbiol.* **2012**, *20*, 175–183. [CrossRef]

148. Holcik, M.; Sonenberg, N. Translational control in stress and apoptosis. *Nat. Rev. Mol. Cell Biol.* **2005**, *6*, 318–327. [CrossRef]

149. Khong, A.; Matheny, T.; Jain, S.; Mitchell, S.; Wheeler, J.; Parker, R. The Stress Granule Transcriptome Reveals Principles of mRNA Accumulation in Stress Granules. *Mol. Cell* **2017**, *68*, 808–820.e805. [CrossRef]

150. Dang, Y.; Kedersha, N.; Low, W.; Romo, D.; Gorospe, M.; Kaufman, R.; Anderson, P.; Liu, J. Eukaryotic initiation factor 2alpha-independent pathway of stress granule induction by the natural product pateamine A. *J. Biol. Chem.* **2006**, *281*, 32870–32878. [CrossRef]

151. Nakagawa, K.; Narayanan, K.; Wada, M.; Makino, S. Inhibition of Stress Granule Formation by Middle East Respiratory Syndrome Coronavirus 4a Accessory Protein Facilitates Viral Translation, Leading to Efficient Virus Replication. *J. Virol.* **2018**, *92*, e00902-18. [CrossRef] [PubMed]

152. Xu, S.; Chen, D.; Chen, D.; Hu, Q.; Zhou, L.; Ge, X.; Han, J.; Guo, X.; Yang, H. Pseudorabies virus infection inhibits stress granules formation via dephosphorylating eIF2α. *Vet. Microbiol.* **2020**, *247*, 108786. [CrossRef] [PubMed]

153. Brukman, A.; Enquist, L.W. Pseudorabies virus EP0 protein counteracts an interferon-induced antiviral state in a species-specific manner. *J. Virol.* **2006**, *80*, 10871–10873. [CrossRef] [PubMed]

154. Van Lint, A.; Murawski, M.; Goodbody, R.; Severa, M.; Fitzgerald, K.; Finberg, R.; Knipe, D.; Kurt-Jones, E. Herpes simplex virus immediate-early ICP0 protein inhibits Toll-like receptor 2-dependent inflammatory responses and NF-kappaB signaling. *J. Virol.* **2010**, *84*, 10802–10811. [CrossRef]

Article

Pseudorabies Virus Inhibits Expression of Liver X Receptors to Assist Viral Infection

Yi Wang [1,2,3,†], Guo-Li Li [1,2,3,†], Yan-Li Qi [1,2,3], Li-Yun Li [1,2,3], Lu-Fang Wang [1,2,3], Cong-Rong Wang [1,2,3], Xin-Rui Niu [1,2,3], Tao-Xue Liu [1,2,3], Jiang Wang [1,2,3,†], Guo-Yu Yang [2,3,4,5], Lei Zeng [1,2,3,*] and Bei-Bei Chu [1,2,3,4,*]

[1] College of Veterinary Medicine, Henan Agricultural University, Zhengzhou 450046, China; anjxasmwy@sina.com (Y.W.); g2914026145@sina.com (G.-L.L.); lily1570304118@sina.com (Y.-L.Q.); liliyun20221186@yeah.net (L.-Y.L.); wanglufang1@aliyun.com (L.-F.W.); raewang616@outlook.com (C.-R.W.); niuxinrui2022@yeah.net (X.-R.N.); liutaoxue201516@yeah.net (T.-X.L.); wangjiang@henau.edu.cn (J.W.)

[2] Key Laboratory of Animal Biochemistry and Nutrition, Ministry of Agriculture and Rural Affairs of the People's Republic of China, Zhengzhou 450046, China; yangguoyu@henau.edu.cn

[3] Key Laboratory of Animal Growth and Development, Zhengzhou 450046, China

[4] International Joint Research Center of National Animal Immunology, Henan Agricultural University, Zhengzhou 450046, China

[5] College of Animal Science & Technology, Henan University of Animal Husbandry and Economy, Zhengzhou 450047, China

* Correspondence: zenglei2021918@outlook.com (L.Z.); chubeibei@henau.edu.cn (B.-B.C.)

† These authors contributed equally to this work.

Abstract: Pseudorabies virus (PRV) is a contagious herpesvirus that causes Aujeszky's disease and economic losses worldwide. Liver X receptors (LXRs) belong to the nuclear receptor superfamily and are critical for the control of lipid homeostasis. However, the role of LXR in PRV infection has not been fully established. In this study, we found that PRV infection downregulated the mRNA and protein levels of LXRα and LXRβ in vitro and in vivo. Furthermore, we discovered that LXR activation suppressed PRV proliferation, while LXR inhibition promoted PRV proliferation. We demonstrated that LXR activation-mediated reduction of cellular cholesterol was critical for the dynamics of PRV entry-dependent clathrin-coated pits. Replenishment of cholesterol restored the dynamics of clathrin-coated pits and PRV entry under LXR activation conditions. Interestingly, T0901317, an LXR agonist, prevented PRV infection in mice. Our results support a model that PRV modulates LXR-regulated cholesterol metabolism to facilitate viral proliferation.

Keywords: pseudorabies virus; Liver X receptors; clathrin-coated pits; viral entry

Citation: Wang, Y.; Li, G.-L.; Qi, Y.-L.; Li, L.-Y.; Wang, L.-F.; Wang, C.-R.; Niu, X.-R.; Liu, T.-X.; Wang, J.; Yang, G.-Y.; et al. Pseudorabies Virus Inhibits Expression of Liver X Receptors to Assist Viral Infection. *Viruses* **2022**, *14*, 514. https://doi.org/10.3390/v14030514

Academic Editors: Yan-Dong Tang and Xiangdong Li

Received: 21 January 2022
Accepted: 25 February 2022
Published: 3 March 2022

Publisher's Note: MDPI stays neutral with regard to jurisdictional claims in published maps and institutional affiliations.

1. Introduction

Pseudorabies (PR), also called Aujeszky's disease, is a highly infectious disease caused by the PR virus (PRV), which is a member of the subfamily Alphaherpesvirinae of the family Herpesviridae. The genome of PRV is approximately 143 kb encoding at least 70 open reading frames [1]. PRV can infect a wide variety of mammals, including pigs, sheep, and cattle, causing severe clinical symptoms and death [2]. Although scientists have been trying to develop diagnostic approaches and vaccines in recent years, PR remains an important infectious disease that is prevalent in many countries. Several recent reports have suggested that PRV can cause human endophthalmitis and encephalitis [3–5]. These findings indicate that PRV infection is a potential public health risk and not limited to the swine industry. Therefore, new methods are urgently needed to prevent PRV infection.

The liver X receptors (LXRs), a family of transcription factors in the nuclear receptor superfamily, are ligand-activated transcription factors and pivotal regulators of cholesterol and lipid metabolism [6,7]. Two LXR subtypes, LXRα and LXRβ, have been identified. While LXRβ is expressed ubiquitously, LXRα is expressed highly in the liver, spleen, intestine, heart, and macrophages [8]. Ligand binding to LXR results in the formation of a heterodimer between LXR and the retinoid X receptor (RXR). The LXR/RXR complex then binds to the LXR-response elements in the promoter, initiating the transcription of target genes [9,10]. The natural ligands of LXR have been identified as oxysterols, such as 22(R)-hydroxycholesterol (22R-HC). LXR plays important roles in lipid metabolism. A number of small molecules of LXR agonists and reverse agonists, such as LXR-623, T0901317, GW3965, and SR9243, have been developed to treat lipid disorders and atherosclerosis in clinical trials [11].

Lipids are essential components for cellular and viral membranes. It has been revealed that lipids are required for the entire life cycle of a virus, including attachment, entry, genome replication, assembly, and release [12]. LXR is required for efficient replication of a number of viruses, including Newcastle disease virus (NDV) [13], human immunodeficiency virus (HIV) [14], hepatitis C virus (HCV) [15], hepatitis B virus (HBV) [16], murine gammaherpesvirus 68 [17], coxsackie B3 virus [18], and chikungunya virus [19]. However, the role of LXR in regulating PRV replication has not been documented. Here, we examined the effects of LXR on PRV replication. We demonstrated that LXR activation, which led to reduced cellular cholesterol levels, suppressed PRV proliferation. LXR agonists, such as T0901317, dramatically decreased PRV entry via interference with the cholesterol-dependent dynamics of clathrin-coated pits (CCPs). Our data suggest that LXR agonists have potential as antivirals for the control of PRV infection.

2. Materials and Methods

2.1. Mice

We purchased female 6–8-week-old BALB/c mice from the Center of Experimental Animal of Zhengzhou University (Zhengzhou, China). Mice were housed in a specific pathogen-free animal facility at Henan Agricultural University. Animal experiments were performed in accordance with protocols approved by the Use of National Research Center for Veterinary Medicine (Permit 20180521047).

2.2. Cells, Viruses, and Plasmids

Porcine kidney epithelial PK-15 (CCL-33, ATCC), porcine alveolar macrophages 3D4/21 (CRL-2843, ATCC), and human cervical cancer HeLa (CL-82, ATCC) cells were grown in monolayers at 37 °C under 5% CO_2 in DMEM (Gibco, Waltham, MA, USA) supplemented with 10% FBS (Gibco), 100 U/mL penicillin, and 100 μg/mL streptomycin sulfate (Sangon, Shanghai, China). Virus titers were determined by the 50% tissue culture infective dose ($TCID_{50}$) assay, which was calculated with the Reed–Muench method.

The virulent PRV isolate QXX (PRV-QXX) was kindly donated by Yong-Tao Li from the College of Veterinary Medicine, Henan Agricultural University [20]. The recombinant PRV strain of PRV-GFP, derived from the PRV Hubei strain with the TK gene replaced by a GFP expression cassette from the pEGFP-N1 plasmid, was kindly donated by Han-Zhong Wang from Wuhan Institute of Virology, Chinese Academy of Sciences [21].

Full-length porcine AP2B1 cDNA was cloned into the mCherry-N1 expression plasmid using the BamHI and KpnI restriction sites.

2.3. Chemicals and Antibodies

Cholesterol, T0901317, GW3965, LXR-623, and SR9243 were ordered from MedChemExpress (Monmouth Junction, NJ, USA). Filipin complex and 22-R-hydroxycholesterol were ordered from Sigma-Aldrich (St. Louis, MO, USA). Anti-LXRα, anti-LXRβ, anti-AP2B, and anti-β-actin were ordered from Proteintech (Rosemont, IL, USA); anti-ABCA1 was ordered from Novus Biologicals (Littleton, CO, USA); Alexa-Fluor-488-conjugated goat anti-mouse IgG, Alexa-Fluor-568-conjugated goat anti-mouse IgG, and Alexa-Fluor-568-conjugated goat anti-rabbit IgG were ordered from Thermo Fisher Scientific (Waltham, MA, USA). Antiserum against PRV glycoprotein gB and gE was generated by immunization of mice with purified recombinant gB and gE.

2.4. Cell Viability Assays

PK-15 cells were seeded at 1×10^4 per well in 96-well plates. On the next day, the medium was changed to DMEM/10% FBS supplemented with various concentrations of LXR agonists for 24–48 h. CCK-8 (10 μL, DingGuo, Beijing, China) was then added to each well, and the cells were incubated for 3 h at 37 °C. Absorbance was detected at 450 nm with a microplate reader (Varioskan Flash; Thermo Fisher Scientific).

2.5. Flow Cytometry Assay

For green fluorescent protein (GFP) reporter assays, PK-15 cells were infected with PRV-GFP [multiplicity of infection (MOI) = 0.01] for 36 h. Cells were digested with trypsin-EDTA (Gibco), collected by centrifugation, and suspended in PBS. The percentage of GFP-positive cells was measured by flow cytometry on a Beckman CytoFLEX instrument (Brea, CA, USA). All data were analyzed with CytExpert software 2.0.

2.6. Immunoblotting Analysis

Cells were collected by centrifugation and lysed in RIPA buffer (Beyotime, Shanghai, China) in the presence of protease and phosphatase inhibitor cocktail (MedChemExpress). The protein concentration was determined by a BCA Protein Assay Kit (DingGuo). Equivalent amounts of total protein (30 μg) were subjected to SDS-PAGE for immunoblotting. The target proteins were detected with specific primary antibodies and appropriate horseradish peroxidase (HRP)-conjugated secondary antibodies. Visualization was performed with Luminata Crescendo Western HRP Substrate (Millipore, Billerica, MA, USA) on a GE AI600 imaging system.

2.7. Cell Surface Biotinylation Assay

Cells were incubated in ice-cold PBS-CM (0.1 mM $CaCl_2$, 1 mM $MgCl_2$) in the presence of 1 mg/mLEZ-Link NHS-biotin (Thermo Fisher Scientific) for 30 min at 4 °C. After two washes with ice-cold PBS-CM, cells were cultured in medium containing 10 mM glycine for 30 min at 37 °C. Cytosolic and membrane fractions were prepared as previously described [22]. Briefly, the cells were homogenized by passing through a #7 needle 30 times in 0.5 mL of homogenization buffer (10 mM HEPES [pH 7.4], 10 mM KCl, 1.5 mM $MgCl_2$, 5 mM sodium EDTA, 5 mM sodium EGTA, 250 mM sucrose) supplemented with protease and phosphatase inhibitors (MedChemExpress) and centrifuged at $1000 \times g$ at 4 °C for 7 min. The pellet, containing the crude nuclear fraction, was discarded, and the supernatant was centrifuged at $12,000 \times g$ at 4 °C for 15 min. The resulting supernatant was the cytosol, and the pellet containing the membrane fraction was dissolved in lysis buffer (10 mM Tris-HCl [pH 6.8], 100 mM NaCl, 1% SDS, 1 mM EDTA, 1 mM EGTA) supplemented with protease and phosphatase inhibitors (MedChemExpress). Each fraction was incubated with NeutrAvidin-agarose (Thermo Fisher Scientific) and rotated for 2 h at 4 °C. After three washes with homogenization buffer (10 mM HEPES pH 7.4, 10 mM KCl, 1.5 mM $MgCl_2$, 5 mM EDTA, 5 mM EGTA, 250 mM sucrose), biotinylated proteins on NeutrAvidin–agarose were eluted with SDS-PAGE sample buffer and subjected to immunoblotting analysis.

2.8. Quantitative Real-Time PCR (qRT-PCR)

Total RNA was extracted using TRIzol Reagent (TaKaRa, Shiga, Japan) and then reverse-transcribed with a PrimeScript RT reagent Kit (TaKaRa). qRT-PCR was performed in triplicate using SYBR Premix Ex Taq (TaKaRa). Data were normalized to the expression of the control gene encoding *β-actin*. The relative expression changes were calculated by the The $2^{-\Delta\Delta CT}$ method. Quantification of the genome copy number of PRV was performed as previously described [23]. Primers used for qRT-PCR analysis were as follows: porcine *β-actin*-Fw: 5′-GCACAGAGCCTCGCCTT-3′, porcine *β-actin*-Rv: 5′-CCTTGCACATGCCGGAG-3′; porcine *Lxra*-Fw: 5′-CGTCCACTCAGAGCAAGTGT-3′, porcine *Lxra*-Rv: 5′-CAGATCTCAGAGAGCAGCGG-3′; porcine *Lxrb*-Fw: 5′-ACGCTACAACCACGAGACAG-3′, porcine *Lxrb*-Rv: 5′- CGGTGGAAGTCATCCTTGCT-

3'; porcine *Abca1*-Fw: 5'-ATGGATCACTGCCCCAGTTC-3', porcine *Abca1*-Rv: 5'-ATGTCCGCGGTGTTCTGTTT-3'; porcine *Abcg1*-Fw: 5'-GTGTACTGGATGACGTCGCA-3', porcine *Abcg1*-Rv: 5'-CGAAGCTGACGAAGAACCCT-3'; PRV *gB*-Fw: 5'-CTCGCCATCGTCAGCAAPRV-3', PRV *gB*-Rv: 5'-GCTGCTCCTCCATGTCCTT-3'; mouse *β-actin*-Fw: 5'-CCCCATTGAACATGGCATTG-3', mouse *β-actin*-Rv: 5'-ACGACCAGAGGCATACAGG-3'; mouse *Lxrα*-Fw: 5'-CTGATTCTGCAACGGAGTTGT-3', mouse *Lxrα*-Rv: 5'-GACGAAGCTCTGTCGGCTC-3'; mouse *Lxrb*-Fw: 5'-GCCTGGGAATGGTTCTCCTC-3', mouse *Lxrb*-Rv: 5'-AGATGACCACGATGTAGGCAG-3'.

2.9. RNA Interference (RNAi)

Cells were transfected with the indicated siRNAs (GenePharma, Shanghai, China) using Lipofectamine RNAiMAX Reagent (Invitrogen, Waltham, MA, USA) according to the manufacturer's instructions. The medium was replaced with DMEM containing 10% FBS at 8 h post-transfection. The knockdown efficacy was assessed by immunoblotting analysis at 48 h post-transfection. The siRNA sequences were as follows: negative control (NC): 5'-UUCUCCGAACGUGUCACGU-3'; siLXRα: 5'-CCCACGGAUGCUAAUGAAAUU-3'; siLXRβ: 5'-UCCCGCGAAUGCUGAUGAAUU-3'.

2.10. Fluorescence Recovery after Photobleaching (FRAP)

HeLa cells were transfected with the AP2B1-mCherry plasmid for 24 h. After 8 h of pre-treatment with the indicated compound, the cells were bleached at maximum laser intensity for 30 s in a region of 7×7 μm^2 and then imaged for 5 min at 37 °C. CCP dynamics was defined by the fluorescence recovery of AP2B1-mCherry, which was performed on a Zeiss LSM 800 confocal microscope.

2.11. Histological Analysis

Animal tissues were fixed in 4% paraformaldehyde and embedded in paraffin. The paraffin blocks were sectioned (7 μm) for hematoxylin-eosin staining.

2.12. Statistical Analysis

All data were obtained from three independent experiments for quantitative analyses and expressed as the mean ± standard error. All statistical analyses were performed with a two-tailed Student's *t* test. Significant differences relative to the corresponding controls were accepted at * $p < 0.05$. For mouse survival studies, Kaplan–Meier survival curves were generated and analyzed for statistical significance.

3. Results

3.1. PRV Infection Inhibits LXRα and LXRβ Expression In Vitro and In Vivo

To determine the role of LXR in PRV infection, we evaluated the expression of LXR under PRV challenge in vitro. Cells were infected with PRV-QXX for 0–24 h, and cells were processed to measure the mRNA and protein levels of LXRα and LXRβ. PRV infection caused a reduction in *Lxra* and *Lxrb* mRNA in PK-15 and 3D421 cells (Figure 1A,B). Consistent with the mRNA levels, LXRα and LXRβ protein levels in PK-15 and 3D421 cells were all downregulated in response to PRV infection (Figure 1C,D). These results suggested that PRV inhibited LXRα and LXRβ expression in vitro.

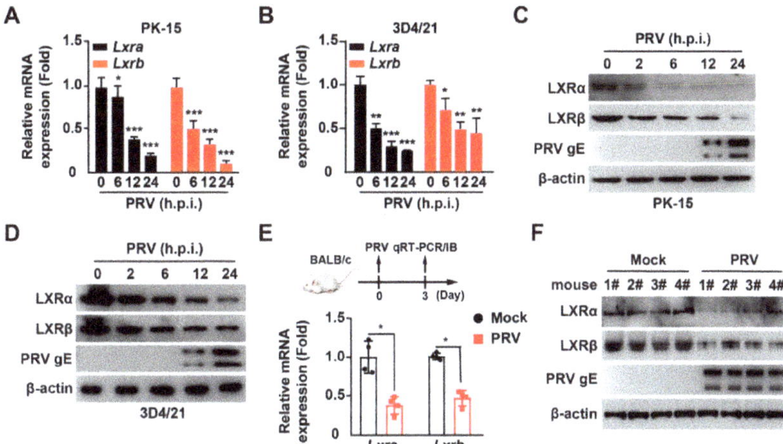

Figure 1. PRV infection downregulates LXR expression. (**A,B**) PK-15 (**A**) and 3D4/21 (**B**) cells were infected with PRV-QXX (MOI = 0.1) for 0–24 h. The mRNA levels of *Lxra* and *Lxrb* were assessed by qRT-PCR analysis. * $p < 0.05$, ** $p < 0.01$, *** $p < 0.001$. (**C,D**) PK-15 (**C**) and 3D4/21 (**D**) cells were infected with PRV-QXX (MOI = 0.1) for 0–24 h. LXRα, LXRβ, and PRV gE were assessed by immunoblotting analysis. (**E**) Mice were mock infected or intranasally infected with PRV-QXX (5×10^3 TCID$_{50}$/50 μL per mouse) for 3 days. The mRNA levels of *Lxra* and *Lxrb* in the lung were assessed by qRT-PCR analysis ($n = 4$ per group). * $p < 0.05$. (**F**) Mice were treated as in G. LXRα, LARβ, and PRV gE in the lung were assessed by immunoblotting analysis ($n = 4$ per group).

We verified whether PRV reduced LXR expression in vivo. Mice were mock infected or intranasally infected with PRV-QXX for 3 days, and the lungs were assessed for mRNA and protein levels of LXRα and LXRβ by qRT-PCR and immunoblotting analysis. PRV infection resulted in a >two-fold decrease in *Lxra* and *Lxrb* mRNAs as compared to that in mock-infected lungs, as well as the protein levels of LXRα and LXRβ (Figure 1E,F). These data indicated that PRV infection suppressed LXR expression both in vitro and in vivo.

3.2. Inhibition of LXR Increases PRV Infection

We aimed to determine whether LXR was involved in PRV infection. SR9243 is an inverse agonist of LXR that induces LXR–co-repressor interaction and downregulates LXR-mediated gene expression [24]. As expected, treatment of PK-15 cells with SR9243 inhibited the transcription of LXR target genes, such as *Abca1* and *Abcg1* (Figure 2A). The mRNA levels of PRV *gB* in SR9243-treated cells were significantly higher than those in control cells (Figure 2B). This indicated that SR9243 promoted transcription of PRV genes. PRV gE expression was enhanced in an SR9243 dose-dependent manner (Figure 2C). We next detected the multiplication of PRV progeny virus in response to SR9243 using a viral titer assay. PK-15 cells were infected with PRV-QXX (MOI = 0.1 and 1.0) and treated with SR9243 (0–10 μM) for 24 h. SR9243 significantly promoted the production of PRV progeny virus (Figure 2D). To gain further insight into the effect of SR9243 on PRV infection, we assessed the growth kinetics of PRV under SR9243 treatment. TCID$_{50}$ assay of viral titer indicated that SR9243 enhanced the production of the PRV progeny virus at 8 h post-treatment (Figure 2E). We also determined whether knockdown of LXRα and LXRβ improved PRV infection. Simultaneous interference LXRα and LXRβ expressions increased the production of PRV progeny virus (Figure 2F,C). These data demonstrated that inhibition of LXR benefited PRV infection.

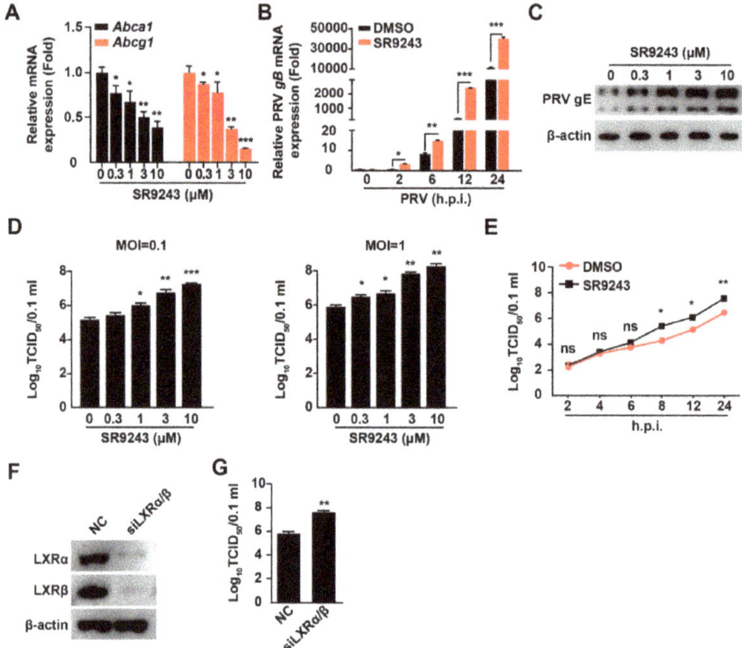

Figure 2. LXR inverse agonist SR9243 promotes PRV proliferation. (**A**) PK-15 cells were treated with SR9243 (0–10 μM) for 24 h. *Abca1* and *Abcg1* mRNA was assessed by qRT-PCR analysis. * $p < 0.05$, ** $p < 0.01$, *** $p < 0.001$. (**B**) PK-15 cells were infected with PRV-QXX (MOI = 0.1) and simultaneously treated with DMSO or SR9243 (10 μM) for 0–24 h. PRV *gB* mRNA was assessed by qRT-PCR analysis. * $p < 0.05$, ** $p < 0.01$, *** $p < 0.001$. (**C**) PK-15 cells were infected with PRV-QXX (MOI = 0.1) and simultaneously treated with SR9243 (0–10 μM) for 24 h. PRV gE was assessed by immunoblotting analysis. (**D**) PK-15 cells were infected with PRV-QXX (MOI = 0.1 and 1) and simultaneously treated with SR9243 (0–10 μM) for 24 h. Viral titers were assessed by a $TCID_{50}$ assay. * $p < 0.05$, ** $p < 0.01$, *** $p < 0.001$. (**E**) PK-15 cells were infected with PRV-QXX (MOI = 0.1) and simultaneously treated with SR9243 (10 μM) for 2–24 h. One-step growth curves of PRV-QXX were assessed using a $TCID_{50}$ assay of viral titers. * $p < 0.05$, ** $p < 0.01$. ns, no significance. (**F**) PK-15 cells were transfected with NC and siLXRα/β for 48 h. LXRα and LXRβ were assessed by immunoblotting analysis. (**G**) PK-15 cells were transfected with NC and siLXRα/β. At 24 h post-transfection, cells were infected with PRV-QXX (MOI = 1) for another 24 h. Viral titers were assessed by a $TCID_{50}$ assay. ** $p < 0.01$.

3.3. Activation of LXR by Their Agonists Inhibits PRV Infection

To confirm the negative role of LXR in PRV replication, we utilized four agonists of LXR (LXR-623, T0901317, 22R-HC, and GW3965) [25]. We first performed cell viability assays to examine the cytotoxicity of LXR agonists, and 20–60 μM of LXR-623 and T0901317 was harmful to PK-15 cells at 36–48 h post treatment (Figure 3A). 22R-HC (20 μM) and GW3965 (10 μM) showed cytotoxicity at 24–48 h post treatment (Figure 3A). LXR-623, T0901317, 22R-HC, and GW3965 resulted in decreased PRV-GFP proliferation, as indicated by flow cytometry analysis of GFP-positive cells (Figure 3B). We verified the inhibitory effect of LXR agonists on PRV infection by a viral titer assay. PK-15 cells were infected with PRV-QXX (MOI = 0.1 and 1) and treated with LXR-623 (0–6 μM), T0901317 (0–6 μM), 22R-HC (0–6 μM), and GW3965 (0–3 μM) for 24 h. Multiplication of the PRV progeny virus decreased with an increased concentration of LXR agonists (Figure 3C). PRV gB and gE expression was inhibited by LXR agonists (Figure 3D). The growth kinetics of PRV assessed by $TCID_{50}$ assay of viral titer indicated that LXR agonists decreased the production of PRV progeny virus at 8 h post-treatment (Figure 3E). These data suggested that LXR played a negative role in PRV infection.

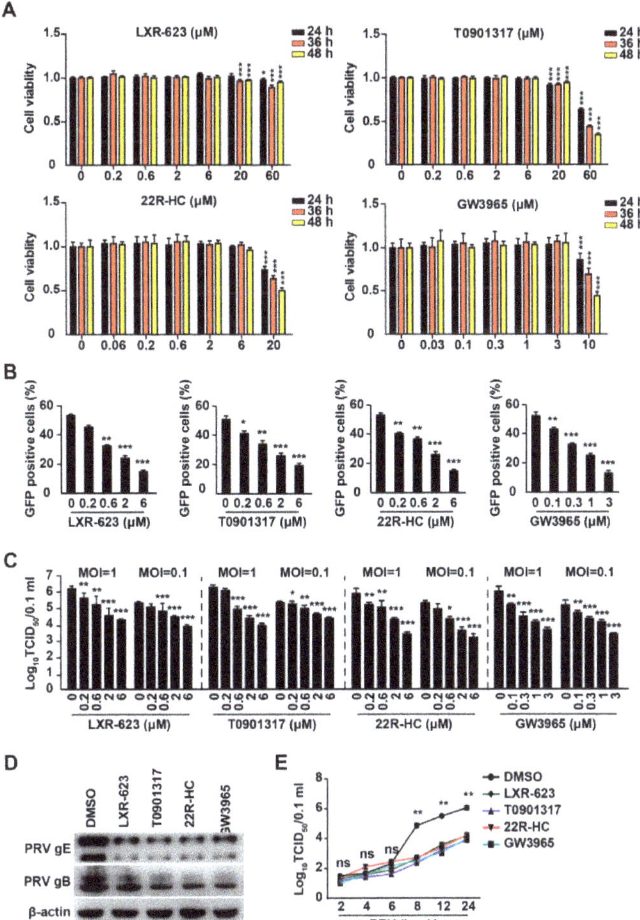

Figure 3. LXR agonists inhibit PRV infection. (**A**) PK-15 cells were treated with LXR-623, T0901317, 22R-HC, and GW3965 at indicated concentrations for 24–48 h. Cell viability was assessed with CCK-8 cell counting assays. * $p < 0.05$, ** $p < 0.01$, *** $p < 0.001$. (**B**) PK-15 cells were infected with PRV-GFP (MOI = 0.01) and simultaneously treated with LXR-623 (0–6 μM), T0901317 (0–6 μM), 22R-HC (0–6 μM), and GW3965 (0–3 μM) for 36 h. GFP-positive cells were analyzed by flow cytometry. * $p < 0.05$, ** $p < 0.01$, *** $p < 0.001$. (**C**) PK-15 cells were infected with PRV-QXX (MOI = 0.1 and 1) and simultaneously treated with LXR-623 (0–6 μM), T0901317 (0–6 μM), 22R-HC (0–6 μM) and GW3965 (0–3 μM) for 24 h. Viral titers were assessed by a TCID$_{50}$ assay. * $p < 0.05$, ** $p < 0.01$, *** $p < 0.001$. (**D**) PK-15 cells were infected with PRV-QXX (MOI = 0.1) and simultaneously treated with DMSO, LXR-623 (6 μM), T0901317 (6 μM), 22R-HC (6 μM), and GW3965 (3 μM) for 24 h. PRV gB and gE were assessed by immunoblotting analysis. (**E**) PK-15 cells were infected with PRV-QXX (MOI = 0.1) and simultaneously treated with LXR-623 (6 μM), T0901317 (6 μM), 22R-HC (6 μM) and GW3965 (3 μM) for 2–24 h. One-step growth curves of PRV-QXX were assessed using a TCID$_{50}$ assay of viral titers. ** $p < 0.01$. ns, no significance.

3.4. Activation of LXR Inhibits PRV Entry

Next, we sought to determine which stage of viral life cycle was influenced by LXR agonists and inverse agonists in a time-of-addition assay. We examined whether activation of LXR could influence PRV attachment to cells. We pretreated PK-15 cells with LXR agonists for 8 h, and infected cells with PRV-QXX combined with LXR agonists for 1 h at 4 °C. After three washes with ice-cold PBS, we analyzed viral attachment by quantification of the PRV genome copy number by qRT-PCR analysis. LXR-623, T0901317, 22R-HC, and GW3965 did not affect PRV attachment to cells (Figure 4A). We next performed a viral entry assay by quantification of the PRV genome copy number in cells. qRT-PCR analysis indicated that LXR agonists inhibited PRV entry (Figure 4B). We also assessed PRV entry by immunoblotting analysis of PRV gE. PRV gE was decreased in cells treated with LXR agonists, which further indicated that LXR agonists inhibited PRV entry (Figure 4C).

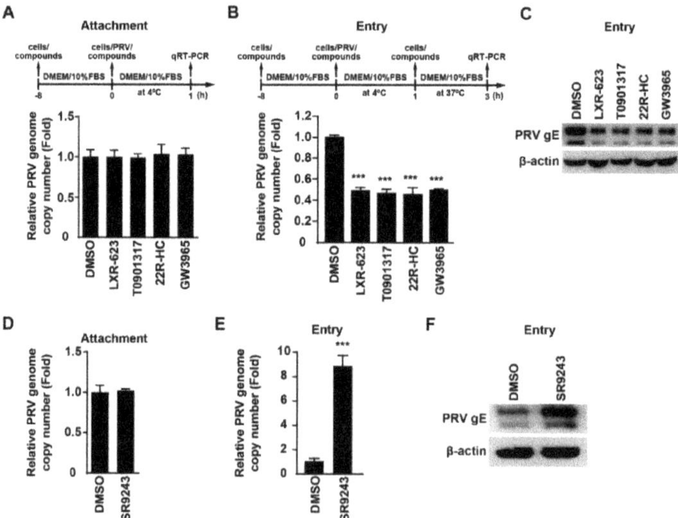

Figure 4. Activation of LXR inhibits PRV entry. (**A**) PK-15 cells were pretreated with DMSO, LXR-623 (6 μM), T0901317 (6 μM), 22R-HC (6 μM) and GW3965 (3 μM) for 8 h at 37 °C. Cells were incubated with PRV-QXX (MOI = 0.1) combined with DMSO, LXR-623 (6 μM), T0901317 (6 μM), 22R-HC (6 μM), and GW3965 (3 μM) for 1 h at 4 °C. After three washes with ice-cold PBS, the viral genome was isolated. PRV genome copy numbers on cells were assessed by qRT-PCR analysis. (**B**) PK-15 cells were pretreated with DMSO, LXR-623 (6 μM), T0901317 (6 μM), 22R-HC (6 μM), and GW3965 (3 μM) for 8 h at 37 °C. Cells were incubated with PRV-QXX (MOI = 0.1) combined with DMSO, LXR-623 (6 μM), T0901317 (6 μM), 22R-HC (6 μM), and GW3965 (3 μM) for 1 h at 4 °C. After three washes with ice-cold PBS, cells were cultured in prewarmed medium containing DMSO, LXR-623 (6 μM), T0901317 (6 μM), 22R-HC (6 μM), and GW3965 (3 μM) for 2 h at 37 °C. PRV genome copy numbers in cells were assessed by qRT-PCR analysis. *** $p < 0.001$. (**C**) PK-15 cells were treated as in (**B**). PRV gE in cells was assessed by immunoblotting analysis. (**D**) PK-15 cells were pretreated with DMSO and SR9243 (10 μM) for 8 h at 37 °C. Cells were then incubated with PRV-QXX (MOI = 0.1) combined with DMSO and SR9243 (10 μM) for 1 h at 4 °C. After three washes with ice-cold PBS, the viral genome was isolated. PRV genome copy numbers on cells were assessed by qRT-PCR analysis. (**E**) PK-15 cells were pretreated with DMSO and SR9243 (10 μM) for 8 h at 37 °C. Cells were incubated with PRV-QXX (MOI = 0.1) combined with DMSO and SR9243 (10 μM) for 1 h at 4 °C. After three washes with ice-cold PBS, cells were cultured in prewarmed medium containing DMSO and SR9243 (10 μM) for 2 h at 37 °C. PRV genome copy numbers in cells were assessed by qRT-PCR analysis. *** $p < 0.001$. (**F**) PK-15 cells were infected and treated as in (**E**). PRV gE in cells was assessed by immunoblotting analysis.

In addition, we used LXR inverse agonists to examine whether SR9243 could promote PRV attachment and entry. qRT-PCR analysis indicated that SR9243 had no inhibitory effect on PRV attachment to cells, but it could promote PRV entry (Figure 4D,E). Immunoblotting analysis of PRV gE indicated that SR9243 increased gE in SR9243-treated cells, suggesting that SR9243 boosted PRV entry (Figure 4F). These results demonstrated that LXR was related to PRV entry.

3.5. PRV Infection Increases Cellular Cholesterol Content That Is Inhibited by LXR Activation

Cholesterol is critical for PRV entry [26], so we examined cellular cholesterol content by filipin staining in PRV-infected and T0901317-treated cells. PRV infection significantly increased cellular cholesterol content (Figure 5A). However, activation of LXR by T0901317 abrogated PRV-induced enhancement of cellular cholesterol, which was restored by cholesterol replenishment (Figure 5A). This phenomenon was verified by quantification of cellular cholesterol (Figure 5B). We examined whether cholesterol replenishment rescued PRV entry during LXR activation. qRT-PCR indicated that PRV genome copy number in T0901317-treated PK-15 cells was gradually increased with the concentration of cholesterol (Figure 5C). Exogenous supplementation of cholesterol in T0901317-treated PK-15 cells restored the internalization of PRV gE, as indicated by immunoblotting analysis (Figure 5D). These data suggested that LXR influenced cellular cholesterol to inhibit PRV entry.

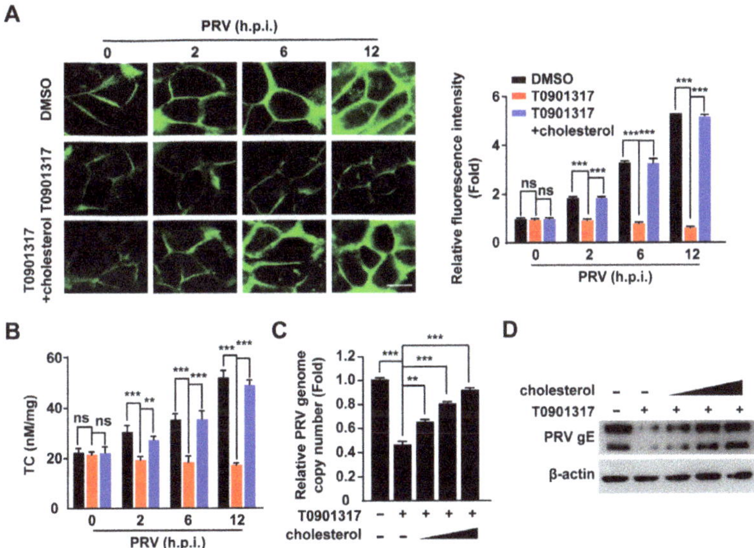

Figure 5. Activation of LXR decreases cellular cholesterol to inhibit PRV entry. (**A**) PK-15 cells were infected with PRV-QXX (MOI = 0.1) and simultaneously treated with DMSO, T0901317 (6 μM), and T0901317 (6 μM) + cholesterol (0.003 μg/mL) for 0–12 h. Cholesterol was detected by filipin staining (left). Quantification of the relative fluorescence intensity of filipin is shown on the right. *** $p < 0.001$. ns, no significance. Scale bar: 10 μm. (**B**) PK-15 cells were treated as in (**A**). Quantification of cellular cholesterol was performed by biochemical determination. ** $p < 0.01$, *** $p < 0.001$. ns, no significance. (**C**) PK-15 cells were incubated with PRV-QXX (MOI = 0.1) for 1 h at 4 °C and then in medium containing T0901317 (6 μM) and cholesterol (0, 0.0003, 0.001, and 0.003 μg/mL) as indicated for 2 h at 37 °C. PRV genome copy numbers in cells were assessed by qRT-PCR analysis ** $p < 0.01$, *** $p < 0.001$. (**D**) PK-15 cells were infected and treated as in (**C**). PRV gE in cells was assessed by immunoblotting analysis.

3.6. T0901317 Inhibit CCP Dynamics through Reducing Cellular Cholesterol

Our previous study suggested that Niemann–Pick C1 deficiency attenuates PRV entry by decreasing cholesterol abundance and by inhibiting CCP dynamics [26]. Therefore, we examined whether LXR agonists acted via a similar mechanism. To corroborate the role of T0901317 in CCP dynamics, FRAP analysis was carried out with live-cell confocal microscopy imaging. Following the 20-s bleaching laser pulse, AP2B1-mCherry fluorescence rapidly recovered to about 50% of its initial value at approximately 180 s in control cells, while T0901317-treated cells did not show recovery of AP2B1-mCherry fluorescence (Figure 6A,B). Exogenous cholesterol was able to complement the inhibitory effect of T0901317 on the recovery of AP2B1-mCherry fluorescence (Figure 6A,B). Additionally, using a cell surface biotinylation assay, we confirmed that T0901317 had no effect on the interaction between PRV virions and AP2B1, but inhibited viral entry, which could be restored by the addition of cholesterol (Figure 6C). These data demonstrated that activation of LXR blocked PRV entry by interfering with cholesterol-dependent CCP dynamics.

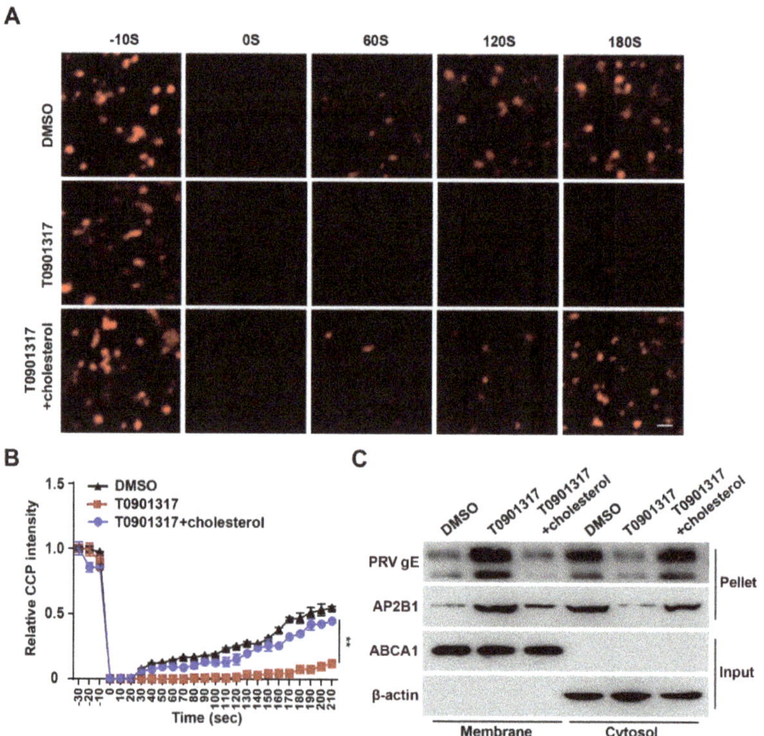

Figure 6. T0901317 blocks CCP dynamics-dependent viral entry. (**A**) HeLa cells were transfected with AP2B1-mCherry plasmid for 24 h followed by treatment with DMSO, T0901317 (6 μM) and T0901317 (6 μM) + cholesterol (0.003 μg/mL) for a further 8 h. The CCP dynamics were assessed by FRAP analysis. Scale bar: 1 μm. (**B**) Quantification of the relative fluorescent intensity of AP2B1 puncta in the FRAP region over time from (**A**) ($n = 10$). ** $p < 0.01$. (**C**) PK-15 cells were incubated with PRV-QXX (MOI = 0.1) for 1 h at 4 °C and then in medium containing DMSO T0901317 (6 μM) and T0901317 (6 μM) + cholesterol (0.003 μg/mL) at 37 °C. The internalization of PRV gE and AP2B1 was assessed by cell surface biotinylation assay after viral entry for 15 min.

3.7. T0901317 Prevents PRV Infection In Vivo

To determine whether LXR agonists can be used as antivirals in vivo, we examined the protective effect of T0901317 against PRV infection. Mice were intraperitoneally injected with T0901317 twice every 2 days before the PRV challenge. Mice were intranasally infected with PRV-QXX for 10 days. We observed that all mice died at 4 days post-infection in the vehicle-treated group (Figure 7A). However, 92% of mice (10/12) survived in the T0901317-treated group (Figure 7A). Analysis of PRV gE by immunoblotting and immunohistochemistry showed that T0901317 decreased PRV gE expression in lungs, suggesting that T0901317 inhibited PRV proliferation in vivo (Figure 7B,C). Less infiltration of inflammatory cells was observed in the lungs in T0901317-treated mice than in vehicle-treated mice (Figure 7D). All the results indicated that activation of LXR prevented PRV infection in vivo.

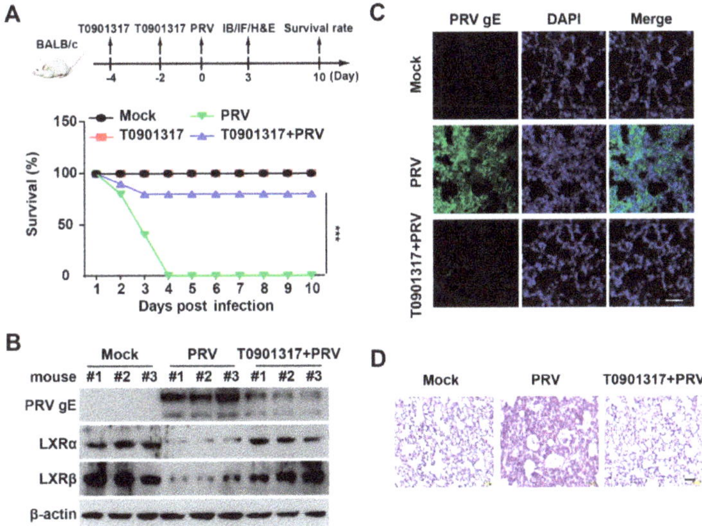

Figure 7. T0901317 inhibits PRV infection in vivo. (**A**) Mice were intraperitoneally injected with vehicle or T0901317 (30 mg/kg) on days 4 and 2. On day 0, the mice were mock infected or intranasally infected with PRV-QXX (5×10^3 TCID$_{50}$/50 μL per mouse). The survival rate was monitored daily for 10 days ($n = 12$ per group). *** $p < 0.001$. (**B**) PRV gE, LXRα, and LXRβ in the lungs were assessed by immunoblotting analysis at 3 days post-infection ($n = 3$ per group). (**C**) PRV gE in the lungs was assessed by immunofluorescence at 3 days post-infection ($n = 3$ per group). Scale bar: 100 μm. (**D**) The lung injury was assessed by H&E staining at 3 days post-infection ($n = 3$ per group). Scale bar: 100 μm.

4. Discussion

Viruses can manipulate lipid metabolism to facilitate their replication [27]. During infection, some viruses induce changes in cell membrane structures or utilize lipid synthetic enzymes to build a suitable microenvironment for different stages of the infection [12]. LXR controls cellular lipid homeostasis [28]. Oxysterols are endogenous ligands of LXR [29], and they promote HBV gene expression through activation of LXR [30]. Murine gammaherpesvirus 68 infection increases LXR expression, and there is not a corresponding increase in LXR target genes [17]. NDV infection activates LXR and its downstream lipogenic gene expression [13]. In this study, we found that PRV infection downregulated LXR expression both in vitro and in vivo, and LXR activation inhibited PRV infection. Our data indicated that different viruses used LXR for optimal replication through diverse mechanisms throughout their entire life cycle. Other physiological functions of LXR may be required for virus replication, as well as their roles in lipid metabolism.

It has been demonstrated that synthetic LXR agonists restrict replication of NDV [13], HCV [15], vector-borne flaviviruses [31], chikungunya virus, and HIV [14], by altering cholesterol homeostasis. Recent evidence suggests that activation of LXR can induce cholesterol 25-hydroxylase (CH25H) mRNA and protein expression [32]. 25-hydroxycholesterol (25HC) is the enzymatic product of CH25H and exerts broad antiviral functions by inhibiting viral entry [33,34]. We have indicated that porcine CH25H acts as a host restriction factor on PRV infection by 25HC [33], so we speculated that LXR-activated CH25H expression may be responsible for inhibiting PRV infection.

Cholesterol is a critical component that determines membrane fluidity and architecture. For some viruses, cholesterol plays an essential role in the process of virus entry into cells, such as PRV [33], foot-and-mouth disease virus [35], and human rhinovirus type 2 [36], and depletion of cholesterol significantly inhibits viral entry and infection. Our data indicated that PRV enhanced cellular cholesterol levels, which were abrogated by LXR agonists. We further demonstrated that LXR activation decreased cellular cholesterol levels to inhibit PRV entry-dependent CCP dynamics, which could be rescued by cholesterol replenishment. Our study demonstrated a mechanism by which PRV disturbed LXR expression to promote viral entry through modulation of cholesterol homeostasis. This is in accordance with our previous report that cholesterol is critical for CCP dynamics in PRV entry [26]. These data provide novel insights into the prevention and control of diverse viruses that require cholesterol-regulated CCP dynamics for viral entry.

5. Conclusions

Viruses hijack cellular metabolism for their optimal replication. A better understanding of the interaction between virus and cellular metabolism will provide insights into an antiviral strategy. Lipids are the key component of cellular membrane compartments that participate in viral replication. We previously reported that Niemann–Pick C1-mediated intracellular cholesterol transport is essential for CCP dynamics and for viral entry. Here, we reported that PRV modulated cholesterol metabolism by downregulating the expression of LXR to assist viral entry through clathrin-mediated endocytosis. Therefore, our data further validated that cholesterol-regulated CCP dynamics are pivotal for PRV infection, suggesting that pharmacological reduction of cellular cholesterol levels has the potential to prevent PRV infection.

Author Contributions: Conceptualization, B.-B.C. and L.Z.; methodology, Y.W., G.-L.L. and Y.-L.Q.; software, L.-Y.L., L.-F.W. and C.-R.W.; validation, X.-R.N. and T.-X.L.; formal analysis, G.-Y.Y.; investigation, Y.W. and G.-L.L.; data curation, J.W.; writing—original draft preparation, B.-B.C.; writing—review and editing, B.-B.C.; supervision, B.-B.C. and L.Z.; funding acquisition, B.-B.C. All authors have read and agreed to the published version of the manuscript.

Funding: This work was supported by grants from the National Natural Science Foundation of China (32072858), the Ten Thousand Talents Program for Young Talents (W03070106), and Outstanding Talents of Henan Agricultural University (30600773).

Institutional Review Board Statement: Not applicable.

Data Availability Statement: All available data are presented in the article.

Conflicts of Interest: The authors declare that they have no conflict of interest.

References

1. Pomeranz, L.E.; Reynolds, A.E.; Hengartner, C.J. Molecular biology of pseudorabies virus: Impact on neurovirology and veterinary medicine. *Microbiol. Mol. Biol. Rev.* **2005**, *69*, 462–500. [CrossRef] [PubMed]
2. Wozniakowski, G.; Samorek-Salamonowicz, E. Animal herpesviruses and their zoonotic potential for cross-species infection. *Ann. Agric. Environ. Med.* **2015**, *22*, 191–194. [CrossRef]
3. Ai, J.W.; Weng, S.S.; Cheng, Q.; Cui, P.; Li, Y.J.; Wu, H.L.; Zhu, Y.M.; Xu, B.; Zhang, W.H. Human Endophthalmitis Caused By Pseudorabies Virus Infection, China, 2017. *Emerg. Infect. Dis.* **2018**, *24*, 1087–1090. [CrossRef]
4. Yang, X.; Guan, H.; Li, C.; Li, Y.; Wang, S.; Zhao, X.; Zhao, Y.; Liu, Y. Characteristics of human encephalitis caused by pseudorabies virus: A case series study. *Int. J. Infect. Dis.* **2019**, *87*, 92–99. [CrossRef] [PubMed]
5. Wong, G.; Lu, J.; Zhang, W.; Gao, G.F. Pseudorabies virus: A neglected zoonotic pathogen in humans? *Emerg. Microbes Infect.* **2019**, *8*, 150–154. [CrossRef]

6. Lewis, G.F.; Rader, D.J. New insights into the regulation of HDL metabolism and reverse cholesterol transport. *Circ. Res.* **2005**, *96*, 1221–1232. [CrossRef] [PubMed]

7. Wang, B.; Tontonoz, P. Liver X receptors in lipid signalling and membrane homeostasis. *Nat. Rev. Endocrinol.* **2018**, *14*, 452–463. [CrossRef]

8. Chen, M.; Beaven, S.; Tontonoz, P. Identification and characterization of two alternatively spliced transcript variants of human liver X receptor alpha. *J. Lipid Res.* **2005**, *46*, 2570–2579. [CrossRef]

9. Repa, J.J.; Turley, S.D.; Lobaccaro, J.A.; Medina, J.; Li, L.; Lustig, K.; Shan, B.; Heyman, R.A.; Dietschy, J.M.; Mangelsdorf, D.J. Regulation of absorption and ABC1-mediated efflux of cholesterol by RXR heterodimers. *Science* **2000**, *289*, 1524–1529. [CrossRef]

10. Costet, P.; Luo, Y.; Wang, N.; Tall, A.R. Sterol-dependent transactivation of the ABC1 promoter by the liver X receptor/retinoid X receptor. *J. Biol. Chem.* **2000**, *275*, 28240–28245. [CrossRef]

11. Hong, C.; Tontonoz, P. Liver X receptors in lipid metabolism: Opportunities for drug discovery. *Nat. Rev. Drug Discov.* **2014**, *13*, 433–444. [CrossRef] [PubMed]

12. Lorizate, M.; Krausslich, H.G. Role of lipids in virus replication. *Cold Spring Harb. Perspect. Biol.* **2011**, *3*, a004820. [CrossRef] [PubMed]

13. Sheng, X.X.; Sun, Y.J.; Zhan, Y.; Qu, Y.R.; Wang, H.X.; Luo, M.; Liao, Y.; Qiu, X.S.; Ding, C.; Fan, H.J.; et al. The LXR ligand GW3965 inhibits Newcastle disease virus infection by affecting cholesterol homeostasis. *Arch. Virol.* **2016**, *161*, 2491–2501. [CrossRef]

14. Jiang, H.; Badralmaa, Y.; Yang, J.; Lempicki, R.; Hazen, A.; Natarajan, V. Retinoic acid and liver X receptor agonist synergistically inhibit HIV infection in CD4+ T cells by up-regulating ABCA1-mediated cholesterol efflux. *Lipids Health Dis.* **2012**, *11*, 69. [CrossRef]

15. Bocchetta, S.; Maillard, P.; Yamamoto, M.; Gondeau, C.; Douam, F.; Lebreton, S.; Lagaye, S.; Pol, S.; Helle, F.; Plengpanich, W.; et al. Up-regulation of the ATP-binding cassette transporter A1 inhibits hepatitis C virus infection. *PLoS ONE* **2014**, *9*, e92140. [CrossRef]

16. Zeng, J.; Wu, D.; Hu, H.; Young, J.A.T.; Yan, Z.; Gao, L. Activation of the Liver X Receptor Pathway Inhibits HBV Replication in Primary Human Hepatocytes. *Hepatology* **2020**, *72*, 1935–1948. [CrossRef]

17. Lange, P.T.; Schorl, C.; Sahoo, D.; Tarakanova, V.L. Liver X Receptors Suppress Activity of Cholesterol and Fatty Acid Synthesis Pathways To Oppose Gammaherpesvirus Replication. *mBio* **2018**, *9*, e01115-18. [CrossRef]

18. Papageorgiou, A.P.; Heggermont, W.; Rienks, M.; Carai, P.; Langouche, L.; Verhesen, W.; De Boer, R.A.; Heymans, S. Liver X receptor activation enhances CVB3 viral replication during myocarditis by stimulating lipogenesis. *Cardiovasc. Res.* **2015**, *107*, 78–88. [CrossRef] [PubMed]

19. Hwang, J.; Wang, Y.; Fikrig, E. Inhibition of Chikungunya Virus Replication in Primary Human Fibroblasts by Liver X Receptor Agonist. *Antimicrob. Agents Chemother.* **2019**, *63*, e01220-19. [CrossRef]

20. Li, Y.; Chang, H.; Yang, X.; Zhao, Y.; Chen, L.; Wang, X.; Liu, H.; Wang, C.; Zhao, J. Antiviral Activity of Porcine Interferon Regulatory Factor 1 against Swine Viruses in Cell Culture. *Viruses* **2015**, *7*, 5908–5918. [CrossRef]

21. Xu, N.; Zhang, Z.F.; Wang, L.; Gao, B.; Pang, D.W.; Wang, H.Z.; Zhang, Z.L. A microfluidic platform for real-time and in situ monitoring of virus infection process. *Biomicrofluidics* **2012**, *6*, 34122. [CrossRef]

22. Wang, J.; Chu, B.; Du, L.; Han, Y.; Zhang, X.; Fan, S.; Wang, Y.; Yang, G. Molecular cloning and functional characterization of porcine cyclic GMP-AMP synthase. *Mol. Immunol.* **2015**, *65*, 436–445. [CrossRef] [PubMed]

23. Wang, J.; Li, G.L.; Ming, S.L.; Wang, C.F.; Shi, L.J.; Su, B.Q.; Wu, H.T.; Zeng, L.; Han, Y.Q.; Liu, Z.H.; et al. BRD4 inhibition exerts anti-viral activity through DNA damage-dependent innate immune responses. *PLoS Pathog.* **2020**, *16*, e1008429. [CrossRef] [PubMed]

24. Griffett, K.; Solt, L.A.; El-Gendy Bel, D.; Kamenecka, T.M.; Burris, T.P. A liver-selective LXR inverse agonist that suppresses hepatic steatosis. *ACS Chem. Biol.* **2013**, *8*, 559–567. [CrossRef]

25. Xu, P.; Li, D.; Tang, X.; Bao, X.; Huang, J.; Tang, Y.; Yang, Y.; Xu, H.; Fan, X. LXR agonists: New potential therapeutic drug for neurodegenerative diseases. *Mol. Neurobiol.* **2013**, *48*, 715–728. [CrossRef]

26. Li, G.; Su, B.; Fu, P.; Bai, Y.; Ding, G.; Li, D.; Wang, J.; Yang, G.; Chu, B. NPC1-regulated dynamic of clathrin-coated pits is essential for viral entry. *Sci. China Life Sci.* **2022**, *65*, 341–361. [CrossRef]

27. Ketter, E.; Randall, G. Virus Impact on Lipids and Membranes. *Annu. Rev. Virol.* **2019**, *6*, 319–340. [CrossRef] [PubMed]

28. Oosterveer, M.H.; Grefhorst, A.; Groen, A.K.; Kuipers, F. The liver X receptor: Control of cellular lipid homeostasis and beyond Implications for drug design. *Prog. Lipid Res.* **2010**, *49*, 343–352. [CrossRef] [PubMed]

29. Edwards, P.A.; Kennedy, M.A.; Mak, P.A. LXRs: Oxysterol-activated nuclear receptors that regulate genes controlling lipid homeostasis. *Vascul. Pharmacol.* **2002**, *38*, 249–256. [CrossRef]

30. Kim, H.Y.; Cho, H.K.; Kim, H.H.; Cheong, J. Oxygenated derivatives of cholesterol promote hepatitis B virus gene expression through nuclear receptor LXRalpha activation. *Virus Res.* **2011**, *158*, 55–61. [CrossRef]

31. Mlera, L.; Offerdahl, D.K.; Dorward, D.W.; Carmody, A.; Chiramel, A.I.; Best, S.M.; Bloom, M.E. The liver X receptor agonist LXR 623 restricts flavivirus replication. *Emerg. Microbes Infect.* **2021**, *10*, 1378–1389. [CrossRef] [PubMed]

32. Liu, Y.; Wei, Z.; Ma, X.; Yang, X.; Chen, Y.; Sun, L.; Ma, C.; Miao, Q.R.; Hajjar, D.P.; Han, J.; et al. 25-Hydroxycholesterol activates the expression of cholesterol 25-hydroxylase in an LXR-dependent mechanism. *J. Lipid Res.* **2018**, *59*, 439–451. [CrossRef]

33. Wang, J.; Zeng, L.; Zhang, L.; Guo, Z.Z.; Lu, S.F.; Ming, S.L.; Li, G.L.; Wan, B.; Tian, K.G.; Yang, G.Y.; et al. Cholesterol 25-hydroxylase acts as a host restriction factor on pseudorabies virus replication. *J. Gen. Virol.* **2017**, *98*, 1467–1476. [CrossRef]

34. Liu, S.Y.; Aliyari, R.; Chikere, K.; Li, G.; Marsden, M.D.; Smith, J.K.; Pernet, O.; Guo, H.; Nusbaum, R.; Zack, J.A.; et al. Interferon-inducible cholesterol-25-hydroxylase broadly inhibits viral entry by production of 25-hydroxycholesterol. *Immunity* **2013**, *38*, 92–105. [CrossRef] [PubMed]
35. Martin-Acebes, M.A.; Gonzalez-Magaldi, M.; Sandvig, K.; Sobrino, F.; Armas-Portela, R. Productive entry of type C foot-and-mouth disease virus into susceptible cultured cells requires clathrin and is dependent on the presence of plasma membrane cholesterol. *Virology* **2007**, *369*, 105–118. [CrossRef] [PubMed]
36. Snyers, L.; Zwickl, H.; Blaas, D. Human rhinovirus type 2 is internalized by clathrin-mediated endocytosis. *J. Virol.* **2003**, *77*, 5360–5369. [CrossRef] [PubMed]

Article

A Nectin1 Mutant Mouse Model Is Resistant to Pseudorabies Virus Infection

Xiaohui Yang [1,2], Chuanzhao Yu [1,2], Qiuyan Zhang [1,2], Linjun Hong [1,2], Ting Gu [1,2], Enqin Zheng [1,2], Zheng Xu [1,2], Zicong Li [1,2], Changxu Song [1,2], Gengyuan Cai [1,2], Zhenfang Wu [1,2,*] and Huaqiang Yang [1,2,*]

[1] National Engineering Research Center for Breeding Swine Industry, College of Animal Science, South China Agricultural University, Guangzhou 510642, China; yangxiaohui@stu.scau.edu.cn (X.Y.); yuchuanzhao@stu.scau.edu.cn (C.Y.); qiuyan@stu.scau.edu.cn (Q.Z.); linjun.hong@scau.edu.cn (L.H.); tinggu@scau.edu.cn (T.G.); eqzheng@scau.edu.cn (E.Z.); stonezen@scau.edu.cn (Z.X.); lizicong@scau.edu.cn (Z.L.); cxsong@scau.edu.cn (C.S.); cgy0415@scau.edu.cn (G.C.)

[2] Guangdong Provincial Key Laboratory of Agro-Animal Genomics and Molecular Breeding, South China Agricultural University, Guangzhou 510642, China

* Correspondence: wzf@scau.edu.cn (Z.W.); yangh@scau.edu.cn (H.Y.)

Abstract: The present study generated nectin1-mutant mice with single amino acid substitution and tested the anti-pseudorabies virus (PRV) ability of the mutant mice, with the aim to establish a model for PRV-resistant livestock. A phenylalanine to alanine transition at position 129 (F129A) of nectin1 was introduced into the mouse genome to generate nectin1 (F129A) mutant mice. The mutant mice were infected with a field-isolated highly virulent PRV strain by subcutaneous injection of virus. We found that the homozygous mutant mice had significantly alleviated disease manifestations and decreased death rate and viral loading in serum and tissue compared with heterozygous mutant and wild-type mice. In addition to disease resistance, the homozygous mutant mice showed a defect in eye development, indicating the side effect on animals by only one amino acid substitution in nectin1. Results demonstrate that gene modification in nectin1 is an effective approach to confer PRV resistance on animals, but the mutagenesis pattern requires further investigation to increase viral resistance without negative effect on animal development.

Keywords: antiviral breeding; genetic modification; nectin1; pig; PRV; disease resistance

Citation: Yang, X.; Yu, C.; Zhang, Q.; Hong, L.; Gu, T.; Zheng, E.; Xu, Z.; Li, Z.; Song, C.; Cai, G.; et al. A Nectin1 Mutant Mouse Model Is Resistant to Pseudorabies Virus Infection. *Viruses* **2022**, *14*, 874. https://doi.org/10.3390/v14050874

Academic Editors: Yan-Dong Tang and Xiangdong Li

Received: 21 March 2022
Accepted: 20 April 2022
Published: 22 April 2022

Publisher's Note: MDPI stays neutral with regard to jurisdictional claims in published maps and institutional affiliations.

1. Introduction

Pseudorabies virus (PRV) is an economically important pathogen causing severe losses to pig production. PRV causes reproductive and respiratory problems in breeding and finishing pigs and central nervous system signs and high mortality in piglets [1–4]. Although the natural reservoir is pig, PRV has a broad host range and can lethally infect a wide variety of animal species [5–9]. PRV infection in pigs is controlled using inactivated and attenuated live vaccines [10]. However, PRV outbreak in vaccinated pig populations still occur widely in many countries mainly due to variations of virus, usually compromising vaccine effectiveness and increasing virulence to pigs [1,11,12].

PRV belongs to *herpesvirinae* family and *alphaherpesvirinae* subfamily [13,14]. The representative alphaherpesviruses, such as PRV and herpes simplex virus (HSV), share some common molecular machinery and mechanism to infect host cells [15,16]. The host membrane-bound protein, nectin1, plays a pivotal role in alphaherpesvirus infection. Global nectin1 gene knockout in mice can confer resistance to HSV1 and HSV2 infection, resulting in a milder disease after HSV1 and HSV2 infection in mice [17–21]. Our previous work found that nectin1 knockout in pig cells reduce PRV growth by impairing cell-to-cell spread step of virus [22]. In this respect, nectin1 gene modification in animals provides a practical route for PRV antiviral breeding given that genetic engineering approach can be used for pig breeding.

We established a mouse model carrying specific mutation in nectin1 to investigate the anti-PRV effect of nectin1 gene modification in animals. Previous work reported that the extracellular N-terminal variable region-like (V) domain of nectin1, which interacts with viral glycoprotein D (gD), is important for alphaherpesvirus infection. The critical residues in V domain for PRV-gD/nectin1 engagement include N77, I80, M85, R110, and F129 [23,24]. Amino acid substitution in some of the residues can severely reduce PRV-gD/nectin1 binding and PRV entry activity. The in vitro cell-based results showed that mutation in F129 has fewer side effects than other key residues on homotypic and heterotypic nectin–nectin interactions, which are important for cell adhesion [23]. Considering these factors, we introduced an F129A single amino acid substitution in mouse nectin1 gene. The homozygous F129A mutant mice were studied with respect to their infection levels in PRV challenge experiments. We also investigated the impact of mutation on the physiologic function of nectin1 in vivo.

2. Materials and Methods

2.1. Animals

The animals and procedures in this study were in accordance with the guidelines and approval of the Institutional Animal Care and Use Committees at South China Agricultural University (approval ID: 2020B032). We used CRISPR-mediated homology-directed repair method for amino acid substitution in mouse genome. The guide RNA (gRNA) sequence targeting mouse nectin1 gene was ACGGTTGCCCGTAGGGAAGG, and a donor oligo containing F129A (TTC to GCT) mutation was designed as follows: ATCCGCCTCTCCGGTCTGGAGCTGGAGGACGAGGGCATGTACATCTGTGAATTTGC-CACC<u>GCT</u>CCTACGGGCAACCGTGAAAGCCAGCTCAATCTCACTGTGATGGGTAAG-CTGCCCTGGGCC; the mutation site is underlined. The gRNA, donor oligo, and Cas9 mRNA were co-injected into fertilized eggs of C57BL/6 mice to generate targeted knock-in offspring. F1 founder animals were identified by PCR followed by sequence analysis, and then bred to test germline transmission and generate F2 mutant animals.

2.2. Analysis of Nectin1 Expression

Mouse brains were minced and lysed in Pierce RIPA buffer (Thermo Fisher Scientific, Rockford, IL, USA) supplemented with protease inhibitor cocktail (Thermo Fisher Scientific, Rockford, IL, USA). Tissue lysate was collected by centrifugation and quantified using Pierce BCA protein assay kit (Thermo Fisher Scientific, Rockford, IL, USA). An equal amount of lysate was boiled in loading buffer and subjected to SDS-PAGE. Subsequently, the separated protein in gel was transferred onto the PVDF membrane, which was then blocked in 5% milk and incubated with anti-nectin1 mouse monoclonal antibody (sc-21722, Santa Cruz Biotechnology, Dallas, TX, USA) at 4 °C overnight. The membrane was washed and further incubated with HRP-conjugated goat-mouse IgG secondary antibody. After thorough washing, the target protein was imaged using a SuperSignal West Pico enhanced chemiluminescence kit (Thermo Fisher Scientific, Carlsbad, CA, USA). The same blots were probed with a β-actin rabbit monoclonal antibody (13E5, Cell Signaling Technology, Danvers, MA, USA) as a loading control.

2.3. Viral Culture and Titration

A field-isolated PRV strain was propagated and titered in PK15 cells [22]. Virus diluted in DMEM was inoculated in PK15 cell monolayer and cultured to reach full cytopathic effect (CPE) in 5% CO_2 incubator at 37 °C. The culture was collected and aliquoted as viral stock. To examine the viral titers, TCID50 assay was performed on PK15 cells grown in 96-well plates. Virus was serially diluted from 10^{-1} to 10^{-10} in DMEM, and 0.1 mL of viral dilution was added per well; eight wells were infected per dilution. Virus was allowed to adsorb to cells for 2 h, followed by replacing the viral dilution with 0.1 mL of fresh DMEM for each well. The plates were cultured in 5% CO_2 incubator at 37 °C for monitoring CPE for one week. The titer was calculated using the method of Muench and Reed [25].

2.4. Viral Infection in Mice

Homozygous mutant mice and age-matched controls (wild-type (WT) and heterozygous) were injected subcutaneously in the neck at a single dose of 100 µL virus prediluted to the indicated titers ($10^{5.43}$ and $10^{7.42}$ TCID50 for two viral challenge experiments). Signs of disease and survival rate of infected mice were recorded for 96 h. The symptoms of PRV infection were scored using a 3-point system: 0 = normal posture; 1 = attempt to scratch or slight scratching; 2 = frequent scratching and abnormal posture such as hunchback; and 3 = scratching with biting and bleeding of the wound. All mice were sacrificed at 96 h to collect serum and various tissues for viral titer quantification. Viral DNA in serum and brain tissues in the same amount was extracted with RaPure Viral RNA/DNA Kit (Magen, Guangzhou, China) and subjected to quantitative PCR (qPCR) by using Premix ExTaq (Probe qPCR) (Takara, Dalian, China) and the following primers and probe, PRV-gE-Forward, CCCACCGCCACAAAGAACACG, PRV-gE-Reverse, GATGGGCATCG-GCGACTACCTG, and PRV-gE-Probe, FAM-CAGCGCGAGCCGCCCATCGTCAC-BHQ1 in a QuantStudio 7 Flex Real-Time PCR System (Thermo Fisher Scientific, Foster City, CA, USA). A PRV gE gene plasmid was serially diluted as templates to generate a reference curve between Ct value and DNA copies in the same qPCR reaction of viral DNA samples. Viral DNA was quantified using the standard curve and expressed as genome equivalents (GE) in extracted DNA solution.

2.5. Histological Analysis

The eyes of mice were collected and soaked in FAS eye fixative (Servicebio, Wuhan, China) for 24 h at room temperature. The fixed tissues were dehydrated with gradient concentrations of ethanol, embedded in paraffin wax, and sectioned to tissue slice of 5 µm thickness. For Hematoxylin and Eosin (H&E) staining, the tissue sections were deparaffinized in the xylene and rehydrated by passing through decreasing concentrations of ethanol baths and water. The rehydrated tissue sections were stained in hematoxylin for 5 min at room temperature. The sections were rinsed in tap water and differentiated in 1% HCl in 70% alcohol for 5 min. After rinsing in tap water, the sections were treated with ammonia water to convert the hematoxylin to a dark blue color. The sections were then rinsed and stained in 1% Eosin Y for 10 min at room temperature. After staining, the sections were washed in tap water for 5 min, dehydrated in increasing concentrations of ethanol, cleared in xylene, and mounted in mounting media for microscopy assay.

2.6. Statistical Analysis

All statistical analyses were performed using Prism (GraphPad, v8, LaJolla, CA, USA). Mean ± standard deviation was calculated for replicate data. Means comparisons in body weight and viral copies were conducted using unpaired *t* tests or ANOVA, and survival rates were compared with Gehan–Breslow–Wilcoxon test, as indicated in the figure legends.

3. Results

3.1. Generation of Nectin1 (F129A) Mutant Mice

Nectin1 F129 residue is conserved across different mammalian species (Figure S1). Its functional significance for engagement of gD of multiple alphaherpesviruses has been found in nectin1 of mouse, pig, and cattle [23,24,26]. Nectin1 (F129A) mutant mice were produced by CRISPR-mediated knock-in using zygote injection of CRISPR system cleaving nectin1 and a DNA template donor carrying F129A mutation (TTC→GCT in DNA sequence) (Figure 1A). The founder mice were genotyped by PCR amplification of nectin1 target region and Sanger sequencing identifying the presence of mutation. The positive mice were bred to generate homozygous and heterozygous mutant offspring (Figure 1B). To detect if the nectin1 protein expression was affected by amino acid substitution, the brain tissues from WT, heterozygous, and homozygous mice were lysed to extract total protein. Western blot assay showed that all three genotypes of mice had nectin1 expression in the similar levels, indicating no impact of F129A mutation on protein expression in brains (Figure 1C). The homozygous mutant mice generally had similar growth rate to WT and heterozygous mutants within one month after birth. The increase in body weight of homozygous mutants was slightly slower than WT and heterozygous mutants after one month, and the difference was more significant in female mice (Figure 1D,E).

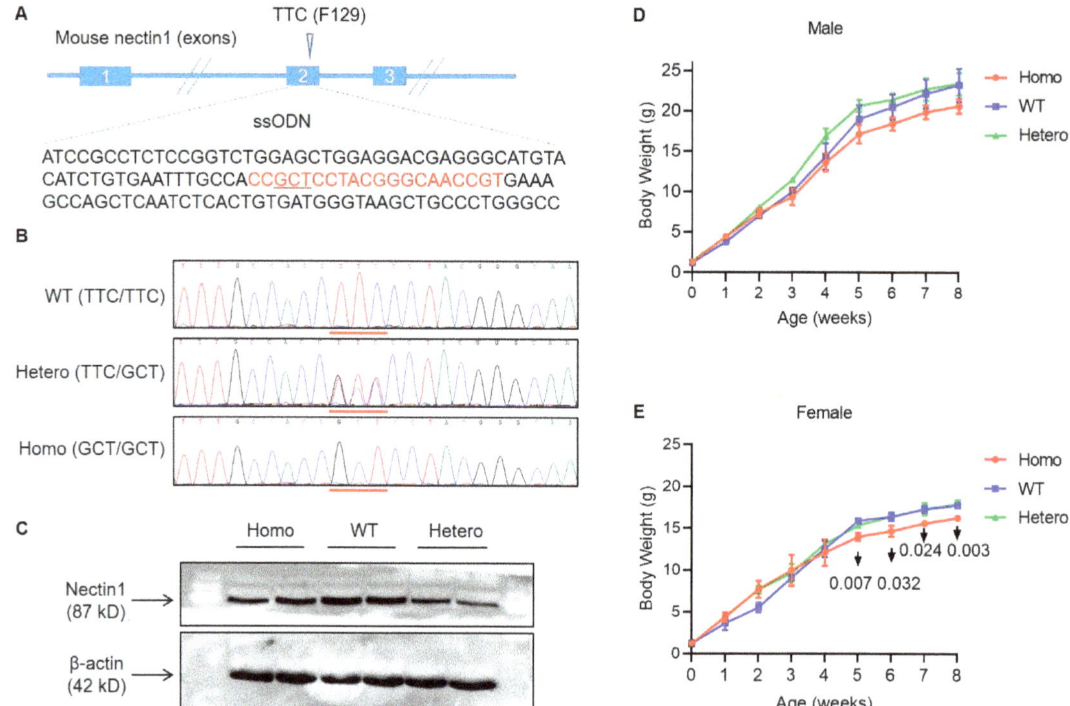

Figure 1. Characterization of nectin1 (F129A) mutant mice. (**A**) Gene targeting strategy to introduce F129A mutation in mouse nectin1. Sequences in red letters are gRNA-recognized target harboring F129A mutation (underlined sequences). (**B**) Sequences of nectin1 mutation site in homozygous, heterozygous, and WT mice. The underlined sequences are F129 mutation sites. (**C**) Nectin1 expression status in brain tissues of homozygous, heterozygous, and WT mice. (**D,E**) Growth curves of male (**D**) and female (**E**) F129A mutant mice. Female homozygous mutant mice had significantly reduced body weight compared with WT littermates at five, six, seven, and eight weeks. Data are presented in mean ± standard deviation. *p*-values represent homozygous versus WT and were analyzed with one-way ANOVA followed by Dunnett's multiple comparison test. ssODN, single-stranded oligodeoxynucleotides; Homo, homozygous nectin1 (F129A) mutant mice; Hetero, heterozygous nectin1 (F129A) mutant mice; WT, wild-type mice.

3.2. Anti-PRV Ability of Nectin1 (F129A) Mutant Mice

We conducted two viral challenge experiments to test the PRV infection status in the mutant mice. Prior to experiments, we evaluated the susceptibility of the mouse strain we used (C57BL/6) to PRV. Our pre-experiments showed that the attenuated PRV vaccine strain Bartha-K61 was mildly pathogenic to C57BL/6 mice, but a field-isolated highly virulent strain of PRV [22] can cause acute symptoms with a high mortality in 3–10 days after viral inoculation in different doses. Infected mice showed viremia and tissue lesion, and high viral load in infected tissues can be detected by qPCR (data not shown). The susceptibility of C57BL/6 mice to PRV has also been comprehensively investigated in previous reports [27–29]. Upon confirming the PRV susceptibility of mice, we performed experiment 1, in which seven WT and seven homozygous at six weeks were given subcutaneous injection with $1 \times 10^{5.43}$ TCID50 PRV field strain (Figure 2A). After PRV inoculation, three mice of each genotype were sacrificed to collected tissues and serum to detect tissue morphology and viral loads at 36 h. The four other mice in each group were maintained to observe disease development for 96 h. Among them, two WT mice had symptoms of scratching and body incoordination starting at 48 h after inoculation and died at 72 h and 96 h, respectively (Figures 2B,C and S2). All WT mice displayed acute itch symptom during the challenge period, but only two homozygous mutant mice had slight symptom starting at 84 h (Figure S2). qPCR results showed that viral copies in brain and serum of homozygous mice were lower than WT (Figure 2D).

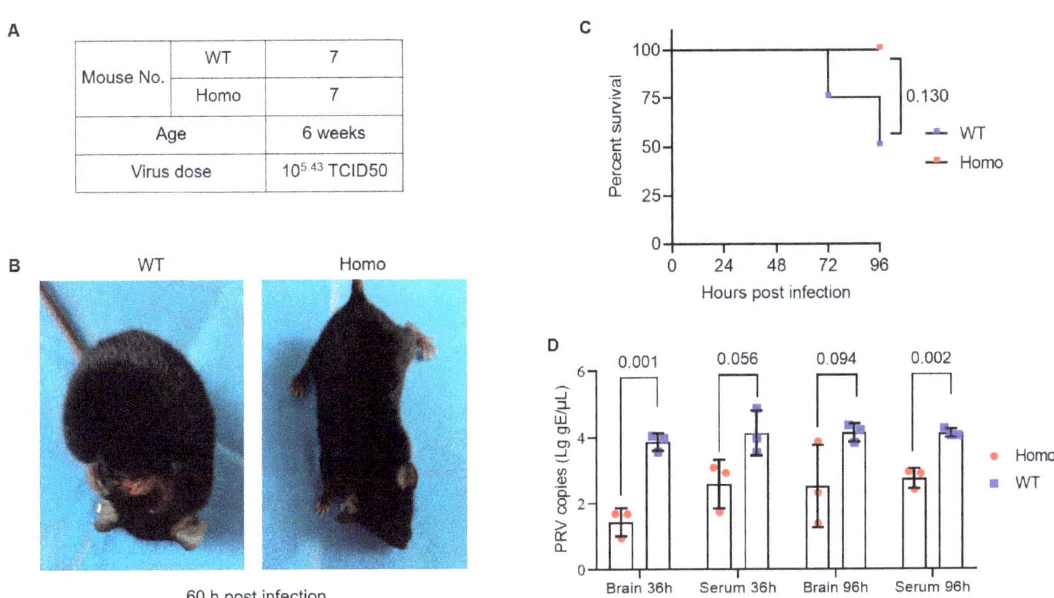

Figure 2. Viral challenge experiment 1 in nectin1 (F129A) mutant mice. (**A**) Experimental groups and virus dose for PRV challenge in mice. (**B**) Typical symptoms present in PRV-infected mice. A WT mouse showed body incoordination and severe wound in the neck because of scratching of the viral injection site at 60 h after challenge (left), whereas no symptoms were observed in nectin1 (F129A) homozygous mutant mouse (right). (**C**) Survival curve for nectin1 (F129A) homozygous mutant and WT mice after PRV challenge within 96 h. Four mice in each group were used for analysis of survival curve. The other three mice in each group were sacrificed to analyze viral infection level in various tissues at 36 h. (**D**) PRV viral loads at 36 h and 96 h in brain and serum by qPCR quantification of PRV gE gene. All data are presented in mean ± standard deviation. Statistically significant differences between mutant and WT groups are indicated by *p*-values, analyzed with Gehan–Breslow–Wilcoxon test and unpaired *t*-test for survival curve and viral copies, respectively. Homo, homozygous nectin1 (F129A) mutant mice; WT, wild-type mice.

We further repeated viral challenge experiment (experiment 2) by using a high dose of virus and including heterozygous mice as subjects. A total of nine WT, nine heterozygous, and seven homozygous at the age of eight weeks were subcutaneously injected with $1 \times 10^{7.42}$ TCID50 PRV field strain (Figure 3A). Follow-up observations showed that six WT and seven heterozygous mice displayed neurological symptoms, such as scratching and biting, starting from 36 h, and finally died within 96 h. In the homozygous group, one showed neurological symptoms at 72 h and died at 96 h; three had slight scratching behavior at 96 h (Figures 3B and S3). All infected mice were sacrificed at 96 h for serum and tissue sampling. Viral DNA extracted from half brain and serum in the same amount showed a significantly reduced viral copies in homozygous mice compared with WT and heterozygous. The viral copies in serum and brain between WT and heterozygous were not different significantly (Figure 3C).

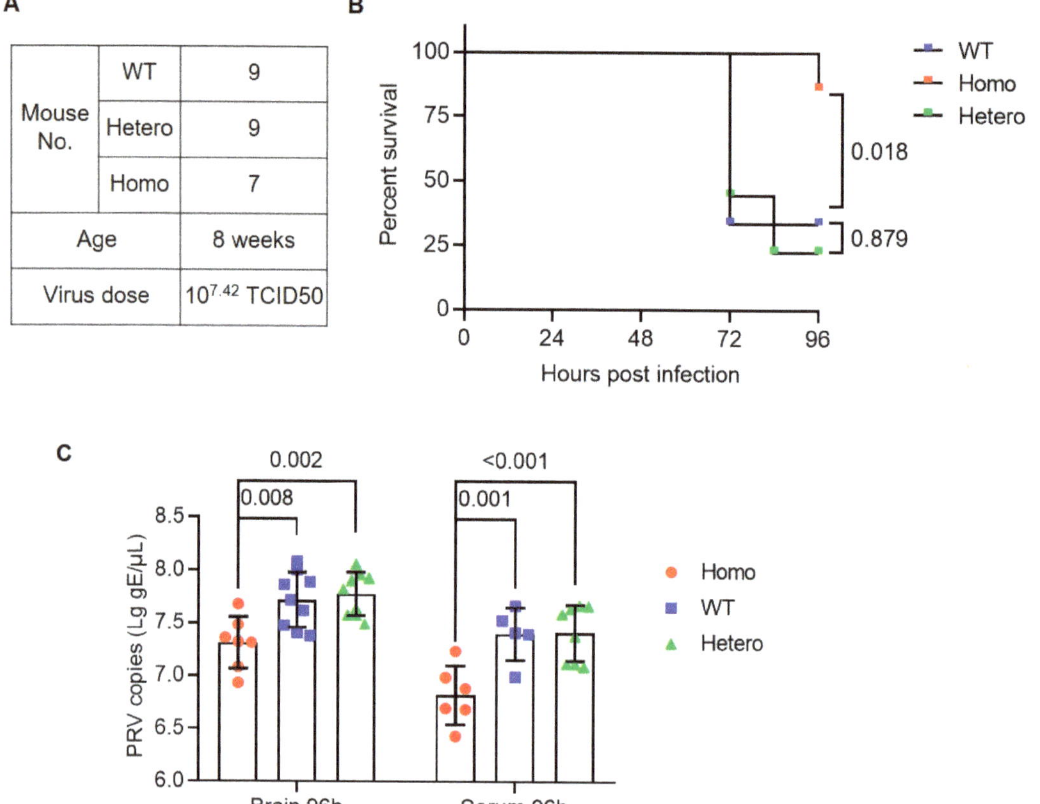

Figure 3. Viral challenge experiment 2 in nectin1 (F129A) mutant mice. (**A**) Experimental groups and virus dose for PRV challenge. (**B**) Survival curve for nectin1 (F129A) homozygous mutant, heterozygous mutant, and WT mice after PRV challenge within 96 h. (**C**) PRV viral loads in brain and serum by qPCR quantification of PRV gE gene. All surviving mice were sacrificed to analyze viral load in brains and serum at 96 h after viral challenge. Data are presented in mean ± standard deviation. Statistically significant differences are indicated by *p*-values, analyzed with Gehan–Breslow–Wilcoxon test and two-way ANOVA for survival curve and viral copies, respectively. Homo, homozygous nectin1 (F129A) mutant mice; Hetero, heterozygous nectin1 (F129A) mutant mice; WT, wild-type mice.

3.3. Defect in Eye Development in Nectin1 (F129A) Mutant Mice

We noted an abnormal eye development in nectin1 (F129A) mutant mice. Homozygous mutant mice showed microphthalmia (Figure 4A). Histological analysis of the eyes showed severe deformation of eye structure in homozygous mice. The vitreous body totally disappeared, and the lenses adhered to the ciliary epithelia and retinal layers (Figure 4B). Furthermore, we observed a deformed ciliary body, which loses ciliary processes. Ciliary processes include the double epithelial layer consisting of pigment and non-pigment epithelia. As shown in Figure 4C, WT mice had the ciliary processes with the contacted pigment and non-pigment epithelia. However, such double-layer structure was not observed in ciliary body of homozygous mutant mice, which thus cannot make the ciliary processes (Figure 4C). A similar defect in eye development was also found in global nectin1 knockout mice or transgenic mice expressing the first V domain of nectin1, in which the main feature is microphthalmia with disappeared vitreous body and ciliary processes in eyes [30,31]. Western blot assay showed that nectin1 protein level in eyes of homozygous mutant mice was greatly less than that in WT littermates, however, nectin1 protein level did not differ greatly in brains between them (Figure 4D). Severe eye abnormality may be partly explained by specifically reduced nectin1 expression in eyes in homozygous mutant mice.

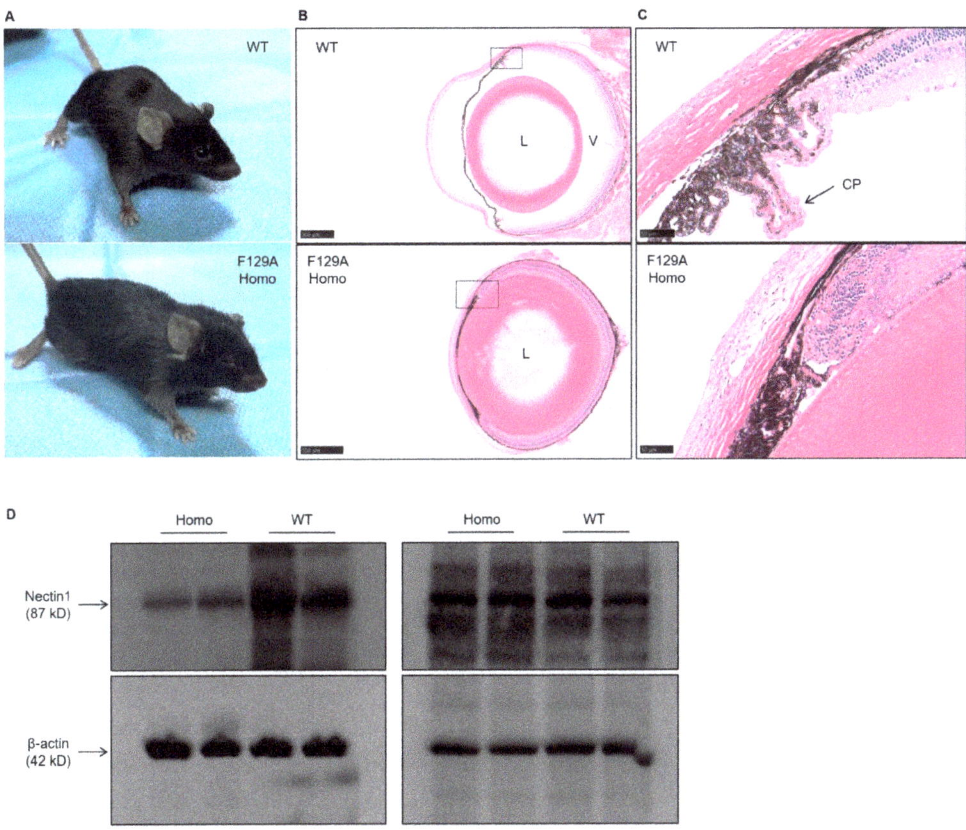

Figure 4. Defect in eye development in nectin1 (F129A) mutant mice. (**A**) Nectin1 (F129A) homozygous mutant mice show bilateral microphthalmia. Representative images of the eyes of nectin1 (F129A) homozygous mutant and age-matched WT mice are shown. (**B**) Histological analysis of the eyes of nectin1 (F129A) homozygous mutant and age-matched WT mice. The absent vitreous body and abnormal lenses can be found in homozygous mutant mice. (**C**) Magnified views of the ciliary body in the boxed areas in the panel (**B**). The ciliary body of WT mice displays the double cell layer structure of the ciliary epithelia composed of pigment (black-colored cells) and non-pigment cells (stained only by eosin in cytoplasm), whereas the homozygous mutant mice have deformed ciliary body, in which the ciliary processes with double cell layer structure of the ciliary epithelia are absent. (**D**) Nectin1 expression levels in eyes and brains of homozygous mutant and WT mice. Tissues of two mice from the same litter for each group were used for western blot assay. Homo, homozygous nectin1 (F129A) mutant mice; WT, wild-type mice; L, lens; V, vitreous body; CP, ciliary processes. Scale bars: 500 μm in (**B**), 50 μm in (**C**).

4. Discussion

4.1. Nectin1 Modification Resists PRV Infection in Hosts

As a representative member of alphaherpesvirus, PRV prevails in pig population and remains a serious threat to the current pig industry in many countries. The emerging animal genetic engineering technology provides an alternative strategy to modify host gene instead of treating the virus itself for viral prevention and control in pigs. The host gene modification antiviral strategy has been proven effective to control viral infection in animals, offering a simple pathway for antiviral breeding to benefit livestock production [32,33]. Nectin1 or nectin2 have long been recognized as the key host factor mediating infection of alphaherpesviruses [15,16]. Although a similar mode by interaction of viral gD and host nectin1 is proposed for entry and spread of alphaherpesviruses, different amino acids in gD/nectin1 binding interface are utilized for nectin1 engagement for different alphaherpesviruses. Structural data of pig nectin1-bound PRV gD revealed N77, I80, M85, and F129 of pig nectin1 as the key residues for their contacting. A mutagenesis study of the key interface residues in nectin1 further confirmed the functional necessity of F129 in PRV-gD engagement [23,24]. Our in vivo viral challenge results were in agreement with the in vitro structural and functional data, that is, homozygous F129A mutant mice harbored enhanced resistance to PRV infection, manifested by reduced viral load in serum and brain, alleviated tissue lesion, and enhanced survival rate compared with WT and heterozygous mutant mice. The mice model-based results lay the groundwork for anti-PRV study in pigs by genetic engineering approaches.

4.2. Correlation of PRV Infection in Mice and Pigs

PRV can infect a wide range of host animals. Similar host factors and machinery are employed by the virus for cell entry and replication in hosts. As a common coreceptor for multiple human and animal alphaherpesviruses, nectin1 may serve a similar role for alphaherpesvirus infection in both natural and non-natural hosts [16]. Therefore, the mouse model of PRV infection could be a valuable reference to study virus–host interaction in pigs. Some previous publications have also reported using transgenic mice as a model for pseudorabies-resistant livestock [34,35]. However, mice and pigs display different clinical manifestations after PRV infection [27–29]. PRV always causes a severe neuropathic itch followed by acute death. The virus replicates in the skin and peripheral nervous system (PNS) neurons, but few infections exist in the brain. In adult pigs, PRV infects mucosal epithelium and spreads to PNS neurons, where a quiescent, latent infection is established. Viral replication can be re-activated to spread back to mucosal surfaces, mainly causing respiratory disease. Acute itch and death rarely occur in adult pigs [27–29]. Investigations indicate that different immune responses, such as type I IFN or specific inflammation, mainly contribute to the difference in pathogenesis and clinical outcomes of PRV infection between mice and pigs [28,29]. The immunity-controlled differential disease manifestations could not affect the disease-resistance phenotype conferred by nectin1 modification, which is a host factor controlling PRV entry or spread in natural and non-natural hosts. Nectin1-mutation could be a universal anti-PRV even anti-alphaherpesvirus strategy in susceptible hosts.

4.3. Nectin1 (F129A) Mutation Impairs Nectin1 Physiologic Function

A previous work reported presence of deleterious effects of global nectin1 knock-out on eye development in mice. Thus, we used single amino acid substitution other than global knockout for anti-PRV mouse preparation. However, only single amino acid mutation seemed to severely impair the normal physiological role of nectin1 in mice. A similar impairment to eye development to that in global nectin1 knockout mice was observed in F129A mutant mice [30]. Our result provides in vivo evidence that F129 is the residue critical for nectin–nectin homophilic or heterophilic trans-interactions; it is essential to form a functional ciliary body with double cell layer structure. This result also implies that PRV infection and nectin1-mediated cell adhesion utilize the same molecular area or machinery to exert their functions. Further work needs to precisely identify the different regions/residues of nectin1 separately hijacked by alphaherpesviruses for cell entry and working in homophilic or heterophilic trans-interactions to maintain normal cellular functions.

4.4. Exploration of Ideal Anti-PRV Gene Targets with Breeding Values in Pigs

The impaired eye function and other undefined abnormality caused by nectin1 mutation could affect the application of the same gene modification for pig antiviral breeding. The genetic engineering technology in agricultural animals should be safely utilized to confer desired phenotypes without penalties to economic traits or animal welfare. The ideal anti-PRV gene targets can be exploited by identifying the safe nectin mutant genotypes or other host factors essential for PRV infection. First, more cell-based works should be performed to define the specific regions in nectin1 or 2 that are critical for viral infection, but minimally interfering with normal cellular functions. Second, other reported gene targets, such as PVR, which may play a key role in PRV cell entry, can be investigated in animals with respect their antiviral ability in vivo [16,36]. Third, the recently developed CRISPR library can effectively screen host genes essential for viral infection [37,38]. A genome-wide screen has been performed for some alphaherpesviruses including PRV and Bovine Herpes Virus Type 1 (BHV-1) and revealed more potential gene targets, which are promising resources facilitating antiviral study [39,40].

Supplementary Materials: The following supporting information can be downloaded at: https://www.mdpi.com/article/10.3390/v14050874/s1, Figure S1: Conservation of Nectin1 F129 residues across different mammalian species; Figure S2: Clinical score for WT and homozygous (Homo) nectin1 F129A mutant mice following PRV infection in experiment 1; Figure S3: Clinical score for WT, heterozygous (Hetero), and homozygous (Homo) nectin1 F129A mutant mice following PRV infection in experiment 2.

Author Contributions: Conceptualization, Z.W. and H.Y.; Data curation, X.Y. and H.Y.; Formal analysis, X.Y., C.Y., Q.Z., L.H., T.G., E.Z., Z.X., Z.L., G.C., Z.W. and H.Y.; Funding acquisition, Z.W. and H.Y.; Investigation, X.Y., C.Y. and Q.Z.; Resources, C.S.; Supervision, H.Y.; Writing—original draft, H.Y.; Writing—review and editing, X.Y. and H.Y. All authors have read and agreed to the published version of the manuscript.

Funding: This research was funded by the Department of Science and Technology of Guangdong Province, grant number 2019B1515210030.

Institutional Review Board Statement: The animals and procedures used in this study were in accordance with the guidelines and approval of the Institutional Animal Care and Use Committees at South China Agricultural University (approval ID: 2020B032).

Informed Consent Statement: Not applicable.

Data Availability Statement: All original data are available upon request.

Conflicts of Interest: The authors declare no conflict of interest.

References

1. Tan, L.; Yao, J.; Yang, Y.; Luo, W.; Yuan, X.; Yang, L.; Wang, A. Current status and challenge of pseudorabies virus infection in China. *Virol. Sin.* **2021**, *36*, 588–607. [CrossRef] [PubMed]
2. Zuckermann, F.A. Aujeszky's disease virus: Opportunities and challenges. *Vet. Res.* **2000**, *31*, 121–131. [CrossRef] [PubMed]
3. Mettenleiter, T.C. Aujeszky's disease (pseudorabies) virus: The virus and molecular pathogenesis—State of the art, June 1999. *Vet. Res.* **2000**, *31*, 99–115. [CrossRef] [PubMed]
4. Mettenleiter, T.C. Pathogenesis of neurotropic herpesviruses: Role of viral glycoproteins in neuroinvasion and transneuronal spread. *Virus Res.* **2003**, *92*, 197–206. [CrossRef]
5. Cheng, Z.; Kong, Z.; Liu, P.; Fu, Z.; Zhang, J.; Liu, M.; Shang, Y. Natural infection of a variant pseudorabies virus leads to bovine death in China. *Transbound. Emerg. Dis.* **2019**, *67*, 518–522. [CrossRef]
6. Jin, H.L.; Gao, S.M.; Liu, Y.; Zhang, S.F.; Hu, R.L. Pseudorabies in farmed foxes fed pig offal in Shandong province, China. *Arch. Virol.* **2016**, *161*, 445–448. [CrossRef]

7. Wong, G.; Lu, J.; Zhang, W.; Gao, G.F. Pseudorabies virus: A neglected zoonotic pathogen in humans? *Emerg. Microbes Infect.* **2019**, *8*, 150–154. [CrossRef]
8. Steinrigl, A.; Revilla-Fernandez, S.; Kolodziejek, J.; Wodak, E.; Bago, Z.; Nowotny, N.; Schmoll, F.; Kofer, J. Detection and molecular characterization of Suid herpesvirus type 1 in Austrian wild boar and hunting dogs. *Vet. Microbiol.* **2012**, *157*, 276–284. [CrossRef]
9. Müller, T.; Hahn, E.C.; Tottewitz, F.; Kramer, M.; Klupp, B.G.; Mettenleiter, T.C.; Freuling, C. Pseudorabies virus in wild swine: A global perspective. *Arch. Virol.* **2011**, *156*, 1691. [CrossRef]
10. Freuling, C.M.; Müller, T.F.; Mettenleiter, T.C. Vaccines against pseudorabies virus (PrV). *Vet. Microbiol.* **2017**, *206*, 3–9. [CrossRef]
11. Tong, W.; Liu, F.; Zheng, H.; Liang, C.; Zhou, Y.J.; Jiang, Y.F.; Shan, T.L.; Gao, F.; Li, G.X.; Tong, G.Z. Emergence of a Pseudorabies virus variant with increased virulence to piglets. *Vet. Microbiol.* **2015**, *181*, 236–240. [CrossRef] [PubMed]
12. An, T.Q.; Peng, J.M.; Tian, Z.J.; Zhao, H.Y.; Li, N.; Liu, Y.M.; Chen, J.Z.; Leng, C.L.; Sun, Y.; Chang, D.; et al. Pseudorabies virus variant in Bartha-K61-vaccinated pigs, China, 2012. *Emerg. Infect. Dis.* **2013**, *19*, 1749–1755. [CrossRef] [PubMed]
13. McGeoch, D.J.; Cook, S. Molecular phylogeny of the alphaherpesvirinae subfamily and a proposed evolutionary timescale. *J. Mol. Biol.* **1994**, *238*, 9–22. [CrossRef] [PubMed]
14. Pomeranz, L.E.; Reynolds, A.E.; Hengartner, C.J. Molecular biology of pseudorabies virus: Impact on neurovirology and veterinary medicine. *Microbiol. Mol. Biol. Rev.* **2005**, *69*, 462–500. [CrossRef]
15. Spear, P.G. Herpes simplex virus: Receptors and ligands for cell entry. *Cell Microbiol.* **2004**, *6*, 401–410. [CrossRef]
16. Geraghty, R.J.; Krummenacher, C.; Cohen, G.H.; Eisenberg, R.J.; Spear, P.G. Entry of alphaherpesviruses mediated by poliovirus receptor-related protein 1 and poliovirus receptor. *Science* **1998**, *280*, 1618–1620. [CrossRef]
17. Petermann, P.; Rahn, E.; Their, K.; Hsu, M.J.; Rixon, F.J.; Kopp, S.J.; Knebel-Mörsdorf, D. Role of nectin-1 and herpesvirus entry mediator as cellular receptors for herpes simplex virus 1 on primary murine dermal fibroblasts. *J. Virol.* **2015**, *89*, 9407–9416. [CrossRef]
18. Petermann, P.; Their, K.; Rahn, E.; Rixon, F.J.; Bloch, W.; Özcelik, S.; Krummenacher, C.; Barron, M.J.; Dixon, M.J.; Scheu, S.; et al. Entry mechanisms of herpes simplex virus 1 into murine epidermis: Involvement of nectin-1 and herpesvirus entry mediator as cellular receptors. *J. Virol.* **2015**, *89*, 262–274. [CrossRef]
19. Kopp, S.J.; Karaba, A.H.; Cohen, L.K.; Banisadr, G.; Miller, R.J.; Muller, W.J. Pathogenesis of neonatal herpes simplex 2 disease in a mouse model is dependent on entry receptor expression and route of inoculation. *J. Virol.* **2013**, *87*, 474–481. [CrossRef]
20. Karaba, A.H.; Kopp, S.J.; Longnecker, R. Herpesvirus entry mediator and nectin-1 mediate herpes simplex virus 1 infection of the murine cornea. *J. Virol.* **2011**, *85*, 10041–10047. [CrossRef]
21. Taylor, J.M.; Lin, E.; Susmarski, N.; Yoon, M.; Zago, A.; Ware, C.F.; Pfeffer, K.; Miyoshi, J.; Takai, Y.; Spear, P.G. Alternative entry receptors for herpes simplex virus and their roles in disease. *Cell Host Microbe* **2007**, *2*, 19–28. [CrossRef] [PubMed]
22. Huang, Y.; Li, Z.; Song, C.; Wu, Z.; Yang, H. Resistance to pseudorabies virus by knockout of nectin1/2 in pig cells. *Arch. Virol.* **2020**, *165*, 2837–2846. [CrossRef] [PubMed]
23. Struyf, F.; Martinez, W.M.; Spear, P.G. Mutations in the N-terminal domains of nectin-1 and nectin-2 reveal differences in requirements for entry of various alphaherpesviruses and for nectin-nectin interactions. *J. Virol.* **2002**, *76*, 12940–12950. [CrossRef] [PubMed]
24. Li, A.; Lu, G.; Qi, J.; Wu, L.; Tian, K.; Luo, T.; Shi, Y.; Yan, J.; Gao, G.F. Structural basis of nectin-1 recognition by pseudorabies virus glycoprotein D. *PLoS Pathog.* **2017**, *13*, e1006314. [CrossRef]
25. Reed, L.J.; Muench, H. A simple method of estimating fifty per cent endpoints. *Am. J. Hyg.* **1938**, *27*, 493–497. [CrossRef]
26. Yue, D.; Chen, Z.; Yang, F.; Ye, F.; Lin, S.; He, B.; Cheng, Y.; Wang, J.; Chen, Z.; Lin, X.; et al. Crystal structure of bovine herpesvirus 1 glycoprotein D bound to nectin-1 reveals the basis for its low-affinity binding to the receptor. *Sci. Adv.* **2020**, *6*, eaba5147. [CrossRef]
27. Brittle, E.E.; Reynolds, A.E.; Enquist, L.W. Two modes of pseudorabies virus neuroinvasion and lethality in mice. *J. Virol.* **2004**, *78*, 12951–12963. [CrossRef]
28. Laval, K.; Vernejoul, J.B.; Van Cleemput, J.; Koyuncu, O.O.; Enquist, L.W. Virulent Pseudorabies Virus Infection Induces a Specific and Lethal Systemic Inflammatory Response in Mice. *J. Virol.* **2018**, *92*, e01614-18. [CrossRef]
29. Laval, K.; Enquist, L.W. The Neuropathic Itch Caused by Pseudorabies Virus. *Pathogens* **2020**, *9*, 254. [CrossRef]
30. Inagaki, M.; Irie, K.; Ishizaki, H.; Tanaka-Okamoto, M.; Morimoto, K.; Inoue, E.; Ohtsuka, T.; Miyoshi, J.; Takai, Y. Roles of cell-adhesion molecules nectin 1 and nectin 3 in ciliary body development. *Development* **2005**, *132*, 1525–1537. [CrossRef]
31. Yoshida, K.; Tomioka, Y.; Kase, S.; Morimatsu, M.; Shinya, K.; Ohno, S. Transgenic mice generating group, Ono E. Microphthalmia and lack of vitreous body in transgenic mice expressing the first immunoglobulin-like domain of nectin-1. *Graefes Arch. Clin. Exp. Ophthalmol.* **2008**, *246*, 543–549. [CrossRef] [PubMed]
32. Xu, K.; Zhou, Y.; Mu, Y.; Liu, Z.; Hou, S.; Xiong, Y.; Fang, L.; Ge, C.; Wei, Y.; Zhang, X.; et al. CD163 and pAPN double-knockout pigs are resistant to PRRSV and TGEV and exhibit decreased susceptibility to PDCoV while maintaining normal production performance. *eLife* **2020**, *9*, e57132. [CrossRef] [PubMed]
33. Whitworth, K.M.; Rowland, R.R.; Ewen, C.L.; Trible, B.R.; Kerrigan, M.A.; Cino-Ozuna, A.G.; Samuel, M.S.; Lightner, J.E.; McLaren, D.G.; Mileham, A.J.; et al. Gene-edited pigs are protected from porcine reproductive and respiratory syndrome virus. *Nat. Biotechnol.* **2016**, *34*, 20–22. [CrossRef] [PubMed]

34. Ono, E.; Amagai, K.; Taharaguchi, S.; Tomioka, Y.; Yoshino, S.; Watanabe, Y.; Cherel, P.; Houdebine, L.M.; Adam, M.; Eloit, M.; et al. Transgenic mice expressing a soluble form of porcine nectin-1/herpesvirus entry mediator C as a model for pseudorabies-resistant livestock. *Proc. Natl. Acad. Sci. USA* **2004**, *101*, 16150–16155. [CrossRef]
35. Ono, E.; Tomioka, Y.; Taharaguchi, S.; Cherel, P. Comparison of protection levels against pseudorabies virus infection of transgenic mice expressing a soluble form of porcine nectin-1/HveC and vaccinated mice. *Vet. Microbiol.* **2006**, *114*, 327–330. [CrossRef]
36. Baury, B.; Geraghty, R.J.; Masson, D.; Lustenberger, P.; Spear, P.G.; Denis, M.G. Organization of the rat Tage4 gene and herpesvirus entry activity of the encoded protein. *Gene* **2001**, *265*, 185–194. [CrossRef]
37. Zhao, C.; Liu, H.; Xiao, T.; Wang, Z.; Nie, X.; Li, X.; Qian, P.; Qin, L.; Han, X.; Zhang, J.; et al. CRISPR screening of porcine sgRNA library identifies host factors associated with Japanese encephalitis virus replication. *Nat. Commun.* **2020**, *11*, 5178. [CrossRef]
38. Yu, C.; Zhong, H.; Yang, X.; Li, G.; Wu, Z.; Yang, H. Establishment of a pig CRISPR/Cas9 knockout library for functional gene screening in pig cells. *Biotechnol. J.* **2021**, *in press*. [CrossRef]
39. Hölper, J.E.; Grey, F.; Baillie, J.K.; Regan, T.; Parkinson, N.J.; Höper, D.; Thamamongood, T.; Schwemmle, M.; Pannhorst, K.; Wendt, L.; et al. A genome-wide CRISPR/Cas9 screen reveals the requirement of host sphingomyelin synthase 1 for infection with pseudorabies virus mutant gD(-)Pass. *Viruses* **2021**, *13*, 1574. [CrossRef]
40. Tan, W.S.; Rong, E.; Dry, I.; Lillico, S.G.; Law, A.; Whitelaw, C.B.A.; Dalziel, R.G. Genome-wide CRISPR knockout screen reveals membrane tethering complexes EARP and GARP important for Bovine Herpes Virus Type 1 replication. *bioRxiv* **2020**. [CrossRef]

Article

Isolation and Characterization of Two Pseudorabies Virus and Evaluation of Their Effects on Host Natural Immune Responses and Pathogenicity

Qiongqiong Zhou [1,†], Longfeng Zhang [1,2,†], Hongyang Liu [1], Guangqiang Ye [1], Li Huang [1,3,*] and Changjiang Weng [1,3,*]

1   State Key Laboratory of Veterinary Biotechnology, Division of Fundamental Immunology, Harbin Veterinary Research Institute, Chinese Academy of Agricultural Sciences, Harbin 150069, China; qqhenanly@163.com (Q.Z.); yicunsangzhe@126.com (L.Z.); lhyyyds1234@163.com (H.L.); ygqyyds123@163.com (G.Y.)
2   College of Animal Science, Yangtze University, Jingzhou 434000, China
3   Key Laboratory of Veterinary Immunology of Heilongjiang Provincial, Harbin 150069, China
*   Correspondence: highlight0315@163.com (L.H.); wengchangjiang@caas.cn (C.W.)
†   These authors contributed equally to this work.

Abstract: Pseudorabies, caused by the pseudorabies virus (PRV), is an acute fatal disease, which can infect rodents, mammals, and other livestock and wild animals across species. Recently, the emergence of PRV virulent isolates indicates a high risk of a variant PRV epidemic and the need for continuous surveillance. In this study, PRV-GD and PRV-JM, two fatal PRV variants, were isolated and their pathogenicity as well as their effects on host natural immune responses were assessed. PRV-GD and PRV-JM were genetically closest to PRV variants currently circulating in Heilongjiang (HLJ8) and Jiangxi (JX/CH/2016), which belong to genotype 2.2. Consistently, antisera from sows immunized with PRV-Ea classical vaccination showed much lower neutralization ability to PRV-GD and PRV-JM. However, the antisera from the pigs infected with PRV-JM had an extremely higher neutralization ability to PRV-TJ (as a positive control), PRV-GD and PRV-JM. In vivo, PRV-GD and PRV-JM infections caused 100% death in mice and piglets and induced extensive tissue damage, cell death, and inflammatory cytokine release. Our analysis of the emergence of PRV variants indicate that pigs immunized with the classical PRV vaccine are incapable of providing sufficient protection against these PRV isolates, and there is a risk of continuous evolution and virulence enhancement. Efforts are still needed to conduct epidemiological monitoring for the PRV and to develop novel vaccines against this emerging and reemerging infectious disease.

Keywords: pseudorabies virus; virus isolation; pathogenicity; mortality; inflammatory response

Citation: Zhou, Q.; Zhang, L.; Liu, H.; Ye, G.; Huang, L.; Weng, C. Isolation and Characterization of Two Pseudorabies Virus and Evaluation of Their Effects on Host Natural Immune Responses and Pathogenicity. *Viruses* **2022**, *14*, 712. https://doi.org/10.3390/v14040712

Academic Editor: Xiangdong Li

Received: 16 February 2022
Accepted: 26 March 2022
Published: 29 March 2022

Publisher's Note: MDPI stays neutral with regard to jurisdictional claims in published maps and institutional affiliations.

1. Introduction

Pseudorabies (PR, Aujeszky's disease), caused by the pseudorabies virus (Aujeszky's disease virus or Suid alphaherpesvirus 1) (PRV), brings substantial economic losses to swine factories in some countries. The PRV belongs to the family Herpesviridae, subfamily Alphaherpesvirinae (https://talk.ictvonline.org/ictv-reports/ictv_online_report/dsdna-viruses/w/herpesviridae/1609/subfamily-alphaherpesvirinae, accessed on 10 Mar 2022). The PRV genome is a double-stranded DNA of about 143 kb in length, and the G+C content is more than 70% [1]. The genome of the PRV encompasses a unique long (UL) segment and a unique short (US) region flanked by the internal and terminal repeat sequences (IRS and TRS, respectively), encoding more than 70 proteins [2].

Since the early 1980s, PR disease had spread nearly globally. In 1947, PRV infection was reported firstly in eastern China. Because of the lack of accurate detection technology and poor biosafety measures, PR was widely prevalent in most of China. Owing to the wide usage of the traditional vaccine strain Bartha-K61 (a live attenuated vaccine based on the PRV Bartha-K61 strain), PR diseases were controlled in China [3]. However, in 2011, the outbreaks of PR were reported and rapidly spread in China. In some farms, the PRV seropositive occurred in a very short time. Research confirmed that the current PR outbreaks were caused by PRV variants and the Bartha-K61 vaccine could not provide adequate protection against these variant strains [3,4]. Subsequently, several PRV isolates were isolated in China, which were in the same evolutionary branch as the JS strain (KP257591), and in two independent branches with the European and American strains (Kapla, Becker) [5]. The variation regions of PRV isolates are mainly concentrated in IRS, TRS and SSR sequences in the promoter region, which can affect the gene expression and pathogenicity of PRV [6]. In addition, it was reported that the pathogenicity of PRV isolates to mice and pigs was significantly higher. For instance, the LD_{50} of SC strain was 10-fold and 8.5-fold higher than that of PRV-TJ (KJ789182) and PRV-HeN1 strains (KP098534.1), respectively [4]. In 2016, Yang et al. illustrated that PRV variant isolate HN1201 (KP722022) could cause extensive tissue injury and 100% death of piglets, but the Fa isolate (KM189913) only induced weak respiratory symptoms [7]. To better control and prevent PR in China, the Chinese government has issued a series of policies with the intent to eradicate PR in pig breeding farms by the end of 2020. The rational use of vaccines (including inactivated vaccines, live attenuated vaccines, live virus-vectored vaccines) and other novel viral inhibitors are the main strategies.

Without specific host tropism, the PRV infects a wide variety of animal species, including ruminants, carnivores, rodents, and lagomorphs. The genus sus scrofa are the only natural hosts for the PRV, which cause severe clinical symptoms [8]. The PRV can infect pigs at different ages, causing 100% death of piglets, abortion of sows, and respiratory diseases in adult pigs [9]. Aside from pigs, cattle, sheep, cats, dogs, raccoons, minks, and skunks, can all be infected by the PRV, causing "mad itch", neurological symptoms, or death [10,11]. Moreover, it was demonstrated that the PRV infected host cells via both human and swine nectin-1, and that PRV glycoprotein D exhibited similar binding affinities for nectin-1 of two species [1]. Symptomatically, human PRV infection cases in China indicated that PRV infection could induce prominent central nervous system disorders and encephalitis [12,13], which posed a significant threat to public health in China [12–15]. Therefore, it is important to improve the phylogenetic analysis and virulence monitoring of PRV variant isolates.

The immune system is the most significant line of host defense against virus infection. Viral infection is defended by hosts using multiple strategies, including innate immune responses and adaptive immunity. The inflammatory response is a complex mechanism, consisting of immune cells and inflammatory cytokines (e.g., IL-1β, IL-6, TNF-α, MCP-1, IP-10, MIP-1, etc.), which remove invading viruses and promote repair at the sites of damage [16,17]. The type I Interferon (IFN-I) response is the most prominent antiviral response, which plays a central role in innate immunity against viral infection. IFN-I-targeting cells maintain a potent antiviral state by inducing the synthesis of hundreds of antiviral proteins encoded by IFN-stimulated genes (ISGs). Increasing evidence indicates that the PRV has evolved multiple strategies to inhibit type I IFN signaling and establish persistent infection [18,19].

The PRV is an emerging and reemerging infectious disease consistently threatening the pig industry worldwide. As the largest pork producer and consumer, China has experienced two pseudorabies outbreaks [20]. Since 2011, more and more PRV variant isolates, whose gene types belong to Clade 2 (variant PRV), were identified in China [13]. It was reported that the sero-prevalence rate of PR was 34.2% in China from 2016 to 2018. Furthermore, the sero-prevalence in northern China, eastern China, and central and southern China are higher than those in northeastern China, northwestern China, and southwestern China [20]. Notably, increasing number of PRV-infected human cases were reported [13–15]. Given the current global epidemic of PRV variant strains, herein, two fatal PRV (PRV-GD and PRV-JM) were isolated, and the etiological as well as genetic characteristics of these PRV isolates were investigated. Moreover, the pathogenicity to mice and pigs and the natural immune responses induced by these PRV isolates were also explored in vitro and in vivo.

2. Materials and Methods

2.1. Isolation of PRV-GD, PRV-JM and Genome Sequencing

PRV-GD and PRV-JM were isolated from the aborted piglet samples of PRV-positive pig farms at Foshan and Jinmen, Guangdong Province, respectively. The PRV-GD and PRV-JM isolates, purified by plaque, were inoculated onto PK-15 cells. To confirm the occurrence of virus multiplication, the one-step growth curve of the PRV-TJ, PRV-GD, and PRV-JM isolates in PK-15 cells and the *gB* gene expression were assessed.

The genomic DNA of PRV-GD and PRV-JM isolates were extracted from the infected PK-15 cells. DNA quality, integrity, and concentration were assessed, and sequencing was performed by Harbin Biotech Gene Company (Harbin, China). Then, the nucleotide sequences of two virus were compared with known PRV strains retrieved from the GenBank database. Phylogenetic tree analysis was constructed using MEGA 7.0 software.

2.2. Virus Neutralization Assay

The positive antisera, collected from different pigs immunized with classic PRV-Ea live attenuated vaccines, were used for the neutralization assay as previously described [21]. Briefly, the antisera were heat-inactivated for 30 min at 56 °C and serially diluted from 2^0–2^{-7} in 2-fold. The neutralization tests were conducted by adding 50 μL (containing 100 $TCID_{50}$) of virus suspension into 50 μL of the diluted antisera in 96-well plate in quadruplets, and then incubated at 37 °C for 40 min. Subsequently, the 100 μL mixtures were added into 96-well plate and incubated with PK-15 cells at 37 °C incubator to observe the cytopathic effect (CPE). The neutralizing titer was expressed as the highest dilution that reduced the CPE by 50% as compared to the control. The neutralizing titer was calculated based on the Reed–Muench method [16].

2.3. Immune Responses in Mouse Peritoneal Macrophages Induced by PRV Strains

The PRV can infect a wide variety of mammals, including pigs, wild boars, rodents, bears, ruminants, and carnivores. In general, PRV-variant infections are fatal to rodents [8]. To evaluate the pathogenicity and immune responses of PRV isolates more comprehensively, the mice and piglets were used to perform the experimental infection study.

Firstly, we detected the innate immune responses in mouse peritoneal macrophages induced by PRV-GD, JM, and TJ strains. Primary peritoneal macrophages were isolated from C57BL/6J mice and infected with PRV-GD, JM and TJ strains for 0, 3, 6, 9, 12, and 24 h. The total mRNA and the cell culture supernatant were harvested. The transcription levels and the protein levels of multiple pro-inflammatory cytokines, such as interleukin-1β (IL-1β), IL-6, tumor necrosis factor-α (TNF-α), and interferon-β (IFN-β) were detected.

2.4. Experimental Mice Infection Study

Then, a total of 40 six-week-old SPF C57BL6/J mice were purchased from Liaoning Changsheng Biotechnology Co., Ltd. (Liaoning, China). All mice were generated and housed in specific pathogen-free (SPF) barrier facilities at the Harbin Veterinary Research Institute (HVRI) of the Chinese Academy of Agricultural Sciences (CAAS) (Harbin, China) (the ethical approval number: 210608-01). All mice were randomly divided into eight groups (n = 5, respectively). Three groups were challenged with PRV-GD (n = 5) and another three groups were challenged with PRV-JM (n = 5) isolates at different doses by intraperitoneal injection (i.p.). The remaining two groups received PBS (50 mM, pH 7.4) injections. Mortality in each group was recorded and LD_{50} was calculated based on the Reed–Muench method [16]. The liver, spleen, lung, kidney, and brain samples were collected and used for H&E staining.

Furthermore, to detect the effects of PRV-GD and PRV-JM isolates on the natural immune response of mice, nine six-week-old SPF C57BL6/J mice were randomly divided into 3 groups ($n = 3$, respectively). Three groups of mice were challenged with 20,000 PFU of PRV-GD, PRV-JM isolates, or PBS via tail vein injection. Two days post-infection, all mice were executed by a cervical vertebrae luxation after CO_2 inhalation anesthesia. The serum, liver, spleen, lung, kidney, and brain samples were collected and detected for IL-1β, IL-6, TNF-α, IFN-β by ELISA or RT-qPCR, respectively.

2.5. Pathogenicity in Piglets

Eight crossbred healthy SPF piglets (Landrace × large white, 2-month-old, male, 9~11 kg) were purchased and housed in SPF barrier facilities at the Harbin Veterinary Research Institute (HVRI) of the Chinese Academy of Agricultural Sciences (CAAS) (Harbin, China) (the ethical approval number: 211026-02). All piglets were maintained at an ambient temperature of 20–25 °C in an environmentally controlled room by air conditioning and illumination (12 h light and dark cycles). Each cage was equipped with a feeder and water nipple to allow free access to food and drinking water.

Eight two-month-old piglets were randomly divided into three groups. Pigs in groups 1 and 2 ($n = 3$, respectively) were inoculated intranasally with PRV-GD or PRV-JM at 10^6 TCID$_{50}$. Pigs in group 3 ($n = 2$) received PBS (50 mM, pH 7.4) as the control. Rectal temperature and clinical signs (including diet, water intake, mental status, and neurological symptoms) were recorded daily. Six days post-infection, all PRV-infected pigs died and were necropsied. The PBS-changed pigs were euthanized (electric shock anesthesia). The serum, liver, spleen, lung, kidney, tonsil, and brain samples were collected and detected for IL-1β, IL-6, TNF-α, IFN-β, IFN-α by ELISA or RT-qPCR, respectively. The samples also were used for H&E and virus detection using *gB*-specific TaqMan qPCR.

2.6. Histological Analysis

The tissues (liver, spleen, lung, kidney, tonsil, and brains) of mice and piglets (the tonsils were taken from pigs only) infected with PRV isolates were fixed in 10% formalin neutral buffer solution overnight. Histological analysis of tissue damage was assessed by standard hematoxylin and eosin (H&E) staining. Tunnel solution (Beyotime) was used to detect the damaged nucleus. The results were analyzed by light microscopy. Representative views of the tissue sections are shown.

2.7. Statistical Analysis

Data were analyzed as mean ± SEM of at least three independent replicates. Differences among groups were performed by one-way ANOVA using GraphPad Prism software. p values of < 0.05 were considered to be statistically significant for each test. The significance level for all analyses was set as * $p < 0.05$, ** $p < 0.01$ and *** $p < 0.001$.

3. Results

3.1. Identification of PRV-GD and PRV-JM Isolates

Cytopathic effects (CPE) induced by PRV-GD and PRV-JM, including cell fusion, syncytium, detachment, and numerous rounded and floated cells, were observed at 24 h post-infection (hpi). The CPE in PRV-TJ-infected PK-15 cells were considered as the positive and no CPE were observed in the control (Figure 1A). As shown in Figure 1B, PRV glycoprotein B (*gB*) and glycoprotein E (*gE*) genes were detected by polymerase chain reaction (PCR) amplification to confirm the existence of PRV in cell culture. The results of Figure 1C showed that there was no significant difference in the growth of the three PRV strains on PK-15 cells.

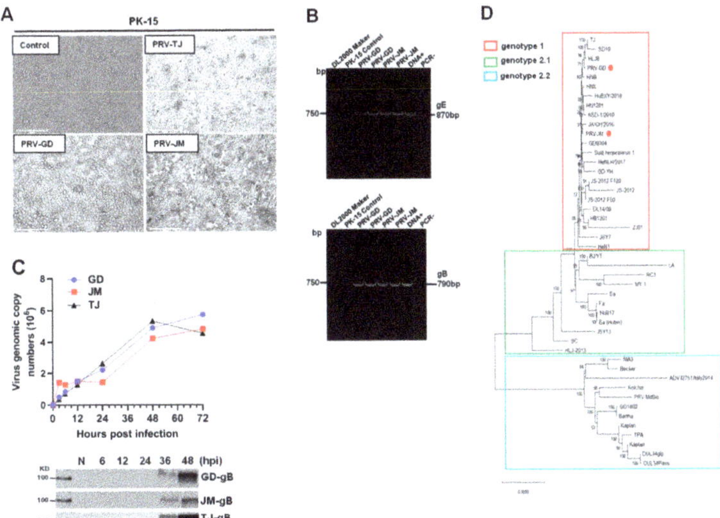

Figure 1. Isolation and identification of PRV-GD and PRV-JM isolates. (**A**), The cytopathic effects (CPEs) of PK-15 cells infected by PRV-GD and PRV-JM for 24 h. The CPEs of PK-15 cells infected with PRV-TJ strain were considered as the positive control. The arrowhead indicates the CPEs observed of PRV-GD and PRV-JM-inoculated PK-15 cells. The CPEs were characterized with rounded and floated cells. (**B**), PCR amplification of PRV *gE* (870 bp) and *gB* (790 bp) fragments from PRV-GD and PRV-JM-inoculated PK-15 cells. DNA+ was the positive control and PCR− was the negative control during PCR amplification. (**C**), One-step growth curves and *gB* protein assessment of 3 PRV strains on PK-15 cells at a multiplicity of infection of 0.1. (**D**), Phylogenetic tree based on genomic nucleotide sequence. PRV-GD and PRV-JM isolates in this work were indicated with a red dot. The phylogenetic tree was constructed by the adjacency method in MEGA 7 (http://www.megasoftware.net, accessed on 14 July 2021).

3.2. Genome Sequencing and Phylogenetic Analysis

The genome of PRV-GD and PRV-JM isolates were extracted and sequenced by second generation sequencing technology. The complete genomes of PRV-GD (GeneBank accession numbers OK338076) and PRV-JM (GeneBank accession numbers OK338077) isolates are 144.05 kb and 142.47 kb, respectively. To further understand the evolutionary relationship between the two PRV isolates and other PRV variants, a phylogenetic tree was constructed based on their genomes (Figure 1D). The results indicated that PRV-GD and PRV-JM isolates, belonging to genotype 2.2, were novel PRV variants. These results demonstrated that the population size of PRV clade 2.2 was increasing, indicating that PRV variants may be still circulating in swine herds and result in a risk in relation to interspecies transmission.

3.3. Proliferation Characteristics of PRV-GD and PRV-JM in PAMs, THP-1 and Mouse Peritoneal Macrophages

One-step growth curves were performed to evaluate the proliferation characteristics of PRV-GD and PRV-JM in PAMs, THP-1, and mouse peritoneal macrophages. The PRV-TJ-infected cells were considered as the positive control. Firstly, virus-induced CPEs in three cells were compared. PAMs, THP-1, and mouse peritoneal macrophages were challenged with PRV-TJ, PRV-GD, and PRV-JM. PAMs, THP-1, and mouse peritoneal macrophages showed the similar CPEs, characterized by shrinkage, fragmentation, and cell death, but mouse peritoneal macrophages showed fibrosis and death. Compared with PRV-GD and PRV-TJ, PRV-JM induced stronger CPEs in the test cell lines (Figure 2A). In one-step growth analysis, PRV-GD and PRV-JM displayed similar growth curves as the PRV-TJ strain in both PK-15 cells and PAMs (Figures 1C and 2B). However, in THP-1 and mouse peritoneal macrophages, PRV-GD and PRV-JM isolates showed a replication superiority compared with the PRV-TJ strain (Figure 2C,D). These observations together supported our proposal that PRV-GD and PRV-JM isolates were the novel PRV variants and behaved differently from other classical PRV variants.

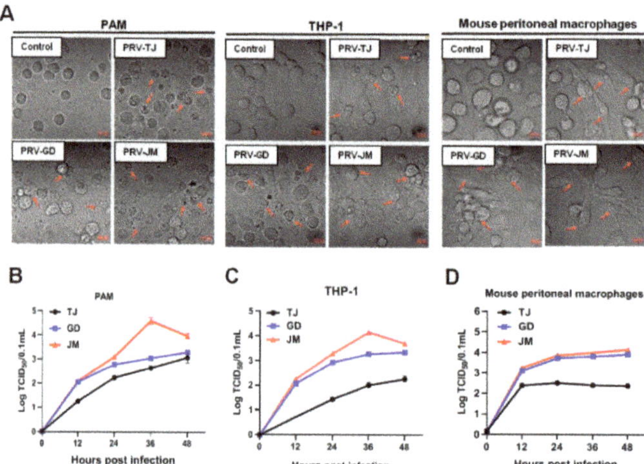

Figure 2. In vitro proliferation properties of PRV-TJ strain, PRV-GD and PRV-JM isolates. (**A**), CPEs in PAMs, THP-1 cells and mouse peritoneal macrophages, caused by PRV-TJ strain, PRV-GD and PRV-JM isolates, were characterized by the disintegrated cells (red arrows). Scale bars: 10 μm. (**B–D**), One-step growth curves of PRV-TJ, PRV-GD and PRV-JM on PAMs, THP-1 cells and mouse peritoneal macrophages at a multiplicity of infection of 0.1.

3.4. Cross-Neutralization Assays

Next, antibody neutralization assay was performed to determine the immunogenicity of PRV-GD and PRV-JM isolates. As shown in Table S1, the antisera from sows immunized with classical Ea vaccination showed strong seropositive against PRV gB protein. Subsequently, the antisera exhibited higher neutralization activity to PRV-TJ but weaker neutralization ability to PRV-GD and PRV-JM (Figure 3A). However, antisera from pigs infected with PRV-JM showed extremely high neutralization activity to PRV-TJ, PRV-GD and PRV-JM (Figure 3B). Collectively, these results suggest that the traditional vaccine prepared with classical PRV strain may not provide effective protection against the challenge of PRV variants, suggesting there is a potential risk of increasing virulence of PRV variants from the perspective of immunogenicity.

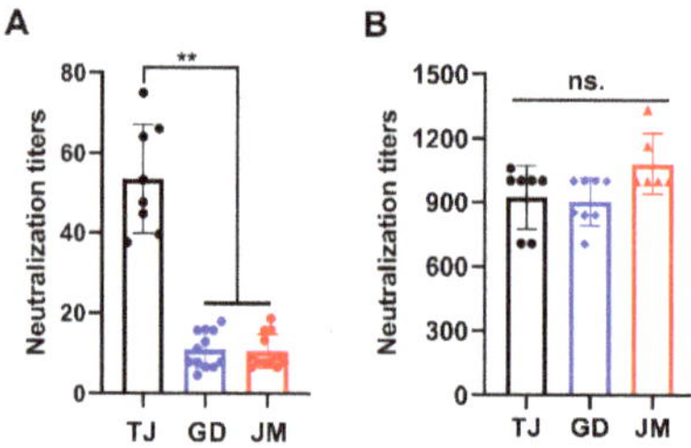

Figure 3. Antisera from pigs infected with PRV-JM isolate has broader spectrum of neutralizing ability. Neutralizing titers of antisera from classical PRV-Ea-vaccinated sows (**A**) and antisera from pigs immunized with PRV-JM isolate (**B**) against PRV-TJ, PRV-GD and PRV-JM. The significance of differences between the PRV-TJ, PRV-GD and PRV-JM were analyzed with t test. ** $p < 0.01$. ns., no significant difference.

3.5. Immune Responses Induced by PRV Strains

The mRNA levels of multiple pro-inflammatory cytokines, such as interleukin-1β (IL-1β), IL-6, tumor necrosis factor-α (TNF-α), and interferon-β (IFN-β), were up-regulated during the three PRV variants infection (Figure 4A). Consistently, ELISA results illustrated that the protein levels of these pro-inflammatory cytokines and IFN-β were also increased (Figure 4B). Additionally, after infection with three PRV strains, the transcriptional levels of IL-1β, IL-6, TNF-α, and IFN-β were highest at 3 h, 6 h, 9 h, and 12 h, respectively, and then gradually decreased. The protein levels of IL-1β, IL-6, and IFN-β were highest after infection of 12 h, but TNF-α was maintained in the higher levels after 24 h (Figure 4). It was noteworthy that PRV-GD and PRV-JM induced higher levels of the pro-inflammatory cytokines and IFN-β than PRV-TJ strain, eventually leading to stronger immune responses.

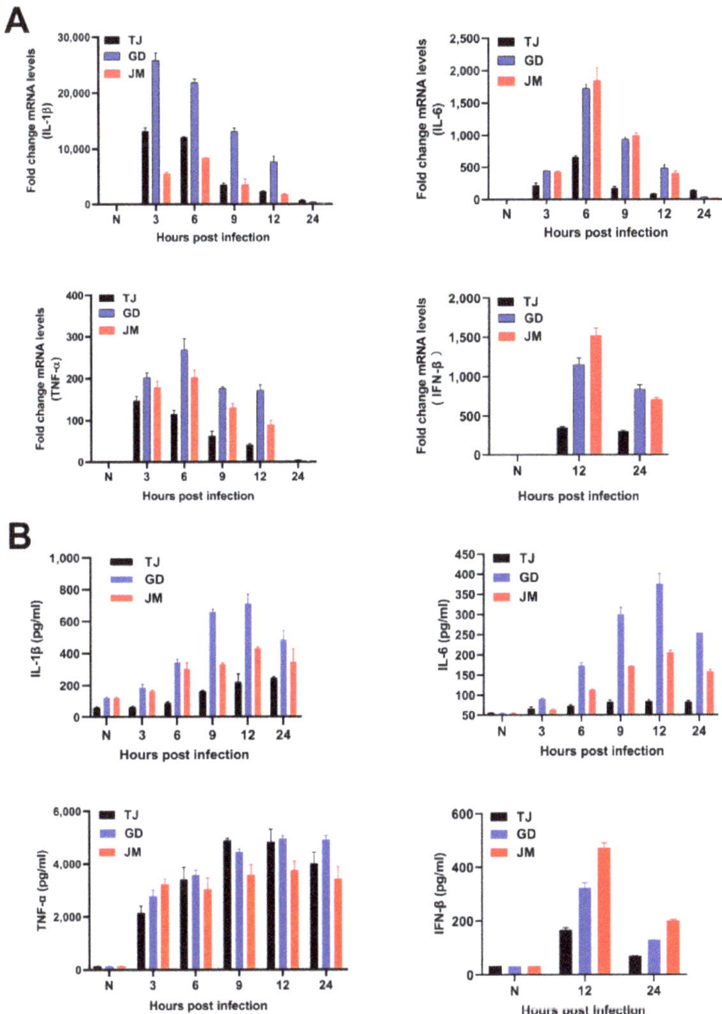

Figure 4. PRV-GD and PRV-JM isolates induced the higher immune responses than PRV-TJ strain. (**A**,**B**), Detection of the mRNA and protein levels of the IL-1β, IL-6, TNF-α, and IFN-β in mouse peritoneal macrophages induced by PRV-TJ, PRV-GD, and PRV-JM after infection for 3, 6, 9, 12 and 24 h, respectively.

3.6. Pathogenicity of PRV-GD and PRV-JM Isolates in Mice

To gain the pathological insight into the overall changes induced by PRV-GD and PRV-JM, pathological examinations in mice were firstly performed. The SPF C57BL/6J mice were infected with two isolates by intraperitoneal injection. As shown in Figure 5A,B, both PRV-GD and PRV-JM infection could induce itch, eventually leading mice to death with a comparable LD_{50} (50% lethal dose): 57.5 and 66.1 $TCID_{50}$ (50% tissue culture infective dose), respectively (Figure 5A,B). Subsequently, dissection of the mice immediately after death was performed to analyze the pathological characteristics by H&E staining. The results showed that PRV-GD and PRV-JM isolates could significantly cause lung congestion, thickening of alveolar septa, lymphocyte infiltration in the spleen, disintegration of hepatocytes, cellular necrosis in the liver and kidney, and microglia proliferation in the brain (Figure 5C).

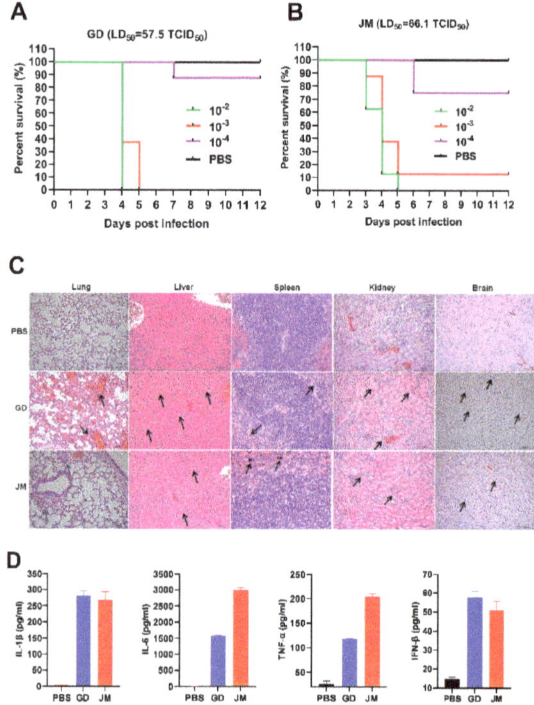

Figure 5. The pathogenicity and the immune responses of PRV-GD and PRV-JM in experimental mice. (**A,B**), LD_{50} of the PRV-GD (**A**) and PRV-JM (**B**) to six−week−old C57BL/6J mice. (**C**), Pathological lesions of mice (H&E staining, black arrow) that died following experimental infection with PRV-GD and PRV-JM isolates. (**D**), Detection of the protein levels of IL-1β, IL-6, TNF-α and IFN-β in serum challenged by PRV-GD and PRV-JM isolates.

In addition, three groups of SPF mice were infected with PRV-GD and PRV-JM isolates at 20,000 PFU/mL to examine the innate immune responses. The results were summarized in Figure S1A and Figure 5D. Among the examined tissues (liver, spleen, lung, and brain), both PRV-GD and PRV-JM could enhance the expression of pro-inflammatory cytokines (IL-1β, IL-6, TNF-α) and IFN-β after 2 days post-infection (dpi). The lung possessed the highest PRV genomic copy number among the examined tissues (Figure S1). Of note, the two isolates induced similar patterns of changes of the pro-inflammatory cytokines and IFN-β in serum by ELISA (Figure 5D). Overall, our results indicate that PRV-GD and PRV-JM are the virulent isolates with high pathogenicity and can induce robust inflammatory responses in mice.

3.7. Pathogenicity of PRV-GD and PRV-JM Isolates in Pigs

To further gain insight into the pathogenicity and inflammatory responses of PRV-GD and PRV-JM isolates in pigs, two-month-old piglets were treated with PRV-GD or PRV-JM at 10^6 TCID$_{50}$, and the innate immune responses and cell death were monitored. Indeed, piglets treated with PRV-GD or PRV-JM initially (48 h) exhibited labored breathing, hydrostomia and reduced feed intake, accompanied by a high fever (>40.5 °C) (Figure 6A). Subsequently, piglets became severely dyspnea, lost weight (Figure 6B), and exhibited neurological symptoms, including convulsion, ataxia, and paddling, and all piglets died after 6 dpi (Figure 6C). The pathological change results showed that PRV-GD and PRV-JM caused hyperemia in the brain, tonsils and kidneys, and necrosis in the lungs and liver (Figure 6D).

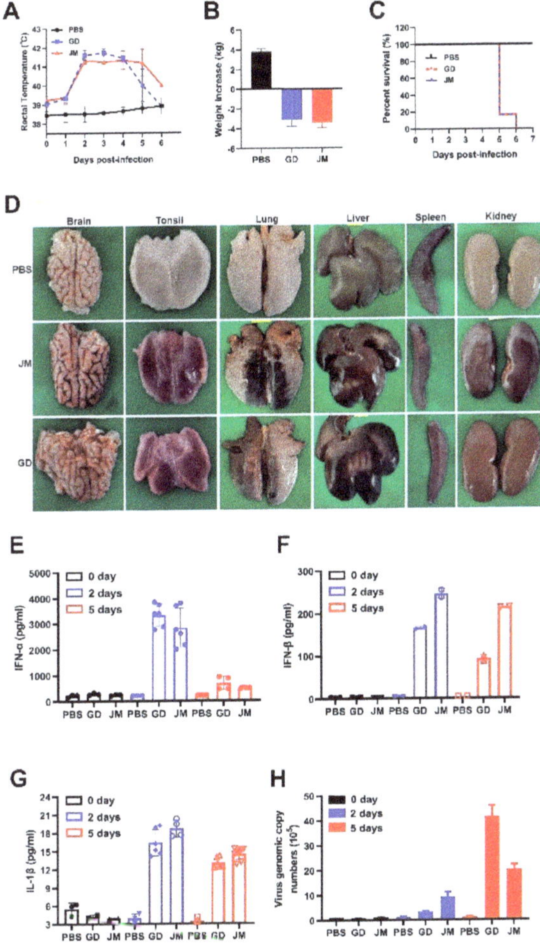

Figure 6. In vivo pathogenicity of PRV-GD and PRV-JM isolates in piglets. (**A**), Rectal temperature of the challenged piglets during the observation. (**B**), The changes of weight of the experimental piglets under the same raised conditions from pre-infection to death. (**C**), Survival rates of piglets infected with PRV-GD or PRV-JM isolates. (**D**), The tissue pathological changes of piglets. (**E–G**), Detection of the protein level of IFN-α, IFN-β and IL-1β in serum from piglets challenged by PRV-GD or PRV-JM. (**H**), Detection of the PRV genomic copy number in blood after infection for 0 days, 2 days and 5 days.

Additionally, qPCR and ELISA were executed to detect the mRNA and protein levels of inflammatory cytokines in piglets infected with PRV-GD and PRV-JM. As structured in Figure S2A–E, the transcription level of pro-inflammatory cytokines (IL-1β, IL-6) and IFN-β were enhanced in PRV-GD and PRV-JM-infected tissues; nevertheless, a weaker change happened in TNF-α. The highest viral load was in the tonsils, while the spleen showed the lowest viral load (Figure S2), suggesting that the tonsils may be the main organ for PRV proliferation in pigs. The ELISA results demonstrated that PRV-infection induced a high level of IFN-α and IFN-β, but a weak release of IL-1β after PRV infection for 2 d and 5 d in pigs (Figure 6E–G). It was noted that there was no IL-6 in serum after PRV infection for 2 d and 5 d (data not shown). Furthermore, the PRV genomic copy number in the blood was significantly enhanced after 2 dpi and then gradually increased (Figure 6H).

Moreover, pathological examinations demonstrated that PRV-GD and PRV-JM infection resulted in glial cell proliferation, neuron degeneration, inflammatory cell infiltration in brain, cell necrosis, congestion in the tonsils, kidneys and spleen, as well as extensive serous exudation, inflammatory cell infiltration, and alveolar epithelial cell abscission in the lung (Figure 7A). Tunnel labeling results suggested that PRV-GD and PRV-JM infection induced cell death (red signal) in the tonsils and brain, and no signal was observed in the PBS-treated piglets (Figure 7B,C).

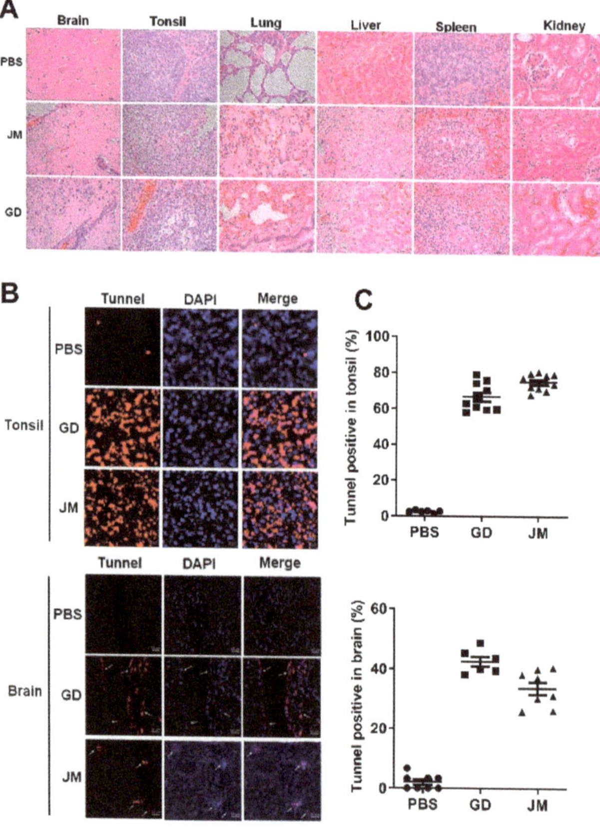

Figure 7. PRV-GD and PRV-JM infection induced tissue injury and cell death. (**A**), Pathological lesions of piglets (H&E staining) that died following infection with PRV-GD or PRV-JM isolate at 10^6 TCID$_{50}$. (**B**), Tunnel solution was used to label the dead cells in the tonsils and brains of piglets. (**C**), The percentage of tunnel-labeled cells was quantified.

4. Discussion

Since the early 1980s, PR had spread almost globally, mainly as sequelae of the international movement of animals and animal products. Due to strict animal disease control measures and eradication programs, including the extensive usage of the traditional vaccine strain Bartha-K61, PR has virtually disappeared from domestic pigs in several countries, e.g., Austria, Germany, Switzerland, Great Britain, Canada, New Zealand, and the United States [22]. However, PR is still endemic in areas with dense pig populations, i.e., some regions in eastern and south eastern Europe, Latin America, Africa, and Asia [22,23]. In addition, wild boar was a potential and persistent reservoir for PRV, since PRV-infected wild boar represent an increasingly obvious threat for the reemergence of PRV into free regions [8,23]. As the largest swine breeding country, highly pathogenic PRV variant epidemics have caused huge economic losses and great difficulties in the prevention and control of PR in China since 2011 [24,25]. In our study, we successfully isolated and identified two new PRV variant strains from the PRV seropositive pig farms (Figure 1). Phylogenetic trees were constructed to reveal the origin and genetic relationships between the two PRV isolates and other strains (Figure 1D). Animal experiments indicated that PRV-GD and PRV-JM isolates could lead to a high mortality in mice and piglets, as well as adult pigs (data not shown). This study provides new information about the prevalence of the PRV variants currently circulating in China.

Symptomatically, PRV infection results in prominent central nervous system disorders and acute encephalitis in both humans and animals [26]. It was reported that PRV-infected pigs could spread PRV to healthy pigs, people, sheep, raccoons, and vice versa. However, the horizontal spread within non-natural hosts might not exist [13]. Therefore, pigs are the only reservoir host of PRV and are recognized as the central link of PRV cross-species transmission. Recombination of PRV strains in pigs may result in changes of antigenicity, virulence, and thus immune failure, which could be the source of continuing epidemics. Therefore, strengthening the monitoring of the prevalence of PRV variants in pigs has great significance in preventing and controlling the spread of PRV variants among other species.

Previous studies showed that PRV strains can be divided into two main clades with frequent interclade and intraclade recombination. Clade 2.2 (PRV variant) is currently the most prevalent genotype worldwide that is most frequently involved in interspecies transmission events (including humans) [13]. According to recent studies, novel PRV variants showed enhanced pathogenicity. Evidence revealed that only 40% of human genes was found to utilize translation initiation sites (TISs) [27]. In comparison, PRV could highly efficiently utilize TISs and integrate the genes of hosts or other viruses, providing more additional possibilities for PRV genetic variation. Indeed, PRV variants delineated a more complex transcriptome and identified an unexpectedly large number of potential novel genes [28]. Additionally, mounting research suggests that overlapping transcription may be the novel strategies of the PRV to regulate its gene expression, escape host innate antiviral immunity, and fulfill genetic evolution at different infection stages [27,29]. In this study, although in the same genetic branch, PRV-GD and PRV-JM isolates had stronger virulence and proliferation rate than that of PRV-TJ on THP-1 cells and mouse peritoneal macrophages (Figure 2). Notably, pigs immunized with classical Ea vaccine were incapable of providing sufficient protection against the two novel PRV-GD and PRV-JM isolates, whereas the antisera from pigs infected with PRV-JM had a high cross-neutralization activity to PRV-TJ, PRV-GD, and PRV-JM (Figure 3). These results suggest that PRV-GD and PRV-JM are virulent isolates and further research is needed for the enhanced virulence. In general, there is an urgent need to strengthen epidemiologic surveillance and develop new effective vaccines to control PR by targeting variant isolates.

In 1970, Jentzsch reported that PR could not occur in humans [30]. Recently increasing human viral infection cases involved in PRV have been reported in China, indicating that PRV can spread from pigs to humans and result in pruritus, throat pain, disturbance of consciousness, or respiratory failure [13]. It makes sense that the virus can spread between humans and swine, because there is a 96% homology of nectin-1 receptor between humans and swine, which mediate several viruses (PRV, HSV-1 and BHV-1) to enter cells [31,32]. In our study, compared with PRV-TJ strain, PRV-GD and PRV-JM isolates had a similar multiplication rate in the PK-15 cells and PAMs but showed enhanced proliferation in THP-1 cells and mouse peritoneal macrophages (Figure 2), suggesting PRV-GD and PRV-JM may have the stronger appetency to these cells. Furthermore, pigs have considerable impacts on human health because of the high similarity of the anatomical structures and immune systems, as well as promising medical resources in xenotransplantation [33]. At present, there are no effective drugs to prevent the progression of the disease caused by PRV infection. Therefore, although human PRV infection cases were rare, it is not be ignored that PRV pose a significant threat to public health, especially in people in close contact with sick pigs and/or related pork products/contaminants. In addition, whether PRV variant strains have enhanced virulence to humans and whether PRV-infection induce a cross-protection against others herpes viruses are also unknown and need to be further studied.

A higher level of pro-inflammatory cytokines (IL-1β, IL-6, TNF-α) were induced in serum from the PRV-infected mice, while only a small amount of IL-1β, and no IL-6 were detected in the serum from the PRV-challenged pigs (Figures 5D and 6G). In addition, PRV infection causes meningitis and conjunctivitis in humans, severe pruritus in mice, and severe respiratory symptoms in adult pigs. These investigations indicated that there were great differences to the hosts' innate immune responses induced by PRV among different species, which may be due to the differences in the immune systems of rodents and pigs. Thus, for pigs or mice, it is worth exploring which is the best model animal to study the pathogenicity of PRV in humans. The detailed molecular mechanisms also need to be further explored.

In summary, two fatal PRV strains, PRV-GD and PRV-JM, were isolated and characterized as the novel variant isolates according to their etiological features and phylogenetic relationships, as well as the lethal rate to mice and pigs. Given the current global epidemic of PRV variant strains in pigs, our analysis of the PRV variants emergence illustrates the need for continuous monitoring and the development of vaccines against specific variants of PRV.

Supplementary Materials: The following supporting information can be downloaded at: https://www.mdpi.com/article/10.3390/v14040712/s1, Figure S1: qPCR analysis of the mRNA level of IL-1β, IL-6, TNF-α, IFN-β and PRV genomic copy number in the liver (A), spleen (B), lung (C) and brain (D) from the mice challenged by PRV-GD and PRV-JM. The significance of differences was analyzed with t test. * $p < 0.05$, ** $p < 0.01$, *** $p < 0.001$; Figure S2: qPCR analysis of the mRNA level of IL-1β, IL-6, TNF-α, IFN-β and PRV genomic copy number in the brain (A), liver (B), tonsil (C), spleen (D) and lung (E) from the piglets challenged by PRV-GD or PRV-JM isolates. The significance of differences were analyzed with t test. * $p < 0.05$, ** $p < 0.01$, *** $p < 0.001$, ns., no significant difference; Table S1: Detection the antibody level of PRV-gB protein in antisera; Table S2: Primers used in this study for the PRV detection; Table S3: Primers used for RT-qPCR amplification; Table S4: The primers and TaqMan probe sequences [34,35].

Author Contributions: Conceptualization, L.H. and C.W.; data curation, Q.Z. and L.Z.; funding acquisition, L.H.; investigation, Q.Z., L.Z., H.L. and G.Y.; methodology, Q.Z., L.Z. and H.L.; project administration, C.W.; software, L.Z.; writing—original draft, Q.Z.; writing—review & editing, C.W. All authors have read and agreed to the published version of the manuscript.

Funding: This study was supported by the National Natural Science Foundation of China (grants No. 31941002), Natural Science Foundation of Heilongjiang Province of China (grants No. YQ2020C022), State Key Laboratory of Veterinary Biotechnology Program (SKLVBP202101).

Institutional Review Board Statement: All animal experiments were performed according to animal protocols approved by the Subcommittee on Research Animal Care at Harbin Veterinary Research Institute (HVRI) of the Chinese Academy of Agricultural Sciences (the ethical approval number: 210608-01 and 211026-02). This study was carried out in strict accordance with the recommendations in the Guide for the Care and Use of Laboratory Animals of the Ministry of Science and Technology of the People's Republic of China.

Informed Consent Statement: Not applicable.

Data Availability Statement: Data supporting the reported results are available in this article and in the Supplementary Materials.

Conflicts of Interest: The authors declare no conflict of interest.

References

1. An, L.; Lu, G.; Qi, J.; Wu, L.; Tian, K.; Luo, T.; Shi, Y.; Yan, J.; Gao, G.F.; Hong, Z.Z. Structural basis of nectin-1 recognition by pseudorabies virus glycoprotein D. *PLoS Pathog.* **2017**, *13*, e1006314.
2. Wu, X.M.; Chen, Q.Y.; Chen, R.J.; Che, Y.L.; Wang, L.B.; Wang, C.Y.; Shan, Y.; Liu, Y.T.; Xiu, J.S.; Zhou, L.J. Pathogenicity and Whole Genome Sequence Analysis of a Pseudorabies Virus Strain FJ-2012 Isolated from Fujian, Southern China. *Can. J. Infect. Dis. Med. Microbiol.* **2017**, *2017*, 9073172. [CrossRef] [PubMed]
3. An, T.Q.; Peng, J.M.; Tian, Z.J.; Zhao, H.Y.; Li, N.; Liu, Y.M.; Chen, J.Z.; Leng, C.L.; Sun, Y.; Chang, D. Pseudorabies Virus Variant in Bartha-K61–Vaccinated Pigs, China, 2012. *Emerg. Infect. Dis.* **2013**, *19*, 1749–1755. [CrossRef] [PubMed]
4. Luo, Y.; Li, N.; Cong, X.; Wang, C.H.; Du, M.; Li, L.; Zhao, B.; Yuan, J.; Liu, D.D.; Li, S. Pathogenicity and genomic characterization of a pseudorabies virus variant isolated from Bartha-K61-vaccinated swine population in China. *Vet. Microbiol.* **2014**, *174*, 107–115. [CrossRef]
5. Ye, C.; Zhang, Q.Z.; Tian, Z.J.; Zheng, H.; Zhao, K.; Liu, F.; Guo, J.C.; Tong, W.; Jiang, C.G.; Wang, S.J.; et al. Genomic characterization of emergent pseudorabies virus in China reveals marked sequence divergence: Evidence for the existence of two major genotypes-ScienceDirect. *Virology* **2015**, *483*, 32–43. [CrossRef]
6. Yu, X.; Zhou, Z.; Hu, D.; Zhang, Q.; Han, T.; Li, X.; Gu, X.; Yuan, L.; Zhang, S.; Wang, B. Pathogenic Pseudorabies Virus, China, 2012. *Emerg. Infect. Dis.* **2014**, *20*, 102. [CrossRef]
7. Yang, Q.; Sun, Z.; Tan, F.; Guo, L.; Xiao, Y. Pathogenicity of a currently circulating Chinese variant pseudorabies virus in pigs. *World J. Virol.* **2016**, *5*, 23. [CrossRef]
8. Sehl, J.; Teifke, J.P. Comparative Pathology of Pseudorabies in Different Naturally and Experimentally Infected Species—A Review. *Pathogens* **2020**, *9*, 633. [CrossRef]
9. Tan, L.; Yao, J.; Yang, Y.; Luo, W.; Wang, A. Current Status and Challenge of Pseudorabies Virus Infection in China. *Virol. Sin.* **2021**, *36*, 588–607. [CrossRef]
10. Hanson, R.P. The history of pseudorabies in the United States. *J. Am. Vet. Med. Assoc.* **1954**, *124*, 259–261.
11. Thawley, D.G.; Wright, J.C. PSEUDORABIES VIRUS INFECTION IN RACCOONS: A REVIEW. *J. Wildl. Dis.* **1982**, *18*, 113–116. [CrossRef] [PubMed]
12. Ai, J.W.; Weng, S.S.; Cheng, Q.; Cui, P.; Li, Y.J.; Wu, H.L.; Zhu, Y.M.; Xu, B.; Zhang, W.H. Human Endophthalmitis Caused By Pseudorabies Virus Infection, China, 2017. *Emerg Infect Dis.* **2018**, *24*, 1087–1090. [CrossRef] [PubMed]
13. Qingyun, L.; Xiaojuan, W.; Caihua, X.; Shifang, D.; Hongna, Y.; Shibang, G.; Jixuan, L.; Lingzhi, Q.; Fuguo, B.; Dongfang, W. A novel human acute encephalitis caused by pseudorabies virus variant strain. *Clin. Infect. Dis.* **2021**, *73*, e3690–e3700.
14. Mravak, S.; Bienzle, U.; Feldmeier, H.; Hampl, H.; Habermehl, K.O. Pseudorabies in man. *Lancet* **1987**, *1*, 501–502.
15. Yang, H.N.; Han, H.; Wang, H.; Cui, Y.; Ding, S.F. A Case of Human Viral Encephalitis Caused by Pseudorabies Virus Infection in China. *Front. Neurol.* **2019**, *10*, 534. [CrossRef] [PubMed]
16. Ye, G.; Liu, H.; Zhou, Q.; Liu, X.; Huang, L.; Weng, C. A Tug of War: Pseudorabies Virus and Host Antiviral Innate Immunity. *Viruses* **2022**, *14*, 547. [CrossRef] [PubMed]
17. Takeuchi, O.; Akira, S. Pattern recognition receptors and inflammation. *Cell* **2010**, *140*, 805–820. [CrossRef]
18. Qin, C.; Zhang, R.; Lang, Y.; Shao, A.; Xu, A.; Feng, W.; Han, J.; Wang, M.; He, W.; Yu, C. Bclaf1 critically regulates the type I interferon response and is degraded by alphaherpesvirus US3. *PLoS Pathog.* **2019**, *15*, e1007559. [CrossRef]
19. Zhang, R.; Xu, A.; Qin, C.; Zhang, Q.; Chen, S.; Lang, Y.; Wang, M.; Li, C.; Feng, W.; Zhang, R. Pseudorabies Virus dUTPase UL50 Induces Lysosomal Degradation of Type I Interferon Receptor 1 and Antagonizes the Alpha Interferon Response. *J. Virol.* **2017**, *91*, e01148-17. [CrossRef]
20. Sun, Y.; Luo, Y.; Wang, C.H.; Yuan, J.; Li, N.; Song, K.; Qiu, H.J. Control of swine pseudorabies in China: Opportunities and limitations. *Vet. Microbiol.* **2016**, *183*, 119–124. [CrossRef]
21. Yu, Z.Q.; Tong, W.; Zheng, H.; Li, L.W.; Li, G.X.; Gao, F.; Wang, T.; Liang, C.; Ye, C.; Wu, J.Q. Variations in glycoprotein B contribute to immunogenic difference between PRV variant JS-2012 and Bartha-K61. *Vet. Microbiol.* **2017**, *208*, 97. [CrossRef] [PubMed]
22. Müller, T.; Hahn, E.C.; Tottewitz, F.; Kramer, M.; Klupp, B.G.; Mettenleiter, T.C.; Freuling, C. Pseudorabies virus in wild swine: A global perspective. *Arch. Virol.* **2011**, *156*, 1691–1705. [CrossRef] [PubMed]
23. Müller, T.; Teuffert, J.; Zellmer, R.; Conraths, F.J. Experimental infection of European wild boars and domestic pigs with pseudorabies viruses with differing virulence. *Am. J. Vet. Res.* **2001**, *62*, 252–258. [CrossRef]
24. Yang, H.C. Epidemiological situation of swine diseases in 2014 and the epidemiological trend and control strategies in 2015. *Swine Ind. Sci.* **2015**, *32*, 38–40.
25. He, W.; Zoé, A.; Zhai, X.; Gary, W.; Zhang, C.; Zhu, H.; Xing, G.; Wang, S.; He, W.; Li, K. Interspecies Transmission, Genetic Diversity, and Evolutionary Dynamics of Pseudorabies Virus. *J. Infect. Dis.* **2019**, *219*, 1705–1715. [CrossRef] [PubMed]

26. Reynolds, L.E.; Brittle, E.E.; Enquist, L. Two modes of pseudorabies virus neuroinvasion and lethality in mice. *J. Virol.* **2004**, *78*, 12951–12963.
27. Noderer, W.L.; Flockhart, R.J.; Bhaduri, A.; Diaz de Arce, A.J.; Zhang, J.; Khavari, P.A.; Wang, C.L. Quantitative analysis of mammalian translation initiation sites by FACS-seq. *Mol. Syst. Biol.* **2014**, *10*, 748. [CrossRef]
28. Torma, G.; Tombácz, D.; Csabai, Z.; Gbhardter, D.; Boldogki, Z. An Integrated Sequencing Approach for Updating the Pseudorabies Virus Transcriptome. *Pathogens* **2021**, *10*, 242. [CrossRef]
29. Toledo-Arana, A.; Lasa, I. Advances in bacterial transcriptome understanding: From overlapping transcription to the excludon concept. *Mol. Microbiol.* **2020**, *113*, 593–602. [CrossRef]
30. Jentzsch, K.D.; Apostoloff, E. Human susceptibility to herpesvirus suis (Aujeszky-virus). 4. Serological determinations in persons in infection endangered occupational groups. *Ztschrift Die Gesamte Hyg. Ihre Grenzgeb.* **1970**, *16*, 692.
31. Spear, P.G.; Eisenberg, R.J.; Cohen, G.H. Three Classes of Cell Surface Receptors for Alphaherpesvirus Entry. *Virology* **2000**, *275*, 1–8. [CrossRef] [PubMed]
32. Milne, R.S.B.; Connolly, S.A.; Krummenacher, C.; Eisenberg, R.J.; Cohen, G.H. Porcine HveC, a Member of the Highly Conserved HveC/Nectin 1 Family, Is a Functional Alphaherpesvirus Receptor. *Virology* **2001**, *281*, 315–328. [CrossRef] [PubMed]
33. Ekser, B.; Ezzelarab, M.; Hara, H.; Windt, D.; Cooper, D. Clinical xenotransplantation: The next medical revolution? *Lancet* **2012**, *379*, 672–683. [CrossRef]
34. Zhang, K.; Zhang, Y.; Xue, J.; Meng, Q.; Liu, H.; Bi, C.; Li, C.; Hu, L.; Yu, H.; Xiong, T.; et al. DDX19 Inhibits Type I Inter-feron Production by Disrupting TBK1-IKKε-IRF3 Interactions and Promoting TBK1 and IKKε Degradation -ScienceDirect. *Cell Rep.* **2019**, *26*, 1258–1272. [CrossRef] [PubMed]
35. Zhou, Q.; Yan, B.; Sun, W.; Chen, Q.; Shi, D. Pig Liver Esterases Hydrolyze Endocannabinoids and Promote Inflammatory Response. *Front. Immunol.* **2021**, *12*, 670426. [CrossRef]

Article

Generation of Premature Termination Codon (PTC)-Harboring Pseudorabies Virus (PRV) via Genetic Code Expansion Technology

Tong-Yun Wang [1,†], Guo-Ju Sang [2,†], Qian Wang [1], Chao-Liang Leng [3], Zhi-Jun Tian [1], Jin-Mei Peng [1], Shu-Jie Wang [1], Ming-Xia Sun [1], Fan-Dan Meng [1], Hao Zheng [2,*], Xue-Hui Cai [1,*] and Yan-Dong Tang [1,*]

[1] State Key Laboratory of Veterinary Biotechnology, Harbin Veterinary Research Institute,
   The Chinese Academy of Agricultural Sciences, Harbin 150069, China; sdndwty@163.com (T.-Y.W.);
   wangqian@caas.cn (Q.W.); tianzhijun@caas.cn (Z.-J.T.); pengjinmei@caas.cn (J.-M.P.);
   wangshujie@caas.cn (S.-J.W.); sunmingxia@caas.cn (M.-X.S.); mengfandan@caas.cn (F.-D.M.)
[2] Shanghai Veterinary Research Institute, The Chinese Academy of Agricultural Sciences,
   Shanghai 200241, China; 18438615167@139.com
[3] Henan Provincial Engineering and Technology Center of Animal Disease Diagnosis and Integrated Control,
   Nanyang Normal University, Nanyang 473061, China; lenghan1223@126.com
* Correspondence: haozheng@shvri.ac.cn (H.Z.); caixuehui139@163.com (X.-H.C.);
   tangyandong2008@163.com (Y.-D.T.)
† These authors contributed equally to this work.

Citation: Wang, T.-Y.; Sang, G.-J.; Wang, Q.; Leng, C.-L.; Tian, Z.-J.; Peng, J.-M.; Wang, S.-J.; Sun, M.-X.; Meng, F.-D.; Zheng, H.; et al. Generation of Premature Termination Codon (PTC)-Harboring Pseudorabies Virus (PRV) via Genetic Code Expansion Technology. *Viruses* **2022**, *14*, 572. https://doi.org/10.3390/v14030572

Academic Editor: Elisa Crisci

Received: 17 February 2022
Accepted: 9 March 2022
Published: 10 March 2022

Abstract: Despite many efforts and diverse approaches, developing an effective herpesvirus vaccine remains a great challenge. Traditional inactivated and live-attenuated vaccines always raise efficacy or safety concerns. This study used Pseudorabies virus (PRV), a swine herpes virus, as a model. We attempted to develop a live but replication-incompetent PRV by genetic code expansion (GCE) technology. Premature termination codon (PTC) harboring PRV was successfully rescued in the presence of orthogonal system MbpylRS/tRNA^{Pyl} pair and unnatural amino acids (UAA). However, UAA incorporating efficacy seemed extremely low in our engineered PRV PTC virus. Furthermore, we failed to establish a stable transgenic cell line containing orthogonal translation machinery for PTC virus replication, and we demonstrated that orthogonal tRNA^{Pyl} is a key limiting factor. This study is the first to demonstrate that orthogonal translation system-mediated amber codon suppression strategy could precisely control PRV-PTC engineered virus replication. To our knowledge, this is the first reported PTC herpesvirus generated by GCE technology. Our work provides a proof-of-concept for generating UAAs-controlled PRV-PTC virus, which can be used as a safe and effective vaccine.

Keywords: premature termination codon; pseudorabies virus; genetic code expansion

1. Introduction

Developing the herpesviruses vaccine is challenging because of the immunologically silent nature of its latency, and the virus mediates immune evasion [1,2]. Pseudorabies virus (PRV) is a swine herpesvirus, also known as Aujeszky virus, which belongs to the genus *Varicellovirus* in the subfamily *Alphaherpesvirinae* of the family *Herpesviridae* [3]. PRV is lethal to many domestic and wild animals, and pigs are the natural host [4]. Since 2011, PRV variants have emerged in China, and commercial vaccines fail to provide complete protection against PRV [5,6]. More seriously, PRV variants can spill over into humans and cause severely nerve-related diseases [7,8]. Developing a safe and effective PRV vaccine is one of the best choices for PRV control in related animals and humans.

Inactivated vaccines play a vital role in eradicating PRV in swine farms. However, the inactivated PRV vaccine mainly induces a humoral immune response, lacks effective T cell response, and inactivated PRV vaccines fail to stop viral shedding post-virus challenge [9,10]. Live-attenuated vaccines have shown the best efficacy against PRV; however, this raises safety concerns, e.g., the attenuated PRV strains are lethal to dogs and can spread horizontally [11]. Therefore, developing a safe and effective PRV vaccine faces a dilemma.

Genetic code expansion (GCE) technology is an orthogonal translation system derived from the *Methanosarcina barkeri*. In this microbe, amber (TAG) stop codon can be read-through with the cooperation of Mb pyrrolysyl tRNA synthetase/tRNA^{Pyl} pair (MbpylRS/tRNA^{Pyl}) and unnatural amino acids (UAA) [12–14]. The application of GCE technology in PTC harboring PRV is illustrated in Figure 1. The GCE technology provides a novel strategy to generate a live but replication-defective candidate vaccine. This technology has been successfully applied in the influenza A virus vaccine [14].

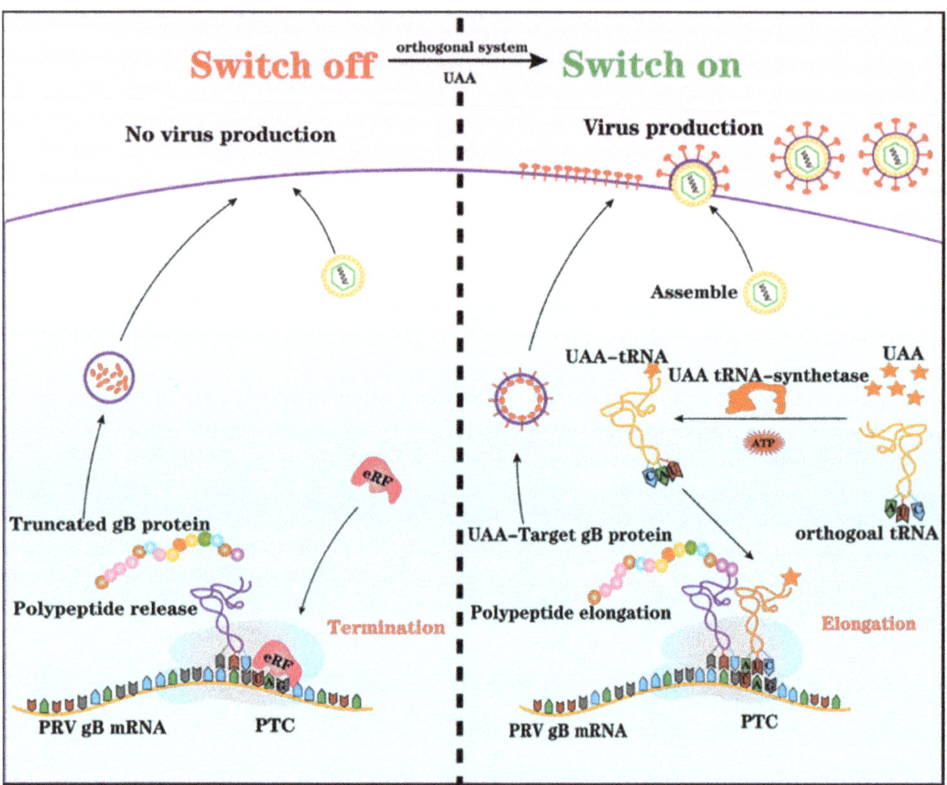

Figure 1. Schematic representation of the construction of PTC harboring PRV via genetic code expansion. PTC harboring PRV failed to replicate in normal cells (**Left**) and can replicate in cells with orthogonal translation machinery system (**Right**). UAA, unnatural amino acid. PTC, premature termination codon.

This study used PRV as a herpesvirus model and attempted to engineer a PTC site in an essential gene of PRV, *gB*, with amber codons (TAG). PTC harboring PRV could be successfully rescued in the orthogonal translation machinery system MbpylRS/tRNA^{Pyl} pair and UAA. However, UAA incorporating efficacy seemed extremely low in our engineered PRV PTC virus. Furthermore, all our attempts to construct cell lines containing orthogonal translation machinery system MbpylRS/tRNA^{Pyl} pair failed. These results suggested that several key issues should be resolved in PTC harboring herpesvirus production in the future.

2. Materials and Methods

2.1. Cells and Plasmids

Human embryonic kidney 293T cells (HEK-293T, ATCC CRL-11268), Rabbit kidney cells (RK13, ATCC CCL-37), Vero cells (ATCC CCL-81), Human embryonic kidney cells (HEK-293A, ATCC CRL-1573), swine testicular cells (ST, ATCC CRL1746), and Porcine kidney cells (PK15, ATCC CCL-33) were maintained in DMEM (Gibco, Waltham, MA, USA) with 10% (v/v) fetal bovine serum (ExCell Bio, Canberra, Australia). As previously described, the pseudorabies virus infectious clone was pBac-JS2012 [15]. Plasmids related to orthogonal translation system, pSD31-pylRS, bjmu-12t-zeo, and pSD31-GFP^{39TAG} were kindly provided by Professor De-Min Zhou of Peking University [14]. PiggyBac transposon plasmids pB513B and PB220A plasmids were described in our previous works [16,17].

2.2. Construction of PTC Harboring gB Mutants

pCAGGS gB plasmid was previously described [18]. The amino acids at positions 149Q, 169K, 171K, 177K, 185W, 206Q, 217K, 221K, 267K, 285W, 319H, 331Q, 362W, 365W, 367W, 370K, 379K, and 413Q of gB protein were mutated to amber codon TAG. Briefly, the indicated amino acid codon was replaced by a TAG amber codon by site-directed mutagenesis PCR. All clones were verified by DNA sequencing. The site-directed mutagenesis primers are listed in Supplementary Table S1.

2.3. Read-Through Efficacy for PTC Harboring gB by GCE

HEK-293T cells in good growth condition were plated in 24 well plates (4×10^5 cells/well). 0.5 µg of MbpylRS/tRNA^{Pyl} plasmids were co-transfected with 0.5 µg of pCAGGS gB PTC plasmid using the jetPRIME transfection reagent (Polyplus, Illkirch, France). Two parallel experiments were conducted. pCAGGS gB was used as a positive control, and non-transfected cells were used as mock. The supernatant was replaced by fresh medium supplemented with 2% FBS in the presence of 1 mM NAEK (TRC, Ottawa, ON, Canada) or not, 6 h post-transfection. At 48 h post-transfection, a Western blot was performed to analyze the read-through efficacy of PTC harboring gB mutants. The cells were lysed by 70 µL lysis buffer (50 mM KCl, 100 mM NaCl, 0.25% NP-40, 1 mM DTT, and 50 mM herpes-NaOH) containing 1% protease inhibitor (Sigma-Aldrich, St. Louis, MO, USA) for 30 min on ice and then centrifuged at $12,000 \times g$ for 10 min at 4 °C. Cell lysates were mixed with $5 \times$ loading buffer and boiled at 100 °C for 10 min. As previously described, the samples were separated by 10% SDS-PAGE and transferred to polyvinylidene fluoride (PVDF) membranes [16].

2.4. Read-Through Efficacy for gB PTC Identified by Cell-to-Cell Fusion Assay

Cell-to-cell fusion assay was performed as previously described [18]. Briefly RK13 cells were placed in 24 well plates (4×10^5 cells/well), and 0.15 µg pCAGGS gB or indicated gB PTCs, 0.15 µg pCAGGS gD, 0.15 µg pCAGGS gH, 0.15 µg pCAGGS gL, 0.15 µg pDC315-GFP, and 0.4 µg MbpylRS/tRNA^{Pyl} were co-transfected. The non-transfected cells were used as control. The supernatant medium was replaced by a fresh medium supplemented with 2% FBS in the presence of 1 mM UAA (NAEK) or not, 6 h post-transfection. 48 h after transfection, fluorescence microscopy was used to analyze the read-through efficacy by indicated gB mediated cell fusion analysis.

2.5. PTC Harboring PRV Construction

PRV-PTC construction was performed as previously described [15]. Briefly, the procedure is as follows. A DNA fragment with a galK expression cassette flanked by 50 bp homologous of *gB* gene was amplified by PCR using the primers gB-galKF/gB-galKR (gB-galKF: 5-GGGACCGCTTCTACGTCTGCCCGCCGCCGTCCGGCTCCACGGTGGT cctgttgacaattaatcatcggca-3; gB-galKR: 5-AGGCGGTCACCTTGTGGTTGTTGCGCACGTAC TCGGCCTTGGAGACGCACTTGCCtcagcactgtcctgctcctt-3) and KOD DNA polymerase (Toyobo, Osaka, Japan). The obtained PCR product was digested with DpnI (Thermo

Fisher, Waltham, MA, USA) at 37 °C for 1 h to remove the original template plasmid, followed by agarose gel purification. To generate SW102-JS-galK, 25 µL SW102-JS electro-competent cells were prepared and electro-transformed with 100 ng galK DNA fragment under the condition of 1.5 kV, 25 µF, 200 Ω. Then, 800 µL SOC medium was added immediately after electro-transformation and incubation at 32 °C, 200 rpm for 1.5 h. The recovered bacteria were washed twice with 1 mL M9 solution and took 150 µL M9 solution to plate the bacteria cells onto M63 plates containing galactose and chloramphenicol. PCR was used to confirm galK positive colonies with primers LgalKup/LgalKdown (LgalKup: 5-TGCTGCGCCTCGACCCCAA-3; LgalKdown: 5-AAGAACTTAACCCGGCACCCT-3). The galK positive colonies were further screened on a MacConkey plate containing chloramphenicol to realize the nucleotide substitution of galK at positions 141 to 187 of gB in pBac-JS2012. Finally, gB fragments harboring PTC points were used to remove the *galK* gene from pBac-JS2012-galK. A DNA fragment with PTC points in gB ORF was amplified from the template gB mutant plasmids with the primers gB-LF/gB-RR (gB-LF: 5-cgacggtatcgataagcttgatCGCTGGTGGCGGTCTTTG-3; gB-RR: 5-ccgggctgcaggaattcgatGAG TCCAGGTCGATGGGGTAG-3). The obtained PCR product was also digested with DpnI (Thermo Fisher, Waltham, MA, USA) at 37 °C for 1 h. The indicated PTC harboring gB fragment was electro-transformed into SW102-JS-galK as described above. After 3 h at 32 °C, the transformed cells were washed and suspended in an M9 medium. The positive clone with mutant gB fragment replacing galK was screened on M63 minimal medium plates containing chloramphenicol, 2-deoxy-galactose, and glycerol. The obtained PRV Bac clone with TAG PTC in gB was termed PRV-PTC.

2.6. Rescue of PRV-PTC Virus

HEK-293T cells and Vero cells (2×10^6) were plated in 6-well plates in DMEM supplemented with 10% FBS. Then, 2 µg MbpylRS/tRNA^{Pyl} plasmid were co-transfected with the 2 µg pPRV-Bac or the indicated pPRV-PTC-Bac using the transfection reagent jetPRIME (Polyplus, Illkirch, France) according to the manufacturer's instructions. At 6 h post-transfection, the supernatant was replaced with DMEM containing 2% FBS and 1 mM UAAs, Nε-2-azidoethyloxycarbonyl-L-lysine (NAEK). To identify the UAA-dependence of PRV-PTC virus, a parallel packaging experiment was conducted in which the medium was not supplemented with UAA. The cells were further incubated at 37 °C in 5% CO_2 until cytopathic effect (CPE) or syncytium was observed.

2.7. Electron Microscopy for PRV-PTC Virus

HEK-293T cells were transfected with plasmids described above for conventional electron microscopy analysis. Then when the cytopathic effect (CPE) or syncytium was observed, cells were fixed with 2.5% (*w/v*) glutaraldehyde in 200 mM HEPES (pH 7.4) for 2 h at room temperature, followed by post-fixation with 1% OsO_4 and 1.5% $K_3Fe(CN)_6$ in H_2O at 4 °C for 30 min. According to standard procedures, samples were dehydrated with acetone and impregnated with epoxy at room temperature and further embedded overnight at 70 °C for polymerization. Then the samples were cut into 70 nm ultrathin sections by ultrathin slicer (Leica, Wetzlar, Germany), stained with 2% uranium acetate for 17 min, lead citrate for 12 min. Specimens were examined using a conventional transmission electron microscope (TEM, h7650, Hitachi, Tokyo, Japan).

2.8. pB513B-Puro-MbpylRS-12 tRNA^{Pyl} Plasmid Construction

MbpylRS gene was optimized to make it more suitable for the mammalian cell system. The original plasmid pB513B was modified to obtain the pB513B plasmid containing MbpylRS and tRNA^{Pyl} simultaneously. Puromycin fragment was amplified by Puro (XbaI) -F: 5-ATTTTCTAGAATGACCGAGTACAAGCCACG-3, puro (NheI-EcoRI-HpaI) -R: 5-GCGTTAACGGTTGAATTCGTCGCTAGCGCGCTTGGGTC-3. *Puromycin* gene and the enzyme sites (NheI/EcoRI/HpaI) were inserted pB513B by enzyme digestion with XbaI/HpaI and named pB513B-Puro. The chicken β-actin promoter of the pCAGGS vector was amplified and digested by NheI/EcoRI to insert into pB513B-Puro. Subsequently, the optimized *MbpylRS* gene was constructed between EcoRI and HpaI to form a pB513B-Puro-MbpylRS plasmid. Furthermore, pB513B-Puro-MbpylRS plasmid was digested by SpeI/SfiI, and a pair of small gene sequences F: 5-GTCTTCCCAATCCTCCCCCTTGGATCCGACGTCAGC GTTCGTCGAC-3, R: 5-CTAGGTCGACGAACGCTGACGTCGGATCCAAGGGGGAGGAT TGGGAAGACTGG-3 were annealed into small fragments and directly inserted into the digested vector pB513B-Puro-MbpylRS. The new vector was named pB513B-Puro-MbpylRS-mid, which contains the homologous arm of 12tRNA. Finally, pB513B-Puro-MbpylRS-mid was digested with BamHI, and 12tRNA fragments were obtained by recovering large fragments by gel electrophoresis after SalI/BbsI digestion of the bjmu-12t-zeo vector. Then, 12 tRNA^{Pyl} copies were constructed into pB513B-puro-MbpylRS-12-tRNA plasmid by homologous recombination. The newly constructed plasmid was named pB513B-puro-MbpylRS-12tRNA. Thus, a plasmid system was constructed simultaneously expressing puromycin, MbpylRS, and multiple tRNA^{Pyl} copies on a transposable plasmid.

2.9. Construction of GFP^{39TAG} Reporter Adenovirus

To generate adenovirus harboring GFP^{39TAG} reporter, HEK293 cells were transiently co-transfected with the pDC-315GFP^{39TAG} (plasmid with the TAG stop codon in *GFP* gene) with pBHGloxΔE1 and E3Cre helper plasmids as previously described [19]. 6 h after transfection, the medium was replaced with a fresh medium containing 2% FBS. The transfected cells were harvested after 7~9 d until plaque was observed.

2.10. Generation of Transgenic Cell Line Containing MbpylRS/tRNA^{Pyl} Orthogonal System

Indicated cells were seeded in 6 well plates and were co-transfected with 3 μg of pB513B-puro-MbpylRS-12tRNA plasmid and 1 μg of pB220A-1 plasmid using the transfection reagent jetPRIME (Polyplus, Illkirch, France). Non-transfected cells were used as control. Then, 6 h later, the transfection medium was replaced by DMEM medium supplemented with 10% FBS and 1 mM UAAs (NAEK). 48 h after transfection, the cells were selected under the pressure of indicated concentrate puromycin (Gibco, Waltham, MA, USA). The medium was replaced every day until the cells in the control group completely died. The resultant cells were stably transfected and continued to cultivate in the presence of 4 μg/mL puromycin. Then these cells were infected with GFP^{39TAG} reporter adenovirus in the presence of UAA, and the single clones were further sorted by fluorescence-activated cell sorting (FACS) according to the GFP reporter.

3. Results

3.1. Evaluation of UAA Site-Specific Incorporation for Potential PRV gB PTC Sites

gB was recognized as an essential gene for PRV replication. Therefore, we selected gB to engineer potential PTC sites. First, 149Q, 169K, 171K, 177K, 185W, 206Q, 217K, 221K, 267K, 285W, 319H, 331Q, 362W, 365W, 367W, 370K, 379K, 413Q of gB was separately engineered into amber codon (TAG). PTC containing gB constructs were confirmed by DNA sequencing (data not shown). Then the indicated gB-PTC constructs were co-transfected with the orthogonal MbpylRS/tRNA^{Pyl} pair plasmid, respectively. The UAA in this study was Nε-2-azidoethyloxycarbonyl-L-lysine (NAEK) as illustrated in Figure 2A. Full-length gB protein can be cleavaged by furin, as demonstrated in Figure 2B. Western blot result showed that gB was successfully expressed in the presence of 1 mM NAEK for indicated gB PTC constructs (Figure 2B,D). There was no gB expression without UAA (Figure 2C,E). This result indicated that some gB PTC sites read-through by GCE technology as expected, although the expression level was lower than wild-type gB (Figure 2B,D). As *gB, gD, gH,* and *gL* are the essential viral genes for virus-mediated cell to-cell fusion, which is an important step for PRV spreading [20]. Next, a cell-to-cell fusion assay was used to test whether these NAEK incorporation sites influenced gB mediated cell-to-cell fusion. As previously described, a transient transfection-based cell-to-cell fusion assay was performed by co-transfection of PTC gB, gD, gH, gL and EGFP plasmids [18,21,22]. The result indicated that some gB PTC sites such as 149Q, 169K, 185W, 206Q, 379K successfully induced syncytia formation in the presence of NAEK, and no syncytia was formed without NAEK (Figure 2F). These results indicated that the effect of substitution by NAEK on function varies with position, and the position of 149 and 185 was labeled in the gB crystal structure (Figure 2G).

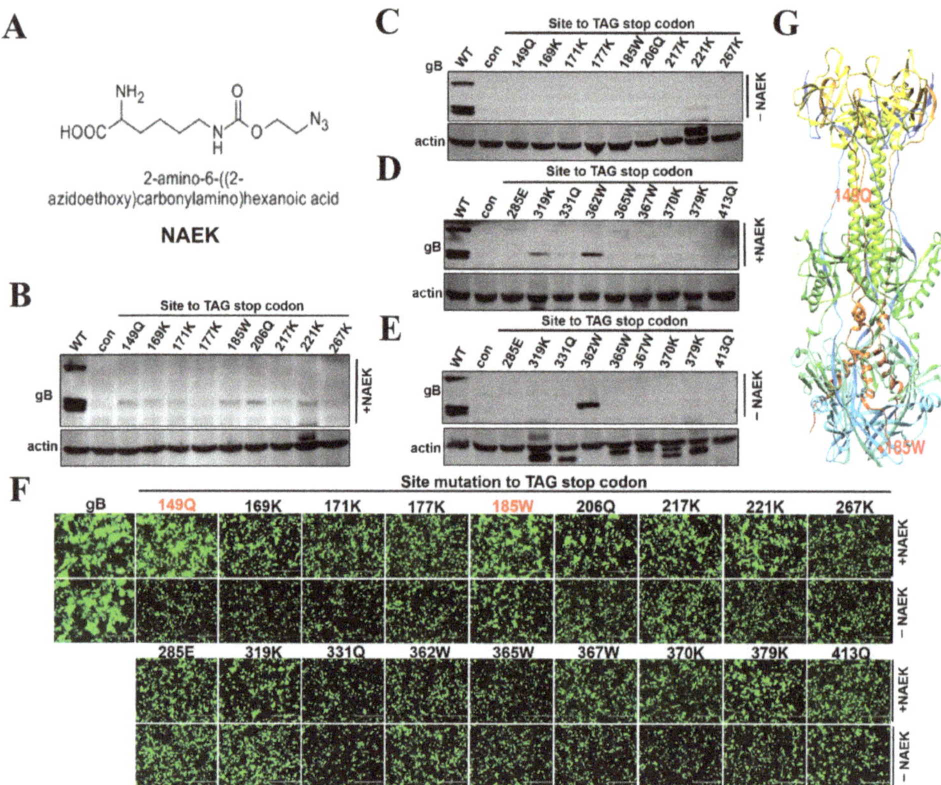

Figure 2. Evaluation of UAA site-specific incorporation for potential PRV gB PTC sites. (**A**) Chemical structure of Nε-2-azideoethyloxycarbonyl-L-lysine (NAEK). (**B,D**) HEK293T cells were co-transfected with GCE machinery and indicated gB-PTC plasmid; 6 h later, the fresh medium replaced the supernatant in the presence of 1 mM UAA (NAEK). Western blot was performed to analyze the UAA-dependent read-through efficacy of PTC harboring gB mutants in the presence of 1 mM UAA (NAEK) or (**C,E**) without UAA. β-actin was used as an internal control. (**F**) Cell-to-cell fusion assay was used to evaluate the read-through efficiency of PTC harboring gB mutants. RK13 cells were co-transfected with MbpylRS/tRNA^Pyl, GFP, gD, gH, gL, gB, or its mutants, and 6 h later the supernatant was replaced by fresh medium supplemented with 2% FBS in the presence of 1 mM UAA (NAEK), or not. Fluorescence microscopy was used to analyze the read-through efficacy by indicating gB mediated cell fusion analysis at 48 h after transfection; the scale bar = 400 μm. The red labeled PTC constructs were used further study. (**G**) The corresponding position of 149 and 185 mutation points in the gB crystal structure.

3.2. Construction and Rescue of the PTC Site Harboring PRV

Next, we selected 149Q and 185W as potential PTC engineering sites in PRV, pPRV-149Q-TAG and pPRV-185W-TAG were subsequently constructed using pBac-JS2012 and the galK selection system as previously described [15]. Vero cells were co-transfected with pPRV-149Q-TAG or pPRV-185W-TAG clones with plasmids containing orthogonal translation systems to rescue the PRV-PTC virus. At 48 h post-transfection, typical PRV-induced syncytia were formed in cells transfected with pPRV-149Q-TAG in the presence of UAA. No syncytia were formed without UAA (Figure 3A). However, pPRV-185W-TAG failed induced syncytia in the presence of UAA (Figure 3A). The same results were obtained in 293T cells (Figure 3B). To test whether infectious PRV particles were produced in the presence of UAA, electron microscopic analysis was performed (Figure 3C). Consistent with our above result, PRV particles were observed only in the pPRV-149Q-TAG transfected group in the presence of UAA. Furthermore, no PRV particles were observed in the control group and pPRV-185W-TAG transfected group (Figure 3C). Taken together, UAA could be used as a precise switch for controlling PRV-PTC virus replication.

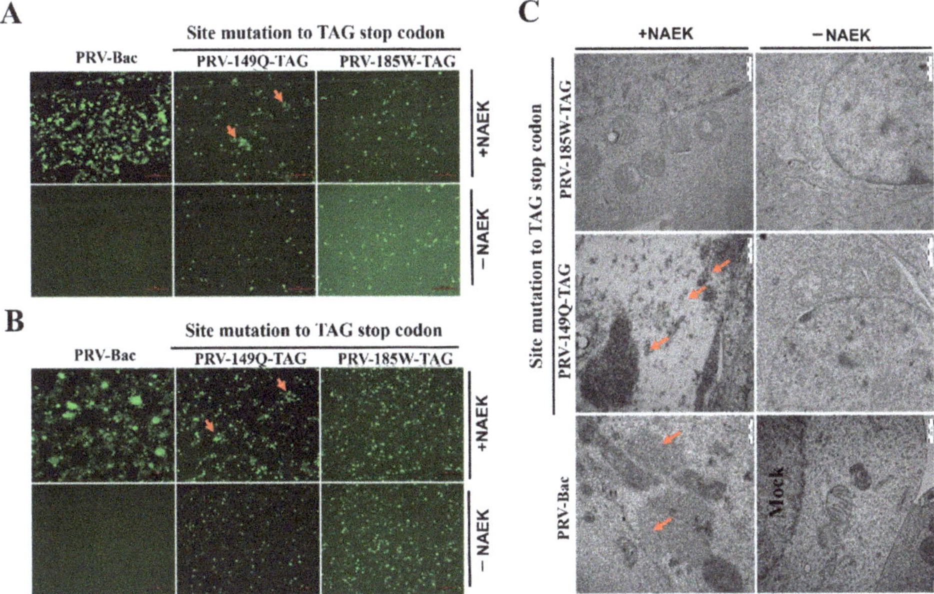

Figure 3. Rescue of PRV-PTC virus on HEK-293T and Vero cells. 2 μg MbpylRS /tRNA^{Pyl} plasmid was co-transfected with the 2 μg pPRV-Bac or the indicated pPRV-PTC-Bac (pPRV-149Q-TAG or pPRV-185W-TAG). At 6 h post-transfection, the supernatant was replaced with DMEM containing 2% FBS in the presence of 1 mM UAA (NAEK), or not. The rescue of PRV-PTC virus on (**A**) Vero cells or (**B**) HEK-293T cells was analyzed by fluorescence microscopy at 48 h post-transfection. The scale bar = 200 μm. (**C**) Electron microscopic analysis was performed to confirm the virus particles of the PRV-PTC virus. The scale bar = 1 μm.

3.3. Generation of MbpylRS/tRNA^{Pyl} Pair Delivery Vector and Reporter Adenovirus

An efficient method to incorporate UAA into the viral PTC site is to generate a stable transgenic cell line harboring MbpylRS/tRNA^{Pyl} pair in the host genome. Lentiviral vector and PiggyBac transposon system are powerful tools to generate stable cell lines [12,14]. However, the PiggyBac transposon system has the advantage of delivering large and complex DNA fragments into the genome of mammalian cells [23]. Therefore, in this study, we used the PiggyBac transposon system to deliver the MbpylRS/tRNA^{Pyl} cassette. First, a PiggyBac transposon vector, pB513B-puro-MbpylRS-12tRNA, was constructed. It contained MbpylRS, which was promoted by chicken β-actin promoter, and 12 tandem tRNA-expression cassettes promoted by U6 or H1 promoters (Figure 4A). To test whether this vector work normally, pB513B-puro-MbpylRS-12tRNA plasmid co-transfected with an amber codon-containing green fluorescent protein (GFP^{39TAG}) reporter plasmid present, with or without UAA (Figure 4B). The results showed that functional GFP was visualized in the presence of UAA (Figure 4C), indicating pB513B-puro-MbpylRS-12tRNA was successfully constructed. To construct transgenic cells without reporter genes, we generate a recombinant adenovirus harboring GFP^{39TAG} (Figure 4D). Recombinant adenovirus was confirmed in RK13 cells, which were first transfected with pB513B-puro-MbpylRS-12tRNA plasmid, then infected with recombinant adenovirus. The results showed that functional GFP was visualized in cells supplemented with UAAs, indicating that recombinant adenovirus was successfully generated (Figure 4E).

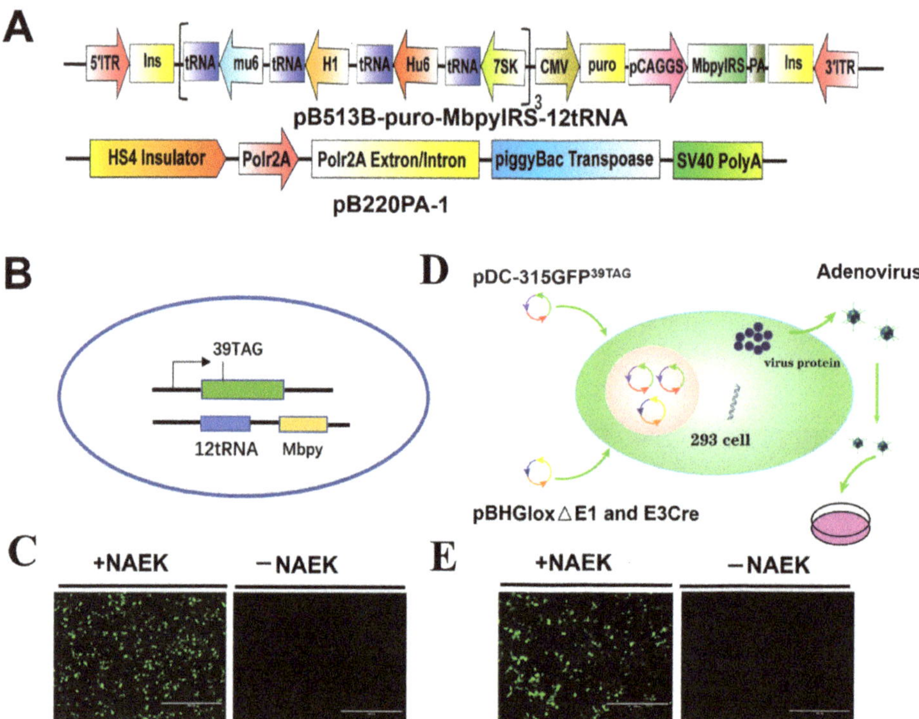

Figure 4. Generation of MbpylRS/tRNA^{Pyl} pair delivery vector and reporter adenovirus. (**A**) A PiggyBac transposon vector, pB513B-puro-MbpylRS-12tRNA, and a helper vector containing the PB transposase expression cassette, PB220A-1, were used to generate orthogonal transgenic stable cell lines. (**B**) *GFP* with an amber codon at position 39 was used as a reporter gene. (**C**) Functional GFP was visualized by fluorescence microscopy in the presence of the MbpylRS/tRNA^{Pyl} pair and UAA. Scale bars, 400 μm. (**D**) Schematic diagram of generating a recombinant adenovirus harboring *GFP^{39TAG}* reporter gene. (**E**) Recombinant adenovirus was confirmed in RK13 cells; RK13 cells were first transfected with pB513B-puro-MbpylRS-12tRNA plasmid, then infected with recombinant adenovirus. The function of adenovirus was verified according to the expression of GFP^{39TAG}. Scale bars, 400 μm.

3.4. Generation of Stable Cell Line Harboring GCE Machinery

To generate stable transgenic RK13, ST, and PK15 cell lines harboring GCE machinery, we co-transfected pB513B-puro-MbpylRS-12tRNA with pB220PA-1 (a vector expressing the PiggyBac transposase) together (Figure 4A). 48 h post-transfection, puromycin (4 μg/mL) was added. Two weeks later, puromycin-resistant cells were infected with reporter adenovirus in the presence of UAA, and GFP expressing single cells were further sorted by fluorescence-activated cell sorting technology (FACS) (Figure 5A). A single-cell of transgenic RK13, ST, PK15 cells was cultivated, to increase over approximately 2–3 weeks. Next, these cell lines were confirmed by infecting with GFP^{39TAG} adenovirus in the presence of UAA, or not. The GFP^{39TAG} expressed well in these cells in the presence of UAA (Figure 5B). The results demonstrated that MbpylRS/tRNA^{Pyl} pair could be successfully delivered by the PiggyBac transposon system.

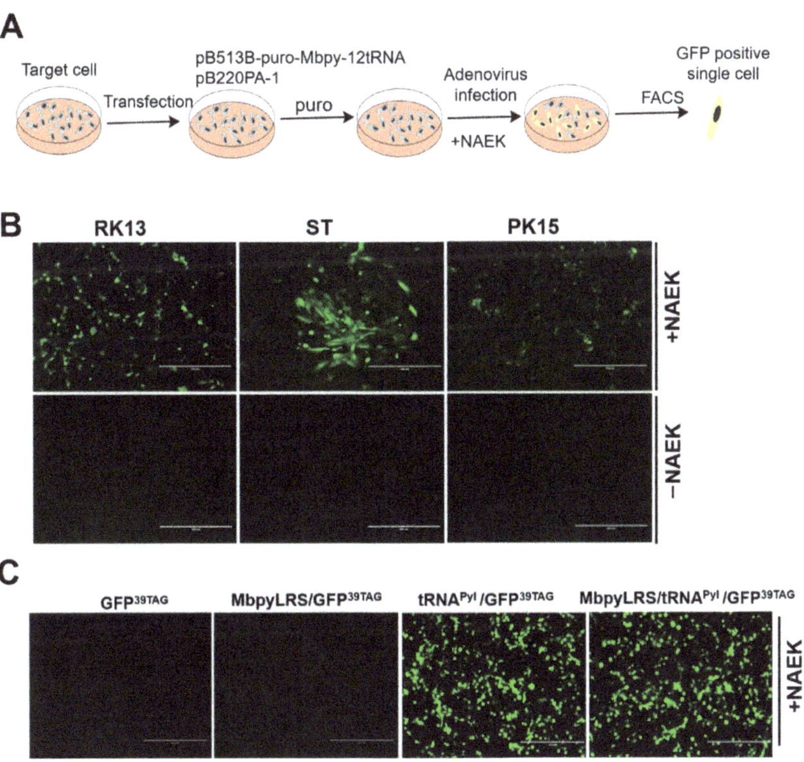

Figure 5. Generation of transgenic cell line containing MbpylRS/tRNA^{Pyl} orthogonal system. (**A**) Schematic representative of the process for generation of transgenic cells containing MbpylRS/tRNA^{Pyl} orthogonal system. (**B**) The well-growth transgenic RK13, ST, PK15 cell lines were infected with packaged adenovirus in the presence of UAA or not. The GFP^{39TAG} expression in these cells was observed by fluorescence microscopy. Scale bars, 400 μm. (**C**) Detection of the key limit factor for orthogonal system fail to translation in RK13 transgenic cell line by transfecting of MbpylRS, GFP^{39TAG} or tRNA^{Pyl} individually or together. The GFP^{39TAG} expression in these groups was observed by fluorescence microscopy. Scale bars, 400 μm.

To our surprise, unlike previous reports [12,14], all our transgenic cell lines are extremely unstable along with the increased passage. An overexpression assay was performed to test which element was lost in these cells. By transfecting MbpylRS, GFP^{39TAG} or tRNA^{Pyl} individually or together, and by co-transfection together, the group was used as a positive control (Figure 5C). Our result revealed that the tRNA^{Pyl} and GFP^{39TAG} co-transfection group restored robust and efficient GFP^{39TAG} expression in the presence of UAA. However, the GFP^{39TAG} transfection group and MbpylRS and GFP^{39TAG} co-transfection group failed to restore efficient GFP^{39TAG} expression in the presence of UAA (Figure 5C). Thus, we concluded that the expression of orthogonal tRNA is the key limiting factor in generating a stable cell line.

4. Discussion

In recent decades, GCE technology has been widely used to engineer PTC sites in essential genes to control different kinds of virus replication, such as HDV, HIV, Zika, FMDV [13,24–27]. Application of GCE technology in Influenza A virus makes it well-known as a novel tool for vaccine development [14]. In this study, we generated PTC harboring PRV, and the results suggested that PRV-PTC virus was successfully rescued in the presence of orthogonal system MbpylRS/tRNA^{Pyl} pair and UAA. This study was the first to demonstrate that UAA was incorporated into PRV gB protein by GCE technology. However, efficiency was generally low. Western blot showed different degrees of weak protein expression in 149Q, 169K, 185W, 206Q, 379K PTC sites, indicating that the incorporation of UAA may have a site preference (Figure 2B). However, only a few of these manifested the UAA-dependent gB mediated cell-to-cell fusion phenotype (Figure 2F). So we speculated that UAA incorporation in target proteins might have deleterious effects on gB function for some PTC sites [28]. More PRV essential genes and potential PTC sites should be screened in the future to obtain ideal PTC sites with efficient, site-specific incorporation of UAAs into PRV.

According to the Western blot and cell-to-cell fusion assay, we chose 149Q and 185W as potential sites to engineer amber codon TAG based on the pPRV-Bac infectious clone and successfully generated pPRV-149Q-TAG or pPRV-185W-TAG PTC virus. pPRV-149Q-TAG PTC virus was successfully rescued in the presence of UAA, demonstrating that GCE technology could control PRV PTC replication in vitro. Unfortunately, in our study the viral titer is extremely low, making it difficult to perform animal experiments in order to evaluate its efficacy as a potential vaccine. pPRV-185W-TAG PTC virus failed to rescue in the presence of UAA. We envisaged that the surrogate of tryptophan at position 185 by UAA might destroy other functions besides a cell-to-cell fusion of gB protein. Therefore, we concluded that the structure of UAA and the incorporation sites might influence the protein function.

Previous reports have demonstrated that the lentivirus vector [13,14] or PiggyBac transposon system successfully constructed stable cell lines harboring the orthogonal translation system [12,13]. Efficient incorporation of UAA requires multiple copies of tRNA^{Pyl} [29], which makes it difficult to package lentivirus efficiently [30]. The PiggyBac transposon system is characterized by rapid and efficient integration of large and complex sequences into mammalian cells' genomes [23,31]. Therefore, this study used the PiggyBac transposon system to generate cell lines containing orthogonal translation machinery. Unfortunately, unlike the previous reports [12–14], all our attempts failed to generate a stable orthogonal MbpylRS/tRNA^{Pyl} system in the current work. The result indicated that insufficient orthogonal tRNA^{Pyl} copies were the limitation steps. Wolfgang H. Schmied et al. developed an optimized pyrrolysyl-tRNA synthetase/tRNA$_{CUA}$ expression system and engineered eukaryotic release factor subunit 1 (eRF1) to efficient incorporate UAA in mammalian cells [29]. Future work will investigate the optimized approaches to efficiently incorporating UAAs at PTC sites in eukaryotic cells.

5. Conclusions

In conclusion, we demonstrated that an orthogonal translation system-mediated amber codon suppression strategy could precisely control PRV-PTC virus replication. To our knowledge, this is the first study reported for herpesvirus generated by GCE technology. However, there are still many challenges remaining to be addressed. Our further work will establish transgenic cell lines with high-efficiency expression of the orthogonal translation system.

Supplementary Materials: The following supporting information can be downloaded at: https://www.mdpi.com/article/10.3390/v14030572/s1, Table S1: primers for constructing PTC harboring gB mutants.

Author Contributions: Conceptualization, T.-Y.W., Z.-J.T., F.-D.M., X.-H.C. and Y.-D.T.; methodology, T.-Y.W., G.-J.S. and Q.W.; software, T.-Y.W. and S.-J.W.; validation, T.-Y.W., Q.W. and J.-M.P.; formal analysis, T.-Y.W. and Q.W.; investigation, T.-Y.W., G.-J.S., C.-L.L. and H.Z.; resources, T.-Y.W., C.-L.L., J.-M.P. and S.-J.W.; data curation, T.-Y.W., S.-J.W. and M.-X.S.; writing—original draft preparation, T.-Y.W. and Y.-D.T.; writing—review and editing, Y.-D.T.; visualization, F.-D.M.; supervision, Z.-J.T., X.-H.C. and Y.-D.T.; project administration, Y.-D.T.; funding acquisition, Q.W. All authors have read and agreed to the published version of the manuscript.

Funding: This research was funded by the Heilongjiang Excellent Youth Fund Project (YQ2019C032 and YQ2019C028).

Informed Consent Statement: Not applicable.

Data Availability Statement: Data is contained within the article or supplementary material. The data presented in this study are available in the insert article or supplementary material here.

Acknowledgments: The authors are grateful to De-Min Zhou in Peking University, who kindly provides plasmids related to orthogonal translation system, pSD31-pylRS, bjmu-12t-zeo, and pSD31-GFP^{39TAG}. The authors are grateful to Chang-Jiang Weng in HVRI for the constructive suggestion.

Conflicts of Interest: The authors declare no conflict of interest.

References

1. Wu, T.T.; Qian, J.; Ang, J.; Sun, R. Vaccine prospect of Kaposi sarcoma-associated herpesvirus. *Curr. Opin. Virol.* **2012**, *2*, 482–488. [CrossRef] [PubMed]
2. Wang, T.Y.; Yang, Y.L.; Feng, C.; Sun, M.X.; Peng, J.M.; Tian, Z.J.; Tang, Y.D.; Cai, X.H. Pseudorabies Virus UL24 Abrogates Tumor Necrosis Factor Alpha-Induced NF-kappaB Activation by Degrading P65. *Viruses* **2020**, *12*, 51. [CrossRef] [PubMed]
3. Pomeranz, L.E.; Reynolds, A.E.; Hengartner, C.J. Molecular biology of pseudorabies virus: Impact on neurovirology and veterinary medicine. *Microbiol. Mol. Biol. Rev.* **2005**, *69*, 462–500. [CrossRef] [PubMed]
4. Mettenleiter, T.C. Molecular biology of pseudorabies (Aujeszky's disease) virus. *Comp. Immunol. Microbiol. Infect. Dis.* **1991**, *14*, 151–163. [CrossRef]
5. Zhang, F.; Wang, M.; Michael, T.; Drabier, R. Novel alternative splicing isoform biomarkers identification from high-throughput plasma proteomics profiling of breast cancer. *BMC Syst. Biol.* **2013**, *7* (Suppl. S5), S8. [CrossRef] [PubMed]
6. Luo, Y.; Li, N.; Cong, X.; Wang, C.H.; Du, M.; Li, L.; Zhao, B.; Yuan, J.; Liu, D.D.; Li, S.; et al. Pathogenicity and genomic characterization of a pseudorabies virus variant isolated from Bartha-K61-vaccinated swine population in China. *Vet. Microbiol.* **2014**, *174*, 107–115. [CrossRef] [PubMed]
7. Ai, J.W.; Weng, S.S.; Cheng, Q.; Cui, P.; Li, Y.J.; Wu, H.L.; Zhu, Y.M.; Xu, B.; Zhang, W.H. Human Endophthalmitis Caused By Pseudorabies Virus Infection, China, 2017. *Emerg. Infect. Dis.* **2018**, *24*, 1087–1090. [CrossRef]
8. Liu, Q.; Wang, X.; Xie, C.; Ding, S.; Yang, H.; Guo, S.; Li, J.; Qin, L.; Ban, F.; Wang, D.; et al. A Novel Human Acute Encephalitis Caused by Pseudorabies Virus Variant Strain. *Clin. Infect. Dis.* **2021**, *73*, e3690–e3700. [CrossRef]
9. Gu, Z.; Dong, J.; Wang, J.; Hou, C.; Sun, H.; Yang, W.; Bai, J.; Jiang, P. A novel inactivated gE/gI deleted pseudorabies virus (PRV) vaccine completely protects pigs from an emerged variant PRV challenge. *Virus Res.* **2015**, *195*, 57–63. [CrossRef]
10. Wang, J.; Guo, R.; Qiao, Y.; Xu, M.; Wang, Z.; Liu, Y.; Gu, Y.; Liu, C.; Hou, J. An inactivated gE-deleted pseudorabies vaccine provides complete clinical protection and reduces virus shedding against challenge by a Chinese pseudorabies variant. *BMC Vet. Res.* **2016**, *12*, 277. [CrossRef]
11. Lin, W.; Shao, Y.; Tan, C.; Shen, Y.; Zhang, X.; Xiao, J.; Wu, Y.; He, L.; Shao, G.; Han, M.; et al. Commercial vaccine against pseudorabies virus: A hidden health risk for dogs. *Vet. Microbiol.* **2019**, *233*, 102–112. [CrossRef] [PubMed]
12. Elsasser, S.J.; Ernst, R.J.; Walker, O.S.; Chin, J.W. Genetic code expansion in stable cell lines enables encoded chromatin modification. *Nat. Methods* **2016**, *13*, 158–164. [CrossRef] [PubMed]
13. Hao, R.; Ma, K.; Ru, Y.; Li, D.; Song, G.; Lu, B.; Liu, H.; Li, Y.; Zhang, J.; Wu, C.; et al. Amber codon is genetically unstable in generation of premature termination codon (PTC)-harbouring Foot-and-mouth disease virus (FMDV) via genetic code expansion. *RNA Biol.* **2021**, *12*, 2330–2341. [CrossRef] [PubMed]
14. Si, L.; Xu, H.; Zhou, X.; Zhang, Z.; Tian, Z.; Wang, Y.; Wu, Y.; Zhang, B.; Niu, Z.; Zhang, C.; et al. Generation of influenza A viruses as live but replication-incompetent virus vaccines. *Science* **2016**, *354*, 1170–1173. [CrossRef]
15. Wang, T.; Tong, W.; Ye, C.; Yu, Z.; Chen, J.; Gao, F.; Shan, T.; Yu, H.; Li, L.; Li, G.; et al. Construction of an infectious bacterial artificial chromosome clone of a pseudorabies virus variant: Reconstituted virus exhibited wild type properties in vitro and in vivo. *J. Virol. Methods* **2018**, *259*, 106–115. [CrossRef]
16. Yang, Y.L.; Liu, J.; Wang, T.Y.; Chen, M.; Wang, G.; Yang, Y.B.; Geng, X.; Sun, M.X.; Meng, F.; Tang, Y.D.; et al. Aminopeptidase N Is an Entry Co-factor Triggering Porcine Deltacoronavirus Entry via an Endocytotic Pathway. *J. Virol.* **2021**, *95*, e0094421. [CrossRef]
17. Yang, Y.L.; Meng, F.; Qin, P.; Herrler, G.; Huang, Y.W.; Tang, Y.D. Trypsin promotes porcine deltacoronavirus mediating cell-to-cell fusion in a cell type-dependent manner. *Emerg. Microbes. Infect.* **2020**, *9*, 457–468. [CrossRef]

18. Wang, Y.; Liu, T.X.; Wang, T.Y.; Tang, Y.D.; Wei, P. Isobavachalcone inhibits Pseudorabies virus by impairing virus-induced cell-to-cell fusion. *Virol. J.* **2020**, *17*, 39. [CrossRef]
19. Tang, Y.D.; Na, L.; Zhu, C.H.; Shen, N.; Yang, F.; Fu, X.Q.; Wang, Y.H.; Fu, L.H.; Wang, J.Y.; Lin, Y.Z.; et al. Equine viperin restricts equine infectious anemia virus replication by inhibiting the production and/or release of viral Gag, Env, and receptor via distortion of the endoplasmic reticulum. *J. Virol.* **2014**, *88*, 12296–12310. [CrossRef]
20. Turner, A.; Bruun, B.; Minson, T.; Browne, H. Glycoproteins gB, gD, and gHgL of herpes simplex virus type 1 are necessary and sufficient to mediate membrane fusion in a Cos cell transfection system. *J. Virol.* **1998**, *72*, 873–875. [CrossRef]
21. Klupp, B.G.; Nixdorf, R.; Mettenleiter, T.C. Pseudorabies virus glycoprotein M inhibits membrane fusion. *J. Virol.* **2000**, *74*, 6760–6768. [CrossRef] [PubMed]
22. Vallbracht, M.; Rehwaldt, S.; Klupp, B.G.; Mettenleiter, T.C.; Fuchs, W. Functional Relevance of the N-Terminal Domain of Pseudorabies Virus Envelope Glycoprotein H and Its Interaction with Glycoprotein, L. *J. Virol.* **2017**, *91*, e00061-17. [CrossRef]
23. Ding, S.; Wu, X.; Li, G.; Han, M.; Zhuang, Y.; Xu, T. Efficient transposition of the piggyBac (PB) transposon in mammalian cells and mice. *Cell* **2005**, *122*, 473–483. [CrossRef] [PubMed]
24. Wang, N.; Li, Y.; Niu, W.; Sun, M.; Cerny, R.; Li, Q.; Guo, J. Construction of a live-attenuated HIV-1 vaccine through genetic code expansion. *Angew. Chem. Int. Ed. Engl.* **2014**, *53*, 4867–4871. [CrossRef]
25. Yuan, Z.; Wang, N.; Kang, G.; Niu, W.; Li, Q.; Guo, J. Controlling Multicycle Replication of Live-Attenuated HIV-1 Using an Unnatural Genetic Switch. *ACS Synth. Biol.* **2017**, *6*, 721–731. [CrossRef]
26. Zhang, R.R.; Ye, Q.; Li, X.F.; Deng, Y.Q.; Xu, Y.P.; Huang, X.Y.; Xia, Q.; Qin, C.F. Construction and characterization of UAA-controlled recombinant Zika virus by genetic code expansion. *Sci. China. Life Sci.* **2021**, *64*, 171–173. [CrossRef] [PubMed]
27. Lin, S.; Yan, H.; Li, L.; Yang, M.; Peng, B.; Chen, S.; Li, W.; Chen, P.R. Site-specific engineering of chemical functionalities on the surface of live hepatitis D virus. *Angew. Chem. Int. Ed. Engl.* **2013**, *52*, 13970–13974. [CrossRef]
28. Wilkerson, J.W.; Smith, A.K.; Wilding, K.M.; Bundy, B.C.; Knotts, T.A., IV. The Effects of p-Azidophenylalanine Incorporation on Protein Structure and Stability. *J. Chem. Inf. Modeling* **2020**, *60*, 5117–5125. [CrossRef]
29. Schmied, W.H.; Elsasser, S.J.; Uttamapinant, C.; Chin, J.W. Efficient multisite unnatural amino acid incorporation in mammalian cells via optimized pyrrolysyl tRNA synthetase/tRNA expression and engineered eRF1. *J. Am. Chem. Soc.* **2014**, *136*, 15577–15583. [CrossRef]
30. Shen, B.; Xiang, Z.; Miller, B.; Louie, G.; Wang, W.; Noel, J.P.; Gage, F.H.; Wang, L. Genetically encoding unnatural amino acids in neural stem cells and optically reporting voltage-sensitive domain changes in differentiated neurons. *Stem Cells* **2011**, *29*, 1231–1240. [CrossRef]
31. Wang, X.; Wei, R.; Li, Q.; Liu, H.; Huang, B.; Gao, J.; Mu, Y.; Wang, C.; Hsu, W.H.; Hiscox, J.A.; et al. PK-15 cells transfected with porcine CD163 by PiggyBac transposon system are susceptible to porcine reproductive and respiratory syndrome virus. *J Virol. Methods* **2013**, *193*, 383–390. [CrossRef] [PubMed]

Article

Histamine Is Responsible for the Neuropathic Itch Induced by the Pseudorabies Virus Variant in a Mouse Model

Bing Wang [1,†], Hongxia Wu [1,†], Hansong Qi [1,†], Hanglin Li [2], Li Pan [1], Lianfeng Li [1], Kehui Zhang [1], Mengqi Yuan [1], Yimin Wang [2,*], Hua-Ji Qiu [1,*] and Yuan Sun [1,*]

[1] State Key Laboratory of Veterinary Biotechnology, Harbin Veterinary Research Institute, Chinese Academy of Agricultural Sciences, 678 Haping Road, Harbin 150069, China; wangbing970112@163.com (B.W.); whx450650@163.com (H.W.); qhs199509@foxmail.com (H.Q.); lipan616@163.com (L.P.); lilianfeng@caas.cn (L.L.); zkhzhangkehui@163.com (K.Z.); yuanmengqi2019@163.com (M.Y.)

[2] Henan Institute of Science and Technology, College of Animal Science and Veterinary Medicine, Xinxiang 453003, China; hanglinLi@hotmail.com

* Correspondence: yiminwang1@hotmail.com (Y.W.); qiuhuaji@caas.cn (H.-J.Q.); sunyuan@caas.cn (Y.S.); Tel.: +86-130-1901-1305 (H.-J.Q.)

† These authors contributed equally to this work.

Abstract: Pseudorabies virus (PRV) is the causative agent of pseudorabies (PR). It can infect a wide range of mammals. PRV infection can cause severe acute neuropathy (the so-called "mad itch") in nonnatural hosts. PRV can infect the peripheral nervous system (PNS), where it can establish a quiescent, latent infection. The dorsal root ganglion (DRG) contains the cell bodies of the spinal sensory neurons, which can transmit peripheral sensory signals, including itch and somatic pain. Little attention has been paid to the underlying mechanism of the itch caused by PRV in nonnatural hosts. In this study, a mouse model of the itch caused by PRV was elaborated. BALB/c mice were infected intramuscularly with 10^5 $TCID_{50}$ of PRV TJ. The frequency of the bite bouts and the durations of itch were recorded and quantified. The results showed that the PRV-infected mice developed spontaneous itch at 32 h postinfection (hpi). The frequency of the bite bouts and the durations of itch were increased over time. The mRNA expression levels of the receptors and the potential cation channels that are relevant to the itch-signal transmission in the DRG neurons were quantified. The mRNA expression levels of tachykinin 1 (TAC1), interleukin 2 (IL-2), IL-31, tryptases, tryptophan hydroxylase 1 (TPH1), and histidine decarboxylase (HDC) were also measured by high-throughput RNA sequencing and real-time reverse transcription PCR. The results showed that the mean mRNA level of the HDC in the DRG neurons isolated from the PRV-infected mice was approximately 25-fold higher than that of the controls at 56 hpi. An immunohistochemistry (IHC) was strongly positive for HDC in the DRG neurons of the PRV-infected mice, which led to the high expression of histamine at the injected sites. The itch of the infected mice was inhibited by chlorphenamine hydrogen maleate (an antagonist for the histamine H1 receptor) in a dose-dependent manner. The mRNA and protein levels of the HDC in the DRG neurons were proportional to the severity of the itch induced by different PRV strains. Taken together, the histamine synthesized by the HDC in the DRG neurons was responsible for the PRV-induced itch in the mice.

Keywords: pseudorabies virus; itch; mouse; histamine; dorsal root ganglion

Citation: Wang, B.; Wu, H.; Qi, H.; Li, H.; Pan, L.; Li, L.; Zhang, K.; Yuan, M.; Wang, Y.; Qiu, H.-J.; et al. Histamine Is Responsible for the Neuropathic Itch Induced by the Pseudorabies Virus Variant in a Mouse Model. *Viruses* **2022**, *14*, 1067. https://doi.org/10.3390/v14051067

Academic Editors: Yan-Dong Tang and Xiangdong Li

Received: 13 April 2022
Accepted: 13 May 2022
Published: 17 May 2022

1. Introduction

Pseudorabies virus (PRV), which is closely related to varicella-zoster virus (VZV) and herpes simplex virus type 1 (HSV-1), is a member of the *Alphaherpesvirinae* subfamily within the *Herpesviridae* family [1]. Aujeszky's disease (AD) is caused by PRV, which was described and demonstrated by Aladár Aujeszky in 1902 [2]. However, the disease was first reported as "mad itch", and it showed clinical severe itch in cattle [3]. Currently, the distinct natural reservoir of PRV is swine; however, PRV can also infect a wide range of mammals, such as mice, rabbits, sheep, goats, and even humans [4,5]. More than 100 years have passed since the "mad itch" disease was first reported in nonnatural hosts. However, the PRV TJ strain, which is a variant PRV currently prevalent in China, caused unusual pruritus in pigs, which are the natural hosts of PRV [6]. The mechanism that underlies the PRV-induced itch in mouse models can provide a basis for the understanding of the PRV-induced pruritus in pigs.

Itch is an "unpleasant" sensation that leads to the scratch reflex, and it has many similarities to pain. There are four categories of itch: the pruriceptive itch, the neurogenic itch, the neuropathic itch, and the psychogenic itch [7]. The pruriceptive itch originates in the skin and is caused by inflammatory disorders. The inflammatory disorders activate the pruriceptive primary afferent, and the itch signals are transmitted from the skin into the sensory neurons in the dorsal root ganglion (DRG). The neurogenic itch results from central-nervous-system (CNS) activation, without the necessary activation of the sensory nerve fibers. The neuropathic itch can originate at any point along the afferent pathway because of the damage to the nervous system caused by viral disease and/or the traumatic nerve injury of the PNS or the CNS, such as peripheral neuropathies (e.g., postherpetic itch), multiple sclerosis, and nerve compression or irritation. The psychogenic itch is related to psychological or psychiatric disorders, such as itch-associated with delusions of parasitosis, stress, and depression.

Itch is mediated by peripheral somatosensory neurons that are termed "pruriceptors", which sense and respond to pruritogens. Tachykinin 1 (TAC1), interleukin 2 (IL-2), IL-31, and tryptases [tryptase alpha/beta 1 (TPSAB1), tryptase beta 2 (TPSB2), and tryptase gamma 1 (TPSG1)] have been identified as pruritogens, which can evoke itch signals [8]. TAC1 is a kind of neuropeptide that elicits biting and scratching in mice [9]. Moreover, IL-2, IL-31, and tryptase (a kind of serine protease) can also elicit itch [10,11]. Histamine or 5-hydroxytryptamine (5-HT) is synthesized by histidine decarboxylase (HDC) and tryptophan hydroxylase 1 (TPH1), respectively, which can evoke scratching. Histamine is a well-established pruritogen, and it has been regarded as the main target for antipruritic therapies. Histamine produces itch in humans accompanied by skin reactions (wheal and flare) [12,13]. Itch can also be elicited in the mice that are injected intradermally with histamine [14,15].

The DRG neurons are composed of the cell bodies of genetically distinguishable primary afferent neurons, and they are crucial structures in sensory transduction and modulation, including in itch transmission. There is a wide range of molecules that are involved in the transmission of itch signals in the DRG neurons [7]. Histamine receptor H1 (HRH1) is expressed in itch-sensing DRG neurons and it mediates histaminergic itch. The family of Mas-related G-protein-coupled receptors (Mrgprs) mediates nonhistaminergic itch, and it includes MrgprA3, MrgprD, and MrgprX1 [7,12]. Additionally, transient receptor potential cation channel subfamily V member 1 (TRPV1), TRPA1, TRPV4, and TRPM8 are also involved in the transmission of itch signals in the DRG. Because of their important roles in signal transduction, the DRG neurons are widely used in itch research.

Some bacterial and viral infections, including PRV, can cause itch. However, the underlying mechanisms that lead to the activation of pruriceptors are not well understood. In this study, a mouse model of PRV TJ-induced itch was established, and the severity of the itch in the mice infected with the PRV TJ, SC, or Bartha-K61 strains was also compared. We showed that the histamine synthesized by the HDC in the DRG neurons was responsible for the PRV-induced itch in the mice.

2. Materials and Methods

2.1. Viruses and Cells

The PRV Bartha-K61 strain (GenBank accession No. JF797217.1) is a widely used attenuated PRV vaccine [16,17]. The PRV SC strain (GenBank accession No. KT809429.1) is a classical virulent PRV strain [18,19]. The PRV TJ strain (GenBank accession No. KJ789182.1) was isolated from a Bartha-K61-vaccinated pig farm in Tianjin, China, in 2012, and it is more virulent than PRV SC. All the PRV strains are saved at the Harbin Veterinary Research Institute (HVRI), at the Chinese Academy of Agricultural Sciences (CAAS). PK-15 cells (ATCC, CCL-33) were cultured in Dulbecco's modified Eagle's medium (DMEM) (Gibco, Carlsbad, CA, USA) supplemented with 5% fetal bovine sera (FBS) (Hyclone, Logan, UT, USA), 100 mg/mL streptomycin, and 100 IU/mL penicillin at 37 °C and 5% CO_2. All the PRV strains were propagated and titrated on PK-15 cells.

2.2. Virus Titration

PK-15 cells were seeded into 96-well plates, infected with 10-fold serially diluted PRV suspensions, and cultured at 37 °C, 5% CO_2, for 72 h. The PRV-infected cells were detected by indirect immunofluorescence assay (IFA), and the number of fluorescent wells was counted, as described previously [19]. Then, the viral titers were calculated by using the Reed and Muench method, and they were expressed as the median tissue culture infectious dose ($TCID_{50}$) [20].

2.3. Behavioral Observation of the PRV-Infected Mice

All the animal experiments were conducted under a protocol (210603-02) approved by HVRI.

Six-week-old specific pathogen-free (SPF) BALB/c mice were used in this study, and all the mice were housed in a pathogen-free environment at 22 to 25 °C and with an ad libitum water and food supply. To observe the behaviors of the PRV-infected mice, twelve mice were divided into two groups, and one group was intramuscularly (i.m.) injected with 100 μL of inoculum containing 10^6 $TCID_{50}$ of PRV TJ into the left hindlimb muscle, and the other group was mock inoculated (100 μL of medium only). To avoid behavior changes caused by different surroundings, mice were videotaped in an isolator from 0 h postinfection (hpi) to the moribund state of 56 hpi [21–23]. Additionally, high-resolution video and analysis techniques at slow playback speeds were used to discriminate between biting and licking. The severity of the itch in the mice was determined by the frequency of bite bouts and by the durations of itch. When the mice started itching, the bite bouts were counted for 30 min at 1.5 h intervals, and the durations of itch were also counted for 30 min at 3.5 h intervals [21–23]. The frequency of the bite bouts and the durations of itch were counted by a blindly trained observer on the basis of videos.

2.4. Isolation of the DRG Neurons and Skin from the PRV-Infected Mice

A total of 33 6-week-old SPF BALB/c mice were i.m. injected with 100 μL of 10^6 $TCID_{50}/_{mL}$ PRV TJ suspensions into the left hindlimb muscle, and the same number of mice was injected with 100 μL of DMEM as the mock. To isolate the DRG neurons located on the left side of the spinal cord, three mice from the PRV TJ group and the mock group were euthanized at 0, 2, 8, 14, 20, 26, 32, 38, 44, 50, and 56 hpi. After removing the fur, muscles, and the dorsal portion of the spine, the spinal cord was exposed. The spinal cord was removed with forceps, and the DRG neurons were collected [24]. Skin at the injection sites was also collected.

2.5. RNA-Seq Analysis

For the whole transcriptomic analysis, total RNAs were extracted from the DRG neurons isolated from the mock or PRV TJ-infected mice by using RNAiso Plus reagent. All the sample analyses were carried out in triplicates. Genome-wide differential gene expression was analyzed by using RNA deep sequencing by BGI Genomics (Shenzhen, China). Poly(A) plus RNAs were sequenced by using the Illumina HiSeq 2500 platform. The high-quality reads were further mapped to the reference genome of the mice by using Hisat2 (https://daehwankimlab.github.io/hisat2/, accessed on 8 June 2021), and the transcripts were assembled by using Stringtie (http://ccb.jhu.edu/software/stringtie/, accessed on 8 June 2021).

2.6. Reverse Transcription-Quantitative PCR (RT-qPCR)

After isolation from the spinal cords of the mice at 0, 2, 8, 14, 20, 26, 32, 38, 44, 50, and 56 hpi, the DRG neurons were ground in liquid nitrogen. Total RNAs from the DRG were extracted by using RNAiso Plus (catalog No. 9108Q; TaKaRa, Beijing, China), according to the manufacturer's protocols. Total RNAs were also extracted from the skin in the same way. Then, the total RNAs were reverse transcribed into cDNA by using the HiScript II Reverse Transcriptase (catalog No. R201-02; Vazyme, Nanjing, China), according to the instructions. GADPH was amplified as a normalization control. RT-qPCR was performed by using the $2 \times$ ChamQ SYBR qPCR Master Mix (catalog No. Q311-03; Vazyme, Nanjing, China). Forty cycles of amplification were performed, which included sequential denaturation at 95 °C for 10 s, annealing at 60 °C for 30 s, and extension at 72 °C for 1 min. All the primers used in this experiment are listed in Table 1, and all the samples were analyzed in triplicates.

Table 1. Primers for RT-qPCR.

Primers	Sequences (5'-3')
HRH1-F	ACTTGAACCGAGAGCGGAAG
HRH1-R	TTGCACAGCGGGTAGATGAG
TRPV4-F	TCACCCTCCTGAATCCGTGC
TRPV4-R	TCTCACCCATGAGGGCGAT
TRPA1-F	GGAAGTAATTCCTTTTCAGAGTGTC
TRPA1-R	ACTCCTCAACCACCCTGTGT
TRPV1-F	ACCACGGCTGCTTACTAT
TRPV1-R	AACTCTTGAGGGATGGTC
TRPM8-F	TACTCTGGCAGCCTTGGG
TRPM8-R	TCGCAGGAGTAGACCAGTAG
MrgprD-F	ATGAACTCCACTCTTGAC
MrgprD-R	AGCACATAGACACAGAAG

2.7. Effects of Chlorphenamine Hydrogen Maleate Treatment on the PRV TJ-Infected Mice

The chlorphenamine hydrogen maleate powder (catalog No. BP081; Sigma-Aldrich, Darmstadt, Germany) was dissolved in DMEM. Thirty 6-week-old SPF BALB/c mice were divided into five groups ($n = 6$): four groups were i.m. injected with 100 μL of inoculum containing 10^6 TCID$_{50}$ of PRV TJ into the left hindlimb muscle, and one group was injected with 100 μL of DMEM as a control. The mice of the four PRV TJ-infected groups were i.m. injected with 0, 10, 30, or 60 mg/kg chlorphenamine hydrogen maleate solution in the left hindlimb muscle at 44 and 50 hpi. Twenty minutes later, the itching behavior was recorded.

2.8. Hematoxylin and Eosin Staining and Immunohistochemistry (IHC)

At 56 hpi, all the mice were euthanized for pathological examinations. The left legs and the lumbosacral DRG neurons of the mice were freshly collected by using scissors and forceps, were fixed in 4% paraformaldehyde for 48 h, followed by 70% ethanol, and were embedded in paraffin, cut into 5-μm sections, and mounted onto glass slides. The sections were stained with hematoxylin and eosin stain [25]. For the immunohistochemical examination, the sections were blocked in 10% normal goat serum, incubated with a rabbit antihistamine antibody (catalog No. H7403-.2ML; Sigma-Aldrich, Darmstadt, Germany) overnight at 4 °C, washed three times in TBST buffer, incubated in SignalStain Boost IHC detection reagent (HRP, Rabbit) (catalog No. 8114P; Sinopharm, Beijing, China) for 30 min at room temperature, washed 3 times, and stained [26].

2.9. Comparison of Itch in the Mice Infected with Different PRV Strains

Twenty-four mice were divided into four groups ($n = 6$): three groups were i.m. injected with 10^5 TCID$_{50}$ of PRV TJ, SC, or Bartha-K61 into the left hindlimb muscle, respectively. The last group was injected with 100 μL of DMEM as a control. All the mice were videotaped, and the bite bouts and durations of itch were counted at 56 hpi [21–23].

Additionally, the DRG neurons were isolated, and the mRNAs of the HDC were quantified by using RT-qPCR.

2.10. qPCR

The mice infected with 10^5 TCID$_{50}$ of PRV TJ, SC, Bartha-K61, or DMEM were euthanized at 56 hpi. The DRG neurons were collected, and DNA was extracted by using the MagaBio plus virus DNA purification kit (catalog No. 9109; BioFlux, Beijing, China), according to the manufacturer's protocols. The genomic copies of PRV were quantified on the basis of a previously described method [27].

2.11. Statistical Analysis

All the experiments were performed in triplicates, and the statistical significance was analyzed by Student's *t* test in Prism v8.0 software (GraphPad, San Diego, CA, USA). Differences were considered significant if the unadjusted *p*-value was less than 0.05.

3. Results

3.1. The Severity of the Itch Caused by PRV TJ Infection Gradually Increased with Time

To quantify the severity of the itch, the mice that were injected with PRV TJ or DMEM were videotaped by using two high-resolution videos, respectively. The behaviors of the mice were recorded constantly from 0 to 56 hpi. As for the biting action, there were no intuitive differences between natural and itchy bites. However, in terms of the biting frequency, natural bites were occasional, while itchy bites were regular and frequent in the PRV TJ-infected mice. In addition, the natural bites were transient, but the durations of the itchy bites increased as the clinical signs progressed. On the basis of this criterion, the frequency of the bite bouts and the durations of itch were quantified. The results showed that the mice infected with PRV TJ did not exhibit spontaneous biting from 0 to 30 hpi; however, from 32 hpi, the mice started itching. The infected mice bit the injection sites about 10 times per 30 min at 32 hpi, spontaneously, and the frequency of the bite bouts increased gradually, reached about 62 times per 30 min at 56 hpi (Figure 1A), and was significantly different between the PRV TJ-infected group and the control group ($p < 0.01$). Additionally, the duration of the biting was counted during the 30 min observation period. We found that the duration of the itch remained at the same level as the mock group until 28 hpi.

None of the PRV TJ-infected mice exhibited itch until 28 hpi, and the average duration of the itching was about 5 s per 30 min at 32 hpi. The duration rose gradually over time, and it peaked at about 90 s per 30 min at 52 hpi (Figure 1B). The mice injected with DMEM did not exhibit spontaneous biting during the whole experiment period.

The mice were also photographed at 0, 32, 44, and 56 hpi. The legs of the infected mice showed no signs of tissue damage from 0 to 32 hpi, although the mice infected with PRV TJ started biting at 32 hpi. At 44 hpi, because of the long-term biting, severe tissue damage appeared at the inoculated sites, and the muscle and skin were detached. At 56 hpi, the left legs of the mice were shriveled. For the control mice, no injuries were found in the legs during the whole experiment period (Figure 1C).

3.2. Different Expression Profiles of the Molecules Relevant to Itch-Signal Transmission Were Noted in the DRG Neurons

The molecules that are relevant to itch-signal transmission were detected by using RNA-seq and RT-qPCR. The DRG neurons from the infected or the control mice were harvested at 56 hpi. On the heat map, the transcription levels of HRH1, MrgprA3, TRPV4, and TRPA1 were comparable between the two groups at 56 hpi. However, the mRNA expression levels of MrgprX1, MrgprD, TrpA1, and TrpV1 were reduced to varying degrees (Figure 2A).

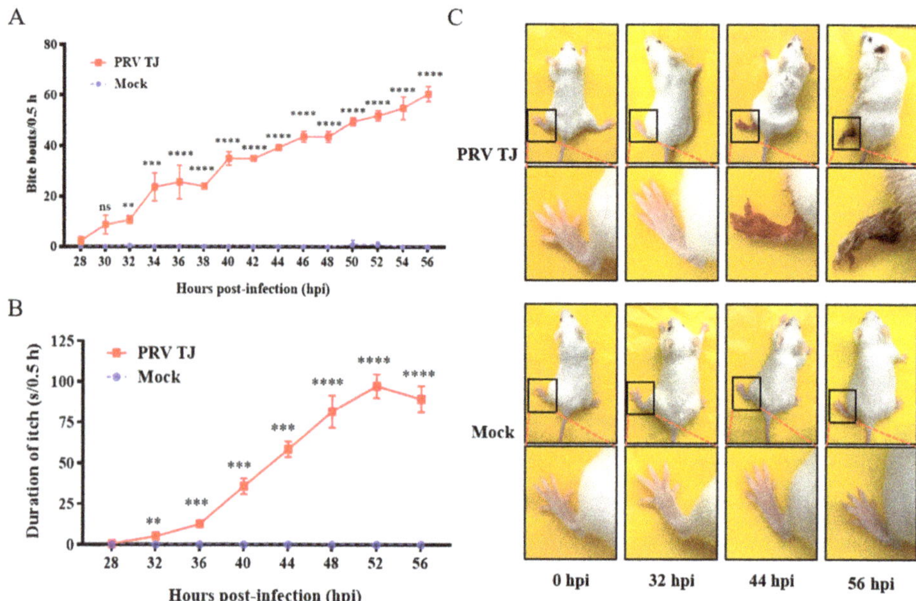

Figure 1. The mice infected with PRV TJ exhibited itch. Twelve SPF BALB/c mice were divided into two groups ($n = 6$): one group was infected intramuscularly with 10^5 TCID$_{50}$ of PRV TJ, and the other group was mock inoculated (100 µL of medium only). (**A**) The behaviors of the mice were recorded from 0 to 56 hpi. The bite bouts of the mice were recorded for 30 min at 1.5 h intervals. (**B**) The durations of biting were recorded for 30 min at 3.5 h intervals. (**C**) Representative photographs of the mice in the two groups were taken at 0, 32, 44, and 56 hpi. Bars represent the means $\pm$ SDs for three independent experiments; ns: not significant ($p \geq 0.05$); ** $p < 0.01$; *** $p < 0.001$; **** $p < 0.0001$.

The mRNA expression levels of the receptors and channels in the DRG neurons were measured by using RT-qPCR. As shown in Figure 2B, the mRNA expression of MrgprX1 did not change significantly from 0 to 50 hpi; however, it decreased at 56 hpi, which was at approximately half of the mock group. The mRNA expression level of MrgprA3 also did not change significantly over the test period. Interestingly, the mRNA level of MrgprD increased gradually, from 0 to 32 hpi, it remained at a plateau from 32 to 38 hpi, its maximum expression in the PRV TJ-infected group was higher than that in the mock group, and it returned to basal level. Finally, the mRNA expression level of MrgprD was reduced by approximately 50%. In addition, the mRNAs of the receptors TRPA1, TRPM8, TRPV1, and TRPV4 were also measured by using RT-qPCR. Among them, TRPV1 and TRPV4 contribute to histamine-mediated neuronal activation. TRPA1 is extensively involved in the neuronal activation in nonhistaminergic itch. TRPM8 is another receptor that modulates or mediates itch. The results showed that the mRNA level of TRPA1 in the PRV TJ-infected group did not change from 0 to 32 hpi, increased transitorily from 32 to 38 hpi, then decreased gradually, and was lower than that of the mock group. The mRNA expression levels of the two TRP cation channels that mediate histaminergic itch were different. The TRPV1 mRNA in the infected group was consistent with that in the mock group from 0 to 20 hpi, then it increased gradually from 20 to 38 hpi, and it decreased from 38 to 56 hpi. The maximal TRPV1 mRNA level was about 2.7-fold higher than that of the control at 38 hpi. In contrast to the TRPA1 and TRPV1, the TRPV4 mRNA remained relatively stable, with changes only from 0 to 56 hpi. The TRPM8 mRNA was unchanged from 0 to 8 hpi, it then increased from 8 to 26 hpi, it remained at a plateau from 26 to 32 hpi (the highest mRNA level of TRPM8 was approximately 3-fold higher than that of the control), and it finally returned to a basal level.

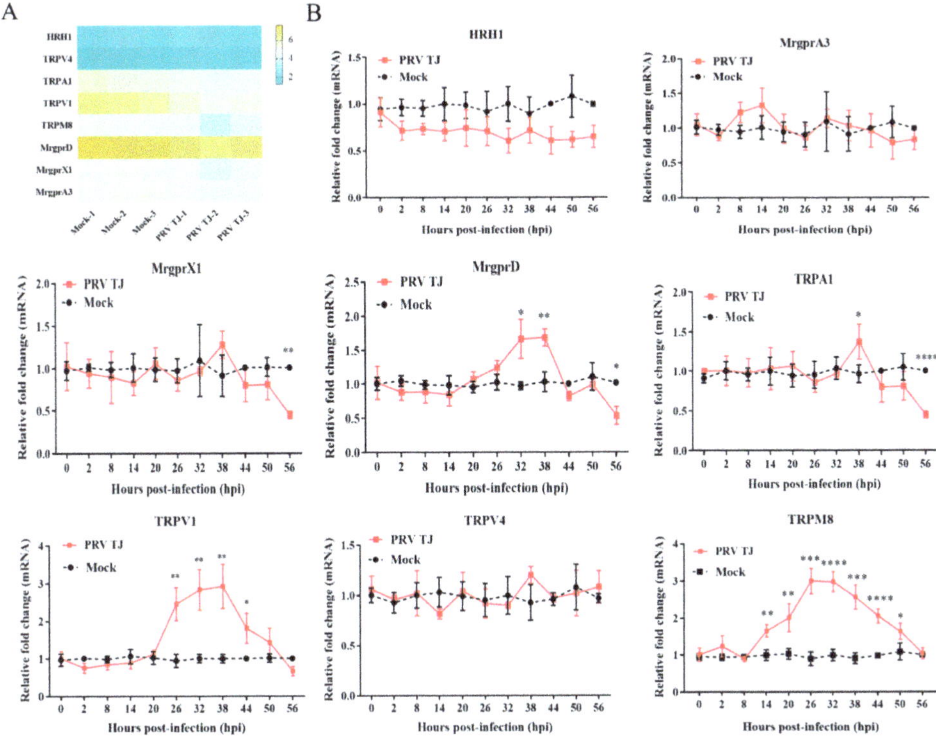

Figure 2. The mRNA expression levels of receptors and the potential cation channels relevant to itch-signal transmission were different. A total of 33 SPF BALB/c mice were infected intramuscularly with 10^5 TCID$_{50}$ of PRV TJ, and the same number of mice was injected with 100 μL of DMEM as a control. The DRG neurons on the left side of the spinal cord were collected, and total RNAs were extracted. (**A**) The mRNAs of HRH1, TRPV4, TRPA1, TRPV4, TRPM8, MrgprD, MrgprX1, and MrgprA3 at 56 hpi were quantified by using RNA-seq. (**B**) The expression kinetics were examined by using RT-qPCR. Bars represent the means ± SDs for three independent experiments; ns: not significant ($p \geq 0.05$); * $p < 0.05$; ** $p < 0.01$; *** $p < 0.001$; **** $p < 0.0001$.

3.3. The Expression of HDC Was Increased in the PRV TJ-Infected DRG Neurons

PRV can replicate and establish quiescent, latent infection in the PNS [28]. Moreover, PRV can induce a severe inflammation response in the DRG neurons. A previous study has shown that PRV in the neurons may trigger itch [29]. TAC1, IL-2, IL-31, TPSAB1, TPSB2, and TPSG1 are known as pruritogens [14,15]. HDC and TPH1 are the key enzymes of pruritogen synthesis. The mRNA levels of pruritogens and their correlated molecules were detected by using RNA-seq. The HDC transcription level of the PRV TJ-infected group was significantly higher than that of the control group (Figure 3A).

HDC mRNA was detected by using RT-qPCR in the DRG neurons of the PRV TJ or mock groups at 0, 2, 8, 14, 20, 26, 32, 38, 44, 50, and 56 hpi (Figure 3B). The mRNA level of the HDC in the PRV TJ group increased 1.7-fold compared to that in the control group at 20 hpi, then it returned to a normal level at 26 hpi, and, notably, it increased rapidly from 32 to 56 hpi. The maximum expression level in the infected group was approximately 25-fold higher than that in the mock group in the moribund state. The gB protein of the PRV TJ was detected in the DRG neurons by using IHC at 56 hpi (red arrows), and the protein expression level of the HDC (blue arrows) was also significantly higher than that of the control (Figure 3C).

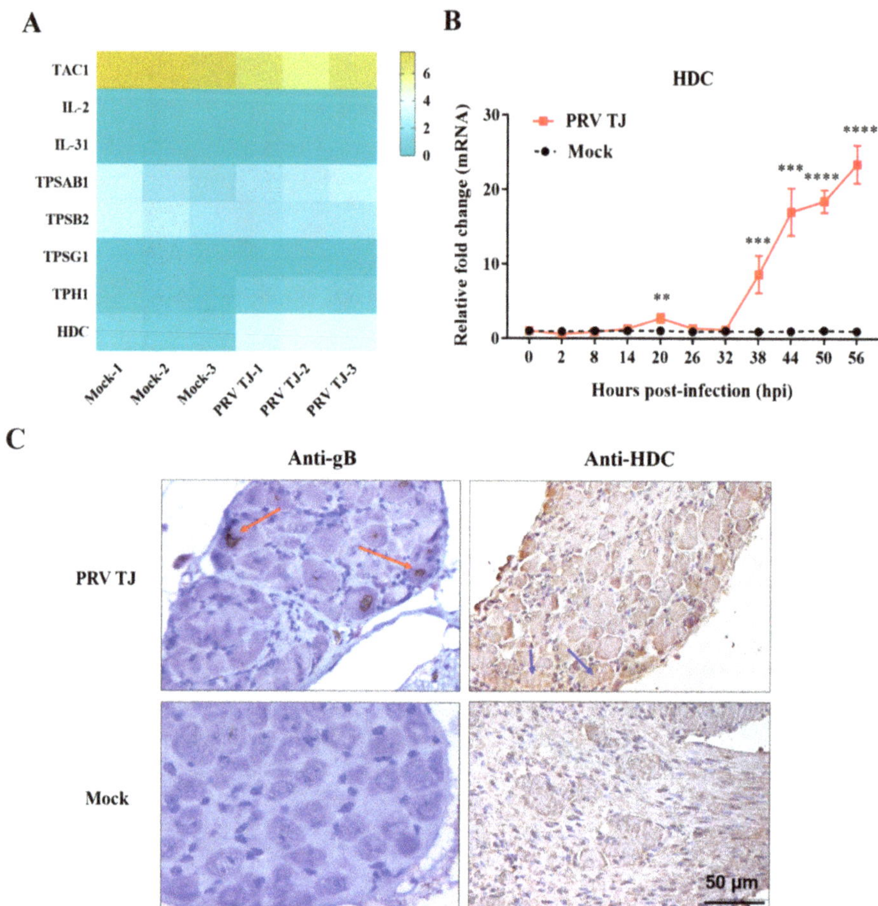

Figure 3. HDC was expressed in the DRG neurons of the PRV TJ-infected mice. Thirty-three SPF BALB/c mice were infected intramuscularly with 10^5 TCID$_{50}$ of PRV TJ, and the same number of mice was injected with 100 μL of DMEM as a control. The DRG neurons on the left side of the spinal cord were isolated, and total RNAs were extracted. (**A**) The mRNA expressions of TAC1, TPH1, IL-31, IL-2, HDC, TPSAB1, TPSB2, and TPSG1 were detected by using RNA-seq at 56 hpi. (**B**) The expression kinetics of HDC was detected by using RT-qPCR. (**C**) The cells expressing the gB protein of PRV (red arrows) and HDC in the lumbosacral DRG neurons (blue arrows) were detected by immunohistochemistry. Bar: 50 μm. Bars represent the means ± SDs for three independent experiments; ns: not significant ($p \geq 0.05$); ** $p < 0.01$; *** $p < 0.001$; **** $p < 0.0001$.

3.4. Histamine Produced in the DRG Neurons Contributed to PRV-Induced Itch and Could Be Inhibited by Chlorpheniramine

HDC catalyzes the conversion of histidine to histamine, which is a "gold standard" itch mediator [7]. A significantly higher expression level of histamine was detected in the PRV TJ-infected group compared to the control group (Figure 4A). The main source of histamine in the body is mast cells; however, several other types of cells can also synthesize histamine, such as neurons and keratinocytes [7]. However, mast cells were not found in either of the injection sites of the mock- and PRV TJ-infected mice. Additionally, the HDC mRNA level of the skin was also quantified. There was no statistical difference between the infected and mock groups ($p \geq 0.05$) (Figure 4B).

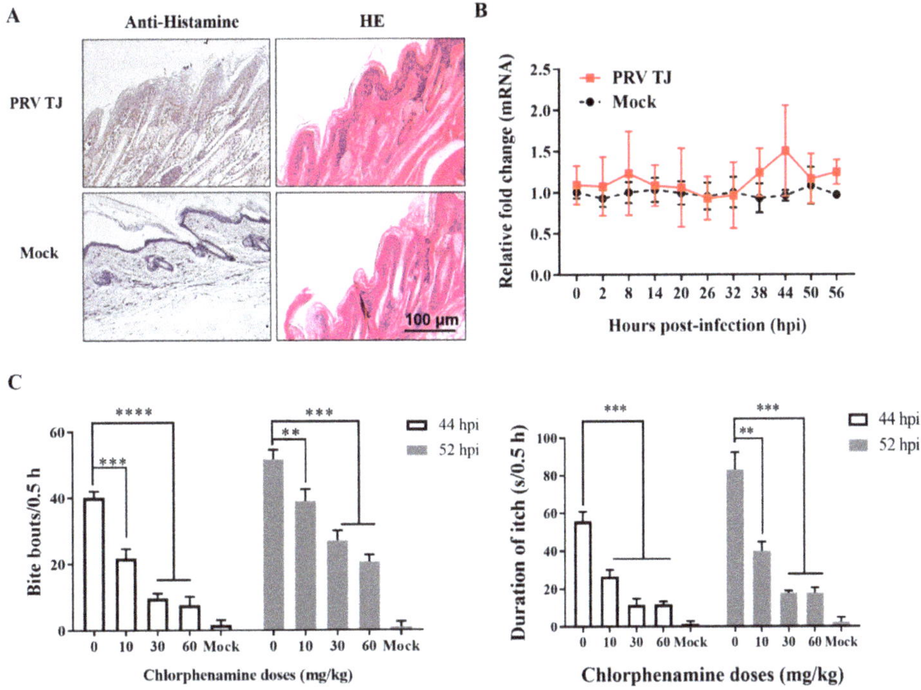

Figure 4. Histamine produced in the DRG neurons contributed to itch and could be inhibited by chlorpheniramine. Twelve SPF BALB/c mice were injected with 10^5 TCID$_{50}$ of PRV TJ or 100 μL of DMEM as a control. The PRV TJ-infected mice were treated with 0, 10, 30, or 60 mg/kg chlorpheniramine at 44 and 56 hpi, respectively, and the behavior of the mice was captured photographically after twenty minutes. The skin at the injection site was collected for pathological detection and total RNA extraction at 56 hpi. (**A**) The histamine was detected by the immunohistochemistry, and the mast cells were detected by using hematoxylin and eosin staining. Bar: 100 μm. (**B**) The mRNA expression trend of the HDC with time was quantified by using RT-qPCR. (**C**) The frequency of bite bouts and the durations of itch were counted after treatment with chlorpheniramine. Bars represent the means $\pm$ SDs of three independent experiments. Ns: not significant ($p \geq 0.05$); ** $p < 0.01$; *** $p < 0.001$; **** $p < 0.0001$.

Chlorpheniramine, which is an H1R antagonist, is widely used as a first-generation sedative antihistamine [30,31]. The mice received an intramuscular injection of either chlorpheniramine (10, 30, or 60 mg/kg) or DMEM at 44 and 52 hpi, and their behaviors were recorded 30 min later. As shown in Figure 4C, the frequency of the bite bouts and the durations of itch of the mice treated with chlorpheniramine were reduced in a dose-dependent manner.

3.5. The Severity of Itch Was Different between the Three PRV Strains and Was Consistent with the HDC Expression

A mouse model of the itch caused by PRV TJ was elaborated, and the increased level of HDC was responsible for the itch. To verify the relationship between HDC and the itch induced by PRV, the mRNAs of the HDC in the mice infected with 10^5 TCID$_{50}$ of PRV Bartha-K61, SC or TJ were quantified. First, the severity of the itch in the mice infected with different PRV strains was evaluated. At 56 hpi, the bite bouts of the mice infected with PRV TJ and SC were about 55 and 56 times per 30 min, respectively. Although the itch severities of the mice infected with PRV TJ and SC were similar in terms of the bite bouts, the bite durations of the mice infected with the two strains were different ($p < 0.05$). The itch duration of the mice infected with PRV TJ was about 91 s per 30 min, which was significantly different from that of the PRV SC (about 67 s per 30 min). Moreover, the mice infected with PRV Bartha-K61 were mildly pruritic, both in terms of the bite bouts and the durations per 30 min (Figure 5A). The mice infected with different PRV strains were photographed in the prone position. As shown in Figure 5B, the mice infected with PRV TJ developed substantial necrotic lesions at the injection sites at 56 hpi. To the contrary, the mice infected with PRV SC were also moribund, but they did not have injuries. Because of the reduced virulence of PRV Bartha-K61, the mice showed no abnormal changes. Overall, the itch severities in the mice infected with the same doses of different PRV strains were in the following order: PRV TJ > PRV SC > PRV Bartha-K61, which presented a positive correlation with the viral virulence. As shown in Figure 5C, the replications of different PRV strains in the DRG neurons were quantified. The genome copies of PRV TJ, PRV SC, and PRV Bartha-K61 in the DRG neurons at 56 hpi were about $10^{4.0}$, $10^{3.6}$, and $10^{4.3}$, respectively. Therefore, the itch severity was not associated with the replication of different PRV strains.

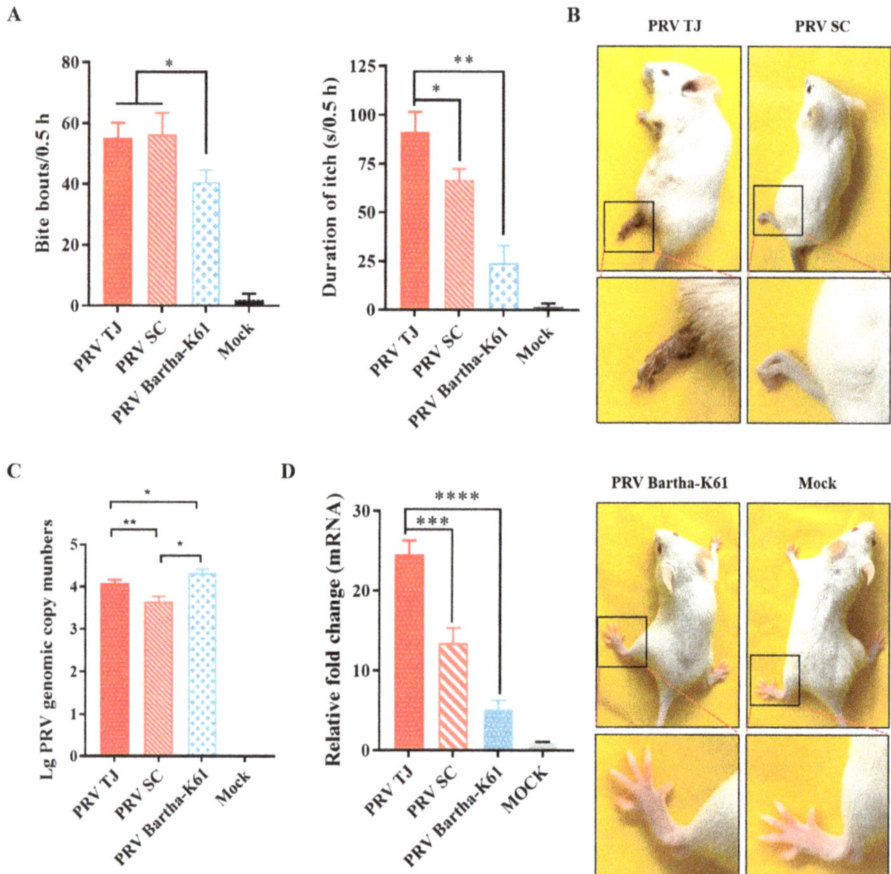

Figure 5. The severities of itch and HDC expressions of the three PRV-strain-infected mice. Twenty-four SPF BALB/c mice were divided into four groups. Six mice were infected with 10^5 $TCID_{50}$ of PRV TJ, PRV SC, PRV Bartha-K61, or DMEM only as a control. The behaviors of the mice infected with the abovementioned strains were recorded at 56 hpi. The DRG neurons of the mice in every group were

isolated at 56 hpi, and the total RNAs were extracted. (**A**) The frequency of bite bouts and the durations of itch were recorded as above. (**B**) Representative photographs of the mice in the above mentioned groups taken at 56 hpi. (**C**) Genomic copies of different PRV strains in the DRG neurons of the infected mice at 56 hpi. (**D**) The DRG neurons of the mice in each group were isolated at 56 hpi, the total RNAs were extracted, and the expressions of HDC in the DRG neurons were quantified by using RT-qPCR. Bars represent the means $\pm$ SDs for three independent experiments; ns: not significant ($p \geq 0.05$); * $p < 0.05$; ** $p < 0.01$; *** $p < 0.001$; **** $p < 0.0001$.

The HDC mRNA levels in the DRG neurons from the mice infected with different PRV strains were quantified (Figure 5D). The mRNA level of the HDC in the DRG neurons that were isolated from the mice infected with PRV TJ, SC, and Bartha-K61 increased 25-, 14-, and 5-fold, respectively, compared to the mock group, and these values were proportional to the itch severities of the mice.

4. Discussion

Since the first case of mad itch was described in 1813, the characteristic itch caused by PRV infection in nonnatural hosts has been frequently reported. Currently, few studies have focused on this aspect of PRV pathogenesis. Self-mutilating behavior, such as biting and scratching, can be stimulated by PRV infection, which is attributed to tissue lesions [32]. It is well established that itch can be induced by nerve injury caused by virus replication in the PNS [29]. However, the exact mechanism remains unknown, and the itch in the mice induced by PRV has not been quantified. Here, we quantified the itch severity of the PRV TJ-infected mice in terms of the bite bouts and durations. In addition, we demonstrated that histamine contributed to the itch caused by the PRV in the mice.

When mice are infected with PRV, itch evokes innate scratching. It was previously shown that viral replication in the skin did not result in necrosis; however, the itch behaviors of the mice were not recorded and quantified [32]. The mice were videotaped after PRV TJ inoculation, they started itching at 32 hpi, and their legs suffered skin injuries at the same time point.

We described the itch phenotype of the mice infected with PRV TJ, and we quantified the transcriptional kinetics of the different receptors and channels expressed in the DRG. The neurotransmitters, receptors, and signal pathways that are involved in acute itch transduction have been well revealed recently. Mrgprs are present in certain kinds of sensory neurons, and they are associated with the transmission of itch signals [33,34].

According to the results, MrgprD and MrgprX1, rather than MrgprA3, expressed differently, compared to the mock group, which may mean that the neural excitation was associated with an increase in the sodium current in the DRG neurons that express MrgprD and MrgprX1 [33,34]. TRPM8, which is a cold-sensitive ion channel, is involved in itch relief [35,36]. Increased TRPM8 expression may inhibit itch signaling in the primary sensory. The TRPV1-gene-expression level significantly increased, which may be due to its important role in histaminergic itch. In conclusion, the transcriptional levels of different molecules were different, which provides a framework for future work on the examination of the neuronal mechanisms that underlie PRV-induced itch and its sensitization.

Although the invasion of PRV into nonnatural hosts has not been well studied, it has been revealed that PRV can invade the PNS and the CNS via anterograde axonal transport. The PRV-induced itch may be triggered by the replication of the virus in the neurons [29]. DRG neurons were isolated and tested for the presence of PRV TJ. As expected, the IHC results demonstrated the presence of PRV TJ in the DRG neurons. Pruritogens, including TAC1, IL-2, IL-31, TPSAB1, TPSB2, and TPSG1, and the key enzymes that are attributed to itch, including HDC and TPH1, were detected by RNA-seq. The mRNA level of HDC was markedly increased, and the HDC protein was strongly positive in the DRG neurons (Figure 3B,C). It was well proven that the PRV TJ replication in the DRG neurons led to the increased expression of the HDC protein.

More importantly, the IHC for histamine was strongly positive in the skin (Figure 4A). Histamine can be secreted by several kinds of cells, including neurons, mast cells, and keratinocytes [7]. The mast cells in the skin were detected by hematoxylin and eosin staining, as well as the mRNA levels of the HDC in the keratinocytes, and all of them presented no differences compared to the mock group (Figure 4A,B). All the results above prove that activated HDC results in increased histamine level in the skin.

Histamine, which is produced in either an autocrine or a paracrine manner, is an inflammatory mediator that is synthesized by HDC. In non-mast cells, HDC is rarely expressed without stimulation; however, its expression is markedly induced by various types of inflammatory stimulants [37,38]. In the mice infected with PRV, neuroinflammation was found in the DRG neurons. Therefore, we speculated that the neuroinflammation induced by PRV resulted in increased expression of HDC, which is regulated by mitogen-activated protein (MAP) kinases and c-Jun N-terminal kinase (JNK), which are important immune-regulatory molecules [39,40]. Upon infection with PRV, the MAPK and JNK expressions are upregulated [41], which leads to the change in normal cellular activities. Moreover, PRV possesses a linear double-stranded DNA with GC-rich sequences [42]. The transcription factor SP1, which binds to GC-rich regions, also regulates the expression of HDC [43]. Thus, it is possible that certain segments of the PRV genome bind to Specificity Protein 1, which results in the upregulation of HDC. In summary, the cellular and metabolic processes in the DRG neurons will be altered upon viral infection. During these processes, multiple viral factors, such as viral DNA, RNA, and proteins, may play important roles, which need to be explored in details.

Histamine exerts various immune-regulatory functions, and it plays an important role in neuroinflammation [44]. The activation of H1R expressed in the neuron by histamine has the ability to induce pruritus and atopic dermatitis [45]. This may explain the itch exhibited by the PRV-infected mice. The H4R-mediated activation of mast cells that is induced by histamine leads to the expression of various proinflammatory cytokines and chemokines, such as IL-6 [46], which have been proven to increase the protein levels in mice after infection with PRV [47,48]. Moreover, the activation of H4R involves several signaling cascades for the release of various allergic inflammatory mediators. Histamine can induce the phosphorylation of ERK in order to mediate the proliferation, differentiation, anti-apoptosis, regulation, and cytokine expression at the gene level. It can also activate NF-κB via the JAK-STAT signaling pathway [49,50], which could be well activated in the DRG neurons from the mice infected with PRV [41]. Taken together, the PRV-induced itch may be due to the histamine-induced inflammation, which presents a positive correlation with PRV virulence, but has no association with the replication of different strains in the DRG neurons (Figure 5B,C).

Many viral proteins are associated with PRV virulence, such as US9, gE, and gI. The loss of these genes usually results in virus attenuation. PRV Becker is a virulent strain; however, when deleting the genes that encode gI, US9, and gE, the virus induced decreased itch in mice. The PRV Bartha-K61, which carries deletions in the above three genes, induced the greatest attenuation of itch, compared with the three PRV isogenic strains [32]. This indicates that the itch severity is related to the virulence-associated proteins of the PRV. In addition, the genes encoding the gI and gE proteins of PRV TJ and SC show some sequence differences [6], which might lead to differences in the severity of the itch [6]. Apart from the three proteins mentioned above, other viral proteins, such as US6, UL27, UL44, etc., are different among the three PRV strains. Therefore, we speculate that these proteins may affect the severity of itch.

In conclusion, we found that the high level of histamine produced in PRV-infected DRG neurons resulted in itch, and that the severity of the itch induced by the different PRV strains was proportional to the HDC mRNA level in the DRG neurons of the PRV-infected mice.

Author Contributions: Conceptualization, Y.S. and H.-J.Q.; methodology, B.W., H.W., H.Q., H.L.; writing—original draft preparation, B.W. and Y.W.; writing—review and editing, H.W., H.Q., L.P., L.L., K.Z., M.Y., H.-J.Q. and Y.S.; funding acquisition, Y.S. and H.-J.Q. All authors have read and agreed to the published version of the manuscript.

Funding: This work was supported by the Natural Science Foundation of China (No. 31870152), and by the Natural Science Foundation of Heilongjiang Province of China (No. QC2018035).

Institutional Review Board Statement: The animal study protocols were approved by the Animal Ethics Committee of the Harbin Veterinary Research Institute (Protocols code 210603-02).

Informed Consent Statement: Not applicable.

Conflicts of Interest: The authors declare no conflict of interest.

References

1. Pomeranz, L.E.; Reynolds, A.E.; Hengartner, C.J. Molecular Biology of Pseudorabies Virus: Impact on Neurovirology and Veterinary Medicine. *Microbiol. Mol. Biol. Rev.* **2005**, *69*, 462–500. [CrossRef] [PubMed]
2. Aujeszky, A. Über Eine Neue Infektionskrankheit Bei Haustieren. *Zentralbl. Bakteriol. Parasitenkd. Infektionskr. Hyg. Abt. 1 Orig.* **1902**, *32*, 353–357.
3. Hanson, R.P. The History of Pseudorabies in the United States. *J. Am. Vet. Med. Assoc.* **1954**, *124*, 259–261. [PubMed]
4. Liu, Q.; Wang, X.; Xie, C.; Ding, S.; Yang, H.; Guo, S.; Li, J.; Qin, L.; Ban, F.; Wang, D.; et al. A Novel Human Acute Encephalitis Caused by Pseudorabies Virus Variant Strain. *Clin. Infect. Dis.* **2021**, *73*, e3690–e3700. [CrossRef] [PubMed]
5. Wang, D.; Tao, X.; Fei, M.; Chen, J.; Guo, W.; Li, P.; Wang, J. Human Encephalitis Caused by Pseudorabies Virus Infection: A Case Report. *J. Neurovirol.* **2020**, *26*, 442–448. [CrossRef]
6. Luo, Y.; Li, N.; Cong, X.; Wang, C.H.; Du, M.; Li, L.; Zhao, B.; Yuan, J.; Liu, D.D.; Li, S.; et al. Pathogenicity and Genomic Characterization of a Pseudorabies Virus Variant Isolated from Bartha-K61-Vaccinated Swine Population in China. *Vet. Microbiol.* **2014**, *174*, 107–115. [CrossRef]
7. Dong, X.; Dong, X. Peripheral and Central Mechanisms of Itch. *Neuron* **2018**, *98*, 482–494. [CrossRef]
8. Akiyama, T.; Carstens, E. Neural Processing of Itch. *Neuroscience* **2013**, *250*, 697–714. [CrossRef] [PubMed]
9. Andoh, T.; Kuraishi, Y. Nitric Oxide Enhances Substance P-Induced Itch-Associated Responses in Mice. *Br. J. Pharmacol.* **2003**, *138*, 202–208. [CrossRef]
10. Wahlgren, C.F.; Tengvall Linder, M.; Hägermark, O.; Scheynius, A. Itch and Inflammation Induced by Intradermally Injected Interleukin-2 in Atopic Dermatitis Patients and Healthy Subjects. *Arch. Dermatol. Res.* **1995**, *287*, 572–580. [CrossRef]
11. Sonkoly, E.; Muller, A.; Lauerma, A.I.; Pivarcsi, A.; Soto, H.; Kemeny, L.; Alenius, H.; Dieu-Nosjean, M.C.; Meller, S.; Rieker, J.; et al. IL-31: A New Link between T Cells and Pruritus in Atopic Skin Inflammation. *J. Allergy Clin. Immunol.* **2006**, *117*, 411–417. [CrossRef] [PubMed]
12. Sutaria, N.; Adawi, W.; Goldberg, R.; Roh, Y.S.; Choi, J.; Kwatra, S.G. Itch: Pathogenesis and Treatment. *J. Am. Acad. Dermatol.* **2021**, *86*, 17–34. [CrossRef] [PubMed]
13. Hosogi, M.; Schmelz, M.; Miyachi, Y.; Ikoma, A. Bradykinin Is a Potent Pruritogen in Atopic Dermatitis: A Switch from Pain to Itch. *Pain* **2006**, *126*, 16–23. [CrossRef]
14. Weisshaar, E.; Ziethen, B.; Gollnick, H. Can a Serotonin Type 3 (5-HT3) Receptor Antagonist Reduce Experimentally-Induced Itch? *Inflamm. Res.* **1997**, *46*, 412–416. [CrossRef]
15. Inagaki, N.; Nakamura, N.; Nagao, M.; Musoh, K.; Kawasaki, H.; Nagai, H. Participation of Histamine H1 and H2 Receptors in Passive Cutaneous Anaphylaxis-Induced Scratching Behavior in ICR Mice. *Eur. J. Pharmacol.* **1999**, *367*, 361–371. [CrossRef]
16. Bartha, A. Experimental Reduction of Virulence of Aujeszky's Disease Virus. *Magy. Allatorv. Lapja* **1961**, *16*, 42–45.
17. Delva, J.L.; Nauwynck, H.J.; Mettenleiter, T.C.; Favoreel, H.W. The Attenuated Pseudorabies Virus Vaccine Strain Bartha K61: A Brief Review on the Knowledge Gathered during 60 Years of Research. *Pathogens* **2020**, *9*, 897. [CrossRef]
18. Yuan, Q.; Li, Z.; Nan, X.; Wu, Y.; Li, Y. Isolation and Identification of Pseudorabies Virus. *Chin. J. Prev. Vet. Med.* **1987**, *3*, 10–11.
19. Qi, H.; Wu, H.; Abid, M.; Qiu, H.J.; Sun, Y. Establishment of a Fosmid Library for Pseudorabies Virus SC Strain and Application in Viral Neuronal Tracing. *Front. Microbiol.* **2020**, *11*, 1168. [CrossRef]
20. Németh, B.; Fasseeh, A.; Molnár, A.; Bitter, I.; Horváth, M.; Kóczián, K.; Götze, Á.; Nagy, B. A Systematic Review of Health Economic Models and Utility Estimation Methods in Schizophrenia. *Expert Rev. Pharmacoecon. Outcomes Res.* **2018**, *18*, 267–275. [CrossRef]
21. Akiyama, T.; Carstens, M.I.; Ikoma, A.; Cevikbas, F.; Steinhoff, M.; Carstens, E. Mouse Model of Touch-Evoked Itch (Alloknesis). *J. Invest. Dermatol.* **2012**, *132*, 1886–1891. [CrossRef] [PubMed]
22. Ohayon, S.; Avni, O.; Taylor, A.L.; Perona, P.; Roian Egnor, S.F. Automated Multi-Day Tracking of Marked Mice for the Analysis of Social Behaviour. *J. Neurosci. Methods* **2013**, *219*, 10–19. [CrossRef] [PubMed]
23. Sakai, K.; Sanders, K.M.; Youssef, M.R.; Yanushefski, K.M.; Jensen, L.; Yosipovitch, G.; Akiyama, T. Mouse Model of Imiquimod-Induced Psoriatic Itch. *Pain* **2016**, *157*, 2536–2543. [CrossRef] [PubMed]

24. Lin, Y.T.; Chen, J.C. Dorsal Root Ganglia Isolation and Primary Culture to Study Neurotransmitter Release. *J. Vis. Exp.* **2018**, *140*, 57569. [CrossRef]
25. Cardiff, R.D.; Miller, C.H.; Munn, R.J. Manual Hematoxylin and Eosin Staining of Mouse Tissue Sections. *Cold Spring Harb. Protoc.* **2014**, *2014*, 655–658. [CrossRef]
26. Zhang, L.; Zhong, C.; Wang, J.; Lu, Z.; Liu, L.; Yang, W.; Lyu, Y. Pathogenesis of Natural and Experimental Pseudorabies Virus Infections in Dogs. *Virol. J.* **2015**, *12*, 44. [CrossRef]
27. Meng, X.Y.; Luo, Y.; Liu, Y.; Shao, L.; Sun, Y.; Li, Y.; Li, S.; Ji, S.; Qiu, H.J. A Triplex Real-Time PCR for Differential Detection of Classical, Variant and Bartha-K61 Vaccine Strains of Pseudorabies Virus. *Arch. Virol.* **2016**, *161*, 2425–2430. [CrossRef]
28. Takahashi, H.; Yoshikawa, Y.; Kai, C.; Yamanouchi, K. Mechanism of Pruritus and Peracute Death in Mice Induced by Pseudorabies Virus (PRV) Infection. *J. Vet. Med. Sci.* **1993**, *55*, 913–920. [CrossRef]
29. Laval, K.; Enquist, L.W. The Neuropathic Itch Caused by Pseudorabies Virus. *Pathogens* **2020**, *9*, 254. [CrossRef]
30. Veglia, E.; Pini, A.; Moggio, A.; Grange, C.; Premoselli, F.; Miglio, G.; Tiligada, K.; Fantozzi, R.; Chazot, P.L.; Rosa, A.C. Histamine Type 1-Receptor Activation by Low Dose of Histamine Undermines Human Glomerular Slit Diaphragm Integrity. *Pharmacol. Res.* **2016**, *114*, 27–38. [CrossRef]
31. Márquez-Valadez, B.; Aquino-Miranda, G.; Quintero-Romero, M.O.; Papacostas-Quintanilla, H.; Bueno-Nava, A.; López-Rubalcava, C.; Díaz, N.F.; Arias-Montaño, J.A.; Molina-Hernández, A. The Systemic Administration of the Histamine H (1) Receptor Antagonist/Inverse Agonist Chlorpheniramine to Pregnant Rats Impairs the Development of Nigro-Striatal Dopaminergic Neurons. *Front. Neurosci.* **2019**, *13*, 360. [CrossRef]
32. Brittle, E.E.; Reynolds, A.E.; Enquist, L.W. Two Modes of Pseudorabies Virus Neuroinvasion and Lethality in Mice. *J. Virol.* **2004**, *78*, 12951–12963. [CrossRef]
33. Liu, Q.; Sikand, P.; Ma, C.; Tang, Z.; Han, L.; Li, Z.; Sun, S.; LaMotte, R.H.; Dong, X. Mechanisms of Itch Evoked by β-Alanine. *J. Neurosci.* **2012**, *32*, 14532–14537. [CrossRef] [PubMed]
34. Qu, L.; Fan, N.; Ma, C.; Wang, T.; Han, L.; Fu, K.; Wang, Y.; Shimada, S.G.; Dong, X.; LaMotte, R.H. Enhanced Excitability of MRGPRA3- and MRGPRD-Positive Nociceptors in a Model of Inflammatory Itch and Pain. *Brain* **2014**, *137*, 1039–1050. [CrossRef]
35. Frölich, M.; Enk, A.; Diepgen, T.L.; Weisshaar, E. Successful Treatment of Therapy-Resistant Pruritus in Lichen Amyloidosis with Menthol. *Acta Derm. Venereol.* **2009**, *89*, 524–526. [CrossRef] [PubMed]
36. Han, J.H.; Choi, H.K.; Kim, S.J. Topical TRPM8 Agonist (Icilin) Relieved Vulva Pruritus Originating from Lichen Sclerosus et Atrophicus. *Acta Derm. Venereol.* **2012**, *92*, 561. [CrossRef]
37. Han, S.K.; Mancino, V.; Simon, M.I. Phospholipase Cbeta 3 Mediates the Scratching Response Activated by the Histamine H1 Receptor on C-Fiber Nociceptive Neurons. *Neuron* **2006**, *52*, 691–703. [CrossRef]
38. Mishra, G.P.; Tamboli, V.; Jwala, J.; Mitra, A.K. Recent Patents and Emerging Therapeutics in the Treatment of Allergic Conjunctivitis. *Recent Pat. Inflamm. Allergy Drug Discov.* **2011**, *5*, 26–36. [CrossRef] [PubMed]
39. Arthur, J.S.; Ley, S.C. Mitogen-Activated Protein Kinases in Innate Immunity. *Nat. Rev. Immunol.* **2013**, *13*, 679–692. [CrossRef] [PubMed]
40. Chuang, H.C.; Wang, X.; Tan, T.H. MAP4K Family Kinases in Immunity and Inflammation. *Adv. Immunol.* **2016**, *129*, 277–314.
41. Laval, K.; Van Cleemput, J.; Vernejoul, J.B.; Enquist, L.W. Alphaherpesvirus Infection of Mice Primes PNS Neurons to an Inflammatory State Regulated by TLR2 and Type I IFN Signaling. *PLoS Pathog.* **2019**, *15*, e1008087. [CrossRef] [PubMed]
42. Zhang, Y.; Liu, S.; Jiang, H.; Deng, H.; Dong, C.; Shen, W.; Chen, H.; Gao, C.; Xiao, S.; Liu, Z.F.; et al. G (2)-Quadruplex in the 3'UTR of IE180 Regulates Pseudorabies Virus Replication by Enhancing Gene Expression. *RNA Biol.* **2020**, *17*, 816–827. [CrossRef] [PubMed]
43. Miltenberger, R.J.; Farnham, P.J.; Smith, D.E.; Stommel, J.M.; Cornwell, M.M. V-Raf Activates Transcription of Growth-Responsive Promoters via GC-Rich Sequences that Bind the Transcription Factor Sp1. *Cell Growth Differ.* **1995**, *6*, 549–556. [PubMed]
44. Buckland, K.F.; Williams, T.J.; Conroy, D.M. Histamine Induces Cytoskeletal Changes in Human Eosinophils via the H(4) Receptor. *Br. J. Pharmacol.* **2003**, *140*, 1117–1127. [CrossRef]
45. Gutzmer, R.; Mommert, S.; Gschwandtner, M.; Zwingmann, K.; Stark, H.; Werfel, T. The Histamine H4 Receptor Is Functionally Expressed on T(H)2 Cells. *J. Allergy Clin. Immunol.* **2009**, *123*, 619–625. [CrossRef]
46. Jemima, E.A.; Prema, A.; Thangam, E.B. Functional Characterization of Histamine H4 Receptor on Human Mast Cells. *Mol. Immunol.* **2014**, *62*, 19–28. [CrossRef]
47. Laval, K.; Vernejoul, J.B.; Van Cleemput, J.; Koyuncu, O.O.; Enquist, L.W. Virulent Pseudorabies Virus Infection Induces a Specific and Lethal Systemic Inflammatory Response in Mice. *J. Virol.* **2018**, *92*, e01614-18. [CrossRef]
48. Desai, P.; Thurmond, R.L. Histamine H4 Receptor Activation Enhances LPS-Induced IL-6 Production in Mast Cells via ERK and PI3K Activation. *Eur. J. Immunol.* **2011**, *41*, 1764–1773. [CrossRef]
49. Morse, K.L.; Behan, J.; Laz, T.M.; West, R.E., Jr.; Greenfeder, S.A.; Anthes, J.C.; Umland, S.; Wan, Y.; Hipkin, R.W.; Gonsiorek, W.; et al. Cloning and Characterization of a Novel Human Histamine Receptor. *J. Pharmacol. Exp. Ther.* **2001**, *296*, 1058–1566.
50. Wang, B.; Wu, H.X.; Li, M.; Gao, Y.; Yuan, M.Q.; Qiu, H.J.; Sun, Y. Transcriptomic Analysis of the Dorsal Root Ganglia of Mice Infected with Pseudorabies Virus. *Chin. J. Prev. Vet. Med.* **2022**, *44*, 1–7. (In Chinese)

Article

Huaier Polysaccharide Interrupts PRV Infection via Reducing Virus Adsorption and Entry

Changchao Huan [1,2,3] , Jingting Yao [1,2,3], Weiyin Xu [1,2,3], Wei Zhang [1,2,3], Ziyan Zhou [1,2,3], Haochun Pan [1,2,3] and Song Gao [1,2,3,*]

1   Institutes of Agricultural Science and Technology Development, College of Veterinary Medicine, Yangzhou University, Yangzhou 225009, China; changchaohuan@yzu.edu.cn (C.H.); yaojingting201609@163.com (J.Y.); xuxuwy@163.com (W.X.); weizhang0918@163.com (W.Z.); yzdxzzy@163.com (Z.Z.); haochunpan0826@163.com (H.P.)
2   Jiangsu Co-Innovation Center for Prevention and Control of Important Animal Infectious Diseases and Zoonoses, Yangzhou 225009, China
3   Key Laboratory of Avian Bioproduct Development, Ministry of Agriculture and Rural Affairs, Yangzhou 225009, China
*   Correspondence: gsong@yzu.edu.cn

Abstract: A pseudorabies virus (PRV) novel virulent variant outbreak occurred in China in 2011. However, little is known about PRV prevention and treatment. Huaier polysaccharide has been used to treat some solid cancers, although its antiviral activity has not been reported. Our study confirmed that the polysaccharide can effectively inhibit infection of PRV XJ5 in PK15 cells. It acted in a dose-dependent manner when blocking virus adsorption and entry into PK15 cells. Moreover, it suppressed PRV replication in PK15 cells. In addition, the results suggest that Huaier polysaccharide plays a role in treating PRV XJ5 infection by directly inactivating PRV XJ5. In conclusion, Huaier polysaccharide might be a novel therapeutic agent for preventing and controlling PRV infection.

Keywords: Huaier polysaccharide; pseudorabies virus; antiviral; infection

Citation: Huan, C.; Yao, J.; Xu, W.; Zhang, W.; Zhou, Z.; Pan, H.; Gao, S. Huaier Polysaccharide Interrupts PRV Infection via Reducing Virus Adsorption and Entry. *Viruses* 2022, 14, 745. https://doi.org/10.3390/v14040745

Academic Editors: Yan-Dong Tang and Xiangdong Li

Received: 17 February 2022
Accepted: 29 March 2022
Published: 1 April 2022

Publisher's Note: MDPI stays neutral with regard to jurisdictional claims in published maps and institutional affiliations.

1. Introduction

Pseudorabies virus (PRV) has been regarded as one of the major causative agents for fatal losses in the swine industry worldwide [1,2]. PRV is a double-stranded linear DNA and enveloped virus that belongs to the subfamily alphaherpesvirinae of the family Herpesviridae [3,4]. PRV can infect various species of mammals, such as swine, wild boars [5], ruminants (e.g., goat, sheep, and cattle) [6,7], carnivores (e.g., hunting dogs, minks, and foxes) [8,9], and rodents. Swine are the unique natural host and reservoir of PRV [10]. PRV infections in swine are often fatal, and the infected swine die from central nervous system disorders and respiratory diseases [3]. Pregnant sows infected with the virus exhibit abortions, stillbirths, and mummified fetuses and are often infertile, with a high rate of rebelliousness [11]. The Bartha-K61 strain is still one of the most widely used live virus vaccines to confront PRV infections worldwide [4]. Pseudorabies (PR) incidence was first reported among cats in 1947 in China; thereafter, PR incidence was reported in other species, particularly in the pig industry [12]. Before 2011, the Bartha-K61 vaccine could prevent and control PR in China; however, the emergence of several variant strains led to the new pseudorabies epidemic among immunized swine [13]. PRV variants exhibit high sequence divergence compared with classical PRV strains [14]. The Chinese triple-gene-deleted (gE/gI/TK) vaccine and TK/gG-deleted vaccine have been extensively used to prevent PR in the Chinese pig population. However, the protection efficacy of the two vaccines has not been reported [3]. More importantly, PRV could infect humans, and it was isolated from an acute human encephalitis case [15–17]. Therefore, identifying new preventive and therapeutic measures for controlling PR, in addition to vaccines, is imperative.

The application of traditional Chinese medicine to treat diseases can be traced back to 200 Anno Domini (AD) [18]. Growing evidence suggests that the traditional Chinese medicine offers numerous compounds having antiviral activities [19]. Trametes robiniophila murr (Huaier), as a traditional Chinese medicine, has been used to treat many diseases for more than one thousand years [20]. Huaier extract can regulate DNA-dependent transcription and the cellular response to hypoxia during breast cancer development and progression [21]; it can also enhance the host immunity and induce apoptosis in breast cancer cells [22]. In addition, studies have confirmed that Huaier granule exhibits good safety in clinical antitumor therapies [23]. In particular, it was considered as a promising adjuvant for the treatment of breast cancer when used in combination with conventional treatment [24]. Huaier polysaccharide also exhibits excellent safety, therapeutic efficacy, and minimal side effects in the clinical treatment of lung cancer, liver cancer, and other solid tumors [25]. However, the effect of Huaier polysaccharide on PRV infection has not been investigated. In this study, we evaluated the mechanism of Huaier polysaccharide in PRV infection; our findings suggested that Huaier polysaccharide inhibits PRV XJ5 adsorption, entry, and replication and thus can dampen PRV infection.

2. Materials and Methods

2.1. Cells and Viruses

Pig kidney (PK15) cells and African green monkey kidney (Vero) cells were stored at the Yangzhou University Infectious Diseases Laboratory and grown in monolayers at 37 °C under 5% CO_2 conditions. PK15 cells and Vero cells were cultured in Dulbecco's modified Eagle medium (DMEM) supplemented with 5% and 6% fetal bovine serum (FBS; Lonsera, Uruguay), 100 units/mL penicillin, 100 μg/mL streptomycin sulfate, and fungizone.

PRV XJ5, PRV NT, and PRV Ra were stored at the Yangzhou University Infectious Diseases Laboratory. We used 0.1 multiplicity of infection (MOI) of PRV-infected PK15 cells in all experiments.

2.2. Reagents and Antibodies

In all subsequent experiments, Huaier polysaccharide (Yangling Ciyuan Biotechnology Co., Ltd., Xian, China) was diluted with PBS to prepare stock solutions of 50 mg/mL and stored at −20 °C. The antibodies for PRV gB were generated by immunization of mice with purified recombinant gB in our laboratory. Actin Ab—T0022 was procured from Affinity Bioscience (Beijing, China). FITC-conjugated goat anti-pig IgG antibody combining PRV-positive sera was purchased from Sigma-Aldrich (St. Louis, MO, USA). DAPI (#C1006) was purchased from Beyotime Biotechnology (Shanghai, China).

2.3. Cell-Based Infectivity Assays

2.3.1. Cells Viability Assay

Cell viability was evaluated with Enhanced Cell Counting Kit-8 (CCK-8) assays, according to the manufacturer's instructions (Beyotime Biotechnology, Shanghai, China). For cell viability assays, cells were seeded at 5000 per well into 96-well plates. On the next day, the medium was changed to DMEM/5% FBS supplemented with different concentrations of Huaier polysaccharide for 24 and 36 h. CCK-8 (10 μL) was then added to each well, and the cells were incubated for 1 h at 37 °C. The absorbance was detected at 450 nm with a microplate reader (Bio-rad, Hercules, CA, USA).

2.3.2. General Effect of Huaier Polysaccharide on PRV Infection

Huaier polysaccharide was added at 37 °C for 24 h to verify its effect on PRV infection. DMEM containing 5% FBS was used to grow the PK15 cells in a 37 °C/5% CO_2 incubator. The PK15 cells were seeded in a 6-well plate, with a density of 5×10^5 cells per well. When the cells grew to approximately 8.5×10^5, the PK15 cells were infected with PRV XJ5 (MOI = 0.1) for 1 h. At 1 h post infection (h.p.i.), DMEM was removed, and the cells were washed with PBS. The cells were incubated with different concentrations (25, 50, 100, or

200 µg/mL) of Huaier polysaccharide in 2% FBS DMEM (2% DMEM). On the next day, the cells were collected to determine PRV XJ5 protein expression through Western blotting and indirect Immunofluorescent Assay (IFA), intracellular viral DNA copies were analyzed by qRT-PCR after 24 h at 37 °C, and cell supernatants were used to measure the virus titer through $TCID_{50}$ assay.

2.3.3. Exploration of Huaier Polysaccharide's Effect on PRV Adsorption and Entry

Huaier polysaccharide was added at 4 °C for 1 h and at 37 °C for 1 h to verify its effect on PRV adsorption and entry. The PK15 cells were cultured using 1 mL of DMEM without FBS to dilute Huaier polysaccharide to 25, 50, 100, 200 µg/mL and then infected with PRV XJ5 (MOI = 0.1) at 4 °C for 1 h. Thereafter, the cells were washed thrice with cold PBS and incubated in 2% DMEM containing appropriate concentrations of Huaier polysaccharide for 1 h at 37 °C. Then, cells were washed with citric acid to remove extracellular virus and again washed with PBS. Adding 2 mL 2% DMEM, intracellular viral proteins were detected through Western blotting and IFA, and the contained viruses' cell supernatants were used to measure virus titers by using the $TCID_{50}$ assay at 24 h.p.i. Intracellular viral DNA copies were analyzed by qRT-PCR after 1 h at 37 °C.

2.3.4. General Effect of Huaier Polysaccharide on PRV Adsorption

Huaier polysaccharide was added at 4 °C for 1 h to verify its effect on PRV adsorption. The PK15 cells were digested with 0.25% trypsin and diluted with DMEM containing 5% FBS, dripped onto a 6-well plate at a concentration of 5×10^5 cells/well in a 37 °C/5% CO_2 incubator. When the cells grew to approximately 70–80%, the cells were treated with different concentrations of Huaier polysaccharide and infected with PRV XJ5 (MOI = 0.1) at 4 °C for 1 h. The cells were cleaned thrice with cold PBS and maintained in 2% DMEM for 24 h. Intracellular viral proteins were detected through Western blotting and IFA. Cell supernatants were used to measure virus titers through $TCID_{50}$ assay. Viral DNA copies of adsorption were analyzed by qRT-PCR after 1 h at 4 °C.

2.3.5. Exploration of Huaier Polysaccharide's Effect on PRV Entry

Huaier polysaccharide was added at 37 °C for 1 h to verify its effect on PRV entry. The PK15 cells were infected with PRV XJ5 (MOI = 0.1) at 4 °C for 1 h without Huaier polysaccharide. The infected cells were then washed with cold PBS three times and cultured in 2% DMEM containing different concentrations of Huaier polysaccharide at 37 °C for 1 h. The cells were washed thrice with citric acid and PBS. Then, cells were cultured in 2% DMEM without Huaier polysaccharide. Intracellular viral proteins were detected through Western blotting and IFA after 24 h at 37 °C. Cell supernatants were used to measure the virus titer through $TCID_{50}$ assay. Intracellular viral DNA copies were analyzed by qRT-PCR after 1 h at 37 °C. The above experiments included three independent experiments.

2.3.6. Effect of Huaier Polysaccharide on PRV Replication

Huaier polysaccharide was added at 37 °C for 3 h or 5 h to verify its effect on PRV replication. When the PK15 cells grew to approximately 70–80%, the cells were incubated with PRV XJ5 (MOI = 0.1) for 1 h at 37 °C and under 5% CO_2. After the cells were washed thrice with PBS, the cells were incubated with 2% DMEM containing 25, 50, 100, and 200 µg/mL Huaier polysaccharide. At 4 h.p.i. and 6 h.p.i., the cells were collected to determine PRV-related protein expression through Western blotting and viral DNA copies were analyzed by qRT-PCR. The above experiments included three independent experiments.

2.3.7. Effect of Huaier Polysaccharide on PRV XJ5

PRV XJ5 (MOI = 0.1) was pretreated with different concentrations of Huaier polysaccharide (25, 50, 100, and 200 µg/mL) for 1 h at 37 °C. Then, the cells were washed thrice with PBS. We co-incubated PRV XJ5 with cells as described above. At 24 h.p.i, the cells were collected to analyzed the expression of PRV-related proteins through Western blotting and viral DNA copies were analyzed by qRT-PCR. PRV XJ5 (MOI = 0.1) was pretreated with different concentrations of Huaier polysaccharide (100 or 200 μg/mL) for 1 h at 37 °C with 100 μL. A transmission electron microscope was used to determine the real destruction of the virions. The above experiments included three independent experiments.

2.3.8. Western Blotting

Cell lysis buffer (Beyotime Biotechnology, Shanghai, China) was used to lyse the cells. The lysed cells were placed in a 1.5 mL EP tube (Eppendorf Micro Test Tubes), to which phosphorylated protease inhibitor and phenylmethanesulfonyl fluoride (PMSF) were added. The concentration of cell lysate protein was determined according to the manufacturer's instructions for the BCA protein quantification kit. Western blotting was performed following the previously described methods [26,27]. The membranes were incubated with PRV gB or actin primary antibodies at 4 °C overnight. On the next day, the membranes were incubated with goat anti-rat secondary antibody (Beyotime Biotechnology, Shanghai, China) at room temperature for 2 h.

2.3.9. Virus Titer Assays

The $TCID_{50}$ assay was used to evaluate virus titers. After trypsinization of Vero cells, these cells were diluted with DMEM containing 6% FBS and added dropwise to a 96-well plate at a concentration of 2×10^3 cells/well. Then, the cells were placed in a 37 °C/5% CO_2 incubator until the cells adhered to the wall. After the monolayer was in the logarithmic growth phase (around 16 h), cells were washed three times with PBS, absorbing the remaining liquid. The cells were inoculated with serially diluted viruses (10^{-1}–10^{-7} fold) for 1 h at 37 °C and eight replicates for each concentration were prepared. Then, 250 μL of maintenance medium was added to each well. At 72 h.p.i., we observed the CPE. Finally, CPE were counted to calculate the $TCID_{50}$ by using the Reed–Muench method.

2.3.10. Indirect Immunofluorescent Assay

After the PK15 cells were infected with PRV for 24 h, the supernatant was discarded, and the cells were washed thrice with PBS. Then, 4% paraformaldehyde that could cover the cell surface was added, and the cells were fixed at 37 °C for 15 min. Thereafter, 0.1% TritonX-100 was used to penetrate the cells for 10 min, and the cells were washed thrice with PBS. The cells were incubated with 5% BSA blocking solution at 37 °C for 2 h or 4 °C overnight. The primary antibody PRV pig-positive serum was diluted to 1:200 and incubated at 37 °C for 2 h, washed three times with PBST, and incubated with FITC-conjugated goat anti-porcine IgG antibody (Sigma-Aldrich, St. Louis, MO, USA) at 37 °C for 1 h; finally, the cells were stained with DAPI for 5 min. The cells were observed under a fluorescence microscope (LAICA, DMi8) with 488nm. All the images were captured at 100× magnification. The fluorescence density of FITC was calculated by Image J (LOCI, Madison, WI, USA).

2.3.11. DNA Extraction and qRT-PCR

Briefly, we extracted DNA for each sample; 10% 92 μL SDS and 8 μL Proteinase K were added to samples and incubated at 58 °C for 1 h, followed by adding 600 μL phenol:chloroform (1:1, *v/v*) to each sample and vortexing, followed by centrifugation for 15 min. Then, 400 μL supernatant was collected and mixed with 800 μL anhydrous ethanol, and kept at −20 °C for 30 min. All samples were centrifuged at 4 °C for 15 min, and the precipitate was washed with anhydrous ethanol. The DNA was dissolved by ddH_2O and incubated at 37 °C for 30 min. Primers used for qRT-PCR are as follows: gB94-F: 5′-ACAAGTTCAAGGCCCACATCTAC-3′, gB94-R: 5′-GTCCGTGAAGCGGTTCGTGAT-3′.

2.3.12. Electron Microscopy

First, we took a small amount of virus liquid or virus containing Huaier polysaccharide (100 or 200 μg/mL) droplets on the copper mesh and used filter paper to absorb excess virus solution, added 2% phosphotungstic acid dropwise on the copper mesh, used filter paper to absorb excess dye, placed the sample in a drying oven, left it to dry, and used a transmission electron microscope (Tecnai 12; Philips, Eindhoven, The Netherlands) to observe it.

2.3.13. Flow Cytometry Assay

For apoptosis assays, PK15 cells were seeded at 1.2×10^6 per well into 6-well plates. On the next day, the medium was changed to 2% DMEM supplemented with Huaier polysaccharide for 24 h. According to the manufacturer's instructions, PI staining was performed with a Cell Apoptosis Kit with PI (Beyotime Biotechnology, Shanghai, China). The percentage of apoptosis cells was measured by flow cytometry on a CytoFLEX instrument (Beckman Coulter, Inc. Brea, CA, USA).

2.3.14. Statistical Analysis

All the experiments were independently repeated at least three times, and all data are presented as the mean $\pm$ SD based on three independent experiments. The data were analyzed using GraphPad Prism software (GraphPad Software, SanDiego, CA, USA). One-way ANOVA, as well as Duncan's multiple range test, was utilized to analyze differences between groups. At a p value of <0.05, the differences were considered to be statistically significant.

3. Results

3.1. Huaier Polysaccharide Has No Inhibitory Effect on PK15 Cells' Growth and Apoptosis

According to a previous study [22], results showed that the vitality of breast cancer cells is more significantly inhibited by Huaier aqueous extract. To explore the effect of Huaier polysaccharide on PK15 cells, we checked the cells' growth using the Enhanced Cell Counting Kit-8. Figure 1A,B reveal that Huaier polysaccharide had no effect on PK15 cells' growth at 24 h.p.i and 36 h.p.i. The results show that cell viability is not affected by Huaier polysaccharide (25, 50, 100, and 200 µg/mL). In addition, we found that Huaier polysaccharide (200 µg/mL) could not induce apoptosis in PK15 cells (Figure 1C).

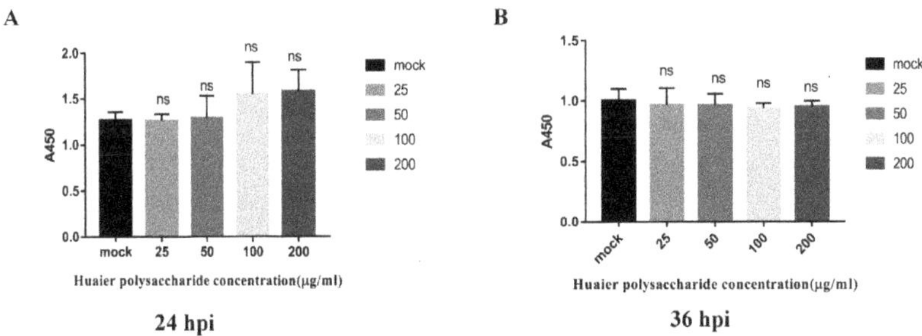

Figure 1. *Cont.*

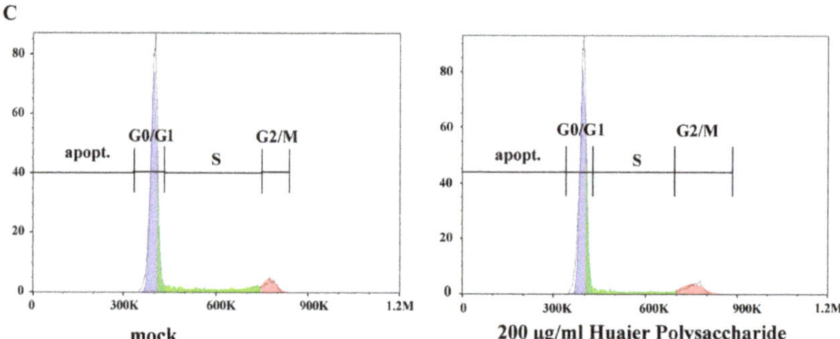

Figure 1. Huaier polysaccharide has no inhibitory effect on PK15 cells' growth and apoptosis. (**A,B**) PK15 cells were pretreated with different concentrations of Huaier polysaccharide at 37 °C for 24 h or 36 h. Cell viability of PK15 cells was assessed with CCK-8 cell counting assays. (**C**) The apoptosis of PK15 cells treated or not with Huaier polysaccharide (200 mg/mL) was checked by flow cytometry. Significance was analyzed using the one-tailed Student's *t*-test. Data are shown as mean ± SD based on three independent experiments. "ns" is not significant.

3.2. Huaier Polysaccharide Inhibis PRV XJ5 Infection in PK15 Cells

Figure 1 reveals that Huaier polysaccharide had no effect on PK15 cells' growth. To investigate the role of Huaier polysaccharides in preventing PRV XJ5 infection, PK15 cells were treated as described in Section 2.3.2. PRV XJ5 (MOI = 0.1) infection resulted in an obvious cytopathic effect (CPE) in the PK15 cells (Figure 2A), which could be inhibited by Huaier polysaccharide, especially at the concentrations of 100 and 200 µg/mL. The antiviral effect of Huaier polysaccharide against PRV XJ5 was further demonstrated through Western blot analysis. gB is a protein that is essential for PRV replication [28]. Thus, we assessed the change in gB expression. The results indicated that the PRV gB protein expression level was reduced after Huaier polysaccharide treatment, with an inhibition rate of 73–94%, at 100 and 200 µg/mL concentrations (Figure 2B,C). The supernatant of the infected PK15 cells was collected to determine the viral titer in terms of the 50% tissue culture infective dose (TCID$_{50}$). As expected, TCID$_{50}$ analysis showed that the Huaier polysaccharide treatment reduced the production of virions; 200 µg/mL Huaier polysaccharide displayed the inhibition rate of 99.9% (Figure 2D). Furthermore, the immunofluorescent assay (IFA) confirmed that Huaier polysaccharide treatment inhibited PRV XJ5 infection in a dose-dependent manner, with the inhibition rate of 13% (25 µg/mL)–99.6% (200 µg/mL) (Figure 2E). qRT-PCR analysis indicated that Huaier polysaccharide decreased the viral DNA copies of PRV XJ5 (Figure 2F). In addition, we explored the effect of Huaier polysaccharide on PRV NT and PRV Ra by qRT-PCR. qRT-PCR analysis revealed that Huaier polysaccharide decreased the viral DNA copies of PRV NT (Figure 2G) and PRV Ra (Figure 2H). The purpose of the infective assay experiment was to emphasize that Huaier polysaccharide has a significant role in the whole process of PRV XJ5 infection of PK15 cells.

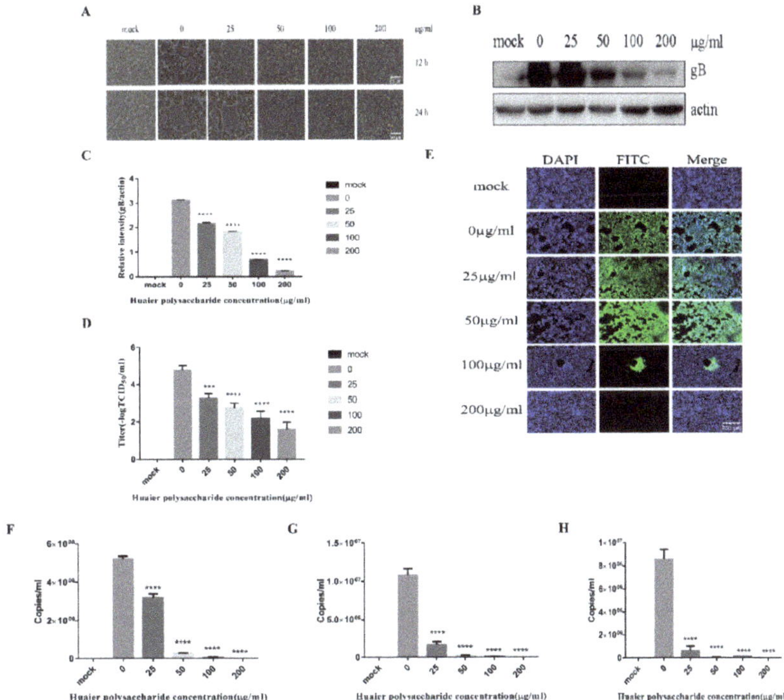

Figure 2. Huaier polysaccharide inhibits PRV XJ5 infection in PK15 cells. (**A**) Infected cells treated with 25, 50, 100, or 200 μg/mL Huaier polysaccharide for 12 or 24 h, showing changes in cell morphology. (**B**) gB and actin protein expression was determined through Western blot assay. (**C**) The relative intensity of intracellular gB to that of actin. Data are presented as means from three independent statistical experiments. The relative intensities of protein were quantified using Image J. Significance was analyzed using a one-tailed Student's *t*-test. (**D**) The viral titers were evaluated through 50% tissue culture infective dose (TCID$_{50}$), and (**E**) immunofluorescent assay (IFA) for internalized virus was performed. (**F–H**) PRV gB were assessed with qRT-PCR analysis in PK15 cells treated with Huaier polysaccharide (25, 50, 100, and 200 μg/mL) at 24 h.p.i. *** $p = 0.0003$, **** $p < 0.0001$. Data are shown as mean $\pm$ SD based on three independent experiments.

3.3. Huaier Polysaccharide Decreased the Adsorption and Entry of PRV XJ5

The PRV life cycle comprises adsorption, entry, viral DNA replication, virion morphogenesis, and viral egress [12,26,27]. In order to further explore the role of Huaier polysaccharides in these processes of PRV XJ5 infected PK15 cells, we firstly studied the effect of Huaier polysaccharide on virus adsorption and entry. The PK15 cells were infected with PRV XJ5 (MOI = 0.1) according to the method described in Section 2.3.3. gB protein was reduced, especially in the presence of 50, 100, and 200 μg/mL Huaier polysaccharide, with inhibition rates of 51%, 75%, and 87%, respectively (Figure 3A,B). Simultaneously, the supernatant was collected, and the viral titer was determined at 24 h.p.i. The results indicated that the PRV XJ5 viral titer was decreased significantly in the cells treated with Huaier polysaccharide (Figure 3C). Furthermore, the IFA demonstrated that Huaier polysaccharide decreased the numbers of cells infected with PRV XJ5 by approximately 48–99.1% (Figure 3D). In addition, we collected cells to quantify the viral DNA copies by qRT-PCR after cells were washed with citric acid and PBS to remove uninternalized virus. qRT-PCR analysis indicated that Huaier polysaccharide decreased viral adsorption and entry in a dose-dependent manner (Figure 3E).

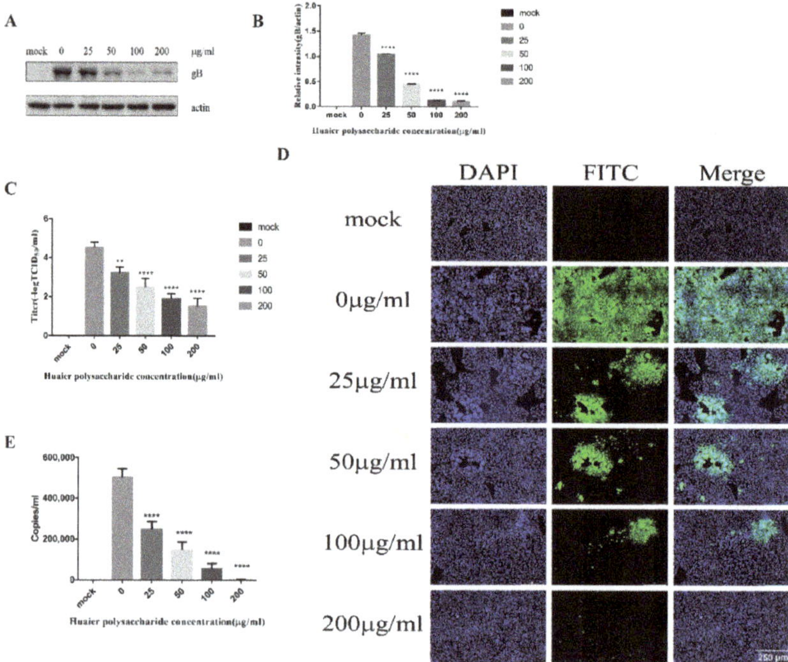

Figure 3. Huaier polysaccharide reduced PRV XJ5 adsorption and entry. The cells were treated according to the method described in Section 2.3.3. (**A**) Western blot showing changes in gB and actin expression. (**B**) The intensity band ratio of intracellular gB to actin. Data are presented as means from three independent statistical experiments. The intensities of protein bands were quantified using Image J. Significance was analyzed using one-tailed Student's *t*-test. **** $p < 0.0001$. (**C**) The viral titers were evaluated through $TCID_{50}$ assay. (**D**) The numbers of cells infected with PRV were observed through IFA. (**E**) The virus DNA copies were observed through qRT-PCR after cells were washed with citric acid and PBS to remove uninternalized virus. ** $p = 0.003$, **** $p < 0.0001$. Data are shown as mean ± SD based on three independent experiments.

3.3.1. Huaier Polysaccharide Attenuates PRV Adsorption

This experiment emphasized PRV XJ5 adsorption. The adsorption of PRV occurred for 1 h at 4 °C. We performed the experiment to study virus adsorption, as described in Section 2.3.4. Figure 4A,B reveal that the Huaier polysaccharide treatment reduced PRV gB protein expression levels, with the inhibition rate of 86–99%, compared with those in untreated cells. To further assess the effect of Huaier polysaccharide on PRV XJ5 adsorption, inhibition of virus adsorption was observed through $TCID_{50}$ assay in the cell supernatant (Figure 4C), and the IFA assay was used to observe the virus-infected cells (Figure 4D). In addition, we collected cells to quantify the viral DNA copies by qRT-PCR after cells were washed with PBS to remove unabsorbed virus. qRT-PCR analysis indicated that Huaier polysaccharide decreased viral adsorption (Figure 4E). The results indicated that Huaier polysaccharide attenuated virus adsorption by the PK15 cells.

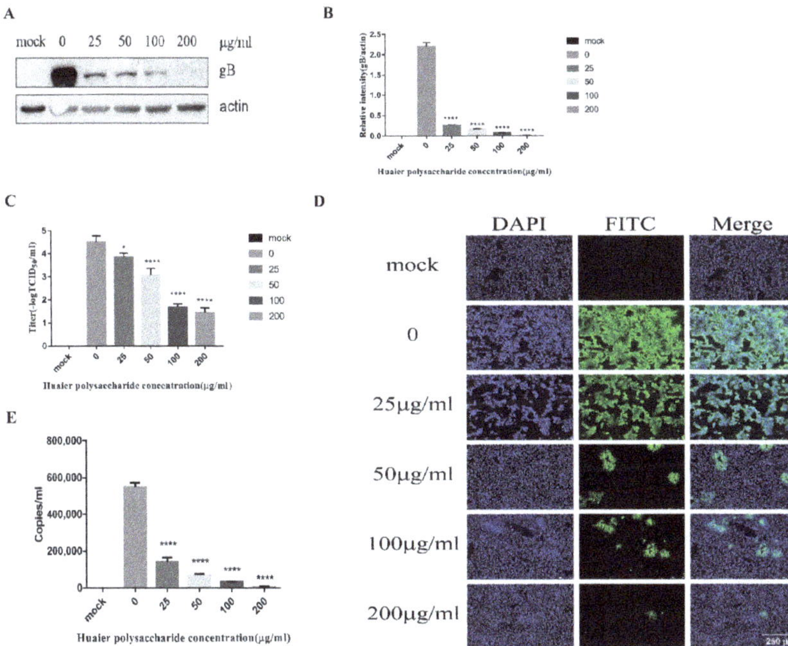

Figure 4. Huaier polysaccharide attenuates PRV XJ5 adsorption. PK15 cells were treated according to the method described in Section 2.3.4. (**A**) At 24 h.p.i., PRV gB and actin protein expression were analyzed through the Western blot assay. (**B**) The intensity band ratio of intracellular gB to actin. The intensities of protein bands were quantified using Image J. (**C**) The viral titers were evaluated through $TCID_{50}$ assay. (**D**) Infected PRV XJ5 cells were observed through IFA. (**E**) The virus DNA copies were observed through qRT-PCR after cells were washed with PBS to remove unadsorbed virus. * $p = 0.0219$, **** $p < 0.0001$. Data are shown as mean ± SD based on three independent experiments.

3.3.2. Huaier Polysaccharide Influences PRV Entry into PK15 Cells

This experiment explored separately the entry of PRV at 1 h and 37 °C. We performed the experiment to study PRV entry according to the method described in Section 2.3.5. Western blot analysis showed that Huaier polysaccharide decreased the expression of PRV gB, with the PRV entry inhibition rate being 13–47% (Figure 5A,B). The $TCID_{50}$ assay revealed that the PRV XJ5 titers were decreased (Figure 5C). The cells infected with PRV were examined through IFA (Figure 5D). In addition, we collected cells to quantify the viral DNA copies by qRT-PCR after cells were washed with citric acid and PBS to remove uninternalized virus. qRT-PCR analysis indicated that Huaier polysaccharide decreased viral entry in a dose-dependent manner (Figure 5E). The results confirmed that Huaier polysaccharide influences PRV XJ5 entry. However, the inhibition rate of PRV entry by Huaier polysaccharide was lower than the observed inhibition rate of PRV adsorption. The results indicate that Huaier polysaccharide might have a prominent role in PRV adsorption.

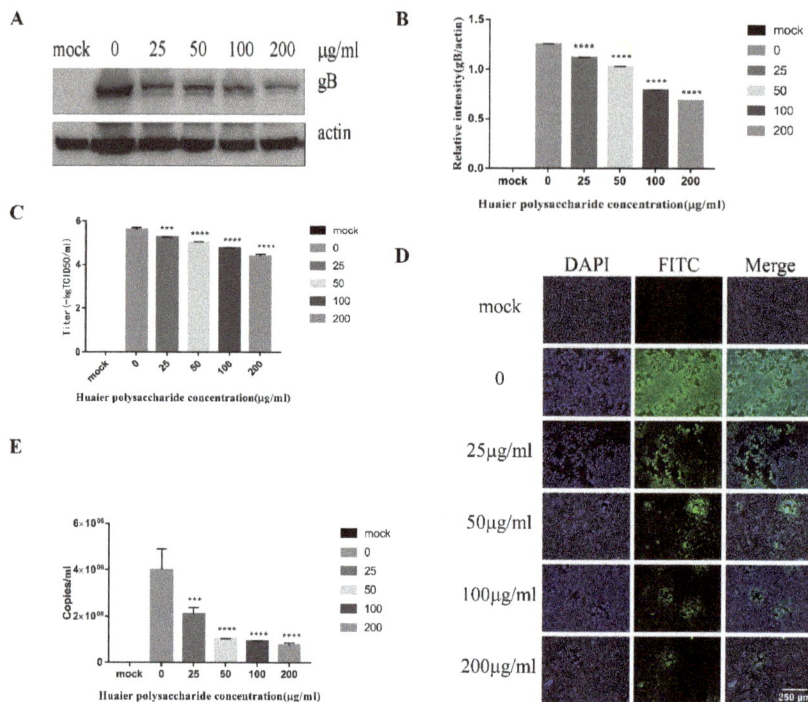

Figure 5. Huaier polysaccharide influences entry of pseudorabies virus in PK15 cells. The PK15 cells were treated according to the method described in Section 2.3.5. (**A**) At 24 h.p.i, PRV gB and actin protein expression were quantified through Western blot assay. (**B**) The relative intensity of intracellular gB to that of actin. Data are presented as the mean value from three independent statistical experiments. The relative intensities of protein were quantified using Image J. Significance was analyzed using the one-tailed Student's *t*-test. (**C**) The viral titers were evaluated through $TCID_{50}$ assay. (**D**) Immunofluorescence assay (IFA) for internalized virus was performed. (**E**) Intracellular viral DNA copies were analyzed by qRT-PCR after cells were washed with citric acid and PBS to remove uninternalized virus. *** $p = 0.0001$, **** $p < 0.0001$. Data are shown as mean $\pm$ SD based on three independent experiments.

3.4. Huaier Polysaccharide Mildly Reduces PRV XJ5 Replication in PK15 Cells

The adsorption of PRV requires 1 h and occurs at 4 °C; entry of PRV requires 1 h and occurs at 37 °C [29]. The life cycle of PRV is completed in approximately 6 h. Hence, we chose the time periods of 4 and 6 h to study PRV replication. The expression of virus protein gB and actin was analyzed through the Western blot assay at 4 and 6 h.p.i. The DNA copies were analyzed by qRT-PCR. The results indicated that Huaier polysaccharide has a slight effect on gB expression (Figure 6A–F), indicating that Huaier polysaccharide mildly affects virus replication.

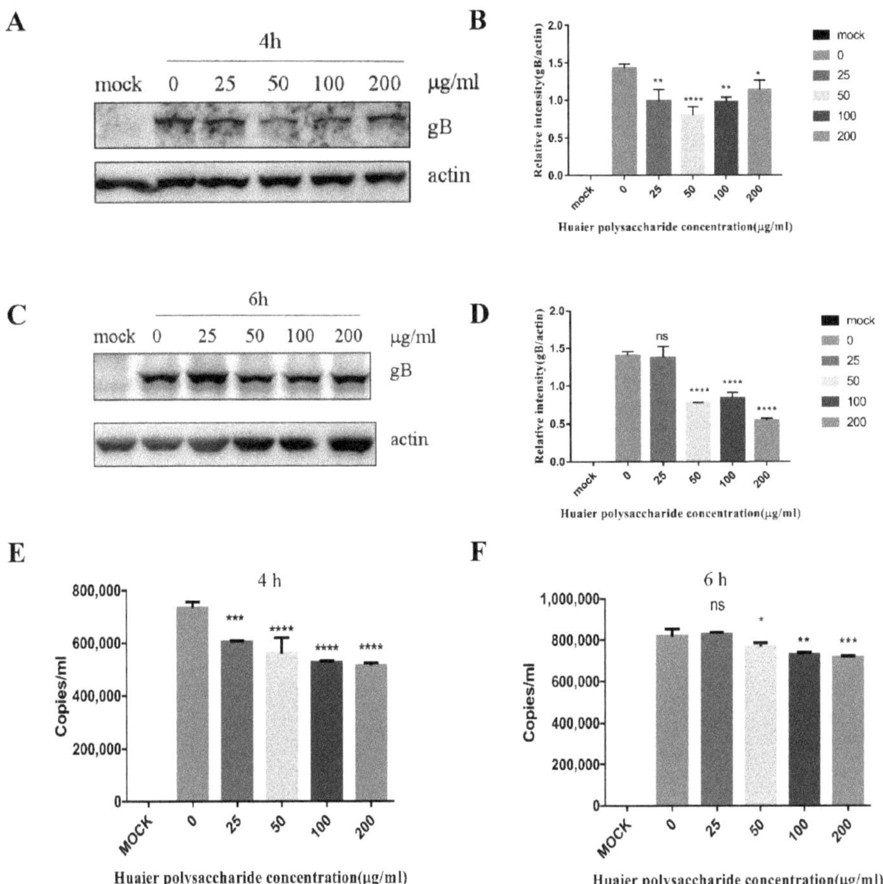

Figure 6. Huaier polysaccharide affects PRV XJ5 replication in PK15 cells. The PK15 cells were treated according to the method described in Section 2.3.6. At 4 and 6 h.p.i., the cells were collected to evaluate the expression of PRV gB and actin proteins through Western blotting (**A,C**). (**B,D**) The relative intensity of intracellular gB to that of actin. The relative intensities of protein were quantified using Image J. (**E,F**) PRV viral DNA copies were analyzed by qRT-PCR at 4 and 6 h.p.i. Significance was analyzed using the one-tailed Student's *t*-test. NS, * $p = 0.0299$, ** $p = 0.0017$, *** $p = 0.0001$, **** $p < 0.0001$. Data are shown as mean $\pm$ SD based on three independent experiments.

3.5. Huaier Polysaccharide May Directly Inactivate PRV XJ5

PRV XJ5 (MOI = 0.1) was pretreated with different concentrations of Huaier polysaccharide for 1 h at 37 °C, and the PK15 cells were infected for 24 h. The cells were collected to assess PRV gB expression through Western blotting. Figure 7A,B illustrate that Huaier polysaccharide reduced PRV gB expression. Figure 7C indicated that virus copies were reduced significantly. In addition, the transmission electron microscope experiment revealed that the PRV envelope was destroyed by Huaier polysaccharide (Figure 7D). These results indicate that Huaier polysaccharide may directly inactivate PRV XJ5.

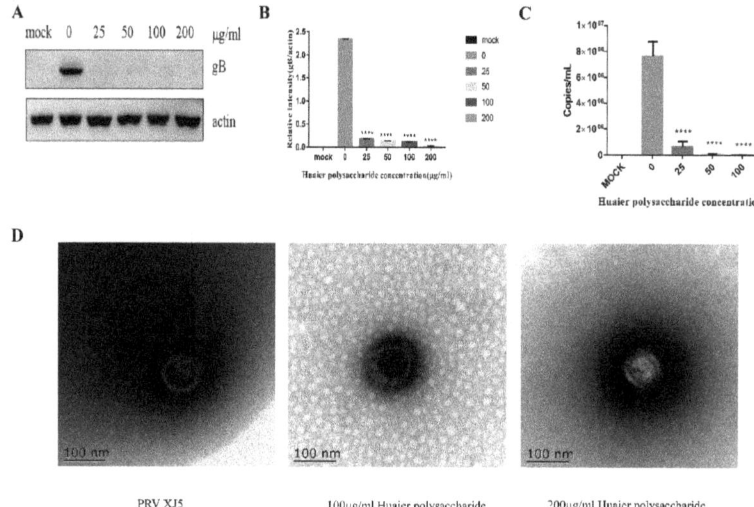

Figure 7. Huaier polysaccharide may directly inactivate PRV XJ5. PRV XJ5 (MOI = 0.1) was treated according to the method described in Section 2.3.8. (**A**) PRV gB was analyzed through Western blotting. (**B**) The relative intensity of intracellular gB to that of actin. (**C**) PRV viral DNA copies were analyzed by qRT-PCR at 24 h.p.i (**D**) PRV particle morphology was observed by electron microscopy. The relative intensities of protein were quantified using Image J. Significance was analyzed using the one-tailed Student's *t*-test. **** $p < 0.0001$. Data are shown as mean ± SD based on three independent experiments.

4. Discussion

Huaier has been used as a traditional Chinese herbal medicine for the treatment of various diseases for thousands of years. In clinical treatment, Huaier polysaccharide has demonstrated a promising auxiliary effect in the treatment of breast cancer, liver cancer, and gastric cancer [30]. Increasing evidence suggests that Huaier polysaccharide can inhibit cell proliferation [31,32], cause cell cycle arrest [33,34], and induce apoptosis [35,36] in cancer cells. In a study, the bi-directional solid fermentation product extract of Huaier with Radix Isatidis (TIF) could upregulate the expression of p53 and caspase-3 in both SK-BR-3 and MDA-MB-231 cell lines [37]. In another study, researchers found that Huaier polysaccharide could inhibit proliferation and promote apoptosis by increasing miR-26b-5p expression in pulmonary cancer cells [38]. Additionally, Huaier polysaccharide extract was found to impair genes related to cell division, the cell cycle, cell cycle phases, and DNA repair [39]. Moreover, in vitro assays confirmed that Huaier polysaccharide could markedly increase the persistence of γ-H2A.X foci and interfere with the homologous recombination pathway [39].

The present study focused on determining the effect of Huaier polysaccharide on virus infection and confirmed that Huaier polysaccharide can affect multiple life cycle stages of PRV XJ5 virus to exert its antiviral effects. PK15 cells are generally used in in vitro PRV research [40]. Huaier polysaccharide mainly affects the adsorption of PRV XJ5 on PK15 cells, and the effect is dose-dependent. The results of qRT-PCR further illustrated this phenomenon. In a study by Zhang et al., the number of successfully invading/migrating cells and the wound closure rate established through the transwell assay and scratch assay, respectively, were reported to be decreased significantly by Huaier polysaccharide, which proved the anti-metastasis effect of Huaier polysaccharide [20]. Thus, we speculate that Huaier polysaccharide inhibits PRV adsorption through its anti-metastasis action; however, the confirmation of this assumption warrants further study. In addition, we revealed that Huaier polysaccharide inhibited PRV entry into PK15 cells by qRT-PCR. Moreover, we found that Huaier polysaccharide prevents virus infection, possibly by binding to a certain receptor on the cell surface, and it protects the cells probably by directly inactivating the virus. The polysaccharide might exert a weak effect on PRV XJ5 replication.

Studies have been focusing on one of the main pathways of programmed cell death after viral infection [41]. The host cell can destroy virus-infected cells through apoptosis, thereby preventing virus infection [42]. Huaier polysaccharide plays a central role in activating the caspase-3 signaling pathway for apoptosis induction [43]. The Huaier polysaccharide component SP1 (a type of purified Huaier polysaccharide) could increase the proportion of Bax/Bcl-2 through the MTDH signaling pathway, which could be another potential mechanism of Huaier in apoptosis induction in breast cancer cells [22]. Moreover, 18β-glycyrrhetinic acid inhibits apoptosis of the cells infected with rotavirus SA11 to decrease rotavirus SA11 infection by Fas (CD95) and FasL (CD178). Similarly, Huaier polysaccharide may depend on Fas (CD95) and FasL (CD178) to regulate apoptosis to inhibit PRV infection [44]. We aim to investigate this assumption in our future studies.

5. Conclusions

In this study, we revealed that Huaier polysaccharide acts against PRV in PK15 cells by blocking PRV adsorption and entry. Hence, Huaier polysaccharide could be further developed as an antiviral agent against PRV infection. Although some studies have investigated the molecular mechanisms, pharmacokinetic and pharmacodynamic models should be developed for an in-depth understanding of the mechanism.

Author Contributions: S.G. and C.H. conceived study, designed the experiments, and analyzed the data. J.Y., W.Z., W.X. and H.P. performed the experiments and analyzed the data. C.H. and J.Y. wrote the manuscript. S.G., C.H., J.Y. and Z.Z. supervised the experiments and edited the manuscript. All authors have read and agreed to the published version of the manuscript.

Funding: This research was funded by the National Natural Science Foundation of China (31902253), the Natural Science Foundation of Jiangsu Province (BK20180921), the Individual Technology Research and Development of Modern Agricultural Industry of Jiangsu Province (CX(19)3024), the China Postdoctoral Science Foundation (2018M632399), the Priority Academic Program Development of Jiangsu Higher Education Institutions (PAPD), and the earmarked fund for Jiangsu Agricultural Industry Technology System, and the Top-level Talents support Program of Yangzhou University.

Institutional Review Board Statement: Not applicable.

Informed Consent Statement: Not applicable.

Conflicts of Interest: The authors declare no conflict of interest.

References

1. Lee, J.Y.; Wilson, M.R. A review of pseudorabies (Aujeszky's disease) in pigs. *Can. Vet. J.* **1979**, *20*, 65–69. [PubMed]
2. Xia, L.M.; Sun, Q.Y.; Wang, J.J.; Chen, Q.; Liu, P.H.; Shen, C.J.; Sun, J.H.; Tu, Y.P.; Shen, S.F.; Zhu, J.C.; et al. Epidemiology of pseudorabies in intensive pig farms in Shanghai, China: Herd-level prevalence and risk factors. *Prev. Vet. Med.* **2018**, *159*, 51–56. [CrossRef] [PubMed]
3. Sun, Y.; Luo, Y.Z.; Wang, C.-H.; Yuan, J.; Li, N.; Song, K.; Qiu, H.-J. Control of swine pseudorabies in China: Opportunities and limitations. *Vet. Microbiol.* **2016**, *183*, 119–124. [CrossRef] [PubMed]
4. Klupp, B.G. Pseudorabies Virus Infections. *Pathogens* **2021**, *10*, 719. [CrossRef]
5. Minamiguchi, K.; Kojima, S.; Sakumoto, K.; Kirisawa, R. Isolation and molecular characterization of a variant of Chinese gC-genotype II pseudorabies virus from a hunting dog infected by biting a wild boar in Japan and its pathogenicity in a mouse model. *Virus Genes* **2019**, *55*, 322–331. [CrossRef]
6. Cheng, Z.; Kong, Z.; Liu, P.; Fu, Z.; Zhang, J.; Liu, M.; Shang, Y. Natural infection of a variant pseudorabies virus leads to bovine death in China. *Transbound. Emerg. Dis.* **2020**, *67*, 518–522. [CrossRef]
7. Kong, H.; Zhang, K.; Liu, Y.; Shang, Y.; Wu, B.; Liu, X. Attenuated live vaccine (Bartha-K16) caused pseudorabies (Aujeszky's disease) in sheep. *Vet. Res. Commun.* **2013**, *37*, 329–332. [CrossRef]
8. Kaneko, C.; Kaneko, Y.; Sudaryatma, P.E.; Mekata, H.; Kirino, Y.; Yamaguchi, R.; Okabayashi, T. Pseudorabies virus infection in hunting dogs in Oita, Japan: Report from a prefecture free from Aujeszky's disease in domestic pigs. *J. Vet. Med. Sci.* **2021**, *83*, 680–684. [CrossRef]

9. Jin, H.-L.; Gao, S.-M.; Liu, Y.; Zhang, S.-F.; Hu, R.-L. Pseudorabies in farmed foxes fed pig offal in Shandong province, China. *Arch. Virol.* **2016**, *161*, 445–448. [CrossRef]

10. Freuling, C.M.; Mueller, T.F.; Mettenleiter, T.C. Vaccines against pseudorabies virus (PrV). *Vet. Microbiol.* **2017**, *206*, 3–9. [CrossRef]

11. Mettenleiter, T.C. Aujeszky's disease (pseudorabies) virus: The virus and molecular pathogenesis—State of the art, June 1999. *Vet. Res.* **2000**, *31*, 99–115. [CrossRef]

12. Tan, L.; Yao, J.; Yang, Y.; Luo, W.; Yuan, X.; Yang, L.; Wang, A. Current Status and Challenge of Pseudorabies Virus Infection in China. *Virol. Sin.* **2021**, *36*, 588–607. [CrossRef]

13. Szpara, M.L.; Tafuri, Y.R.; Parsons, L.; Shamim, S.R.; Verstrepen, K.J.; Legendre, M.; Enquist, L.W. A Wide Extent of Inter-Strain Diversity in Virulent and Vaccine Strains of Alphaherpesviruses. *PLoS Pathog.* **2011**, *7*, e1002282. [CrossRef]

14. Ye, C.; Guo, J.-C.; Gao, J.-C.; Wang, T.-Y.; Zhao, K.; Chang, X.-B.; Wang, Q.; Peng, J.-M.; Tian, Z.-J.; Cai, X.-H.; et al. Genomic analyses reveal that partial sequence of an earlier pseudorabies virus in China is originated from a Bartha-vaccine-like strain. *Virology* **2016**, *491*, 56–63. [CrossRef]

15. Ai, J.-W.; Weng, S.-S.; Cheng, Q.; Cui, P.; Li, Y.-J.; Wu, H.-L.; Zhu, Y.-M.; Xu, B.; Zhang, W.-H. Human Endophthalmitis Caused by Pseudorabies Virus Infection, China, 2017. *Emerg. Infect. Dis.* **2018**, *24*, 1087–1090. [CrossRef]

16. Liu, Q.; Wang, X.; Xie, C.; Ding, S.; Yang, H.; Guo, S.; Li, J.; Qin, L.; Ban, F.; Wang, D.; et al. A novel human acute encephalitis caused by pseudorabies virus variant strain. *Clin. Infect. Dis.* **2020**, *73*, e3690–e3700. [CrossRef]

17. Yang, X.; Guan, H.; Li, C.; Li, Y.; Wang, S.; Zhao, X.; Zhao, Y.; Liu, Y. Characteristics of human encephalitis caused by pseudorabies virus: A case series study. *Int. J. Infect. Dis.* **2019**, *87*, 92–99. [CrossRef]

18. Liu, S.-H.; Chuang, W.-C.; Lam, W.; Jiang, Z.; Cheng, Y.-C. Safety Surveillance of Traditional Chinese Medicine: Current and Future. *Drug Saf.* **2015**, *38*, 117–128. [CrossRef]

19. Kannan, S.; Kolandaivel, P. Antiviral potential of natural compounds against influenza virus hemagglutinin. *Comput. Biol. Chem.* **2017**, *71*, 207–218. [CrossRef]

20. Zhang, N.; Kong, X.; Yan, S.; Yuan, C.; Yang, Q. Huaier aqueous extract inhibits proliferation of breast cancer cells by inducing apoptosis. *Cancer Sci.* **2010**, *101*, 2375–2383. [CrossRef]

21. Kong, X.; Ding, X.; Yang, Q. Identification of multi-target effects of Huaier aqueous extract via microarray profiling in triple-negative breast cancer cells. *Int. J. Oncol.* **2015**, *46*, 2047–2056. [CrossRef]

22. Luo, Z.; Hu, X.; Xiong, H.; Qiu, H.; Yuan, X.; Zhu, F.; Wang, Y.; Zou, Y. A polysaccharide from Huaier induced apoptosis in MCF-7 breast cancer cells via down-regulation of MTDH protein. *Carbohydr. Polym.* **2016**, *151*, 1027–1033. [CrossRef]

23. Chen, Q.; Shu, C.; Laurence, A.D.; Chen, Y.; Peng, B.G.; Zhen, Z.J.; Cai, J.Q.; Ding, Y.T.; Li, L.Q.; Zhang, Y.B.; et al. Effect of Huaier granule on recurrence after curative resection of HCC: A multicentre, randomised clinical trial. *Gut* **2018**, *67*, 2006–2016. [CrossRef]

24. Yao, X.; Wu, W.; Qu, K.; Xi, W. Traditional Chinese biomedical preparation (Huaier Granule) for breast cancer: A PRISMA-compliant meta-analysis. *Biosci. Rep.* **2020**, *40*, BSR20202509. [CrossRef]

25. Wang, M.; Hu, Y.; Hou, L.; Pan, Q.; Tang, P.; Jiang, J. A clinical study on the use of Huaier granules in post-surgical treatment of triple-negative breast cancer. *Gland Surg.* **2019**, *8*, 758–765. [CrossRef]

26. Enquist, L.W.; Husak, P.J.; Banfield, B.W.; Smith, G.A. Infection and spread of alphaherpesviruses in the nervous system. *Adv. Virus Res.* **1998**, *51*, 237–347.

27. Mettenleiter, T.C. Budding events in herpesvirus morphogenesis. *Virus Res.* **2004**, *106*, 167–180. [CrossRef]

28. Ren, J.; Wang, H.; Zhou, L.; Ge, X.; Guo, X.; Han, J.; Yang, H. Glycoproteins C and D of PRV Strain HB1201 Contribute Individually to the Escape From Bartha-K61 Vaccine-Induced Immunity. *Front. Microbiol.* **2020**, *11*, 323. [CrossRef] [PubMed]

29. He, W.; Zhai, X.; Su, J.; Ye, R.; Zheng, Y.; Su, S. Antiviral Activity of Germacrone against Pseudorabies Virus in Vitro. *Pathogens* **2019**, *8*, 258. [CrossRef] [PubMed]

30. Qi, J.; Xie, F.-j.; Liu, S.; Yao, C.-y.; Liu, W.-h.; Cai, G.-q.; Liao, G.-q. Huaier Granule Combined with Tegafur Gimeracil Oteracil Potassium Promotes Stage IIb Gastric Cancer Prognosis and Induces Gastric Cancer Cell Apoptosis by Regulating Livin. *Biomed. Res. Int.* **2020**, *2020*. [CrossRef] [PubMed]

31. Yang, A.; Zhao, Y.; Wang, Y.; Zha, X.; Zhao, Y.; Tu, P.; Hu, Z. Huaier suppresses proliferative and metastatic potential of prostate cancer PC3 cells via downregulation of Lamin B1 and induction of autophagy. *Oncol. Rep.* **2018**, *39*, 3055–3063.

32. Wang, Y.; Lv, H.; Xu, Z.; Sun, J.; Ni, Y.; Chen, Z.; Cheng, X. Huaier n-butanol extract suppresses proliferation and metastasis of gastric cancer via c-Myc-Bmi1 axis. *Sci. Rep.* **2019**, *9*, 447. [CrossRef]

33. Yan, L.; Liu, X.; Yin, A.; Wei, Y.; Yang, Q.; Kong, B. Huaier aqueous extract inhibits cervical cancer cell proliferation via JNK/p38 pathway. *Int. J. Oncol.* **2015**, *47*, 1054–1060. [CrossRef]

34. Hu, Z.; Yang, A.; Su, G.; Zhao, Y.; Wang, Y.; Chai, X.; Tu, P. Huaier restrains proliferative and invasive potential of human hepatoma SKHEP-1 cells partially through decreased Lamin B1 and elevated NOV. *Sci. Rep.* **2016**, *6*, 31298. [CrossRef]

35. Xie, H.-X.; Xu, Z.-Y.; Tang, J.-N.; Du, Y.-A.; Huang, L.; Yu, P.-F.; Cheng, X.-D. Effect of Huaier on the proliferation and apoptosis of human gastric cancer cells through modulation of the PI3K/AKT signaling pathway. *Exp. Ther. Med.* **2015**, *10*, 1212–1218. [CrossRef]

36. Heller, M.J. DNA microarray technology: Devices, systems, and applications. *Annu. Rev. Biomed. Eng.* **2002**, *4*, 129–153. [CrossRef]

37. Liu, Z.; Tang, Y.; Zhou, R.; Shi, X.; Zhang, H.; Liu, T.; Lian, Z.; Shi, X. Bi-directional solid fermentation products of Trametes robiniophila Murr with Radix Isatidis inhibit proliferation and metastasis of breast cancer cells. *J. Chin. Med. Assoc.* **2018**, *81*, 520–530. [CrossRef]
38. Wu, T.; Chen, W.; Liu, S.; Lu, H.; Wang, H.; Kong, D.; Huang, X.; Kong, Q.; Ning, Y.; Lu, Z. Huaier suppresses proliferation and induces apoptosis in human pulmonary cancer cells via upregulation of miR-26b-5p. *Febs Lett.* **2014**, *588*, 2107–2114. [CrossRef]
39. Ding, X.; Yang, Q.; Kong, X.; Haffty, B.G.; Gao, S.; Moran, M.S. Radiosensitization effect of Huaier on breast cancer cells. *Oncol. Rep.* **2016**, *35*, 2843–2850. [CrossRef]
40. Yu, T.; Chen, F.; Ku, X.; Fan, J.; Zhu, Y.; Ma, H.; Li, S.; Wu, B.; He, Q. Growth characteristics and complete genomic sequence analysis of a novel pseudorabies virus in China. *Virus Genes* **2016**, *52*, 474–483. [CrossRef]
41. Danthi, P. Viruses and the Diversity of Cell Death. *Annu. Rev. Virol.* **2016**, *3*, 533–553. [CrossRef]
42. Zhou, X.; Jiang, W.; Liu, Z.; Liu, S.; Liang, X. Virus Infection and Death Receptor-Mediated Apoptosis. *Viruses* **2017**, *9*, 316. [CrossRef]
43. Porter, A.G.; Jänicke, R.U. Emerging roles of caspase-3 in apoptosis. *Cell Death Differ.* **1999**, *6*, 99–104. [CrossRef]
44. Wang, X.; Xie, F.; Zhou, X.; Chen, T.; Xue, Y.; Wang, W. 18β-Glycyrrhetinic acid inhibits the apoptosis of cells infected with rotavirus SA11 via the Fas/FasL pathway. *Pharm. Biol.* **2021**, *59*, 1098–1105. [CrossRef]

Article

Pseudorabies Virus ICP0 Abolishes Tumor Necrosis Factor Alpha-Induced NF-κB Activation by Degrading P65

Xiangbo Zhang [1,†], Jingying Xie [1,2,†], Ming Gao [2], Zhenfang Yan [1], Lei Chen [1], Suocheng Wei [2,*] and Ruofei Feng [1,3,*]

1 Key Laboratory of Biotechnology and Bioengineering of State Ethnic Affairs Commission, Biomedical Research Center, Northwest Minzu University, Lanzhou 730030, China; zhangxiangbowork@163.com (X.Z.); xjy_1314@126.com (J.X.); yzhenfang@163.com (Z.Y.); chenleiwork7@163.com (L.C.)

2 College of Life Science and Engineering, Northwest Minzu University, Lanzhou 730030, China; gaoming001222@163.com

3 Gansu Tech Innovation Center of Animal Cell, Biomedical Research Center, Northwest Minzu University, Lanzhou 730030, China

* Correspondence: weisc668@163.com (S.W.); fengruofei@xbmu.edu.cn (R.F.)

† These authors equally contributed to this work.

Abstract: Nuclear factor κB (NF-κB) is involved in a wide range of innate immune activities in host cells and serves as an important component of a host's immunity system. To survive in infected cells, viruses have evolved intricate strategies to evade the host immune response. Pseudorabies virus (PRV) is a member of the alpha herpesvirus family and is capable of causing reproductive and neurological dysfunction in pigs. PRV has a large DNA genome and therefore has the ability to encode numerous proteins that modulate host innate immune responses. In the present study, we demonstrated that the PRV-encoded immediate early protein ICP0 inhibits the tumor necrosis factor alpha (TNF-α)-mediated NF-κB signaling pathway. An in-depth study showed that ICP0 protein was able to limit NF-κB activation and decreased the expression of inflammatory cytokines interleukin-6 (IL-6) and interleukin 8 (IL-8). In addition, ICP0 blocked the activation of NF-κB through interacting with p65, degrading its protein expression and limiting its phosphorylation. PRV protein ICP0 is shown for the first time to enable escape from innate immune response through the regulation of NF-κB during PRV infection. These results illustrate that PRV ICP0 is able to block NF-κB activation. This mechanism may represent a critical role in the early events leading to PRV infection.

Keywords: pseudorabies virus; ICP0 protein; P65; NF-κB signaling pathway

Citation: Zhang, X.; Xie, J.; Gao, M.; Yan, Z.; Chen, L.; Wei, S.; Feng, R. Pseudorabies Virus ICP0 Abolishes Tumor Necrosis Factor Alpha-Induced NF-κB Activation by Degrading P65. *Viruses* **2022**, *14*, 954. https://doi.org/10.3390/v14050954

Academic Editors: Yan-Dong Tang and Xiangdong Li

Received: 21 March 2022
Accepted: 29 April 2022
Published: 2 May 2022

Publisher's Note: MDPI stays neutral with regard to jurisdictional claims in published maps and institutional affiliations.

1. Introduction

Nuclear factor-κB, also known as NF-κB, is an important component of intrinsic immunity in the body [1–3]. Activation of the NF-κB signaling pathway can be divided into classical and non-classical pathways, with the classical pathway being mainly through the degradation of IκB proteins to allow the release of NF-κB dimers; the non-classical pathway, on the other hand, is processed by p100 and p52, allowing activation of the signaling pathway [4,5]. When cells are subjected to various extracellular stimuli, IκB kinase is activated, leading to phosphorylation of IκB protein and ubiquitination, after which the IκB protein is degraded and NF-κB dimers are released [6]. The NF-κB dimer is then further activated by various post-translational modifications and is translocated to the nucleus to participate in the transcriptional regulation of activated genes and the innate immune response. In the nucleus, it binds to the target gene to facilitate transcription of the target genes [7].

ICP0 is a ring-finger E3 ubiquitin ligase that belongs to the group of early-stage proteins encoded by alpha herpesvirus [8]. The enzyme binds directly to the component proteins of the Ub pathway to inactivate the cellular processes that underlie host immune defense and limit the progression of viral infection [9–11]. Many studies have shown that ICP0 plays a key role in the HSV-1 infection cycle, among many other functions, which are required to facilitate the efficient initiation of lytic infection and the activation of the regeneration of the viral genome from latency [12]. Regarding how HSV-1 counteracts the natural immune response of the host, many studies have shown that ICP0 can inhibit the activation of IFN-β and nuclear factor-κB (NF-κB) [13,14]. Although the HSV-1 ICP0 protein has multiple immune evasion mechanisms, the immune evasion function of the PRV ICP0 protein is not well understood.

Current studies on PRV proteins have focused on their antagonism to type I IFN signaling pathways, especially the cGAS-STING-mediated signaling pathway. As the DNA sensing pathway induced by innate immunity plays a critical role in controlling PRV infection, PRV proteins evolved complex mechanisms to antagonize the innate immune response. Several studies have indicated that PRV UL13 inhibits the IFN-β production by targeting IRF3 in a kinase-dependent manner [15–17]. PRV UL24 efficiently inhibited cGAS-STING-mediated IFN production by interacting with interferon regulatory factor 7 (IRF7) and degrading its expression through the proteasome pathway [18]. In addition, PRV gE is involved in counteracting cGAS-STING-mediated IFN production through degrading CBP [19]. Our previous study also found PRV US3 protein could inhibit cGAS-STING-mediated IFN production by interacting with IRF3 and degrading its expression [20]. There are a few studies on PRV immune evasion and the NF-κB signaling pathway. A recent study showed that PRV UL24 protein abrogated TNF-α-mediated NF-κB activation by interacting with p65 and promoting it for proteasomal degradation [21]. All in all, the study on the evasion of host immune response mediated by PRV viral proteins is very limited.

In this study, we defined the role of PRV ICP0 protein in the inhibition of NF-κB pathway activation. Our results indicated that the PRV ICP0 protein could significantly inhibit TNF-α-mediated NF-κB activation. Additionally, ICP0 prevented the degradation of IκBα and then degraded p65 to inhibit the expression of inflammatory factors IL-6 and IL-8. Co-immunoprecipitation analysis demonstrated that ICP0 interacted with p65 and degraded its expression through proteasome pathway. Meanwhile, ICP0 also inhibited the phosphorylation and nuclear translocation of p65. In conclusion, this study is the first to describe the role of ICP0 in antagonizing the NF-κB pathway. The attenuation of NF-κB activation by PRV ICP0 protein may represent an essential accommodation to enable virus persistence within the host.

2. Materials and Methods

2.1. Cells and Virus

PK15 and HEK293 cells were cultured with Dulbecco's modified Eagle medium (DMEM, BAILING) supplemented with 10% fetal bovine serum, at 37 °C in a 5% CO_2 incubator. PRV Bartha-61 strain was propagated in BHK-21 cells, and the supernatants of infected cells were clarified and stored at −80 °C.

2.2. Antibodies and Reagents

Anti-Flag tag rabbit polyclonal antibody (D191041), horseradish peroxidase (HRP)-conjugated goat anti-rabbit IgG (D110058) and HRP-conjugated goat anti-mouse IgG (D110087) were purchased from Sangon Biotech (Shanghai, China). NF-κB p65 rabbit polyclonal antibody (10745-1-AP), GAPDH mouse monoclonal antibody (60004-1-Ig), Myc tag mouse monoclonal antibody (60003-2-Ig), NFκB1 rabbit polyclonal antibody (14220-1-AP) and TAK1 rabbit polyclonal antibody (12330-2-AP) were purchased from Proteintech (Wuhan, China). Anti-HIST3H3 polyclonal antibody (K106623P) was purchased from Solarbio (Beijing, China). IκBα rabbit polyclonal antibody KO validated (AF5204), phospho-IκBα (Ser32/36) rabbit polyclonal antibody (AF5851) and phospho-NF-κB p65 (Ser276) rabbit polyclonal antibody (AF5875) were purchased from Beyotime (Shanghai, China). *TransStart* Top Green qPCR SuperMix (+Dye II) was purchased from Transgen (Beijing, China). Cell membrane/cytoplasm/nuclear membrane protein step extraction kit (BB-31042) was purchased from BestBio (Shanghai, China). Lipofectamine 3000 was purchased from Invitrogen. Chemical reagents RNase inhibitor (Thermo Fisher, Waltham, MA, USA), MG132 (Beyotime, Nantong, China), chloroquine (CQ) (tlrl-chq, InvivoGen, San Diego, CA, USA), Ac-DEVD-CHO (Beyotime) and TNF-α (InvivoGen) were purchased from indicated manufacturers.

2.3. Plasmids

A plasmid encoding Flag-tagged p65 was constructed by molecular cloning methods. A Myc-tagged ICP0 plasmid was constructed in-house. All plasmids were verified by sequencing. The primer sequences used in this study are available upon request.

2.4. Western Blotting

Cells were harvested and whole-cell extracts were prepared with lysis buffer RIPA (Solarbio, Beijing, China). Cell extracts were subjected to 10% or 15% SDS-PAGE, and the separated proteins were transferred to PVDF membranes (Millipore, Berlington, MA, USA). The PVDF membranes were incubated with specific primary and HRP-conjugated secondary antibodies. GAPDH or β-actin served as loading control. The proteins were detected using ECL Blotting Substrate (Bio-Rad, Hercules, CA, USA).

2.5. Co-Immunoprecipitation Assay

Cells were collected with lysis buffer supplemented with phosphatase inhibitor cocktail and incubated with anti-Flag or anti-p65 antibody for 12 h at 4 °C. Then, 10 μL of Protein G agarose slurry (Beyotime, Nantong, China) was added to each lysate. After incubation for 4 h at 4 °C, the lysates were centrifuged at 2500 rpm for 5 min. The beads were collected and washed 5 times with ice-cold PBS. The precipitates were mixed with SDS buffer and boiled for 5 min at 95 °C. After centrifugation at 6000 rpm for 1 min, the supernatant was collected and used for Western blot analysis.

2.6. RNA Extraction and RT-qPCR

mRNA transcription levels for NF-κB-dependent genes such as IFN-β, IL-6 and IL-8 were determined by relative quantitative PCR (RT-qPCR). Cellular RNA was isolated and reverse-transcribed to cDNA. Methods were performed as previously described [22]. Primers for RT-qPCR are available upon request.

2.7. Transfection

Plasmid DNA was transfected into PK15 cells using Lipofectamine 3000 (Invitrogen, Waltham, MA, USA). All experiments were conducted in accordance with the company's instructions. Cells were then infected with PRV for 24 h at an MOI of 0.01 (except for the cases mentioned in the text) to test the effect of ICP0 on PRV replication. In co-transfection experiments, ICP0 and reporter gene constructs were used in a 1:1 mass ratio.

2.8. Nuclear and Cytoplasmic Extraction

PRV-infected or uninfected PK15 cells were washed in PBS with 400 μL of Extract A, 2 μL of protease inhibitor and 2 μL of phosphatase inhibitor (BestBio, Xi'an, China). The homogenate was centrifuged at 1000 g at 4 °C for 5 min. The supernatant was preserved as cytoplasm and placed on ice for 30 min. Then, 1 μL of protease inhibitor and 1 μL of phosphatase inhibitor (BestBio, Xi'an, China) were added to 200 μL of Extract B and 5 μL of Extract C. The mixture is placed on ice for another 30 min and kept as a nucleus. The precipitates were analyzed by standard immunoblotting procedures.

2.9. Virus Titer

BHK-21 cells grown in 96-well plates were infected with 10-fold serial dilutions of PRV samples. After 2 h at 37 °C, the culture medium was replaced with fresh DMEM.

The plates were incubated for 72–96 h at 37 °C. PRV titers were calculated using the Reed–Muench method.

2.10. ICP0 mRNA Detection

PK15 cells were infected with 1 MOI PRV, and then cells were collected at 1 h, 2 h, 3 h, 4 h and 5 h post-infection. Cellular RNA was extracted, and the expression of ICP0 mRNA expression was detected by RT-qPCR. Primers used were as follows: ICP0-qF, GCGACGCTTCGTTTGTGG; ICP0-qR, GGTTCATCCCGTGCTCCTG.

2.11. Enzyme-Linked Immunosorbent Assay (ELISA)

IFN-β secretion expression levels in the cell supernatants were detected using a swine IFN-β ELISA kit (Jianglaibio, Shanghai, China), according to the manufacturer's instructions.

2.12. P65 Polyubiquitination Assay

PK15 cells were co-transfected with the Myc-tagged ICP0, FLAG-tagged p65 and Myc-tagged Ub expression vector at a 1:1:1 ratio using the Lipofectamine 3000 transfection method. Protein was extracted 30 h post-transfection. The p65–ubiquitin complexes were immunoprecipitated using anti-Flag antibody and immunoblotted with anti-Myc antibody to detect ubiquitinated proteins.

2.13. Statistical Analysis

Measurements were compared using one-way ANOVA. Statistical significance comparisons were calculated using Student's *t*-test in GraphPad Prism 7.0 software (La Jolla, CA, USA). Unless otherwise stated, data are presented as the mean ± standard deviation (SD) of at least three independent experiments. Asterisks indicate statistically significant differences (*** $p < 0.001$, ** $p < 0.01$ and * $p < 0.05$).

3. Results

3.1. ICP0 Promotes PRV Replication in PK15 Cells

To investigate the role of viral protein ICP0 in the process of PRV infection, pCMV-Myc-ICP0 plasmid was transfected into PK15 cells. After 24 h, cells were inoculated with PRV to observe the effect of ICP0 on PRV multiplication. RT-qPCR results show that the amount of viral genomic DNA copies in the ICP0 group grew considerably as compared to the empty vector (EV) group (Figure 1A). Meanwhile, TCID$_{50}$ assay showed the same result: there was a significant upregulation of virus titer in the ICP0 transfection group (Figure 1B). These results suggest that PRV viral protein ICP0 significantly promotes the replication of PRV in PK15 cells.

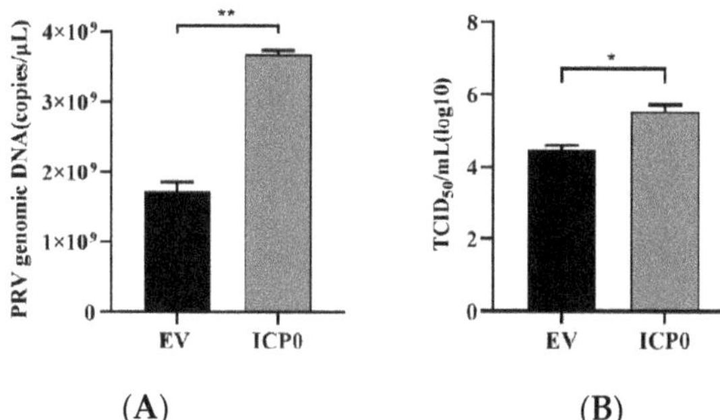

Figure 1. Exogenous expression of ICP0 enhances PRV replication in vitro. PK15 cells were transfected with 1 μg pCMV-Myc plasmid (empty vector, EV) or pCMV-Myc-ICP0 plasmid (ICP0) for 24 h. Then, cells were infected with 0.01 MOI PRV for 24 h before PRV viral copy numbers and titers were measured. Real-time quantitative PCR was used to determine viral copy number (**A**), and TCID50

assay was performed for viral titer detection (Reed–Muench method) (**B**). Data are listed as mean ± SD from three independent experiments. Comparison between two groups was evaluated by unpaired Student's *t*-test. * $p < 0.05$, ** $p < 0.01$.

3.2. ICP0 Inhibits the Transcription of Inflammatory Factors

To investigate whether ICP0 is involved in the regulation of inflammatory factor expression, we assessed the effect of ICP0 on IL-6 and IL-8 mRNA transcription by RT-qPCR. TNF-α was used as a stimulatory factor, and PK15 cells were transfected with an ICP0 expression plasmid for 24 h prior to stimulation. Interestingly, TNF-α significantly increased IL-6 and IL-8 mRNA expression, while gene transcription levels were significantly reduced in ICP0 expression cells (Figure 2A). In addition, IL-6 and IL-8 could be efficiently activated during PRV infection; in contrast, in viral protein ICP0 expression groups, this was accompanied by a significant reduction in the transcription of these genes (Figure 2B). Indeed, these results indicated that ICP0 protein dramatically inhibited TNF-α-mediated NF-κB activation.

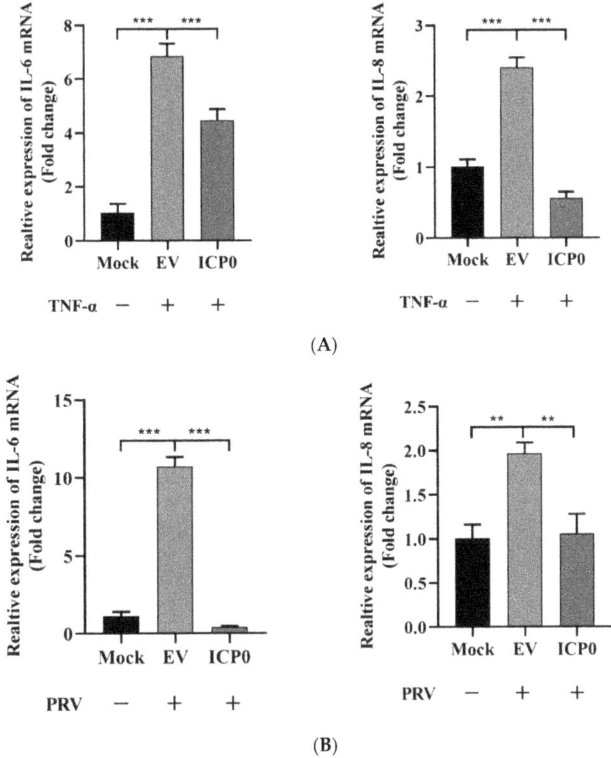

Figure 2. Viral protein ICP0 inhibits IL-6 and IL-8 mRNA transcription. (**A**) PK15 cells were transfected with pCMV-Myc empty vector (EV) (1.0 μg) or pCMV-Myc-ICP0 (1.0 μg) plasmids for 24 h. Then, cells were treated with or without 50 ng/mL recombinant human TNF-α and incubated for an additional 12 h, followed by total RNA extraction. IL-6 and IL-8 mRNA expression levels were measured by RT-qPCR. (**B**) PK15 cells were transfected with pCMV-Myc empty vector (EV) (1.0 μg) or pCMV-Myc-ICP0 (1.0 μg) plasmids for 24 h. Then, cells were infected with 0.01 MOI PRV for another 24 h. Cells were collected for cellular RNA extraction, and RT-qPCR was performed for IL-6 and IL-8 mRNA detection. Data are shown as mean ± SD from three independent experiments. Comparison between two groups was evaluated by unpaired Student's *t*-test. ** $p < 0.01$, *** $p < 0.001$.

3.3. p65 May Be a Target of Pseudorabies Virus ICP0 Protein

During PRV infection, we found PRV infection induced p65 degradation at 5 h post-infection at 1 MOI infection (Figure 3A left). When infected with a lower MOI (0.01 MOI), the degradation of p65 started at 12 h post-infection (data not shown). This indicated there might be other viral proteins participating in degrading p65 in the late infection of PRV. We also detected ICP0 mRNA expression in 1 MOI PRV infected cells at indicated time points. Results showed that ICP0 mRNA expression was increased with infection time (Figure 3A Right). This means p65 degradation may be associated with ICP0 mRNA transcription in infected cells. In order to determine whether p65 of the NF-κB signaling pathway is ICP0's molecular target, we overexpressed the viral protein ICP0 in PK15 cells. Western blot results showed that ICP0 was able to block the phosphorylation of IκBα, as well as the degradation of p65 (Figure 3B). Interestingly, the degradation of p65 by ICP0 was shown to proceed in a dose-dependent manner (Figure 3C). Based on the previous results, we can speculate that the degradation of p65 upon PRV infection may be related to viral protein ICP0.

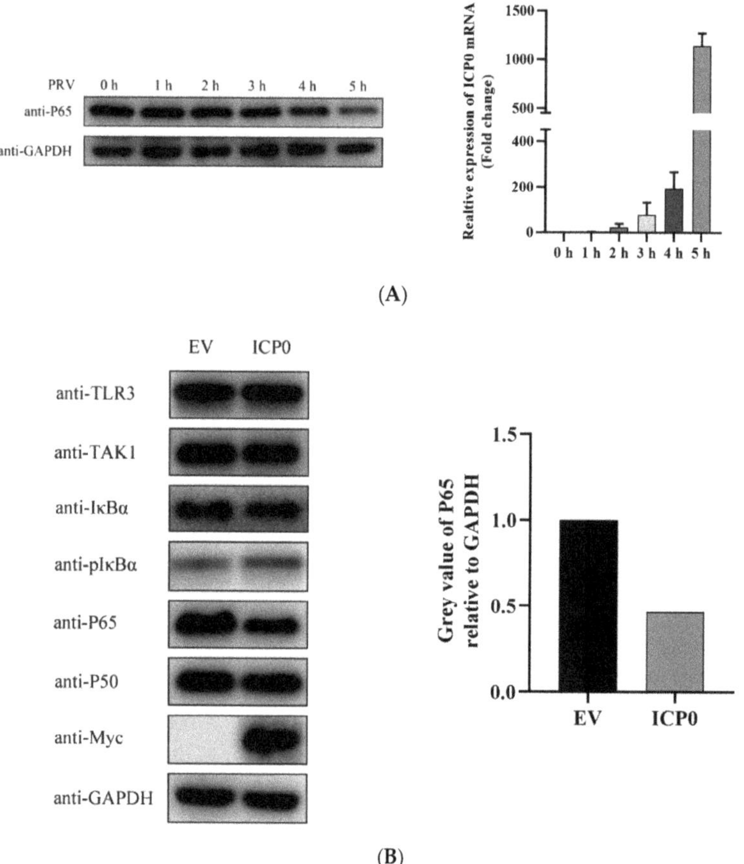

Figure 3. *Cont.*

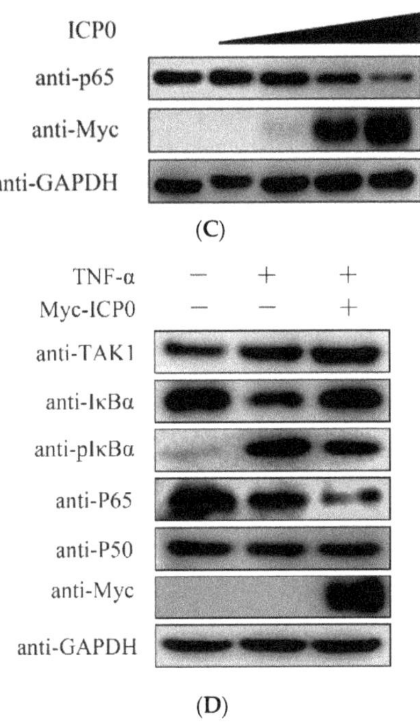

Figure 3. p65 may be a target of pseudorabies virus ICP0 protein. (**A**) PK15 cells were mock infected or infected with 1 MOI PRV for 1 h, 2 h, 3 h, 4 h and 5 h. Then, cells were collected at the indicated time points. Western blot was performed for p65 expression detection (Left). The expression of ICP0 mRNA was detected by RT-qPCR (Right). GAPDH served as loading control. (**B**) PK15 cells were transfected with pCMV-Myc empty vector (EV) (1.0 μg) or Myc-ICP0 (1.0 μg) plasmids for 24 h. Then, cells were harvested and lysed for Western blot detection. Factors included in the NF-κB signaling pathway such as TAK1, IκBα, pIκBα, p50 and p65 were detected. GAPDH served as loading control. The grey value of p65 relative to GAPDH was quantified. (**C**) PK15 cells were transfected with increasing concentrations of expression vectors for ICP0. After 24 h, cells were collected and lysed for p65 detection. GAPDH served as loading control. The grey value of p65 protein expression relative to GAPDH was also quantified. (**D**) PK15 cells were transfected with pCMV-Myc empty vector (EV) (1.0 μg) or Myc-ICP0 (1.0 μg) plasmids for 24 h. Then, cells were treated with or without 50 ng/mL recombinant human TNF-α and incubated for an additional 12 h. Western blot was performed for TAK1, IκBα, pIκBα, p50 and p65 detection with the indicated antibodies. GAPDH served as loading control.

Next, we wanted to verify whether ICP0 played a similar role in the TNF-α-induced NF-κB signaling pathway activation. TNF-α was used as a stimulating factor in ICP0-transfected PK15 cells. The results showed that TNF-α efficiently activated the NF-κB pathway, but ICP0 overexpression significantly inhibited IκBα phosphorylation and degraded p65 (Figure 3D), which is consistent with our previous results. Moreover, we also found when the NF-κB pathway was activated during TNF-α stimulation, ICP0 could inhibit the degradation of IκBα. This result was different from the result shown in Figure 3B, which indicated that ICP0 had no effect on IκBα unless the NF-κB pathway was activated. Based on these results, we inferred that ICP0 may target p65 or its downstream to inhibit NF-κB pathway activation.

3.4. ICP0 Interacts with p65 and Degrades p65 through the Proteasome Pathway

PRV ICP0 exhibited a remarkable inhibitory effect on p65 protein, suggesting it could target p65. To investigate whether there is an interaction between ICP0 and p65, PK15 cells were co-transfected with Flag-p65 and Myc-ICP0 plasmids for 30 h. Cells were collected and a co-immunoprecipitation assay was carried out. The results showed that p65 co-precipitated with ICP0 protein (Figure 4A), suggesting that there is a direct interaction between p65 and ICP0. We have further demonstrated this interaction under physiological conditions. Myc-ICP0 plasmid was transfected into PK15 cells and verified using immuno-precipitation. Interestingly, we obtained similar results (Figure 4B), which would further certify an interaction between p65 and ICP0.

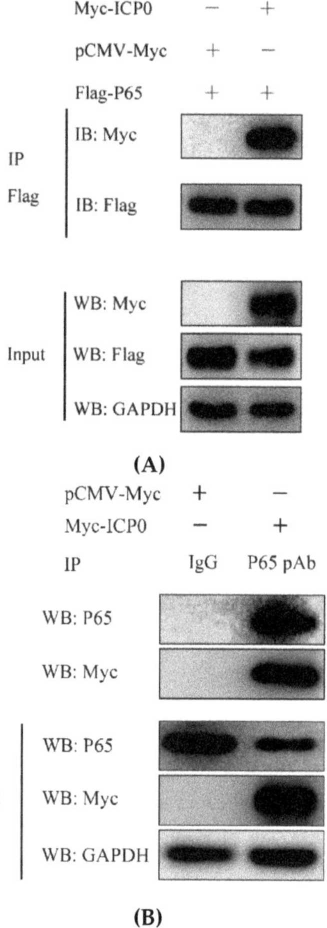

Figure 4. *Cont.*

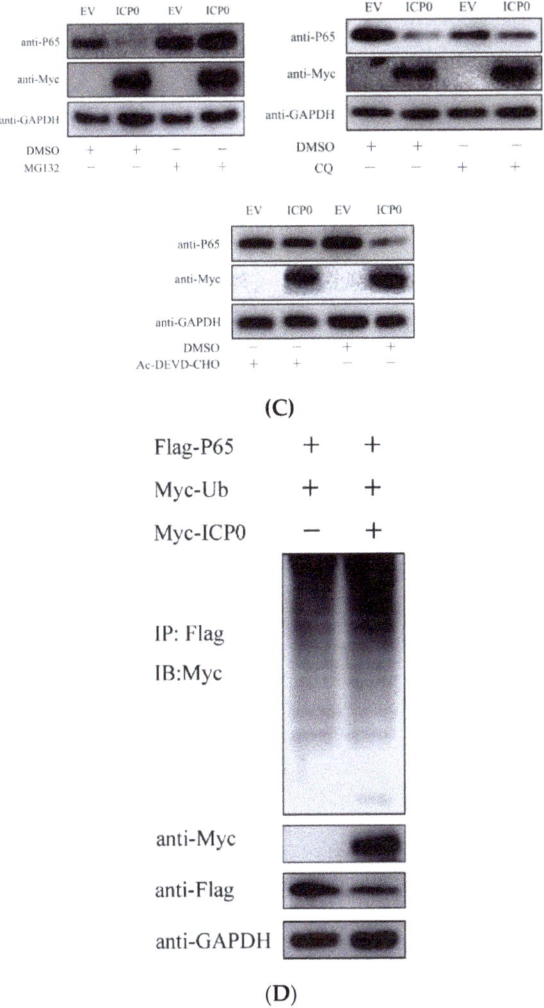

Figure 4. ICP0 targets p65 to inhibit NF-κB signaling. (**A**) PK15 cells were co-transfected with pCMV-Myc empty vector (1.0 µg) or Myc-ICP0 (1.0 µg) plasmids and Flag-p65 (1.0 µg) plasmids for 24 h. The cells were then lysed and immunoprecipitated with an anti-Flag antibody. The whole-cell lysates (input) and immunoprecipitation (IP) complexes were analyzed using an anti-Myc, anti-Flag or anti-GAPDH antibody by Western blotting. (**B**) PK15 cells were transfected with pCMV-Myc (1.0 µg) or Myc-ICP0 (1.0 µg) plasmids for 24 h. The cells were lysed and immunoprecipitated with an anti-p65 antibody. The input and IP complexes were analyzed by Western blotting using anti-p65, anti-Myc or anti-GAPDH antibodies. (**C**) PK15 cells were transfected with Myc-ICP0 (1.0 µg) or pCMV-Myc empty vector (EV) (1.0 µg) for 24 h, then treated with proteasomal inhibitor MG132 (7.5 µM), lysosome inhibitor CQ (50 µM) or caspase 3 inhibitor Ac-DEVD-CHO (50 µM) for 12 h. DMSO-treated cells served as vehicle control. Then, cells were collected and immunoblotted for p65 and Myc-tagged ICP0. GAPDH served as loading control. (**D**) PK15 cells were co-transfected with the Myc-tagged ICP0, FLAG-tagged p65 and Myc-tagged Ub 30 h post-transfection. The p65–ubiquitin complexes were immunoprecipitated using anti-Flag antibody and immunoblotted with anti-Myc antibody to detect ubiquitinated proteins.

Next, to identify the pathway by which ICP0 achieves the degradation of p65, we transfected the Myc-ICP0 plasmid into PK15 cells and treated the cells with MG132 (ubiquitin–proteasome inhibitor), chloroquine (CQ, lysosome pathway inhibitor) or Ac-DEVD-CHO (caspase-3 inhibitor). Western blot results showed that MG132 prevented the degradation of p65 by ICP0 but not by CQ and caspase-3 inhibitor Ac-DEVD-CHO (Figure 4C). It is suggested that the degradation of p65 by ICP0 is achieved through the ubiquitin–proteasome pathway. Proteins degraded via the proteasome pathway must be ubiquitinated first. The above data indicated that ICP0 interacted well with p65; we next examined whether ICP0 affects p65 ubiquitination. The ubiquitination assay showed that ICP0 could increase p65 polyubiquitination (Figure 4D).

3.5. ICP0 Protein Suppresses p65 Phosphorylation

The classical hallmark of NF-κB activation is the degradation of IκBα, releasing two subunits, namely p50 and p65, which then undergo phosphorylation and ubiquitination modifications in the nucleus. ICP0 affects the NF-κB signaling pathway by targeting p65; therefore, it was necessary to verify the effect of ICP0 on p65 phosphorylation. Empty vector (EV) and Myc-ICP0 plasmids were transfected into PK15 cells respectively. Then, cells were treated with MG132 and TNF-α. The results showed that TNF-α induced significant phosphorylation of p65. Meanwhile, in the ICP0 transfection group, the phosphorylation level of p65 was effectively inhibited (Figure 5). These results indicated that ICP0 abrogated p65 phosphorylation.

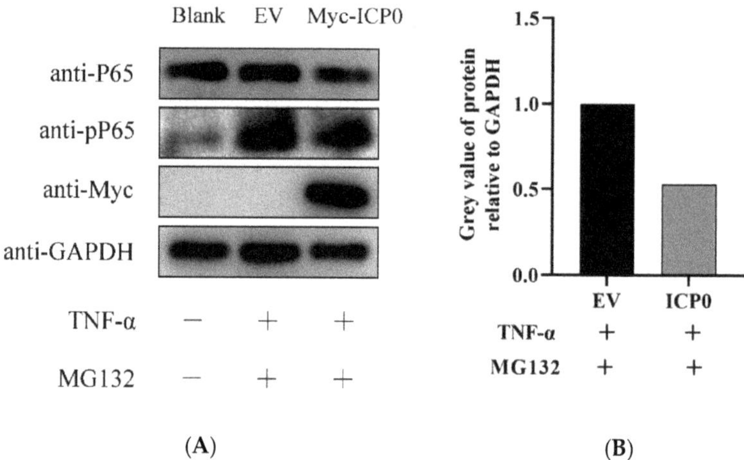

Figure 5. ICP0 protein suppresses p65 phosphorylation. (**A**)PK15 cells were transfected with empty vector (EV) (1.0 µg) or Myc-ICP0 (1.0 µg) plasmids for 24 h and then stimulated with TNF-α (50 µg/mL) for 12 h. Cells were treated with 7.5 µM MG132 for another 12 h before collection. p65, phosphorylated p65 (pP65) and ICP0 protein (Myc) expression were detected by immune blotting. GAPDH served as loading control. (**B**) The grey value of p65 phosphorylation relative to GAPDH was quantified.

3.6. ICP0 Protein Blocks p65 Nuclear Translocation

The phosphorylation of p65 causes its nuclear translocation. To verify whether ICP0 affects the nuclear translocation process of p65, we transfected an ICP0 expression plasmid into PK15 cells and examined the distribution of p65 in the cytoplasm and nucleus by nucleoplasmic separation assay. Compared to the empty vector transfection group, in the ICP0 transfection group, most of the p65 remained in the cytoplasm and only a small amount entered the nucleus (Figure 6). The above results suggest that ICP0 inhibits the

TNF-α-induced nuclear translocation process of p65, thereby limiting the activation of the NF-κB signaling pathway.

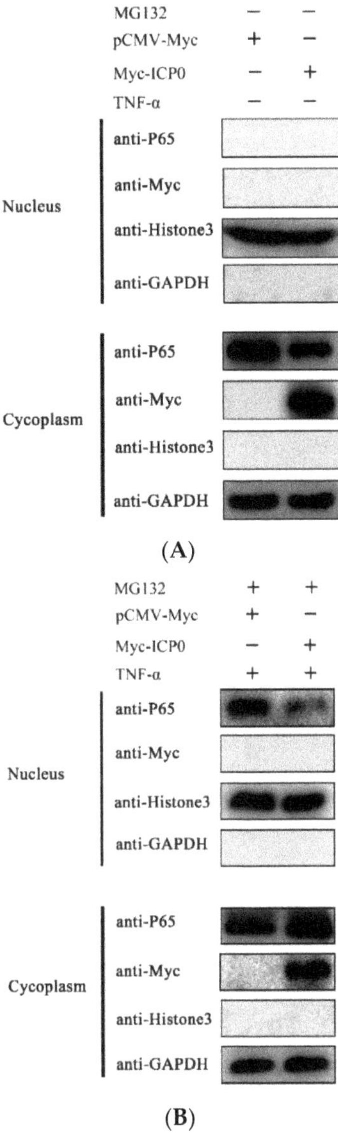

Figure 6. ICP0 protein blocks p65 nuclear translocation. (**A**) PK15 cells were transfected withpCMV-Myc (1.0 µg) or Myc-ICP0 (1.0 µg) for 36 h. Cells were collected without MG132 and TNF-α treatment. Cytoplasmic and nuclear proteins were extracted and subjected to Western blot analysis.(**B**) TNF-α (50 µg/mL) was added into PK15 cells in the presence of either pCMV-Myc (1.0 µg) or Myc-ICP0 (1.0 µg) for 24 h. Before being collected, cells were treated with MG132 and TNF-α for 12 h. Then, cytoplasmic and nuclear proteins were extracted and subjected to Western blot analysis. Expression of p65 and Myc-tagged ICP0 was detected with specific antibodies. Histone 3 was used as a nuclear protein marker. GAPDH served as loading control.

3.7. ICP0 Protein Promotes PRV Proliferation via Decreasing IFN-β Production

In Figure 1, we demonstrate that ICP0 expression enhances PRV replication in PK15 cells, but it is unclear whether this observation is due to the attenuation of cytokine production via NF-κB suppression shown in the above results. As it is well known that NF-κB regulates IFN-β production, we then detected IFN-β expression in empty-vector- or ICP0-transfected cells during PRV infection. Results indicated that ICP0 could abrogate IFN-β mRNA transcription compared to the empty vector group in the condition of viral infection (Figure 7A). ELISA was also used to quantify IFN-β secretion expression. As shown in Figure 7B, ICP0 significantly inhibited antiviral factor IFN-β production. These results demonstrated that ICP0 overexpression could block IFN-β production. Furthermore, the increase in PRV proliferation caused by ICP0 was related to the decrease in antiviral factor IFN-β production.

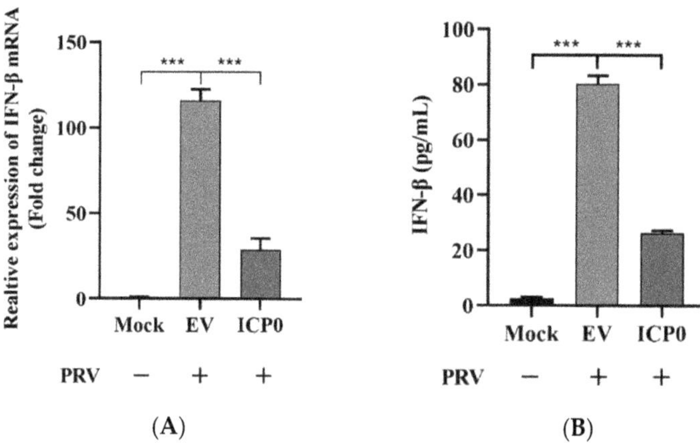

Figure 7. ICP0 protein promotes PRV proliferation via decreasing IFN-β production. (**A**) PK15 cells were transfected with pCMV-Myc and Myc-ICP0 plasmids. Then, cells were infected with PRV. Cellular RNA was extracted, and RT-qPCR was performed to detect IFN-β mRNA transcription. (**B**) PK15 cells were transfected with pCMV-Myc and Myc-ICP0 plasmids. Then, cells were infected with PRV. Cell culture supernatant was collected. The secretion expression of IFN-β was detected using ELISA. Data are shown as mean ± SD from three independent experiments. Comparison between two groups was evaluated by unpaired Student's *t*-test. *** $p < 0.001$.

4. Discussion

The innate immune response of the host is thought to play an important role in resistance to viral infection. Activation of NF-κB as a strategy that protects the host against viral pathogens plays a vital role in the regulation of the intrinsic immune response [23,24]. Therefore, many viruses have evolved different strategies to modulate NF-κB activation and thus evade the host immune response [25–27]. In this study, we have demonstrated that the PRV-encoded immediate early protein ICP0 inhibits the TNF-α-mediated NF-κB signaling pathway activation.

PRV is a swine alphaherpesvirus closely related to the human herpes simplex virus type 1 (HSV-1). PRV infects a broad host range of mammals. PRV infection primarily causes an acute lytic infection in the adult pig, its natural host, characterized by respiratory distress and reproductive failure while resulting in neurological symptoms and high mortality in newborn piglets and non-natural hosts [28]. Although hosts have evolved powerful innate immune mechanisms in response to virus invasion, PRV has evolved strategies to hijack host immune responses for viral replication and the establishment of persistent infection.

In the present study, the PRV-encoded ICP0 protein has been verified to have the ability to prevent TNF-α-stimulated NF-κB signaling pathway activation (Figures 2 and 3). Given that the innate immune response is the first line of host antiviral systems, these results indicate that ICP0 plays an important role in PRV immune evasion of the NF-κB signaling transduction pathway. The NF-κB pathway controls the transcription of many immune molecules required to initiate an immune response to foreign pathogens; as a result, disruption of the NF-κB pathway is likely to inhibit the immune response capacity of the host cell.

PRV uses numerous viral proteins to antagonize the host innate immune system. Previous studies showed that PRV UL24 had an inhibitory role in the NF-κB pathway [21]. In this study, we proved that PRV ICP0 had a similar effect to UL24 in the inhibition of the NF-κB pathway. Using Western blot analysis, we identified p65 as a target of the PRV ICP0 protein, through which it inhibits TNF-α-mediated activation of the NF-κB signaling pathway (Figure 3).

p65 is a key regulator of the NF-κB pathway [29]; it can be phosphorylated by cellular and viral proteins, contributing to the activation or inhibition of the transcriptional activity of p65 and, as a result, leading to an increase or decrease in the production of inflammatory factors.

Here, we found that the PRV ICP0 protein antagonized the NF-κB pathway by targeting p65 and inhibited p65 phosphorylation and nuclear translocation (Figures 5 and 6). Moreover, there was a direct interaction between ICP0 and p65, and endogenous p65 expression was also degraded by ICP0 through the proteasomal pathway (Figure 4).

As there is a direct interaction between ICP0 and p65 (Figure 4), and endogenous p65 levels are affected by ICP0 (Figure 3), we can infer that the nucleus p65 decrease in non-TNF-α/MG132-treated cells is related to the degradation effect of ICP0 on p65. In contrast, in TNF-α/MG132-treated cells, compared with the empty-vector-transfected group, in the ICP0-transfected group, most of the p65 protein remained in the cytoplasm, and only a small amount of p65 entered the nucleus (Figure 6). These results further confirmed that ICP0 could inhibit the nuclear translocation of ICP0. In addition, results in Figures 3 and 5 also indicated that ICP0 could inhibit the phosphorylation level of IκBα and p65 when cells were treated with TNF-α or TNF-α combined with MG132. So, the different results in Figure 6A,B may be due to the inhibitory effect of ICP0 on phosphorylation of IκBα or p65.

In a related study on herpes simplex virus 1 study, scientists found that its ICP0 protein inhibits TNF-α-induced NF-κB activation by interacting with p65 and p50. HSV-1 ICP0 also degraded p50 via its E3 ubiquitin ligase activity. Our results showed that PRV ICP0 had no effect on p50. Although both HSV-1 and PRV belong to Herpesviridae, and ICP0 is a relatively conserved viral protein, the mechanisms by which it plays a role in different viral infections may be different. As mentioned in this study, the ICP0 proteins of the two viruses, which both affect the NF-κB signaling pathway, have different target molecules. This difference may be related to the infectious properties of the viruses, and it also greatly enriches our understanding of the role of ICP0 protein in the natural immune response.

In summary, our data demonstrate a possible mechanism by which ICP0 abolishes the NF-κB signaling pathway (Figure 8). ICP0 inhibits NF-κB activation by targeting p65, and there is a direct interaction between ICP0 and p65. In addition, ICP0 decreases the expression of p65 and blocks the activation of the NF-κB pathway by inhibiting phosphorylation and nuclear translocation of p65. These findings suggest that pseudorabies virus ICP0 can inhibit the TNF-α-mediated NF-κB signaling pathway and provide new insights into the innate immune evasion of the pseudorabies virus.

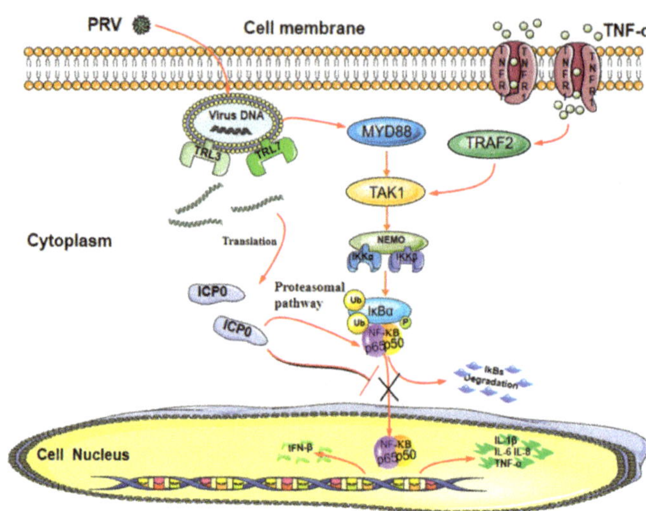

Figure 8. Model of ICP0 interfering with the NF-κB signaling pathway. The PRV protein ICP0 blocks the degradation of IκBα and inhibits its phosphorylation. Meanwhile, ICP0 interacts with p65 and degrades p65 protein expression via the proteasome pathway. In addition, ICP0 inhibits p65 phosphorylation and prevents its nuclear translocation, thereby negatively regulating the NF-κB signaling pathway.

Author Contributions: Methodology, X.Z., J.X., M.G. and Z.Y.; software, L.C.; validation, M.G., Z.Y. and L.C.; resources, R.F.; data curation, X.Z.; writing—original draft preparation, X.Z. and J.X.; writing—review and editing, S.W. and R.F.; supervision, J.X.; project administration, J.X. and R.F.; funding acquisition, X.Z., J.X. and R.F. All authors have read and agreed to the published version of the manuscript.

Funding: This research was funded by Gansu Province Graduate Student Innovation Star Project, grant number [2021CXZX-684], Open Funds of the Biomedical Research Center from Northwest Minzu University, grant number [EB202101] and Young Doctor Fund Project of Gansu Province Education Department, grant number [2021QB-064].

Institutional Review Board Statement: Not applicable.

Informed Consent Statement: Not applicable.

Data Availability Statement: All available data are presented in the article.

Conflicts of Interest: The authors declare no conflict of interest.

References

1. DiDonato, J.A.; Mercurio, F.; Karin, M. NF-KB and the Link between Inflammation and Cancer: NF-KB Links Inflammation and Cancer. *Immunol. Rev.* **2012**, *246*, 379–400. [CrossRef]
2. Hayden, M.S.; Ghosh, S. NF-KB in Immunobiology. *Cell Res.* **2011**, *21*, 223–244. [CrossRef] [PubMed]
3. Zhang, Q.; Lenardo, M.J.; Baltimore, D. 30 Years of NF-KB: A Blossoming of Relevance to Human Pathobiology. *Cell* **2017**, *168*, 37–57. [CrossRef] [PubMed]
4. Sedger, L.M.; McDermott, M.F. TNF and TNF-Receptors: From Mediators of Cell Death and Inflammation to Therapeutic Giants–Past, Present and Future. *Cytokine Growth Factor Rev.* **2014**, *25*, 453–472. [CrossRef] [PubMed]
5. Williams, L.M.; Gilmore, T.D. Looking Down on NF-KB. *Mol. Cell Biol.* **2020**, *40*, e00104-20. [CrossRef]
6. Zhao, X.; Cui, Q.; Fu, Q.; Song, X.; Jia, R.; Yang, Y.; Zou, Y.; Li, L.; He, C.; Liang, X.; et al. Antiviral Properties of Resveratrol against Pseudorabies Virus Are Associated with the Inhibition of IκB Kinase Activation. *Sci. Rep.* **2017**, *7*, 8782. [CrossRef]
7. Oeckinghaus, A.; Hayden, M.S.; Ghosh, S. Crosstalk in NF-KB Signaling Pathways. *Nat. Immunol.* **2011**, *12*, 695–708. [CrossRef]

8. Everett, R.D.; Boutell, C.; McNair, C.; Grant, L.; Orr, A. Comparison of the Biological and Biochemical Activities of Several Members of the Alphaherpesvirus ICP0 Family of Proteins. *J. Virol.* **2010**, *84*, 3476–3487. [CrossRef]

9. Ly, C.Y.; Yu, C.; McDonald, P.R.; Roy, A.; Johnson, D.K.; Davido, D.J. Simple and Rapid High-Throughput Assay to Identify HSV-1 ICP0 Transactivation Inhibitors. *Antivir. Res.* **2021**, *194*, 105160. [CrossRef]

10. Rodríguez, M.C.; Dybas, J.M.; Hughes, J.; Weitzman, M.D.; Boutell, C. The HSV-1 Ubiquitin Ligase ICP0: Modifying the Cellular Proteome to Promote Infection. *Virus Res.* **2020**, *285*, 198015. [CrossRef]

11. Song, K.; Li, S. The Role of Ubiquitination in NF-KB Signaling during Virus Infection. *Viruses* **2021**, *13*, 145. [CrossRef] [PubMed]

12. Boutell, C.; Everett, R.D. Regulation of Alphaherpesvirus Infections by the ICP0 Family of Proteins. *J. Gen. Virol.* **2013**, *94*, 465–481. [CrossRef] [PubMed]

13. Shahnazaryan, D.; Khalil, R.; Wynne, C.; Jefferies, C.A.; Ní Gabhann-Dromgoole, J.; Murphy, C.C. Herpes Simplex Virus 1 Targets IRF7 via ICP0 to Limit Type I IFN Induction. *Sci. Rep.* **2020**, *10*, 22216. [CrossRef] [PubMed]

14. Zhang, J.; Wang, K.; Wang, S.; Zheng, C. Herpes Simplex Virus 1 E3 Ubiquitin Ligase ICP0 Protein Inhibits Tumor Necrosis Factor Alpha-Induced NF-KB Activation by Interacting with P65/RelA and P50/NF-KB1. *J. Virol.* **2013**, *87*, 12935–12948. [CrossRef] [PubMed]

15. Bo, Z.; Miao, Y.; Xi, R.; Zhong, Q.; Bao, C.; Chen, H.; Sun, L.; Qian, Y.; Jung, Y.-S.; Dai, J. PRV UL13 Inhibits CGAS–STING-Mediated IFN-β Production by Phosphorylating IRF3. *Vet. Res.* **2020**, *51*, 118. [CrossRef] [PubMed]

16. Lv, L.; Bai, J.; Gao, Y.; Jin, L.; Wang, X.; Cao, M.; Liu, X.; Jiang, P. Peroxiredoxin 1 Interacts with TBK1/IKKε and Negatively Regulates Pseudorabies Virus Propagation by Promoting Innate Immunity. *J. Virol.* **2021**, *95*, e00923-21. [CrossRef]

17. Lv, L.; Cao, M.; Bai, J.; Jin, L.; Wang, X.; Gao, Y.; Liu, X.; Jiang, P. PRV-Encoded UL13 Protein Kinase Acts as an Antagonist of Innate Immunity by Targeting IRF3-Signaling Pathways. *Vet. Microbiol.* **2020**, *250*, 108860. [CrossRef]

18. Liu, X.; Zhang, M.; Ye, C.; Ruan, K.; Xu, A.; Gao, F.; Tong, G.; Zheng, H. Inhibition of the DNA-Sensing Pathway by Pseudorabies Virus UL24 Protein via Degradation of Interferon Regulatory Factor 7. *Vet. Microbiol.* **2021**, *255*, 109023. [CrossRef]

19. Lu, M.; Qiu, S.; Zhang, L.; Sun, Y.; Bao, E.; Lv, Y. Pseudorabies Virus Glycoprotein GE Suppresses Interferon-β Production via CREB-Binding Protein Degradation. *Virus Res.* **2021**, *291*, 198220. [CrossRef]

20. Xie, J.; Zhang, X.; Chen, L.; Bi, Y.; Idris, A.; Xu, S.; Li, X.; Zhang, Y.; Feng, R. Pseudorabies Virus US3 Protein Inhibits IFN-β Production by Interacting with IRF3 to Block Its Activation. *Front. Microbiol.* **2021**, *12*, 761282. [CrossRef]

21. Wang, T.-Y.; Yang, Y.-L.; Feng, C.; Sun, M.-X.; Peng, J.-M.; Tian, Z.-J.; Tang, Y.-D.; Cai, X.-H. Pseudorabies Virus UL24 Abrogates Tumor Necrosis Factor Alpha-Induced NF-KB Activation by Degrading P65. *Viruses* **2020**, *12*, 51. [CrossRef] [PubMed]

22. Xie, J.; Bi, Y.; Xu, S.; Han, Y.; Idris, A.; Zhang, H.; Li, X.; Bai, J.; Zhang, Y.; Feng, R. Host Antiviral Protein IFITM2 Restricts Pseudorabies Virus Replication. *Virus Res.* **2020**, *287*, 198105. [CrossRef] [PubMed]

23. Hayden, M.S.; Ghosh, S. Shared Principles in NF-KB Signaling. *Cell* **2008**, *132*, 344–362. [CrossRef]

24. Santoro, M.G.; Rossi, A.; Amici, C. NF-KB and Virus Infection: Who Controls Whom. *EMBO J.* **2003**, *22*, 2552–2560. [CrossRef] [PubMed]

25. Barrado-Gil, L.; del Puerto, A.; Galindo, I.; Cuesta-Geijo, M.Á.; García-Dorival, I.; de Motes, C.M.; Alonso, C. African Swine Fever Virus Ubiquitin-Conjugating Enzyme Is an Immunomodulator Targeting NF-KB Activation. *Viruses* **2021**, *13*, 1160. [CrossRef] [PubMed]

26. Neidel, S.; Ren, H.; Torres, A.A.; Smith, G.L. NF-KB Activation Is a Turn on for Vaccinia Virus Phosphoprotein A49 to Turn off NF-KB Activation. *Proc. Natl. Acad. Sci. USA* **2019**, *116*, 5699–5704. [CrossRef] [PubMed]

27. Palmer, W.H.; Joosten, J.; Overheul, G.J.; Jansen, P.W.; Vermeulen, M.; Obbard, D.J.; Van Rij, R.P. Induction and Suppression of NF-KB Signalling by a DNA Virus of *Drosophila*. *J. Virol.* **2019**, *93*, e01443-18. [CrossRef]

28. Sun, Y.; Luo, Y.; Wang, C.-H.; Yuan, J.; Li, N.; Song, K.; Qiu, H.-J. Control of Swine Pseudorabies in China: Opportunities and Limitations. *Vet. Microbiol.* **2016**, *183*, 119–124. [CrossRef]

29. Sumner, R.P.; Maluquer de Motes, C.; Veyer, D.L.; Smith, G.L. Vaccinia Virus Inhibits NF-KB-Dependent Gene Expression Downstream of P65 Translocation. *J. Virol.* **2014**, *88*, 3092–3102. [CrossRef]

Article

Pseudorabies Virus Tegument Protein UL13 Suppresses RLR-Mediated Antiviral Innate Immunity through Regulating Receptor Transcription

Ningning Zhao [1,2,3,†], Fan Wang [1,2,†], Zhengjie Kong [1,2] and Yingli Shang [1,2,3,*]

1   Department of Preventive Veterinary Medicine, College of Veterinary Medicine, Shandong Agricultural University, Taian 271018, China; ningningzhao@163.com (N.Z.); wf201709@163.com (F.W.); kongzj2016@163.com (Z.K.)
2   Shandong Provincial Key Laboratory of Animal Biotechnology and Disease Control and Prevention, Shandong Agricultural University, Taian 271018, China
3   Institute of Immunology, Shandong Agricultural University, Taian 271018, China
*   Correspondence: shangyl@sdau.edu.cn; Tel.: +86-538-8240680
†   These authors contributed equally to this work.

Abstract: Pseudorabies virus (PRV) has evolved various strategies to escape host antiviral immune responses. However, it remains unclear whether and how PRV-encoded proteins modulate the RIG-I-like receptor (RLR)-mediated signals for immune evasion. Here, we show that the PRV tegument protein UL13 functions as an antagonist of RLR-mediated antiviral responses via suppression of the transcription of RIG-I and MDA5, but not LGP2. UL13 overexpression significantly inhibits both the mRNA and protein levels of RIG-I and MDA5, along with RIG-I- or MDA5-mediated antiviral immune responses, whereas overexpression of RIG-I or MDA5 counteracts such UL13-induced suppression. Mechanistically, UL13 suppresses the expression of RIG-I and MDA5 by inhibiting activation of the transcription factor NF-κB. Consequently, overexpression of p65 promotes the activation of *RIG-I* and *MDA5* promoters. Moreover, deletion of the p65-binding sites in the promoters of *RIG-I* or *MDA5* abolishes the suppression role of UL13. As a result, mutant PRV lacking UL13 elicits stronger host antiviral immune responses than PRV-WT. Hence, our results provide a novel functional role of UL13-induced suppression of host antiviral immunity through modulating receptors' transcription.

Keywords: pseudorabies virus (PRV); tegument protein UL13; RIG-I; MDA5; NF-κB

Citation: Zhao, N.; Wang, F.; Kong, Z.; Shang, Y. Pseudorabies Virus Tegument Protein UL13 Suppresses RLR-Mediated Antiviral Innate Immunity through Regulating Receptor Transcription. *Viruses* **2022**, *14*, 1465. https://doi.org/10.3390/v14071465

Academic Editors: Yan-Dong Tang and Xiangdong Li

Received: 30 May 2022
Accepted: 30 June 2022
Published: 2 July 2022

Publisher's Note: MDPI stays neutral with regard to jurisdictional claims in published maps and institutional affiliations.

1. Introduction

Pseudorabies virus (PRV), a member of the Alphaherpesvirus subfamily, is the pathogen of Aujeszky's disease, which causes abortions and stillbirths in sows, central nervous system disorders in young piglets, and respiratory disease in older pigs [1–3], generating considerable economic losses worldwide in the swine industry worldwide [4]. Recent studies report that PRV can also infect humans, thus raising great concern about cross-species PRV transmission [5,6]. Similar to other alphaherpesviruses, PRV has evolved multiple strategies to dismantle the host's innate antiviral response [7–10], such as blocking pattern-recognition receptor (PRR)-induced type-I interferon (IFN-I) and neutralizing the antiviral functions of IFN-stimulated genes (ISGs), resulting in the establishment of lifelong latent infection in the host. Nevertheless, viruses may also be released during latent infection, which causes persistent infection in the host, and poses potential risks to the breeding industry and human health. In addition, the constant mutation of PRV leads to classical attenuated vaccines failing to provide sufficient protection against PRV infection. Therefore, it is desirable to explore the strategies of innate immune escape of PRV for drug target selection.

As the first line of defense against viral invasion, the antiviral innate immune response is activated through the recognition of pathogen-associated molecular patterns (PAMPs) by PRRs, leading to the production of interferons (IFNs), inflammatory cytokines, and chemokines to eliminate pathogens [11–13]. Transmembrane or cytosol PRRs—such as Toll-like receptors (TLRs), RIG-I-like receptors (RLRs), and cGAS-STING—recognize distinct pathogen-derived nucleic acids with different features. Among them, RLRs, including retinoic-acid-inducible gene-I (RIG-I), melanoma-differentiation-associated gene 5 (MDA5), and laboratory of genetics and physiology 2 (LGP2), are prominent intracellular PRRs that normally sense RNA virus signals [12,14]. Interestingly, studies have discovered that DNA virus replication generates RNA intermediates, which could be nucleic acid ligands for RLRs [15–18]. In addition, host-derived RNAs produced by DNA viruses can also initiate the RLR-mediated signaling pathway [19]. These studies indicate that RLR-mediated antiviral innate immune response may also play important roles in defense against DNA virus infection. In response, viruses have evolved numerous strategies to evade it [16]. However, little is known about how PRV escapes from RLR-mediated signaling pathways.

PRV has a large linear double-stranded DNA genome that encodes over 70 functional proteins [20]. Among them, the tegument proteins possess a wide variety of functions in viral entry, secondary envelopment, and viral capsid transportation during infection and immune escape [21]. It has been shown that tegument proteins are the main components of alphaherpesviruses antagonizing RLR signal transduction. For instance, HSV-1 pUS11 interacts with endogenous RIG-I and MDA5 through the RNA-binding domain to block IFN-β production [22]. It has been shown that HSV-1 pUL37—a deaminase protein—suppresses RNA-induced activation by targeting RIG-I [23]. UL13, a serine/threonine protein kinase, is an important immune escape protein of PRV. Recent reports have found that the PRV tegument protein UL13 acts as an antagonist of cGAS-STING-mediated IFN-β production [24,25]. However, whether UL13 regulates the RLR pathway remains unknown. Here, we show that PRV UL13 inhibits transcription of RIG-I and MDA5 through regulating NF-κB activation, resulting in suppression of RLR-mediated IFN-β production. Our study preliminarily explores the suppressive role of the PRV UL13 in the activation of RLR-mediated antiviral innate immune response, and identifies a novel strategy of PRV for immune escape.

2. Materials and Methods

2.1. Cells and Viruses

HEK293T cells, PK-15 cells, and BHK-21 cells were purchased from the American Type Culture Collection (ATCC, Manassas, VA, USA). Stable PK-15 cells ectopically expressing PRV UL13 (UL13-PK-15 cells), along with control cells, were generated previously [24]. Cells were maintained in Dulbecco's minimal essential medium (DMEM) (Gibco, New York, NY, USA) containing 10% fetal bovine serum (FBS) (Biological Industries, Israel) and 1% penicillin–streptomycin (Gibco, New York, NY, USA) at 37 °C in 5% CO_2. All cells tested negative for mycoplasma using a mycoplasma detection kit (TransGen Biotech, Beijing, China). PRV (Bartha-K61 strain) was purchased from the China Veterinary Culture Collection Center (Cat# CVCC AV249, Beijing, China), and was purified in BHK-21 cells. The PRV-ΔUL13 recombinant strain was generated previously [24]. The wild-type PRV (PRV-WT) and PRV-ΔUL13 PRV strains were amplified and titrated in PK-15 cells using standard protocols. Sendai virus (SeV), described previously [26], was propagated in 10-day-old embryonated eggs. SeV titers were then determined via the Reed–Muench method using MDCK cells.

2.2. Antibodies and Reagents

The antibodies used and the sources were as follows: Anti-Flag M2 mouse mAb (1:5000, F1804) was obtained from Sigma (St. Louis, USA). Antibodies against TBK1 (1:1000, 3013), p-TBK1 (Ser172) (1:1000, 5483), p-p65 (Ser536) (1:1000, 3033), RIG-I (1:1000, 3743), MDA5 (1:1000, 5321), and MAVS (1:1000, 3993) were purchased from Cell Signaling Technology.

Antibodies against p65 (1:2000, 10745-1-AP), β-actin (1:10,000, 66009-1-Ig), IκBα (1:2000, 10268-1-AP), and Lamin B1 (1:2000, 12987-1-AP) were purchased from Proteintech Group Inc. Lipofectamine 2000 transfection reagent and Opti-MEM were purchased from Thermo Fisher Scientific. Protease inhibitors were purchased from Roche. PMSF and DAPI were purchased from Solarbio Life Sciences, Beijing, China.

2.3. Lentiviral Infection and Stable Cell Line Generation

Stable cell lines were generated as described previously [24]. Briefly, lentiviral particles were produced in HEK293T cells transfected with two packaging plasmids (psPAX2 and pVSVG) and an empty vector or pCDH-Flag-UL13 plasmid using Lipofectamine 2000 (Invitrogen). After 24 h, the recombinant viruses were filtered, then infected HEK293T cells again and supplemented with Polybrene (6 μg/mL, Cat#H8641, Solarbio, Beijing, China). Cells were selected with puromycin (Cat# IP1160, Solarbio, China) at a final concentration of 6 μg/mL for 5 days. Monoclonal cells were obtained in 96-well plates via the limited dilution method, and UL13-HEK293T cells were identified by immunoblotting with anti-Flag monoclonal antibody (Sigma).

2.4. Reverse Transcription and Quantitative Real-Time PCR (qPCR)

Total RNA isolation was carried out using an RNA purification Kit (Promega, Madison, WI, USA) and reverse-transcribed to cDNA using M-MLV reverse transcriptase with RNase inhibitor (Takara Bio, Beijing, China). qPCR was performed in triplicate with RealStar Green Fast Mixture (A303, GenStar, Beijing, China) and on a StepOnePlus thermal cycler (ABI, Thermo Fisher, MA, USA). Threshold cycle numbers were normalized to triplicate samples amplified with primers specific to glyceraldehyde-3-phosphate dehydrogenase (*GAPDH*). qPCR primers for the target genes are listed in Table 1.

2.5. Immunoblotting

Immunoblotting assay was performed as previously described [27]. Cells were lysed on ice with a lysis buffer (50 mM Tris-Cl at pH 7.4, 150 mM NaCl, 1% Triton X-100, 1% sodium deoxycholate, 1 mM Na_3VO_4, 1 mM EDTA, and 1 mM PMSF) for 60 min. Whole-cell lysates were separated by sodium dodecyl sulfate–polyacrylamide gel electrophoresis (SDS–PAGE) and transferred to a polyvinylidene fluoride membrane (Millipore) for immunoblotting with specific antibodies.

2.6. Dual-Luciferase Reporter Assay

Human *RIG-I*, *MDA5*, or *LGP2* promoter sequences (from positions −1500 to 0) were amplified by DNA Polymerase (Vazyme Biotech Co., Ltd., Nanjing, China) and cloned into pGL3-basic vectors to generate RIG-I-luc, MDA5-luc, or LGP2-luc reporter plasmids. Putative p65-binding sites in *RIG-I*, *MDA5*, or *LGP2* promoters were predicted using the online software JASPAR (http://jaspar.genereg.net/ (accessed on 29 May 2022)). The mutants of *RIG-I* or *MDA5* promoters lacking putative p65-binding sites were amplified by overlap PCR using specific primers, and were cloned into pGL3-basic vectors. For luciferase assays, HEK293T cells were co-transfected with RIG-I-luc, MDA5-luc, and LGP2-luc, or RIG-mut-luc, MDA5-mut-luc reporter plasmids, and expressing plasmids encoding UL13 or empty vectors using Lipofectamine 2000 (Invitrogen, Carlsbad, CA, USA). Twenty-four hours after transfection, cell lysates were prepared and analyzed using the Dual-Luciferase Report Assay System (Promega, Madison, WI, USA), according to the manufacturer's instructions. The Renilla luciferase reporter gene (pRL-TK, Promega) was used as an internal control.

Table 1. Primers or sequences used in this study.

Gene	Forward Sequence (5′–3′)	Reverse Sequence (5′–3′)
	Primer sequences for qPCR	
hGAPDH	ATCAAGAAGGTGGTGAAGCA	GTCGCTGTTGAAGTCAGAGGA
hIFNB1	GCACTGGCTGGAATGAGACT	CCTTGGCCTTCAGGTAATG
hISG56	TTCGGAGAAAGGCATTAGA	TCCAGGGCTTCATTCATAT
hMX1	AGCCACTGGACTGACGACTT	ACCACGGCTAACGGATAAG
hOASL	CCCTGGGGCCTTCTCTTC	TCCTAACAGTGCCATTCCCT
hTNF	AATAGGCTGTTCCCATGTAGC	AGAGGCTCAGCAATGAGTGA
hIL-6	TAATGGGCATTCCTTCTTCT	TGTCCTAACGCTCATACTTTT
h-RIG-I	CTGGTTCCGTGGCTTTTTGG	CACCTGCCATCATCCCCTTA
h-RIG-I-Flag	GACTACAAGGACGACGATGA	AGTGTGGCAGCCTCCATTGG
h-MDA5	AAAGCTCCTACCCGAGTGTG	GCTGCCCACTTAGAGAAGCA
h-MDA5-Flag	GACTACAAGGACGACGATGA	AGGCTCCACCTGGATGTACA
h-LGP2	CAGCTGAGCCGACTTAGGAA	CGCAGCAGCAGTACTTAACC
pGAPDH	TACACTGAGGACCAGGTTGTG	TTGACGAAGTGGTCGTTGAG
pIFNB1	TGCATCCTCCAAATCGCTCT	ATTGAGGAGTCCCAGGCAAC
pISG56	TCCGACACGCAGTCAAGTTT	TGTAGCAAAGCCCTGTCTGG
pMX1	GCTTTCAGATGCTTCGCAGG	TGTCGTATGGCTGATTGCCT
pOASL	CAGGCCAACAGGTTCAGACAG	CAGGAAACCGCAGACGATGT
pTNF	CGACTCAGTGCCGAGATCAA	CTCACAGGGCAATGATCCCA
pIL6	AAGCTGCAGTCACAGAACGA	GGACGGCATCAATCTCAGGT
p-RIG-I	TCCTTCTGACTGCTAACGCT	ACTAAGGAAGGTGTCCAGCAG
p-MDA5	TGAGGACTGATGTTTGATTCCAG	ACCTCTGCCCACCAAGATAGA
pLGP2	CAGCCCTGCAAACAGTACGAC	CACTCCAGTTTCGGGTTCTC
	Primer sequences for regular PCR	
PRV UL13	CCGGAATTCATGGACTACAAAGACGATG	CGCGGATCCTCAGGCATCGAGTTCG
hRIG-I-Luc	CGGGGTACCAAGTTTATCTGTAGGTTCAATG	GGAAGATCTGCCTCACTAGCTTTAAAGCC
hMDA5-Luc	CGGGGTACCCCAAGGTTTCATTTACTTCAAC	GGAAGATCTCCTGACTTTGGTTTCTGTTT
hLGP2-Luc	CGGGGTACCGGAGACCAGGTTTCCTTTCCAG	GGAAGATCTAGAAATGGAAACTGAAACTGAG
hRIG-I-Luc-M1	ATTTGGACAACAGGTTATAAAGCTAAACAT	ATGTTTAGCTTTATAACCTGTTGTCCAAAT
hRIG-I-Luc-M2	TTTGGACAACAGGTTATAAAGCTAAACAT	ATGTTTAGCTTTATAACCTGTTGTCCAAA
hMDA5-Luc-M1	GCCTGGCGGGGATCAGGGAGACGC	GCGTCTCCCTGATCCCCGCCAGGC
hMDA5-Luc-M2	AGCATGTGATTTAAAGGGGAAGTG	CACTTCCCCTTTAAATCACATGCT

2.7. Inhibition of Signaling Pathways

For blocking of NF-κB and MAPK signals, specific chemical inhibitors were applied as described previously [28]. Briefly, cells were pretreated with the NF-κB inhibitor Bay11-7082 (Bay11, 10 μM), JNK inhibitor SP600125 (SP, 10 μM), MEK/ERK inhibitor U0126 (10 μM), or p38 MAPK inhibitor SB203580 (SB, 10 μM) for 1 h prior to poly(I:C) (InvivoGen, San Diego, CA, USA) transfection. All chemical inhibitors were purchased from MedChemExpress (MCE, Monmouth Junction, NJ, USA).

2.8. Cytoplasmic and Nuclear Protein Extraction

Cells were lysed on ice with a lysis buffer (10 mM HEPES at pH 7.9, 50 mM NaCl, 0.5 mM Sucrose, 0.1 mM EDTA, 0.5% Triton X-100, 1 mM DTT, 10 mM Sodium pyrophosphate decahydrate, 0.5 M NaF, 0.2 M Na3VO4, 1 mM PMSF, and protease inhibitor mixture) for 60 min. The supernatant was collected for the cytoplasmic extract after centrifuging for 5 min at 1500 rpm. The pellet was resuspended with Buffer A (10 mM HEPES at pH 7.9, 10 mM KCl, 0.1 mM EGTA, 0.1 mM EDTA, 1 mM DTT, 1 mM PMSF, and protease inhibitor mixtures), centrifuged for 5 min at 1500 rpm, and then the supernatant was removed. Then, four volumes of buffer C (10 mM HEPES at pH 7.9, 500 mM NaCl, 0.1 mM EGTA, 0.1 mM EDTA, 0.1% Nonidet P-40, 1 mM DTT, 1 mM PMSF, and protease inhibitor mixtures) were added, and vortexed for 30 min at 4 °C. The supernatant was collected as the nuclear extract after centrifugation for 10 min at 14,000 rpm. β-actin and Lamin B1 were used as loading controls for the cytoplasm and nucleus, respectively.

2.9. Immunofluorescence

UL13-HEK293T cells or control cells were transfected with poly(I:C) for 3 h and then fixed for 20 min with 4% cold paraformaldehyde. Cells were then permeabilized for 10 min with 0.1% Triton X-100, and then blocked with 5% bovine serum albumin (BSA, Sigma) for 30 min. Cells were incubated with the appropriate primary antibodies for 120 min, followed by staining with Alexa Fluor 594-conjugated secondary antibodies (Proteintech Group Inc. China) or 647-conjugated secondary antibodies (TIANGEN, Beijing, China) for 60 min. Nuclei were stained with DAPI. Images were visualized and acquired using a laser scanning confocal microscope with LAS X software (Leica, Wetzlar, Germany).

2.10. Statistical Analysis

All statistical analyses were carried out using GraphPad Prism v8.0 software; p-values were calculated with a two-tailed paired or unpaired Student's t-test, and p-values ≤ 0.05 were considered significant.

3. Results

3.1. UL13 Inhibits RLR-mediated Antiviral Immune Responses

The role of the PRV tegument protein UL13 has been indicated in the suppression of host antiviral immune responses [24,29]. To confirm the suppressive effect of UL13 on RLR-mediated IFN-I signaling, we generated stable HEK293T cells ectopically expressing Flag-tagged UL13 (UL13-HEK293T cells). Immunoblotting analysis showed that Flag-UL13 was successfully expressed in UL13-HEK293T cells (Figure 1A). Upon poly(I:C) transfection or Sendai virus (SeV) infection, which typically activate RLR-mediated immune responses, UL13-HEK293T cells showed impaired gene expression of *IFNB1* and downstream ISGs— including *ISG56*, *MX1*, and *OASL*—compared with control cells (Figure 1B,C), indicating that UL13 inhibits RLR-mediated expression of type-I IFN and downstream ISGs. This observation was further confirmed in PK-15 cells stably expressing UL13 (UL13-PK-15 cells) (Figure 1D,E). These results suggest that UL13 functions as an antagonist of RLR-mediated type-I IFN responses.

3.2. UL13 Inhibits RLR-Mediated Antiviral Response by Suppressing Transcription of RIG-I and MDA5

The transfected cytosolic RNA is mainly recognized by RLRs, and activates the adaptor protein MAVS to initiate innate antiviral immune response [11]. To know whether UL13 targets the RLR-mediated signaling pathway, we next examined the effect of UL13 on transfected poly(I:C)-mediated immune responses in HEK293T cells or PK-15 cells. The results showed that UL13 expression markedly inhibited the expression of RIG-I and MDA5, as well as phosphorylation of TBK1, in response to poly(I:C) transfection (Figure 2A), indicating that UL13 likely targets the expression of RIG-I-like receptors to suppress RNA-mediated immune response. Similarly, upon SeV infection, UL13 expression also inhibited expression of RLR receptors—including RIG-I and MDA5—and phosphorylation of TBK1 and IRF3 in both HEK293T cells and PK-15 cells (Figure 2B), further confirming the suppressive effect of UL13 on RNA-mediated immune responses. Notably, UL13 did not alter total protein levels of MAVS, TBK1, and IRF3 (Figure 2A,B), suggesting that UL13 possibly targets the upstream RLRs to suppress RLR-mediated signaling activation. Indeed, UL13 expression strikingly suppressed gene transcription of *RIG-I* and *MDA5* under poly(I:C) transfection conditions (Figure 2C). However, expression of UL13 did not affect mRNA expression of *LGP2*—another RIG-I like receptor. To confirm the effect of UL13 on RLR transcription, we first constructed luciferase reporter plasmids by introducing RIG-I, MDA5, or LGP2 promoter sequences into the pGL3-basic vector. Luciferase assays showed that UL13 expression remarkably restrained activation of *RIG-I* or *MDA5* promoters in HEK293T cells, but not that of the *LGP2* promoter (Figure 2D), which is consistent with the suppressive effect of UL13 on transcription of RIG-I and MDA5 (Figure 2C). Together, these data demonstrate that UL13 negatively regulates the RLR signaling pathway via suppression of the transcription of *RIG-I* and *MDA5*.

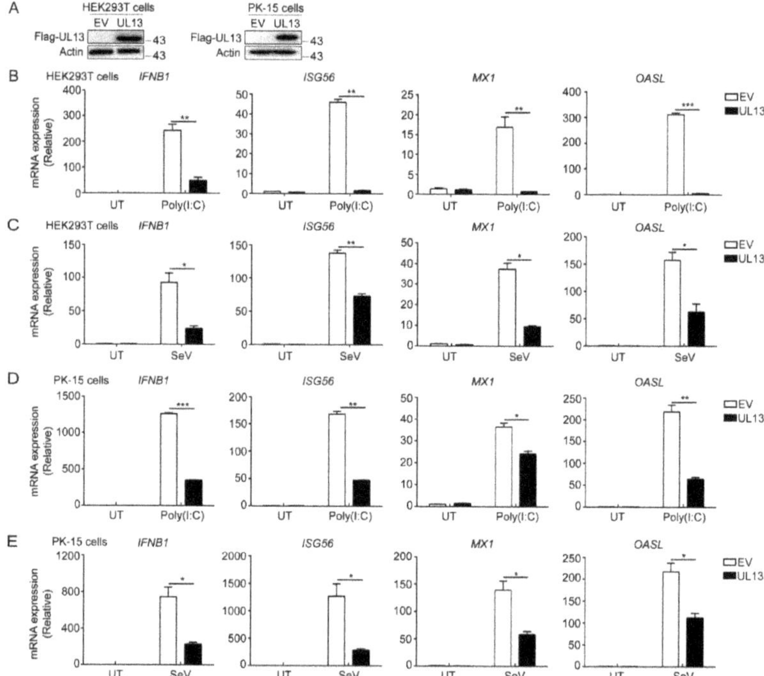

Figure 1. PRV UL13 negatively regulated RLR-mediated expression of type-I IFN and ISGs. (**A**) The expression of Flag-tagged UL13 in HEK293T or PK-15 cell lines was verified by immunoblotting; β-actin served as a loading control. (**B–E**) Quantitative real-time PCR (qPCR) analysis of *IFNB1* and downstream ISG (*ISG56*, *MX1*, and *OASL*) mRNA expression in UL13-HEK293T cells (**B,C**) or UL13-PK-15 cells (**D,E**), and control cells left untreated (UT) or transfected with 1 μg/mL of poly(I:C) (**B,D**) for 6 h or infected with SeV (MOI = 1) (**C,E**) for 12 h. Data are pooled from three independent experiments (mean ± SEM); *, $p < 0.05$; **, $p < 0.01$; ***, $p < 0.001$ (Student's *t*-test).

3.3. Enforced Expression of RIG-I and MDA5 Counteracts UL13-Suppressed Antiviral Immune Responses

Having known that UL13 expression inhibits transcription of RIG-I and MDA5, leading to repression of RNA-triggered induction of IFNs and downstream antiviral responses, we next examined whether UL13 directly targets the mRNA of RLRs. qPCR analysis showed that UL13 overexpression has no effect on exogenous RIG-I/MDA5 transcription (Figure 3A), suggesting that UL13 may repress endogenous transcription of RLRs. Then, we investigated whether overexpression of RIG-I or MDA5 could rescue the phenotype. As expected, enforced expression of RIG-I or MDA5 counteracted UL13-induced suppression of antiviral immune responses, including expression of *IFNB1* and downstream ISGs (*ISG56* and *OASL*), both at the basal level and in response to stimulation of transfected poly(I:C) (Figure 3B), showing that RIG-I and MDA5 are the key targets for UL13 during RLR-mediated immune responses. In line with this observation, overexpression of RIG-I or MDA5 also enhanced phosphorylation of TBK1 induced by transfected poly(I:C) in UL13-HEK293T cells, without altering the expression of total TBK1 (Figure 3C). Collectively, these data demonstrate that UL13 inhibits RLR-mediated antiviral immune responses via downregulation of RLR expression.

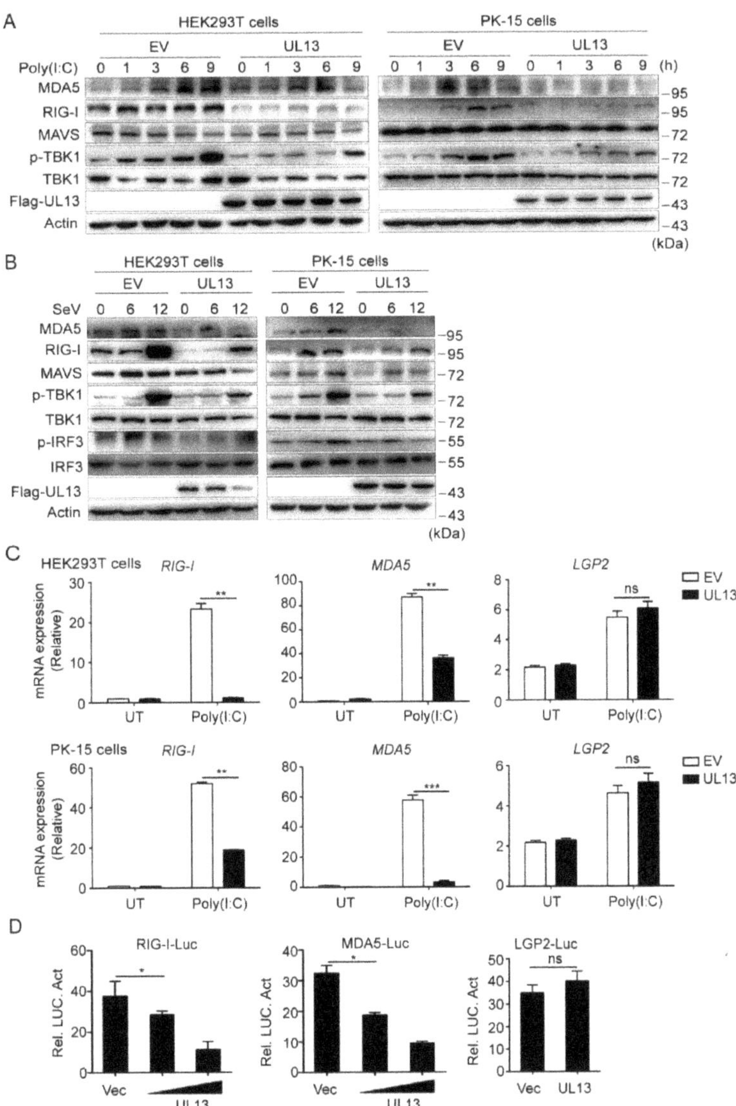

Figure 2. UL13 inhibits transcription of *RIG-I* and *MDA5* to suppress RLR-mediated antiviral responses. (**A,B**) Immunoblotting analysis of RIG-I, MDA5, MAVS, phosphorylated (Ser172) and total TBK1, and phosphorylated (Ser396) and total IRF3 in whole-cell lysates of UL13-HEK293T or UL13-PK-15 cells and control cells stimulated with poly(I:C) (**A**) or infected with SeV (MOI = 1, B) for the indicated times (above lanes); β-actin served as a loading control. (**C**) qPCR analysis of the mRNA levels of *RIG-I*, *MDA5*, and *LGP2* in UL13-HEK293T cells, UL13-PK-15 cells, or control cells left untreated (UT) or transfected with poly(I:C) for 6 h. (**D**) Luciferase activities in HEK293T cells co-transfected with *RIG-I*, *MDA5*, or *LGP2* promoter-driven luciferase reporters (50 ng) and plasmids encoding UL13 (concentration 150 ng, 300 ng) or empty vectors. Twenty-four hours after transfection, cell lysates were analyzed for luciferase activity. Data are representative of three independent experiments (**A,B**), or are pooled from three independent experiments (**C,D**, mean ± SD); *, $p < 0.05$; **, $p < 0.01$; ***, $p < 0.001$; ns, not significant (Student's *t*-test).

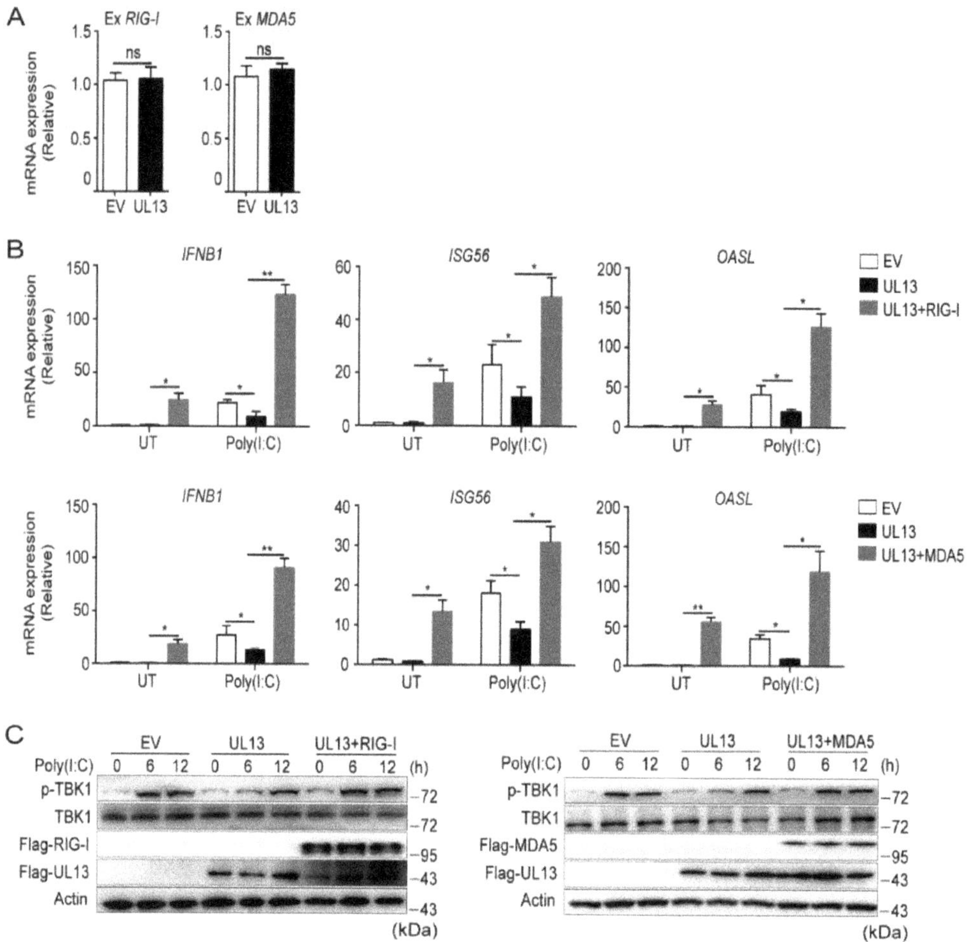

Figure 3. Enforced expression of RIG-I or MDA5 rescues UL13-suppressed antiviral immune responses. (**A**) qPCR analysis of *RIG-I* and *MDA5* mRNA expression in UL13-HEK293T cells or control cells transfected with plasmids encoding Flag-RIG-I or Flag-MDA5 for 24 h. (**B**) qPCR analysis of *IFNB1* and downstream ISG (*ISG56* and *OASL*) mRNA expression in cells transfected with plasmids encoding Flag-RIG-I, Flag-MDA5, or an empty vector for 24 h following poly(I:C) transfection for 6 h. (**C**) Immunoblotting analysis of phosphorylated (Ser172) and total TBK1, Flag-RIG-I or Flag-MDA5, and Flag-UL13 in UL13-HEK293T cells transfected with plasmids encoding Flag-RIG-I, Flag-MDA5, or an empty vector for 24 h following poly(I:C) transfection for the additional indicated periods. Data are pooled from three independent experiments (**A,B**, mean $\pm$ SD) or representative of three independent experiments (**C**); *, $p < 0.05$; **, $p < 0.01$; ns, not significant (Student's *t*-test).

3.4. UL13 Overexpression Restrains NF-κB Activation to Modulate Transcription of RIG-I and MDA5

Next, we sought to investigate the mechanisms by which UL13 suppresses the transcription of RLRs and RLR-mediated antiviral gene expression. It has been established that activation of NF-κB and mitogen-activated protein kinases (MAPKs) modulates gene transcription. We therefore examined whether UL13 modulates the activation of NF-κB and/or MAPKs to regulate IFN responses induced by poly(I:C) transfection. qPCR analysis showed that inhibition of NF-κB by a chemical inhibitor compromised the expression of

IFNB1 and downstream ISGs such as *ISG56* and *MX1*, whereas inhibition of MAPKs did not alter the UL13-mediated suppression of *IFNB1* and ISGs induced by transfected Poly(I:C) in HEK293T cells (Figure 4A), indicating that UL13 regulates NF-κB activation to repress RLR-mediated IFN responses. Consistently, immunoblotting analysis also showed that UL13 overexpression significantly suppressed phosphorylation of p65 and retarded degradation of IκBα in HEK23T cells transfected with poly(I:C) (Figure 4B). Importantly, overexpression of UL13 strikingly decreased the protein expression of RIG-I under both basal and stimulated conditions (Figure 4B), implying that suppression of NF-κB by UL13 is related to RIG-I expression. Moreover, UL13 overexpression impaired poly(I:C)-induced nuclear translocation of p65 (Figure 4C,E) and mRNA expression of *IL6* and *TNF* (Figure 4F)—two key inflammatory cytokines that are dependent on NF-κB. Altogether, these results suggest that UL13 modulates NF-κB activation to suppress RLR expression and RLR-mediated antiviral immune responses.

To further clarify whether UL13 regulates transcription of *RIG-I* or *MDA5* by inhibiting activation of NF-κB, we next examined the effects of p65 on the activation of *RIG-I, MDA5,* and *LGP2* promoters. Luciferase assays showed that p65 significantly promoted the activity of *RIG-I* and *MDA5* promoters, but not the *LGP2* promoter (Figure 4G), demonstrating that NF-κB activation is critical for transcription of *RIG-I* and *MDA5*. Interestingly, *RIG-I* and *MDA5* promoters, but not the *LGP2* promoter, contain several putative p65-binding sites (Figure 4H). To confirm the direct role of NF-κB in the transcription regulation of *RIG-I* or *MDA5*, we constructed two mutant *RIG-I* and *MDA5* luciferase reporter plasmids in which two putative p65-binding sequences were deleted (Figure 4H). The results showed that while UL13 expression dramatically suppressed wild-type *RIG-I*-promoter-driven luciferase activity, UL13 failed to suppress mutant *RIG-I*-promoter-driven luciferase activity when the p65-binding motif between −421 and −412 was deleted (Figure 4I). Similarly, UL13 inhibited wild-type *MDA5* promoter activity but failed to suppress mutant *MDA5* promoter activation when the p65-binding sites were deleted (Figure 4I). Together, the above data suggest that the transcription of *RIG-I* and *MDA5* is dependent on NF-κB, and that UL13 suppresses transcription of *RIG-I* and *MDA5* by inhibiting NF-κB activation.

3.5. UL13 Deficiency Potentiates RLR-Mediated Antiviral Responses during PRV Infection

Next, we investigated the effects of UL13 on *RIG-I* and *MDA5* expression during PRV infection in HEK293T cells. Compared with wild-type PRV (PRV-WT), HEK293T cells showed higher levels of *RIG-I* or *MDA5* when infected with the mutant PRV lacking UL13 (PRV-ΔUL13) (Figure 5A). Consequently, PRV-ΔUL13 infection enhanced the induction of *IFNB1* and downstream ISGs, as well as the key inflammatory mediators, including *IL6* and *TNF* (Figure 5B,C). Consistent with these results, the expression of RIG-I and MDA5, along with the phosphorylation levels of TBK1 and p65—but not the key adaptor protein MAVS—were higher in HEK293T cells infected with PRV-ΔUL13 than in PRV-WT-infected cells (Figure 5D). Collectively, these data suggest that deficiency of UL13 promotes RLR-mediated antiviral immune responses during PRV infection.

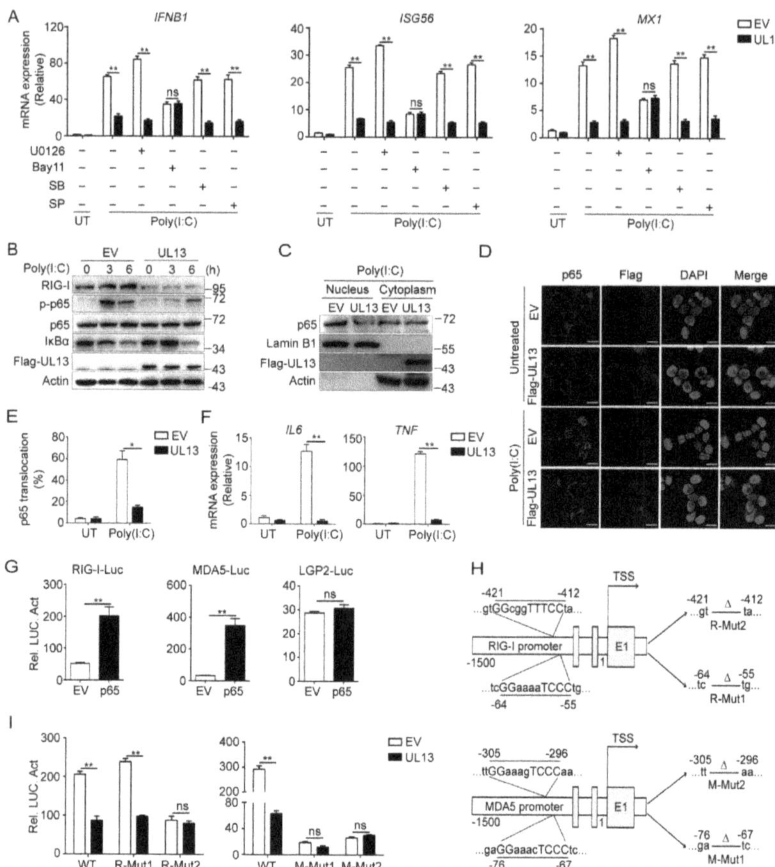

Figure 4. UL13 represses transcription of *RIG-I* and *MDA5* by regulating NF-κB activation. (**A**) qPCR analysis of mRNA expression of *IFNB1* and downstream ISG (*ISG56* and *MX1*) in UL13-HEK293T cells or control cells. Cells were left untreated or transfected with poly(I:C) for 6 h, or pretreated with NF-κB inhibitor (BAY11-7082, 10 μM), JNK inhibitor (SP600125, 10 μM), MEK/ERK inhibitor (U0126, 10 μM), or p38 inhibitor (SB203580, 10 μM) for 1 h, followed by poly(I:C) stimulation for 6 h. (**B**) Immunoblotting analysis of RIG-I, phosphorylated (Ser356) and total p65, IκBα, and Flag-UL13 in whole-cell lysates of UL13-HEK293T cells or control cells transfected with or without poly(I:C) for the indicated times. (**C**) Immunoblotting analysis of p65 protein levels in the nucleus and cytoplasm in UL13-HEK293T cells and control cells transfected with poly(I:C) for 3 h; β-actin and Lamin B1 served as the loading controls for the cytoplasm and nucleus, respectively. (**D**) Immunofluorescence analysis of the nuclear translocation of p65 in UL13-HEK293T cells or control cells left untreated or transfected with poly(I:C) for 3 h. Scale bars = 20 μm. (**E**) Quantification of nuclear localization of p65. (**F**) qPCR analysis of mRNA expression of *IL6* and *TNF* in cells as in (**D**). (**G**) Luciferase activities in UL13-HEK293T cells or control cells co-transfected with *RIG-I-*, *MDA5-*, or *LGP2*-promoter-driven luciferase reporters and plasmids encoding p65 or empty vectors. Twenty-four hours after transfection, cell lysates were analyzed for luciferase activity. (**H**) Annotation of putative p65-binding sites in *RIG-I* promoter or *MDA5* promoter and their mutants. (**I**) Luciferase activities in UL13-HEK293T cells or control cells co-transfected with *RIG-I-* or *MDA5*-promoter-driven luciferase reporters or mutants with plasmid encoding p65 or empty vectors. Twenty-four hours after transfection, cell lysates were analyzed for luciferase activity. Data are pooled from three independent experiments (**A,E–G,I**, mean ± SD) or representative of three independent experiments (**B,D**); *, $p < 0.05$; **, $p < 0.01$; ns, not significant (Student's *t*-test).

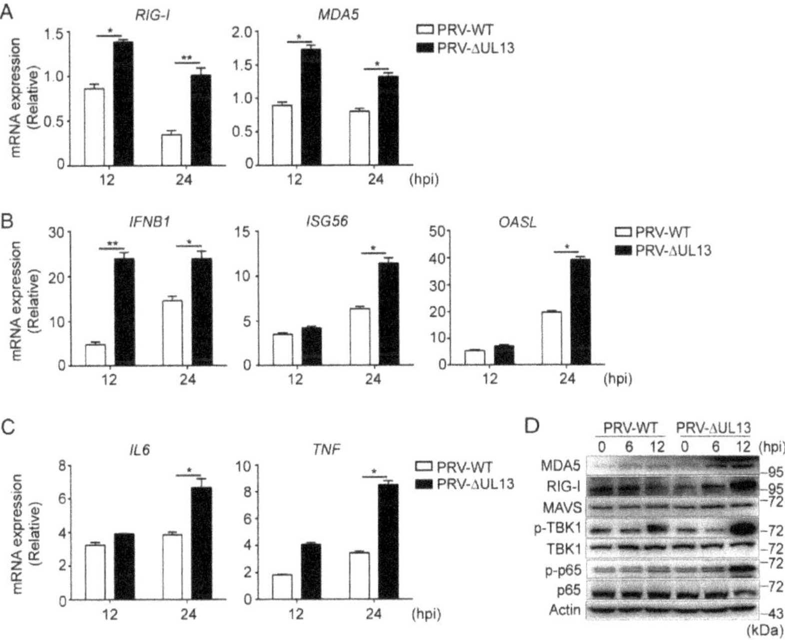

Figure 5. UL13 deficiency increases expression of RLRs and promotes RLR-mediated antiviral responses. (**A**) qPCR analysis of *RIG-I* and *MDA5* mRNA levels in HEK293T cells infected with PRV-WT or PRV-ΔUL13 (MOI = 1) for the indicated times. (**B,C**) qPCR analysis the mRNA levels of *IFNB1*, downstream ISGs (*ISG56* and *OASL*) (**B**), and expression of *IL6* and *TNF* (**C**) in HEK293T cells infected with PRV-WT or PRV-ΔUL13 (MOI = 1) for the indicated periods. (**D**) Immunoblotting analysis of MDA5, RIG-I, MAVS, phosphorylated and total TBK1, and phosphorylated and total p65 in HEK293T cells infected with PRV-WT or PRV-ΔUL13 (MOI = 1). Data are pooled from three independent experiments (A, B, C, mean ± SD) or representative of three independent experiments (**D**); *, $p < 0.05$; **, $p < 0.01$; ns, not significant (Student's *t*-test).

4. Discussion

IFN-mediated innate immune responses are crucial in protecting the host from viral infections and activating adaptive immunity [30,31]. In response to viral infection, host cells activate multiple signaling cascades to disrupt viral replication when specific molecular components of viruses are recognized [32,33]. Generally, cytosolic DNA sensors, such as cGAS, play important roles in recognition of DNA viruses [34–36]. Interestingly, recent studies suggest that RLRs sense RNA intermediates and host-derived RNAs during DNA virus infection [37–40]. As a DNA virus, PRV can establish long-term infection by evading cGAS-STING- and TLR3-induced antiviral innate immune responses [7,24,29]. However, how PRV evades RLR-mediated immune responses remains obscure. Here, we show that the PRV tegument protein UL13 suppresses type-I IFN production and downstream ISGs induced by RNA stimulation, indicating that UL13 functions as a novel antagonist of RLR-mediated antiviral responses. Hence, our study provides evidence that UL13 plays important roles in PRV's evasion of RLR-mediated antiviral innate immune responses.

In recognition of RNAs, RLRs activate the MAVS–TBK1–IRF3 axis to induce type I IFN responses [21]. Viruses, such as alphaherpesviruses, have developed multitudinous strategies for the evasion of RLR-mediated signaling [21]. Among these strategies, targeting RIG-I and MDA5 receptors is an effective way to inhibit RLR-mediated signaling transduction. For example, the HSV-1 UL11 C-terminal RNA-binding domain binds to RIG-I and MDA5, resulting in inhibition of downstream signaling activation [22]. HSV-1 UL37 deamidates RIG-I, rendering it unable to sense viral RNA and, thus, blocking its ability to induce antiviral immune responses [23]. HSV-2 utilizes the virion host shutoff (vhs) protein to suppress *RIG-I* and *MDA5* expression [41]. Consistently, we found that PRV infection could also significantly inhibit the gene transcription of *RIG-I* and *MDA5*, but not *LGP2*. Mechanistically, the PRV tegument protein UL13 inhibits transcription of *RIG-I* or *MDA5* by suppressing activation of NF-κB. Notably, both RIG-I and MDA5 are ISGs, and their expression can be quickly induced upon viral infection and dsRNA stimulation [42,43]. As IFN-inducible genes, increases in RIG-I and MDA5 expression may amplify the effects of IFNs [44]. Therefore, targeting of RIG-I and MDA5 by PRV can be a novel and simple strategy to escape RLR-mediated antiviral innate immune response.

Gene expression is under tight control of transcription factors that bind to unique DNA enhancer/repressor elements [26]. For *RIG-I*, several transcription factors have been reported to be directly involved in its transcriptional regulation. For example, interferon regulatory factor-1 (IRF1) positively regulates interferon- or dsRNA-induced RIG-I transcription [45]. CCAAT/enhancer-binding protein beta (C/EBPβ) acts as a repressor element to bind the *RIG-I* promoter for RIG-I transcriptional inhibition [46]. In this study, we found a key p65-binding site in the *RIG-I* proximal promoter. Deletion of the p65-binding site resulted in dramatic loss of RIG-I promoter activation induced by p65, which is consistent with a previous report [46]. Unlike *RIG-I*, the molecular mechanism that induces *MDA5* expression has not been presented. Here, we identified that p65 expression noticeably induced *MDA5* promoter activity. Deletion of two key p65-binding sites abolished activity of the *MDA5* promoter induced by p65, indicating that p65 is also the key transcriptional factor for *MDA5* transcription. In contrast, no putative p65-binding site was found in the *LGP2* promoter, which is consistent with the finding that UL13 did not affect *LGP2* activation. UL13 is a serine/threonine protein kinase that targets several viral and cellular substrates [47]. Studies have shown that UL13 can participate in various processes depending on its kinase activity. For example, UL13 directly modulates the phosphorylation of viral proteins VP11/12, ICP22, and UL49 [48,49], promotes the assembly and release of mature infectious virions [50], and stabilizes the viral ICP0 protein against degradation. Furthermore, UL13 targets STING or IRF3 to escape IFN-mediated antiviral innate immune response [24,25,29]. Here, we found that UL13 inhibited the expression of RIG-I and MDA5 by blocking the activation of NF-κB. However, the detailed mechanisms of how UL13 prevents NF-κB activation require further investigation.

In summary, this study identifies novel evasion strategies for PRV via suppression of the transcription of *RIG-I* and *MDA5* by UL13 through targeting NF-κB activation, further highlighting the importance of UL13 in the regulation of host innate antiviral immune responses.

Author Contributions: Conceptualization, Y.S.; methodology, Z.K. and F.W.; validation, N.Z. and F.W.; investigation, N.Z. and F.W.; writing—original draft preparation, N.Z. and Y.S.; writing—review and editing, Y.S.; visualization, N.Z. and F.W.; project administration, Y.S.; funding acquisition, Y.S. All authors have read and agreed to the published version of the manuscript.

Funding: This study was supported by the National Natural Science Foundation of China (grants 32072869, 31941015) and the Shandong Modern Technology System of Agricultural Industry (SDAIT-09-06).

Institutional Review Board Statement: Not applicable.

Informed Consent Statement: Informed consent was obtained from all subjects involved in the study.

Data Availability Statement: All analyzed data are contained in the main text. Raw data are available from the authors upon request.

Conflicts of Interest: The authors declare no conflict of interest.

References

1. Laval, K.; Enquist, L.W. The Neuropathic Itch Caused by Pseudorabies Virus. *Pathogens* **2020**, *9*, 254. [CrossRef] [PubMed]
2. Li, H.; Liang, R.; Pang, Y.; Shi, L.; Cui, S.; Lin, W. Evidence for interspecies transmission route of pseudorabies virus via virally contaminated fomites. *Vet. Microbiol.* **2020**, *251*, 108912. [CrossRef] [PubMed]
3. Liu, J.; Chen, C.; Li, X. Novel Chinese pseudorabies virus variants undergo extensive recombination and rapid interspecies transmission. *Transbound. Emerg. Dis.* **2020**, *67*, 2274–2276. [CrossRef] [PubMed]
4. Liu, Q.; Wang, X.; Xie, C.; Ding, S.; Yang, H.; Guo, S.; Li, J.; Qin, L.; Ban, F.; Wang, D.; et al. A novel human acute encephalitis caused by pseudorabies virus variant strain. *Clin. Infect. Dis.* **2020**, *73*, e3690–e3700. [CrossRef] [PubMed]
5. Ai, J.-W.; Weng, S.-S.; Cheng, Q.; Cui, P.; Li, Y.-J.; Wu, H.-L.; Zhu, Y.-M.; Xu, B.; Zhang, W.-H. Human Endophthalmitis Caused By Pseudorabies Virus Infection, China, 2017. *Emerg. Infect. Dis.* **2018**, *24*, 1087–1090. [CrossRef]
6. Fan, S.; Yuan, H.; Liu, L.; Li, H.; Wang, S.; Zhao, W.; Wu, Y.; Wang, P.; Hu, Y.; Han, J.; et al. Pseudorabies virus encephalitis in humans: A case series study. *J. Neurovirol.* **2020**, *26*, 556–564. [CrossRef]
7. Zhang, R.; Tang, J. Evasion of I interferon-mediated innate immunity by pseudorabies virus. *Front. Microbiol.* **2021**, *12*, 801257. [CrossRef]
8. Lu, M.; Qiu, S.; Zhang, L.; Sun, Y.; Bao, E.; Lv, Y. Pseudorabies virus glycoprotein gE suppresses interferon-β production via CREB-binding protein degradation. *Virus Res.* **2020**, *291*, 198220. [CrossRef]
9. Xie, J.; Zhang, X.; Chen, L.; Bi, Y.; Idris, A.; Xu, S.; Li, X.; Zhang, Y.; Feng, R. Pseudorabies virus US3 protein inhibits IFN-β production by interacting with IRF3 to block its activation. *Front. Microbiol.* **2021**, *12*, 761282. [CrossRef]
10. Koyanagi, N.; Kawaguchi, Y. Evasion of the Cell-Mediated Immune Response by Alphaherpesviruses. *Viruses* **2020**, *12*, 1354. [CrossRef]
11. Rehwinkel, J.; Gack, M.U. RIG-I-like receptors: Their regulation and roles in RNA sensing. *Nat. Rev. Immunol.* **2020**, *20*, 537–551. [CrossRef] [PubMed]
12. Stok, J.E.; Vega Quiroz, M.E.; van der Veen, A.G. Self RNA Sensing by RIG-I-like Receptors in Viral Infection and Sterile Inflammation. *J. Immunol.* **2020**, *205*, 883–891. [CrossRef] [PubMed]
13. Zhang, X.; Bai, X.C.; Chen, Z.J. Structures and Mechanisms in the cGAS-STING Innate Immunity Pathway. *Immunity* **2020**, *53*, 43–53. [CrossRef] [PubMed]
14. Onomoto, K.; Onoguchi, K.; Yoneyama, M. Regulation of RIG-I-like receptor-mediated signaling: Interaction between host and viral factors. *Cell Mol. Immunol.* **2021**, *18*, 539–555. [CrossRef]
15. Crill, E.K.; Furr-Rogers, S.R.; Marriott, I. RIG-I is required for VSV-induced cytokine production by murine glia and acts in combination with DAI to initiate responses to HSV-1. *Glia* **2015**, *63*, 2168–2180. [CrossRef]
16. Zhu, H.; Zheng, C. The Race between Host Antiviral Innate Immunity and the Immune Evasion Strategies of Herpes Simplex Virus 1. *Microbiol. Mol. Biol. Rev.* **2020**, *84*, e00099-20. [CrossRef]
17. Zhao, J.; Qin, C.; Liu, Y.; Rao, Y.; Feng, P. Herpes Simplex Virus and Pattern Recognition Receptors: An Arms Race. *Front. Immunol.* **2020**, *11*, 613799. [CrossRef]
18. Yu, H.; Bruneau, R.; Brennan, G.; Rothenburg, S. Battle Royale: Innate Recognition of Poxviruses and Viral Immune Evasion. *Biomedicines* **2021**, *9*, 765. [CrossRef]
19. Chiang, J.J.; Sparrer, K.M.J.; Van Gent, M.; Lässig, C.; Huang, T.; Osterrieder, N.; Hopfner, K.-P.; Gack, M.U. Viral unmasking of cellular 5S rRNA pseudogene transcripts induces RIG-I-mediated immunity. *Nat. Immunol.* **2018**, *19*, 53–62. [CrossRef]
20. Pomeranz, L.E.; Reynolds, A.E.; Hengartner, C.J. Molecular biology of pseudorabies virus: Impact on neurovirology and veterinary medicine. *Microbiol. Mol. Biol. Rev.* **2005**, *69*, 462–500. [CrossRef]
21. Yang, L.; Wang, M.; Cheng, A.; Yang, Q.; Wu, Y.; Jia, R.; Liu, M.; Zhu, D.; Chen, S.; Zhang, S.; et al. Innate Immune Evasion of Alphaherpesvirus Tegument Proteins. *Front. Immunol.* **2019**, *10*, 2196. [CrossRef] [PubMed]
22. Xing, J.; Wang, S.; Lin, R.; Mossman, K.L.; Zheng, C. Herpes simplex virus 1 tegument protein US11 downmodulates the RLR signaling pathway via direct interaction with RIG-I and MDA-5. *J. Virol.* **2012**, *86*, 3528–3540. [CrossRef]
23. Zhao, J.; Zeng, Y.; Xu, S.; Chen, J.; Shen, G.; Yu, C.; Knipe, D.; Yuan, W.; Peng, J.; Xu, W.; et al. A Viral Deamidase Targets the Helicase Domain of RIG-I to Block RNA-Induced Activation. *Cell Host Microbe* **2016**, *20*, 770–784. [CrossRef]
24. Kong, Z.; Yin, H.; Wang, F.; Liu, Z.; Luan, X.; Sun, L.; Liu, W.; Shang, Y. Pseudorabies virus tegument protein UL13 recruits RNF5 to inhibit STING-mediated antiviral immunity. *PLoS Pathog.* **2022**, *18*, e1010544. [CrossRef]
25. Bo, Z.; Miao, Y.; Xi, R.; Zhong, Q.; Bao, C.; Chen, H.; Sun, L.; Qian, Y.; Jung, Y.S.; Dai, J. PRV UL13 inhibits cGAS-STING-mediated IFN-β production by phosphorylating IRF3. *Vet. Res.* **2020**, *51*, 118. [CrossRef]
26. Zhu, Q.; Yu, T.; Gan, S.; Wang, Y.; Pei, Y.; Zhao, Q.; Pei, S.; Hao, S.; Yuan, J.; Xu, J.; et al. TRIM24 facilitates antiviral immunity through mediating K63-linked TRAF3 ubiquitination. *J. Exp. Med.* **2020**, *217*, e20192083. [CrossRef] [PubMed]
27. Ning, F.; Li, X.; Yu, L.; Zhang, B.; Zhao, Y.; Liu, Y.; Zhao, B.; Shang, Y.; Hu, X. Hes1 attenuates type I IFN responses via VEGF-C and WDFY1. *J. Exp. Med.* **2019**, *216*, 1396–1410. [CrossRef] [PubMed]
28. Zhang, W.; Fu, Z.; Yin, H.; Han, Q.; Fan, W.; Wang, F.; Shang, Y. Macrophage Polarization Modulated by Porcine Circovirus Type 2 Facilitates Bacterial Coinfection. *Front. Immunol.* **2021**, *12*, 688294. [CrossRef]
29. Lv, L.; Cao, M.; Bai, J.; Jin, L.; Wang, X.; Gao, Y.; Liu, X.; Jiang, P. PRV-encoded UL13 protein kinase acts as an antagonist of innate immunity by targeting IRF3-signaling pathways. *Vet. Microbiol.* **2020**, *250*, 108860. [CrossRef]

30. Domingo-Calap, P.; Segredo-Otero, E.; Durán-Moreno, M.; Sanjuán, R. Social evolution of innate immunity evasion in a virus. *Nat. Microbiol.* **2019**, *4*, 1006–1013. [CrossRef]
31. Ank, N.; West, H.; Bartholdy, C.; Eriksson, K.; Thomsen, A.R.; Paludan, S.R. Lambda interferon (IFN-lambda), a type III IFN, is induced by viruses and IFNs and displays potent antiviral activity against select virus infections in vivo. *J. Virol.* **2006**, *80*, 4501–4509. [CrossRef] [PubMed]
32. Thaiss, C.A.; Zmora, N.; Levy, M.; Elinav, E. The microbiome and innate immunity. *Nature* **2016**, *535*, 65–74. [CrossRef] [PubMed]
33. Fitzgerald, K.A.; Kagan, J.C. Toll-like receptors and the control of immunity. *Cell* **2020**, *180*, 1044–1066. [CrossRef]
34. Ma, Z.; Ni, G.; Damania, B. Innate sensing of DNA virus genomes. *Annu. Rev. Virol.* **2018**, *5*, 341–362. [CrossRef] [PubMed]
35. Chen, Q.; Sun, L.; Chen, Z.J. Regulation and function of the cGAS-STING pathway of cytosolic DNA sensing. *Nat. Immunol.* **2016**, *17*, 1142–1149. [CrossRef] [PubMed]
36. Beachboard, D.C.; Horner, S.M. Innate immune evasion strategies of DNA and RNA viruses. *Curr. Opin. Microbiol.* **2016**, *32*, 113–119. [CrossRef]
37. Samanta, M.; Iwakiri, D.; Kanda, T.; Imaizumi, T.; Takada, K. EB virus-encoded RNAs are recognized by RIG-I and activate signaling to induce type I IFN. *EMBO J.* **2006**, *25*, 4207–4214. [CrossRef]
38. Minamitani, T.; Iwakiri, D.; Takada, K. Adenovirus virus-associated RNAs induce type I interferon expression through a RIG-I-mediated pathway. *J. Virol.* **2011**, *85*, 4035–4040. [CrossRef]
39. García-Sastre, A. Ten Strategies of Interferon Evasion by Viruses. *Cell Host Microbe* **2017**, *22*, 176–184. [CrossRef]
40. Chan, Y.K.; Gack, M.U. Viral evasion of intracellular DNA and RNA sensing. *Nat. Rev. Microbiol.* **2016**, *14*, 360–373. [CrossRef]
41. Yao, X.D.; Rosenthal, K.L. Herpes simplex virus type 2 virion host shutoff protein suppresses innate dsRNA antiviral pathways in human vaginal epithelial cells. *J. Gen. Virol.* **2011**, *92 Pt 9*, 1981–1993. [CrossRef] [PubMed]
42. Yoneyama, M.; Kikuchi, M.; Natsukawa, T.; Shinobu, N.; Imaizumi, T.; Miyagishi, M.; Taira, K.; Akira, S.; Fujita, T. The RNA helicase RIG-I has an essential function in double-stranded RNA-induced innate antiviral responses. *Nat. Immunol.* **2004**, *5*, 730–737. [CrossRef] [PubMed]
43. Yuzawa, E.; Imaizumi, T.; Matsumiya, T.; Yoshida, H.; Fukuhara, R.; Kimura, H.; Fukui, A.; Tanji, K.; Mori, F.; Wakabayashi, K.; et al. Retinoic acid-inducible gene-I is induced by interferon-gamma and regulates CXCL11 expression in HeLa cells. *Life Sci.* **2008**, *82*, 670–675. [CrossRef]
44. Patel, J.R.; García-Sastre, A. Activation and regulation of pathogen sensor RIG-I. *Cytokine Growth Factor Rev.* **2014**, *25*, 513–523. [CrossRef] [PubMed]
45. Su, Z.-Z.; Sarkar, D.; Emdad, L.; Barral, P.M.; Fisher, P.B. Central role of interferon regulatory factor-1 (IRF-1) in controlling retinoic acid inducible gene-I (RIG-I) expression. *J. Cell Physiol.* **2007**, *213*, 502–510. [CrossRef] [PubMed]
46. Kumari, R.; Guo, Z.; Kumar, A.; Wiens, M.; Gangappa, S.; Katz, J.M.; Cox, N.J.; Lal, R.B.; Sarkar, D.; Fisher, P.B.; et al. Influenza virus *NS1-C/EBPβ* gene regulatory complex inhibits RIG-I transcription. *Antivir. Res.* **2020**, *176*, 104747. [CrossRef] [PubMed]
47. Koyanagi, N.; Kato, A.; Takeshima, K.; Maruzuru, Y.; Kozuka-Hata, H.; Oyama, M.; Arii, J.; Kawaguchi, Y. Regulation of herpes simplex virus 2 protein kinase UL13 by phosphorylation and its role in viral pathogenesis. *J. Virol.* **2018**, *92*, e00807-18. [CrossRef]
48. Eaton, H.E.; Saffran, H.A.; Wu, F.W.; Quach, K.; Smiley, J.R. Herpes simplex virus protein kinases US3 and UL13 modulate VP11/12 phosphorylation, virion packaging, and phosphatidylinositol 3-kinase/Akt signaling activity. *J. Virol.* **2014**, *88*, 7379–7388. [CrossRef]
49. Asai, R.; Ohno, T.; Kato, A.; Kawaguchi, Y. Identification of proteins directly phosphorylated by UL13 protein kinase from herpes simplex virus 1. *Microbes Infect.* **2007**, *9*, 1434–1438. [CrossRef]
50. Tanaka, M.; Nishiyama, Y.; Sata, T.; Kawaguchi, Y. The role of protein kinase activity expressed by the *UL13* gene of herpes simplex virus 1: The activity is not essential for optimal expression of UL41 and ICP0. *Virology* **2005**, *341*, 301–312. [CrossRef]

viruses

MDPI

Article

Metabolomics Analysis of PK-15 Cells with Pseudorabies Virus Infection Based on UHPLC-QE-MS

Panrao Liu [1], Danhe Hu [1], Lili Yuan [1], Zhengmin Lian [1], Xiaohui Yao [1], Zhenbang Zhu [1] and Xiangdong Li [1,2,*]

[1] Jiangsu Co-Innovation Center for Prevention and Control of Important Animal Infectious Diseases and Zoonoses, College of Veterinary Medicine, Yangzhou University, Yangzhou 225009, China; liupanrao@163.com (P.L.); hdh768@163.com (D.H.); yll96213@163.com (L.Y.); lian_zm@126.com (Z.L.); xiaohuiyao1995@163.com (X.Y.); 007583@yzu.edu.cn (Z.Z.)

[2] Joint International Research Laboratory of Agriculture and Agri-Product Safety, The Ministry of Education of China, Yangzhou University, Yangzhou 225009, China

* Correspondence: 007352@yzu.edu.cn; Tel.: +86-514-8797-9036

Abstract: Viruses depend on the metabolic mechanisms of the host to support viral replication. We utilize an approach based on ultra-high-performance liquid chromatography/Q Exactive HF-X Hybrid Quadrupole-Orbitrap Mass (UHPLC-QE-MS) to analyze the metabolic changes in PK-15 cells induced by the infections of the pseudorabies virus (PRV) variant strain and Bartha K61 strain. Infections with PRV markedly changed lots of metabolites, when compared to the uninfected cell group. Additionally, most of the differentially expressed metabolites belonged to glycerophospholipid metabolism, sphingolipid metabolism, purine metabolism, and pyrimidine metabolism. Lipid metabolites account for the highest proportion (around 35%). The results suggest that those alterations may be in favor of virion formation and genome amplification to promote PRV replication. Different PRV strains showed similar results. An understanding of PRV-induced metabolic reprogramming will provide valuable information for further studies on PRV pathogenesis and the development of antiviral therapy strategies.

Keywords: pseudorabies virus; metabolomic analysis; UHPLC-QE-MS; PK-15 cells

Citation: Liu, P.; Hu, D.; Yuan, L.; Lian, Z.; Yao, X.; Zhu, Z.; Li, X. Metabolomics Analysis of PK-15 Cells with Pseudorabies Virus Infection Based on UHPLC-QE-MS. *Viruses* **2022**, *14*, 1158. https://doi.org/10.3390/v14061158

Academic Editor: Douglas Gladue

Received: 8 May 2022
Accepted: 25 May 2022
Published: 27 May 2022

Publisher's Note: MDPI stays neutral with regard to jurisdictional claims in published maps and institutional affiliations.

1. Introduction

The pseudorabies virus (PRV) is a member of the *Herpesviridae* family, which causes Aujeszky's disease (AD) in pigs [1]. Pigs are the main host of PRV, and pigs of different ages can be infected with PRV. AD leads to high mortality and symptoms related to the central nervous system in piglets, respiratory disease in adult pigs, and decreased reproduction in sows, which have resulted in great economic losses for the pig industry [2]. PRV also infects other mammals, such as ruminants, carnivores, rodents, and even humans [3–5], posing a concern for public health. PRV virions are composed of double-stranded DNA genomes, capsids, teguments, and envelopes. The genome is approximately 150 kb and encodes over 100 proteins [1]. PRV was discovered in the 1900s and then widely distributed in the world. Although PRV has been eradicated in some Western countries (such as the United States, Germany, and Canada), it is still prevalent in many countries [6]. The Bartha K61 strain is one of the classically attenuated PRV strains, known as Bartha K61, which was isolated in 1961 and attenuated by a series of passages in embryos and chicken cells [7]. As an attenuated live vaccine, Bartha K61 can induce effective immune responses against PRV in pigs [8,9]. However, in 2011, PRV variant strains emerged in Northern China, then caused appalling outbreaks in swine farms, including in PRV-vaccinated swine farms [10,11]. Although much progress has been made in the research on the pathogenesis of PRV, the detailed mechanisms of the interaction between PRV and host cells remain unclear.

Metabolomics is a new subject developed after genomics and proteomics. Nowadays, metabolomics has been applied to disease diagnosis, pharmaceutical research and development, nutrition science, environmental science, botany, and other fields closely related to human health [12,13]. Metabolomics focuses on low-molecular-weight metabolites (MW < 1 KD, such as sugars, lipids, amino acids, and vitamins) in various metabolic pathways, and it can reflect the changes in the metabolic response of cells or tissues to external stimulation or genetic modifications, which contribute to reveal the mechanism of interaction between host cells and external factors [14,15].

Viruses are intracellular parasites and cannot proliferate independently. They must hijack and rely on the metabolic mechanisms and resources of host cells for their own replication [16,17]. Virus infection remodels the metabolic machineries in host cells to deal with the higher metabolic demands during virus replication [18–20]. Zika virus infection increased the glucose utilization in the tricarboxylic acid (TCA) cycle in HFF-1 cells and elevated the AMP/ATP ratios, which led to cell death [21]. Influenza virus was shown to affect host metabolic pathways to ensure the production of viral particles. Glucose uptake and aerobic glycolysis were increased, while fatty acid β-oxidation were decreased in cells infected with influenza virus [22]. Human cytomegalovirus (HCMV) increased glycolytic flux to replenish the TCA cycle, and herpes simplex virus type-1 (HSV-1) induced the elevation of pyrimidine nucleotide components [23]. In addition, SARS-CoV-2 infection induced sphingolipid metabolism reprogramming, which was required for viral replication. The levels of glycosphingolipid and sphingolipid (sphingosine, GA1, and GM3) were markedly increased in cells and the murine model after SARS-CoV-2 infection [24]. However, a decrease in cholesterol and high- and low-density lipoproteins was induced in the blood of patients with COVID-19, which may be potential markers for monitoring the disease [25]. Therefore, metabolomics is widely used as an important tool to investigate complex virus–host interactions. In this way, these studies provide an insight into the pathogenic mechanisms and novel therapeutic methods of the virus.

The metabolic alterations induced by different viruses are distinct. PRV is one of the main pathogens of pigs, thus understanding the changes of PRV in host cell metabolism is necessary. Few studies have been performed on the host metabolism of PRV. Gou et al. established that the metabolic flux derived from glycolysis, the pentose phosphate pathway, and glutamine metabolism for nucleotide biosynthesis was necessary for PRV replication [26]. The changes of PRV infection in immortalized porcine alveolar macrophages (iPAMs) on glycerolipids, fatty acyls, glycerophospholipids, and sphingolipids have also been determined [27]. In this study, we analyze the metabolic alterations in porcine kidney cells (PK-15) infected with a PRV variant strain and Bartha K61 strain using ultra-high-performance liquid chromatography/Q Exactive HF-X Hybrid Quadrupole-Orbitrap Mass (UHPLC-QE-MS). The results show that plenty of metabolites and metabolic pathways are significantly changed during PRV infection, when compared to uninfected cell groups. It suggests that those alterations may be in favor of better viral replication. These findings may be helpful to understand the host response to PRV infection and development for this disease control.

2. Materials and Methods

2.1. Cell Culture and Virus Infection

PK-15 cells were purchased from the American Type Culture Collection (ATCC) and cultured in Dulbecco's modified Eagle medium (DMEM) (Gibco, Waltham, MA, USA) containing 10% fetal bovine serum (FBS, Thermo Fisher Scientific, Waltham, MA, USA) at 37 °C with 5% CO_2. PRV variant strain JS21 (abbreviation of PRV-G) and PRV Bartha K61 strain (GenBank accession no. JF797217; with abbreviation of PRV-K) were preserved in our laboratory. PRV titers were determined as the median tissue culture infective doses ($TCID_{50}$) on PK-15 cells.

2.2. Virus Infection

PK-15 cells were cultured overnight at 37 °C with 5% CO_2. When the cells' density reached approximately 80%, they were infected with PRV at a multiplicity of infection

(MOI) of 1 and incubated at 37 °C for 1 h. After washing with phosphate-buffered saline (PBS), the cells were incubated in DMEM supplemented with 2% FBS. PK-15 cell samples were harvested at 0, 6, 12, and 24 h post infection (h.p.i.).

2.3. Western Blot and Immunofluorescence Assay

PK-15 cells were infected with PRV (PRV-G or PRV-K) at MOI of 1. Cell samples were harvested at 6, 12, and 24 h.p.i. for immunoblotting analysis, which was performed as previously described [28]. Briefly, cells were lysed with 200 µL lysis buffer (Beyotime, Shanghai, China) for 15 min on ice. Following centrifugation, the supernatant of the lysates was denatured. Then, the samples were subjected to SDS-PAGE and transferred to nitrocellulose membranes (Sigma-Aldrich, Whatman, MA, USA). Following the incubation of antibodies of anti-PRV gB protein mAb (1:1000, preserved in our laboratory) and anti-β-actin (1:1000, Cell Signaling Technology, Danvers, MA, USA), the bands were visualized using an enhanced chemiluminescence reagent kit (Share-bio, Shanghai, China) and analyzed using ImageJ software.

PK-15 cells were inoculated on coverslips in the 6-well plate and infected with PRV (PRV-G or PRV-K) at MOI of 1. Cell samples were harvested at 6, 12, and 24 h.p.i. for immunofluorescence assay (IFA), which was performed as previously described [28]. Finally, the slides were placed on the cover glass with antifade mounting medium and visualized using an LSM 880 Zeiss confocal microscope (Carl Zeiss, Jena, Germany).

2.4. Sample Preparation and Extraction and UHPLC-QE-MS Analysis

PK-15 cells were infected with PRV (PRV-G or PRV-K) at MOI of 1. Cell samples were harvested at 0, 6, 12, and 24 h.p.i. There were approximately 1×10^7 cells per sample. The samples of 0 h were used as control, named Mock group. Other groups were named G6, G12, G24, K6, K12, and K24, respectively. Three replicates per group were set. Cells were washed with precooled PBS, and the supernatant was cleared by centrifugation for 10 min at 13,000 rpm at 4 °C. The cell pellet was frozen in liquid nitrogen for 30 s. Following freeze-drying, the samples were dissolved in sterile water and ultrasound treatment in ice water. Following centrifugation at 12,000 rpm for 15 min at 4 °C, the supernatant was extracted with 1 mL of methanol/acetonitrile/water (2:2:1, $v/v/v$) containing isotope-labeled internal standard mixture. Following ultrasound treatment, the samples were incubated at −40 °C for 1 h, and centrifuged at 12,000 rpm at 4 °C for 15 min. The supernatant was used for the UHPLC-QE-MS analysis. All samples were obtained and mixed in equal amounts as quality control (QC) samples before testing. Then, experimental samples and QC samples were tested on the machine.

A UHPLC system (Vanquish, Thermo Fisher Scientific, Waltham, MA, USA) with a UPLC BEH Amide column (2.1 mm × 100 mm, 1.7 µm) coupled to a Q Exactive HFX mass spectrometer (Orbitrap MS, Thermo Fisher Scientific, Waltham, MA, USA) was used for LC-MS/MS analysis. Liquid chromatography phase A is an aqueous phase, containing 25 mmol/L ammonium acetate and 25 mmol/L ammonia water, and phase B is acetonitrile. The QE HFX mass spectrometer was used for acquiring full scan MS/MS spectra on information-dependent acquisition (IDA) mode under software (Xcalibur, Thermo Fisher Scientific, Waltham, MA, USA) control.

2.5. PCA and OPLS-DA Analyses

The raw data were converted to the mzXML format using ProteoWizard software. Then, the data were processed by R package analysis for peak identification, extraction, alignment, and integration. Principal component analysis (PCA) and orthogonal projection to latent structures discriminant analysis (OPLS-DA) were conducted [29,30]. The data were logarithmically (LOG) transformed and centered (CTR) formatted using SIMCA software (V16.0.2, Sartorius Stedim Data Analytics AB, Umea, Sweden), followed by PCA modeling analysis. OPLS-DA modeling analysis was performed on the first principal component and a 7-fold cross-validation in the SIMCA software was performed throughout the analysis.

The R2X or R2Y (interpretability of the model for the categorical variable) and Q2 (the predictability of the model) were used to evaluate the model validity.

2.6. Total RNA Extraction and Quantitative Real-Time PCR (qPCR) Analysis

Cell samples were harvested at 24 h.p.i. after PRV (PRV-G or PRV-K) infection with different MOI; meanwhile, uninfected cells were used as control. Total RNAs from cell samples were extracted using TRNzol (TIANGEN, Beijing, China) and reverse transcribed to cDNA using a HiScript III 1st Strand cDNA Synthesis Kit (Vazyme, Nanjing, China), according to the manufacturer's instructions. The primer sequences of the target genes to be detected were designed, and are shown in Table S1. ACTB was used as an internal reference gene. qPCR was performed using Universal SYBR qPCR Master Mix (Vazyme, Nanjing, China) and an ABI QuantStudio 3 Real-Time PCR (96-Well) Detection System. The reaction parameters were: 95 °C, 30 s; 95 °C, 10 s, 60 °C, 30 s, 40 cycles.; 95 °C, 15 s, 60 °C, 60 s, 95 °C, 15 s. All experiments were performed in triplicate. The mRNA levels of genes were quantified relative to ACTB using the comparative threshold cycle ($2^{-\Delta\Delta CT}$) method [31].

2.7. Statistical Analysis

The first principal component of variable importance in the projection (VIP > 1) and Student's *t*-test ($p < 0.05$) was set as the standard to screen the differential metabolites. Then, the data was subjected to the KEGG Metabolome Database for identification of metabolites. Additionally, the metabolites were analyzed further via online statistical analysis (MetaboAnalyst, http://www.metaboanalyst.ca/ (accessed on 16 November 2021)) for identifying the altered metabolic pathways caused by PRV infection [32].

GraphPad Prism 7.0 software was used for the statistical analyses. *p*-values less than 0.05 were considered statistically significant. The values were expressed as the mean ± standard error of the mean. The significance in figures was indicated as follows: *, $p < 0.05$; **, $p < 0.01$.

3. Results

3.1. Replication of PRV in PK-15 Cells

To confirm PRV replication in PK-15 cells, the cells were infected with PRV-G or PRV-K strain at MOI = 1, respectively. Additionally, the expression levels of PRV-gB protein were determined by Western blot and IFA. In the uninfected cells, there was no signal of viral protein to be detected. As shown in Figure 1A,B, the gray values of PRV-gB protein notably increase over time. Additionally, the amount and the intensity of fluorescence in PRV-infected cells were progressively strong and reached a high level at 24 h.p.i. (Figure 1C,D). These results indicate that both the PRV-G and PRV-K strains could effectively replicate in PK-15 cells within a 24 h infection.

3.2. Multivariate Analysis of PK-15 Cell Metabolites

The UHPLC-QE-MS used positive and negative ion (POS and NEG) switching modes and full-scan assay to screen and identify the numerous metabolites. After obtaining the data, we performed a series of multivariate pattern recognition analyses to evaluate the differences between the samples. PCA and OPLS-DA were performed to obtain more reliable information on the correlation between group differences of metabolites and experimental groups. The PCA-score scatter plot of all samples (including QC samples) is shown in Figure 2A,B. Each scatter represented a sample, and the color and shape of the scatter signed different groups. The results of the PCA score scatter plot show that all samples are in the 95% confidence interval. The OPLS-DA model for different groups versus the mock group was analyzed, and the R2X, R2Y, and Q2 of samples (POS, NEG) are shown in Figure 2C, in which the values of three parameters are close to 1. These results indicate that these different groups are clearly distinguished, and these models are efficient and reliable.

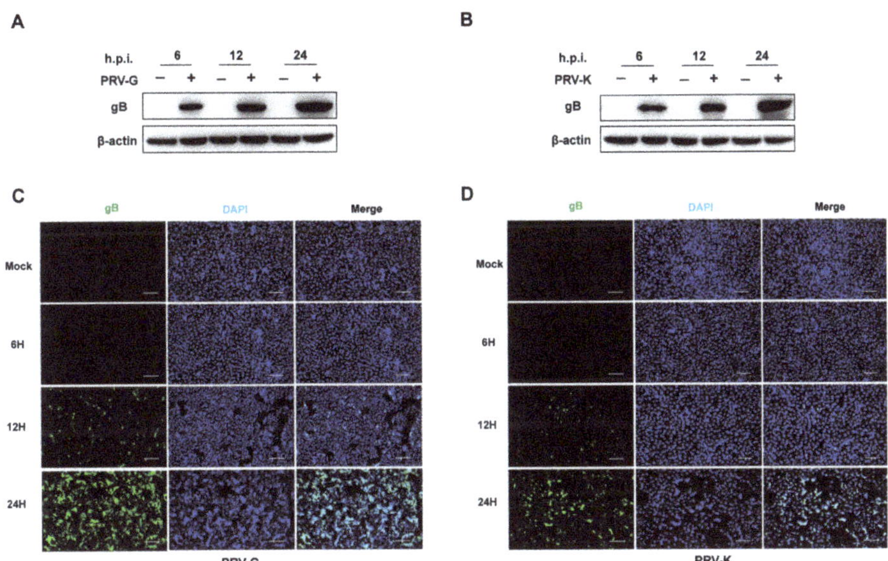

Figure 1. Infections of different PRV strains in PK-15 cells. (**A,B**) The cells were infected with PRV variant strain (abbreviation of PRV-G) and PRV Bartha K61 strain (abbreviation of PRV-K) at MOI = 1, and cell samples were collected at 6, 12, and 24 h for immunoblotting detection. (**C,D**) PK-15 cells were infected with PRV-G and PRV-K at MOI = 1 for 6, 12, and 24 h at 37 °C with 5% CO_2. The expression levels of PRV-gB protein detected by IFA. Scale bars = 200 μm.

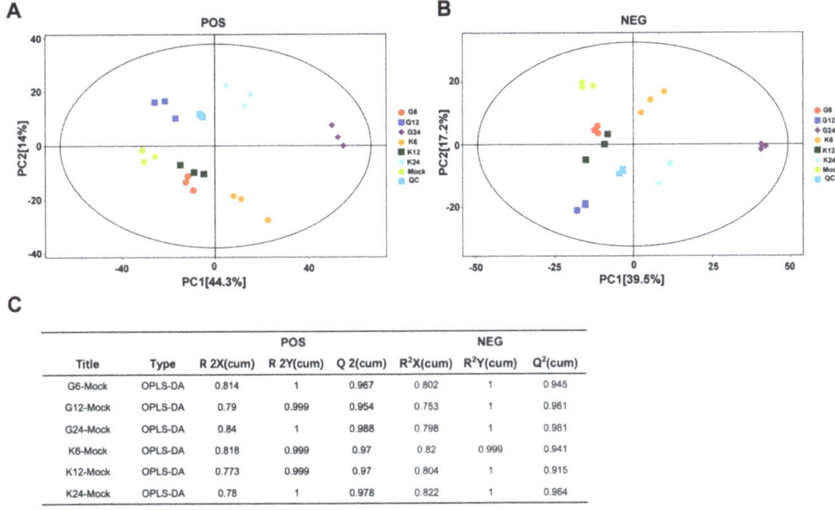

Title	Type	POS			NEG		
		R 2X(cum)	R 2Y(cum)	Q 2(cum)	R^2X(cum)	R^2Y(cum)	Q^2(cum)
G6-Mock	OPLS-DA	0.814	1	0.967	0.802	1	0.945
G12-Mock	OPLS-DA	0.79	0.999	0.954	0.753	1	0.961
G24-Mock	OPLS-DA	0.84	1	0.988	0.798	1	0.981
K6-Mock	OPLS-DA	0.818	0.999	0.97	0.82	0.999	0.941
K12-Mock	OPLS-DA	0.773	0.999	0.97	0.804	1	0.915
K24-Mock	OPLS-DA	0.78	1	0.978	0.822	1	0.964

Figure 2. Score scatter plots of PCA and OPLS-DA of PRV-infected and uninfected cells. (**A,B**) Score scatter plot of the PCA model for the different infection groups versus mock group. Electrospray ionization served as the source of UHPLC-QE-MS, including positive and negative ion modes (POS and NEG). (**A**) was derived from POS and (**B**) from NEG. The lines denote 95% confidence interval Hotelling's ellipses. (**C**) OPLS-DA model for the different PRV-strain infection group versus mock group.

3.3. Differentially Expressed Metabolites during PRV Infection

Based on OPLS-DA analysis, the VIP > 1 and $p < 0.05$ were set as the standards to screen the differential metabolites. We found that a great number of metabolites were altered during PRV infection, which were summarized and shown in Venn diagrams. The numbers of differential metabolites in PK-15 cells infected with PRV-G were 430, 426, and 606 at 6, 12, and 24 h.p.i., respectively. Additionally, the numbers of differential metabolites changed by PRV-K infection were 556, 425, and 535 at different time points (Figure 3A,B). Compared to the mock group, a total of 375 and 194 metabolites were significantly upregulated in PRV-G- and PRV-K-infected cells, respectively (Figure 3C,D). Furthermore, the different changes in metabolites were due to the different times of PRV infection. In addition, these differential metabolites were classified and analyzed. As shown in Figure 3E–H, lipids and lipid-like molecules, organic acids and derivatives, nucleosides, nucleotides, analogues, and organic oxygen compounds account for nearly 80% in the PRV-infected cells. It is worth noting that, among these differential metabolites caused by both PRV strains, lipid metabolites accounted for the highest proportion: around 35%. These results suggest that the lipid metabolism of the host cell may play an important role in PRV replication.

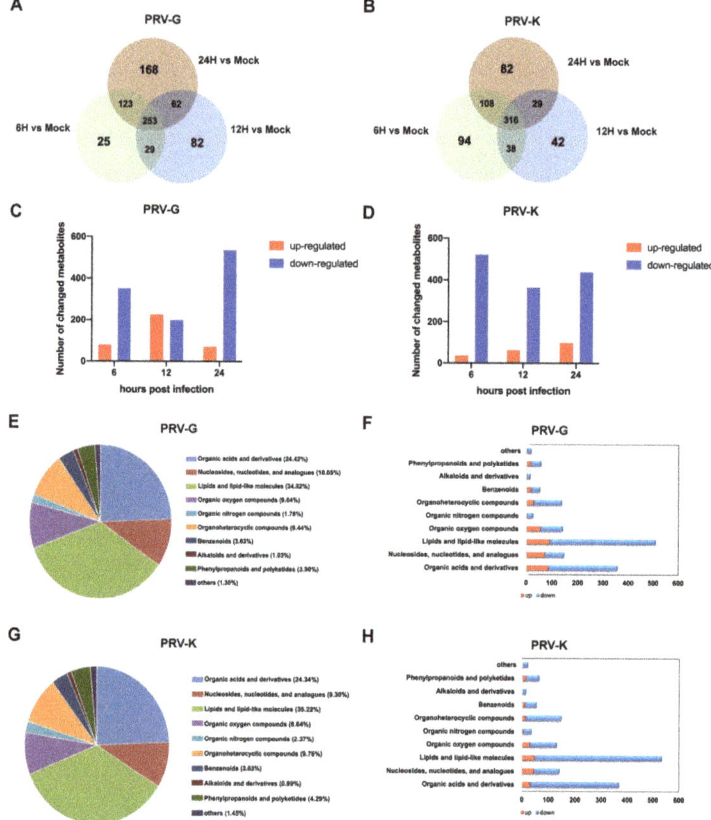

Figure 3. Analysis of differentially expressed metabolites in PK-15 cells infected with different PRV strains. (**A,B**) Venn diagrams between PRV-infected groups (6H, 12H, 24H) and mock group. (**C,D**) Numbers of differentially expressed metabolites upregulated (red) and downregulated (blue) in infected groups. (**E–H**) Pie charts and the histogram graphs showing proportions of different categories among differentially expressed metabolites in PRV-infected PK-15 cells.

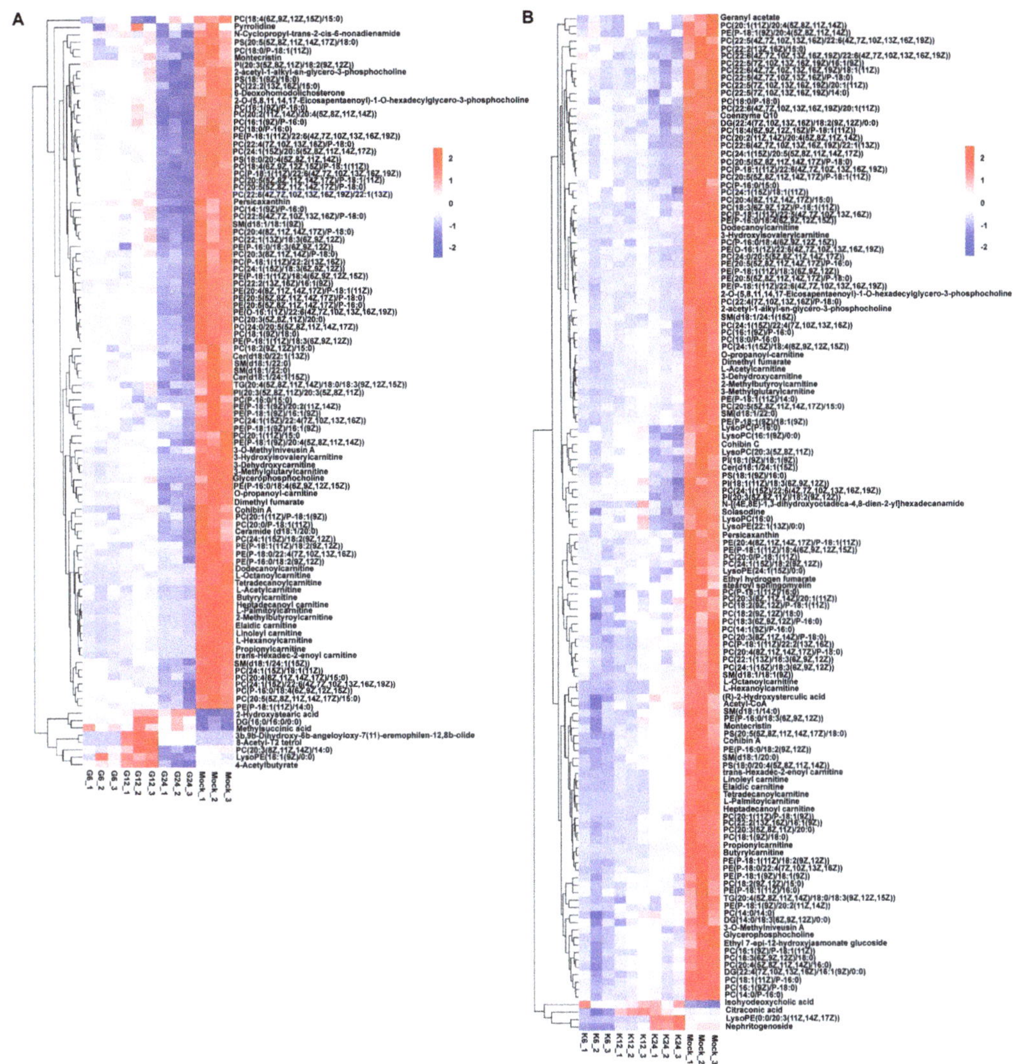

Figure 4. Heatmap analysis of 103 and 136 metabolites among PRV-G, PRV-K, and mock groups. Rows: metabolites; columns: samples. The color of each rectangle represents the relative level of the differential metabolites. Red: upregulated; blue: downregulated. (**A**) PRV-G; (**B**) PRV-K.

To study the crucial metabolites related to the PRV replication process, common differential metabolites among the three comparisons, i.e., 6 h.p.i. vs. mock, 12 h.p.i. vs. mock, and 24 h.p.i. vs. mock were screened. A total of 103 and 136 metabolites were obtained in PRV-G and PRV-K groups, respectively, and presented in the heatmap of hierarchical clustering analysis (Figure 4). As expected, the metabolites changed along with the virus infection. Many metabolites were significantly downregulated after PRV infection, especially lipids and lipid-like molecules. Glycerophospholipids and sphingolipids are important phospholipid molecules. Glycerophospholipids are divided into many categories according to the substitution groups, such as phosphatidylglycerol (PG), phosphatidylcholine (PC), phosphatidylserine (PS), phosphatidylethanolamine (PE), phosphatidylinositol (PI), and cardiolipin (CL). During the early stage of PRV infection, some of the glycerophospholipids increased. At 24 h.p.i., PRV-G infection induced a decrease in the levels of 78 species of glycerophospholipids, including 53 species of PC, 18 species of PE, 3 species of PS, and 2 species of PI and others (Figure 4A and Table S2). Meanwhile, PRV-K infection caused a decrease in the levels of 95 species of glycerophospholipids, including 64 species of PC, 21 species of PE, 3 species of PS, and 3 species of PI and others (Figure 4B and Table S3). In addition, few species of sphingolipids significantly decreased in PRV-G- and PRV-K-infected cells, respectively. These results indicate that, in the late stage of virus infection, PRV needs to consume a large amount of lipids in the host cell to ensure its replication.

3.4. Metabolic Pathway Analysis of Metabolites

These differential metabolites were annotated by using the KEGG Metabolome Database and further comprehensive analysis, including enrichment analysis and topological analysis, was conducted to find the metabolic pathways with high correlations. The results are shown in a bubble plot (Figure 5). In PRV-G vs. mock, differential metabolites were mainly enriched in arginine and proline metabolism; glycerophospholipid metabolism; glycine, serine, and threonine metabolism; purine metabolism; pyrimidine metabolism; and sphingolipid metabolism (Figure 5A–C). For PRV-K vs. mock, the metabolic pathways of the differential metabolites contained thiamine metabolism, purine metabolism, arginine and proline metabolism, glycerophospholipid metabolism, pyrimidine metabolism, and sphingolipid metabolism (Figure 5D–F).

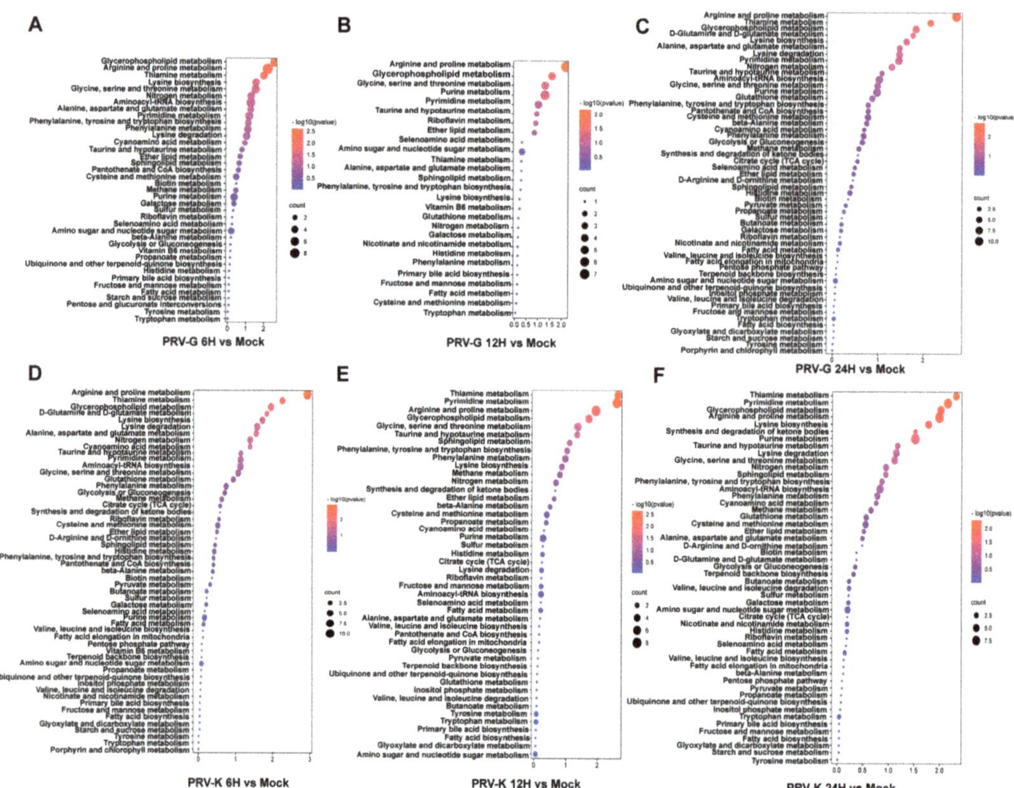

Figure 5. The KEGG-enrichment pathway analysis of differentially expressed metabolites for PK-15 cells infected with different PRV strains in different time courses. (**A–C**) PRV-G (6H, 12H, 24H); (**D–F**) PRV-K (6H, 12H, 24H).

In summary, the results present more visualized profiles of the metabolite changes in PK-15 cells infected with two different PRV strains (Figure 6). The metabolic pathway contained a TCA cycle, lipid metabolism, amino acid metabolism, purine, and pyrimidine metabolism. The levels of adenosine at 12 and 24 h.p.i. in PRV-G/PRV-K-infected cells were more upregulated than that in the mock group as well as the levels of dTMP, which indicated that they may be required in the virus replication cycle. Adenosine is an important intermediate for the synthesis of adenosine triphosphate (ATP), adenine, and adenylate. Additionally, dTMP is a basic unit for deoxyribonucleic acid, which is the material basis for DNA synthesis. The major metabolic pathways of glycerophospholipids and sphingolipids and fatty acids during PRV infection were also reprogrammed, and the levels of PC, PE, ceramide, and sphingomyelin were consumed with the PRV replication process. Additionally, some amino acids were altered after PRV infection. These results indicate that two different PRV-strain infections led to the metabolic reprogramming of PK-15 cells to benefit self-replication.

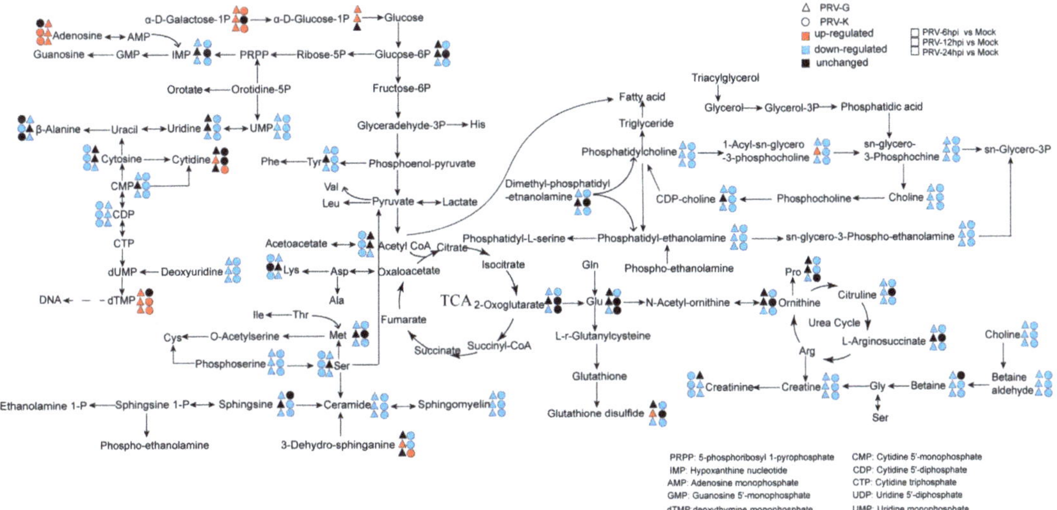

Figure 6. Schematic overview of altered metabolic pathways in PK-15 cells infected with different PRV strains. The metabolites were shown in different colors according to their changes. Black: unchanged; red: upregulated; blue: downregulated.

3.5. Validation of Metabolomic Data by qPCR

Given the possible roles of lipid metabolism during PRV infection, PK-15 samples were harvested at 24 h after PRV (PRV-G/PRV-K) infection with different MOI (1, 5, and 10, respectively) to further validate the metabolomic data. The mRNA levels of enzymes related to glycerophospholipid metabolism and sphingolipid metabolism were analyzed by qRT-PCR (Figure 7), including sphingosine kinase (SPHK1/2), sphingomyelin synthase (SGMS1/2), serine palmitoyltransferase small subunit A/B (SPTSSA/B), sphingomyelin phosphodiesterase 1,2,3 (SMPD1/2/3), fatty acid synthase (FASN), 3-hydroxy-3-methylglutaryl-CoA reductase (HMGCR), phosphate cytidylyltransferase (PCYT1A/2), and phosphatidylserine decarboxylase (PISD). We found that SPTSSB, SMPD3, and PCYT2 were significantly increased in PRV-G-infected PK-15 cells (Figure 7A,B). Moreover, the PRV-K strain could apparently upregulate these three genes (Figure 7C,D). These results suggest that some pathways of lipid metabolism in host cells are promoted during PRV infection to facilitate viral replication.

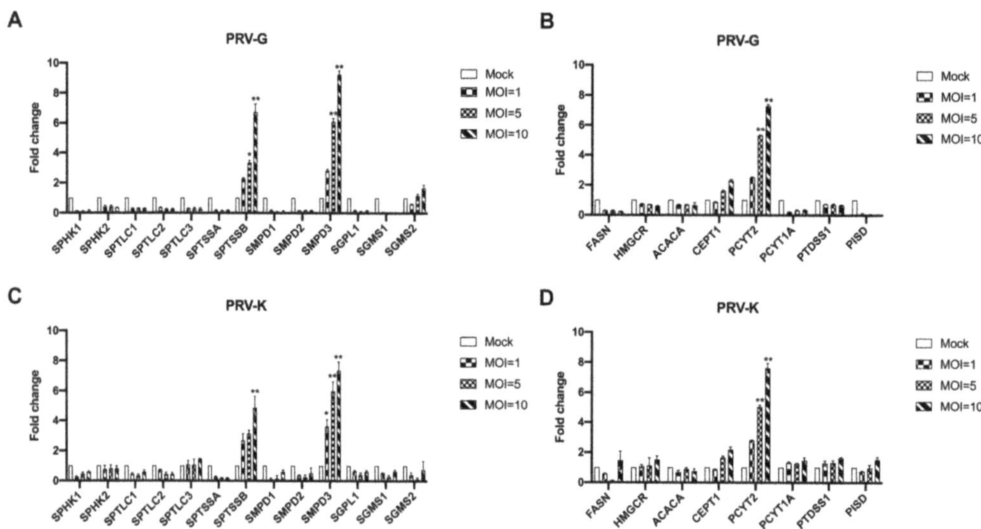

Figure 7. The mRNA levels of sphingolipid- and glycerophospholipid-metabolism-related enzymes after different PRV-strain infections. PK-15 cells were harvested with different MOI (1, 5, 10) at 24 h.p.i after PRV (PRV-G or PRV-K) infection, while non-infected cells were used as control. The mRNA levels of different enzymes were determined by qRT-PCR. β-actin was used as an internal reference gene. (**A,B**) PRV-G; (**C,D**) PRV-K. The significance in the figure was indicated as follows: *, $p < 0.05$; **, $p < 0.01$.

4. Discussion

Recently, a growing number of studies on the combination of metabolite profiling with disease have been reported [33,34]. Through the detection and analysis of endogenous small molecules in cells, researchers evaluated the biochemical differences between healthy and pathological organisms to obtain an insight into the pathology, etiology, and possible treatment options of diseases. In our study, we analyzed the metabolomic profiles of PRV-infected PK-15 cells based on UHPLC-QE-MS, and these findings provide new viewpoints on the interactions between PRV and host cells, which will help future studies on PRV.

This study involved a global metabonomic analysis of PK-15 cells that were infected with two different strains of PRV. Both OPLS-DA and the heatmap of hierarchical clustering analysis indicated that, compared to the mock group, PRV-infected cells showed significantly different metabolic profiles. We determined that PRV infection broke the metabolic homeostasis of PK-15 cells, caused metabolic reprogramming, and significantly affected the metabolism of lipid metabolism and nucleotides metabolism (Figure 6). The alterations of these metabolites and pathways reflected the cellular responses to PRV infection or the nutritional needs in virus replication.

The metabolic alterations caused by the infection of two different strains of PRV were different, which were caused by the characteristics of the virus itself. The sequence homology of the two strains was 96%, but the pathogenicity was significantly different. Following two different strains of PRV infection, PRV-G could cause a more serious typical cytopathic effect than that caused by PRV-K 24 h post infection. Therefore, PRV-infected cells showed strain-specific metabolic characteristics. However, there were some overlaps between these pairwise comparisons in the metabolic pathways (Figure 5), such as glycerophospholipid metabolism, sphingolipid metabolism, purine metabolism, and pyrimidine metabolism. Lipids are the structural basis of the cell biofilm, and are biologically active molecules, which participate in a variety of cellular processes and immune functions. We have known that palmitoyl-oleoyl-phosphatidylglycerol and PI inhibited inflammatory sequelae and the infection of respiratory syncytial virus and influenza A virus, by destroying the binding of virus particles to plasma membrane receptors of host cells [35]. Flaviviruses also increased lipid synthesis to expand the surface area of membranes for better replication [36]. Dengue virus (DENV) infection required the manipulation of cellular fatty acid synthesis and cholesterol biosynthesis and transport [37,38]. It has been previously found that sphingolipids were essential for the successful completion of the viral life cycle, which is involved in attachment, membrane fusion, intracellular replication, assembly, and release of several viruses [39–41]. In our study, as an enveloped virus, PRV also required a large amount of lipids to participate in the formation and release of virions. Consistently, we found that many lipids (including glycerophospholipids and sphingolipids) were consumed in the late stage of virus infection (Figures 4 and 6). Purine metabolism and pyrimidine metabolism are the basic steps for nucleotide synthesis. In COVID-19 patients, some metabolites related to purine metabolism show an upward or downward trend compared to healthy controls, and correlation analysis showed a close correlation between these metabolites and proinflammatory cytokines/chemokines [42]. Previous studies have been reported to inhibit DENV replication with inhibitors (methotrexate and floxuridine) of the thymidine synthesis pathway [43]. Tiwari SK et al. utilized nucleoside metabolic inhibitors fluorouracil and floxuridine to inhibit Zika virus in human microglial cells [44]. Following PRV infection, nucleotide metabolism was markedly changed, suggesting an unknown role in PRV replication (Figure 6). This may expand the novel possibilities for the development of antiviral therapies.

Few studies have been performed on the host metabolism of PRV. Gou et al. explored the metabolic networks in PK-15 cells infected with PRV using gas chromatography-mass spectrometry (GC-MS) analysis. They reported that the metabolic flux derived from glycolysis, the pentose phosphate pathway, and glutamine metabolism for nucleotide biosynthesis was necessary for PRV replication [26]. In addition, Yao et al. indicated changes of PRV infection in iPAM on glycerolipids, fatty acyls, glycerophospholipids, and sphingolipids [27]. Our results are somewhat different from the previous two reports due to differences in the testing methods (GC-MS vs. LC-MS) and cells (iPAM vs. PK-15).

In our results, we found the mRNA level of PCYT2 was significantly increased in PK-15 cells infected with two PRV strains (Figure 7). PCYT2 is an important enzyme for the biosynthesis of PE from ethanolamine and diacylglycerol. Previous studies have shown treatment of PRV-infected PK-15 cells with meclizine, an inhibitor of PCYT2, led to decreased PRV infection and replication [28]. The result showed that glycerophospholipid metabolism was essential for PRV replication, but the mechanism was unclear.

Other herpesviruses, such as HSV-1, HCMV, Kaposi's sarcoma-associated herpesvirus (KSHV), and Epstein–Barr virus (EBV), have shown a notable ability to reprogram the host's metabolism for viral replication. HSV-1 infection led to increased levels of phosphoenolpyruvate, deoxypyrimidines, and pentose phosphate pathway intermediates [23]. Treatment with inhibitor of glucose metabolism or nucleoside analogs decreased the cell-to-cell spread and production of HSV [45,46]. HCMV infection markedly increased glucose uptake and glycolysis flux and promoted flux through the TCA cycle and fatty acid biosynthesis pathway [23,47]. Following the inhibition of fatty acid biosynthesis by drugs, the level of HCMV replication was suppressed [47]. KSHV caused changes in many metabolites of glycolysis, the pentose phosphate pathway, amino acid metabolism, and lipogenesis [48]. Moreover, latent KSHV-infected endothelial cells depended on glutamine and glutaminolysis for survival [49]. This evidence showed that metabolic changes caused by virus infection played an important role in viral replication.

In conclusion, the metabolic profiles of PK-15 cells infected with different PRV strains were analyzed to display the metabolic changes by UHPLC-QE-MS. There were significant differences in lipid metabolism and nucleotide metabolism between the PRV-infected groups and the mock group. The ability of viruses to actively modulate host metabolism is crucial for the successful completion of the viral life cycle. Many inhibitors of lipid metabolism and nucleotide metabolism are already used against some viral infections;

therefore, identifying metabolic targets for antiviral therapy may be a promising strategy. Our study provides much information for a further understanding of PRV pathogenesis and drug intervention for disease control.

Supplementary Materials: The following supporting information can be downloaded at: https://www.mdpi.com/article/10.3390/v14061158/s1: Table S1. Primer sequences used for qPCR; Table S2. Identification and characterization of the metabolites in PK-15 cells infected with PRV-G; Table S3. Identification and characterization of the metabolites in PK-15 cells infected with PRV-K.

Author Contributions: P.L. and X.L. conceived and designed the experiments; P.L. and D.H. performed the experiments, analyzed the data, and drafted the manuscript; L.Y., Z.L., X.Y. and Z.Z. contributed reagents/materials/analysis tools; P.L. and X.L. thoroughly revised the manuscript. All authors have read and agreed to the published version of the manuscript.

Funding: This work was funded by the National Natural Science Foundation of China (Nos. 32102637 and 32172823) and the Project of the Priority Academic Program Development of Jiangsu Higher Education Institutions (PAPD).

Institutional Review Board Statement: Not applicable for studies not involving humans or animals.

Informed Consent Statement: Not applicable.

Data Availability Statement: The metabolomic data is available with the link: https://pan.baidu.com/s/1SRndDrNqckJ7DxofO5zWeA (Password: hv45, accessed on 24 May 2022).

Acknowledgments: We want to express our gratitude to Chan Ding from Shanghai Veterinary Research Institute, CAAS for his valuable suggestions for this project.

Conflicts of Interest: The authors declare no conflict of interest.

References

1. Pomeranz, L.E.; Reynolds, A.E.; Hengartner, C.J. Molecular biology of pseudorabies virus: Impact on neurovirology and veterinary medicine. *Microbiol. Mol. Biol. Rev.* **2005**, *69*, 462–500. [CrossRef] [PubMed]
2. Müller, T.; Hahn, E.C.; Tottewitz, F.; Kramer, M.; Klupp, B.G.; Mettenleiter, T.C.; Freuling, C. Pseudorabies virus in wild swine: A global perspective. *Arch. Virol.* **2011**, *156*, 1691–1705. [CrossRef] [PubMed]
3. He, W.; Auclert, L.Z.; Zhai, X.; Wong, G.; Zhang, C.; Zhu, H.; Xing, G.; Wang, S.; He, W.; Li, K.; et al. Interspecies Transmission, Genetic Diversity, and Evolutionary Dynamics of Pseudorabies Virus. *J. Infect. Dis.* **2019**, *219*, 1705–1715. [CrossRef] [PubMed]
4. Yang, X.; Guan, H.; Li, C.; Li, Y.; Wang, S.; Zhao, X.; Zhao, Y.; Liu, Y. Characteristics of human encephalitis caused by pseudorabies virus: A case series study. *Int. J. Infect. Dis.* **2019**, *87*, 92–99. [CrossRef] [PubMed]
5. Liu, Q.; Wang, X.; Xie, C.; Ding, S.; Yang, H.; Guo, S.; Li, J.; Qin, L.; Ban, F.; Wang, D.; et al. A Novel Human Acute Encephalitis Caused by Pseudorabies Virus Variant Strain. *Clin. Infect. Dis.* **2021**, *73*, e3690–e3700. [CrossRef] [PubMed]
6. Freuling, C.M.; Müller, T.F.; Mettenleiter, T.C. Vaccines against pseudorabies virus (PrV). *Vet. Microbiol.* **2017**, *206*, 3–9. [CrossRef] [PubMed]
7. Delva, J.L.; Nauwynck, H.J.; Mettenleiter, T.C.; Favoreel, H.W. The Attenuated Pseudorabies Virus Vaccine Strain Bartha K61: A Brief Review on the Knowledge Gathered during 60 Years of Research. *Pathogens* **2020**, *9*, 897. [CrossRef]
8. Brukman, A.; Enquist, L.W. Suppression of the interferon-mediated innate immune response by pseudorabies virus. *J. Virol.* **2006**, *80*, 6345–6356. [CrossRef]
9. van Rooij, E.M.; de Bruin, M.G.; de Visser, Y.E.; Middel, W.G.; Boersma, W.J.; Bianchi, A.T. Vaccine-induced T cell-mediated immunity plays a critical role in early protection against pseudorabies virus (suid herpes virus type 1) infection in pigs. *Vet. Immunol. Immunopathol.* **2004**, *99*, 113–125. [CrossRef]
10. Wu, R.; Bai, C.; Sun, J.; Chang, S.; Zhang, X. Emergence of virulent pseudorabies virus infection in northern China. *J. Vet. Sci.* **2013**, *14*, 363–365. [CrossRef]
11. An, T.Q.; Peng, J.M.; Tian, Z.J.; Zhao, H.Y.; Li, N.; Liu, Y.M.; Chen, J.Z.; Leng, C.L.; Sun, Y.; Chang, D.; et al. Pseudorabies virus variant in Bartha-K61-vaccinated pigs, China, 2012. *Emerg. Infect. Dis.* **2013**, *19*, 1749–1755. [CrossRef] [PubMed]
12. Goodacre, R. Making sense of the metabolome using evolutionary computation: Seeing the wood with the trees. *J. Exp. Bot.* **2005**, *56*, 245–254. [CrossRef]
13. Goldansaz, S.A.; Guo, A.C.; Sajed, T.; Steele, M.A.; Plastow, G.S.; Wishart, D.S. Livestock metabolomics and the livestock metabolome: A systematic review. *PLoS ONE* **2017**, *12*, e0177675. [CrossRef] [PubMed]
14. Johnson, C.H.; Ivanisevic, J.; Siuzdak, G. Metabolomics: Beyond biomarkers and towards mechanisms. *Nat. Rev. Mol. Cell Biol.* **2016**, *17*, 451–459. [CrossRef] [PubMed]
15. Parveen, M.; Miyagi, A.; Kawai-Yamada, M.; Rashid, M.H.; Asaeda, T. Metabolic and biochemical responses of *Potamogeton anguillanus* Koidz. (Potamogetonaceae) to low oxygen conditions. *J. Plant Physiol.* **2019**, *232*, 171–179. [CrossRef]

16. Ghazal, P.; González Armas, J.C.; García-Ramírez, J.J.; Kurz, S.; Angulo, A. Viruses: Hostages to the cell. *Virology* **2000**, *275*, 233–237. [CrossRef]
17. Thaker, S.K.; Ch'ng, J.; Christofk, H.R. Viral hijacking of cellular metabolism. *BMC Biol.* **2019**, *17*, 59. [CrossRef]
18. Gualdoni, G.A.; Mayer, K.A.; Kapsch, A.M.; Kreuzberg, K.; Puck, A.; Kienzl, P.; Oberndorfer, F.; Frühwirth, K.; Winkler, S.; Blaas, D.; et al. Rhinovirus induces an anabolic reprogramming in host cell metabolism essential for viral replication. *Proc. Natl. Acad. Sci. USA* **2018**, *115*, E7158–E7165. [CrossRef]
19. Jean Beltran, P.M.; Cook, K.C.; Hashimoto, Y.; Galitzine, C.; Murray, L.A.; Vitek, O.; Cristea, I.M. Infection-Induced Peroxisome Biogenesis Is a Metabolic Strategy for Herpesvirus Replication. *Cell Host Microbe* **2018**, *24*, 526–541.e7. [CrossRef]
20. Hsu, N.Y.; Ilnytska, O.; Belov, G.; Santiana, M.; Chen, Y.H.; Takvorian, P.M.; Pau, C.; van der Schaar, H.; Kaushik-Basu, N.; Balla, T.; et al. Viral reorganization of the secretory pathway generates distinct organelles for RNA replication. *Cell* **2010**, *141*, 799–811. [CrossRef]
21. Thaker, S.K.; Chapa, T.; Garcia, G., Jr.; Gong, D.; Schmid, E.W.; Arumugaswami, V.; Sun, R.; Christofk, H.R. Differential Metabolic Reprogramming by Zika Virus Promotes Cell Death in Human versus Mosquito Cells. *Cell Metab.* **2019**, *29*, 1206–1216.e4. [CrossRef] [PubMed]
22. Keshavarz, M.; Solaymani-Mohammadi, F.; Namdari, H.; Arjeini, Y.; Mousavi, M.J.; Rezaei, F. Metabolic host response and therapeutic approaches to influenza infection. *Cell. Mol. Biol. Lett.* **2020**, *25*, 15. [CrossRef] [PubMed]
23. Vastag, L.; Koyuncu, E.; Grady, S.L.; Shenk, T.E.; Rabinowitz, J.D. Divergent effects of human cytomegalovirus and herpes simplex virus-1 on cellular metabolism. *PLoS Pathog.* **2011**, *7*, e1002124. [CrossRef] [PubMed]
24. Vitner, E.B.; Avraham, R.; Politi, B.; Melamed, S.; Israely, T. Elevation in sphingolipid upon SARS-CoV-2 infection: Possible implications for COVID-19 pathology. *Life Sci. Alliance* **2022**, *5*, e202101168. [CrossRef]
25. Kočar, E.; Režen, T.; Rozman, D. Cholesterol, lipoproteins, and COVID-19: Basic concepts and clinical applications. *Biochim. Biophys. Acta Mol. Cell Biol. Lipids* **2021**, *1866*, 158849. [CrossRef]
26. Gou, H.; Bian, Z.; Li, Y.; Cai, R.; Jiang, Z.; Song, S.; Zhang, K.; Chu, P.; Yang, D.; Li, C. Metabolomics Exploration of Pseudorabies Virus Reprogramming Metabolic Profiles of PK-15 Cells to Enhance Viral Replication. *Front. Cell. Infect. Microbiol.* **2020**, *10*, 599087. [CrossRef]
27. Yao, L.; Hu, Q.; Zhang, C.; Ghonaim, A.H.; Cheng, Y.; Ma, H.; Yu, X.; Wang, J.; Fan, X.; He, Q. Untargeted LC-MS based metabolomic profiling of iPAMs to investigate lipid metabolic pathways alternations induced by different Pseudorabies virus strains. *Vet. Microbiol.* **2021**, *256*, 109041. [CrossRef]
28. Liu, P.; Hu, D.; Yuan, L.; Lian, Z.; Yao, X.; Zhu, Z.; Nowotny, N.; Shi, Y.; Li, X. Meclizine Inhibits Pseudorabies Virus Replication by Interfering With Virus Entry and Release. *Front. Microbiol.* **2021**, *12*, 795593. [CrossRef]
29. Wang, L.; Liu, S.; Yang, W.; Yu, H.; Zhang, L.; Ma, P.; Wu, P.; Li, X.; Cho, K.; Xue, S.; et al. Plasma Amino Acid Profile in Patients with Aortic Dissection. *Sci. Rep.* **2017**, *7*, 40146. [CrossRef]
30. Wiklund, S.; Johansson, E.; Sjöström, L.; Mellerowicz, E.J.; Edlund, U.; Shockcor, J.P.; Gottfries, J.; Moritz, T.; Trygg, J. Visualization of GC/TOF-MS-based metabolomics data for identification of biochemically interesting compounds using OPLS class models. *Anal. Chem.* **2008**, *80*, 115–122. [CrossRef]
31. Schmittgen, T.D.; Livak, K.J. Analyzing real-time PCR data by the comparative C(T) method. *Nat. Protoc.* **2008**, *3*, 1101–1108. [CrossRef]
32. Xia, J.; Sinelnikov, I.V.; Han, B.; Wishart, D.S. MetaboAnalyst 3.0—Making metabolomics more meaningful. *Nucleic Acids Res.* **2015**, *43*, W251–W257. [CrossRef] [PubMed]
33. Wang, T.J.; Larson, M.G.; Vasan, R.S.; Cheng, S.; Rhee, E.P.; McCabe, E.; Lewis, G.D.; Fox, C.S.; Jacques, P.F.; Fernandez, C.; et al. Metabolite profiles and the risk of developing diabetes. *Nat. Med.* **2011**, *17*, 448–453. [CrossRef] [PubMed]
34. Elia, I.; Haigis, M.C. Metabolites and the tumour microenvironment: From cellular mechanisms to systemic metabolism. *Nat. Metab.* **2021**, *3*, 21–32. [CrossRef] [PubMed]
35. Voelker, D.R.; Numata, M. Phospholipid regulation of innate immunity and respiratory viral infection. *J. Biol. Chem.* **2019**, *294*, 4282–4289. [CrossRef] [PubMed]
36. Perera, R.; Riley, C.; Isaac, G.; Hopf-Jannasch, A.S.; Moore, R.J.; Weitz, K.W.; Pasa-Tolic, L.; Metz, T.O.; Adamec, J.; Kuhn, R.J. Dengue virus infection perturbs lipid homeostasis in infected mosquito cells. *PLoS Pathog.* **2012**, *8*, e1002584. [CrossRef]
37. Rothwell, C.; Lebreton, A.; Young Ng, C.; Lim, J.Y.; Liu, W.; Vasudevan, S.; Labow, M.; Gu, F.; Gaither, L.A. Cholesterol biosynthesis modulation regulates dengue viral replication. *Virology* **2009**, *389*, 8–19. [CrossRef]
38. Poh, M.K.; Shui, G.; Xie, X.; Shi, P.Y.; Wenk, M.R.; Gu, F. U18666A, an intra-cellular cholesterol transport inhibitor, inhibits dengue virus entry and replication. *Antivir. Res.* **2012**, *93*, 191–198. [CrossRef]
39. Schneider-Schaulies, J.; Schneider-Schaulies, S. Viral infections and sphingolipids. *Handb. Exp. Pharmacol.* **2013**, *216*, 321–340. [CrossRef]
40. Schneider-Schaulies, S.; Schumacher, F.; Wigger, D.; Schöl, M.; Waghmare, T.; Schlegel, J.; Seibel, J.; Kleuser, B. Sphingolipids: Effectors and Achilles Heals in Viral Infections? *Cells* **2021**, *10*, 2175. [CrossRef]
41. Avota, E.; Bodem, J.; Chithelen, J.; Mandasari, P.; Beyersdorf, N.; Schneider-Schaulies, J. The Manifold Roles of Sphingolipids in Viral Infections. *Front. Physiol.* **2021**, *12*, 715527. [CrossRef]

42. Xiao, N.; Nie, M.; Pang, H.; Wang, B.; Hu, J.; Meng, X.; Li, K.; Ran, X.; Long, Q.; Deng, H.; et al. Integrated cytokine and metabolite analysis reveals immunometabolic reprogramming in COVID-19 patients with therapeutic implications. *Nat. Commun.* **2021**, *12*, 1618. [CrossRef] [PubMed]
43. Fischer, M.A.; Smith, J.L.; Shum, D.; Stein, D.A.; Parkins, C.; Bhinder, B.; Radu, C.; Hirsch, A.J.; Djaballah, H.; Nelson, J.A.; et al. Flaviviruses are sensitive to inhibition of thymidine synthesis pathways. *J. Virol.* **2013**, *87*, 9411–9419. [CrossRef] [PubMed]
44. Tiwari, S.K.; Dang, J.; Qin, Y.; Lichinchi, G.; Bansal, V.; Rana, T.M. Zika virus infection reprograms global transcription of host cells to allow sustained infection. *Emerg. Microbes Infect.* **2017**, *6*, e24. [CrossRef]
45. Birkmann, A.; Zimmermann, H. HSV antivirals—Current and future treatment options. *Curr. Opin. Virol.* **2016**, *18*, 9–13. [CrossRef]
46. McArdle, J.; Schafer, X.L.; Munger, J. Inhibition of calmodulin-dependent kinase kinase blocks human cytomegalovirus-induced glycolytic activation and severely attenuates production of viral progeny. *J. Virol.* **2011**, *85*, 705–714. [CrossRef] [PubMed]
47. Munger, J.; Bennett, B.D.; Parikh, A.; Feng, X.J.; McArdle, J.; Rabitz, H.A.; Shenk, T.; Rabinowitz, J.D. Systems-level metabolic flux profiling identifies fatty acid synthesis as a target for antiviral therapy. *Nat. Biotechnol.* **2008**, *26*, 1179–1186. [CrossRef] [PubMed]
48. Liu, X.; Zhu, C.; Wang, Y.; Wei, F.; Cai, Q. KSHV Reprogramming of Host Energy Metabolism for Pathogenesis. *Front. Cell. Infect. Microbiol.* **2021**, *11*, 621156. [CrossRef]
49. Sanchez, E.L.; Carroll, P.A.; Thalhofer, A.B.; Lagunoff, M. Latent KSHV Infected Endothelial Cells Are Glutamine Addicted and Require Glutaminolysis for Survival. *PLoS Pathog.* **2015**, *11*, e1005052. [CrossRef]

Article

Proteomic Analysis of Vero Cells Infected with Pseudorabies Virus

Xintan Yang [†], Shengkui Xu [†], Dengjin Chen, Ruijiao Jiang, Haoran Kang, Xinna Ge, Lei Zhou, Jun Han, Yongning Zhang, Xin Guo * and Hanchun Yang

Key Laboratory of Animal Epidemiology of the Ministry of Agriculture, College of Veterinary Medicine, China Agricultural University, Beijing 100193, China; s20193050751@cau.edu.cn (X.Y.); skxu0721@163.com (S.X.); chendengjin-cau@foxmail.com (D.C.); wn2jjj@163.com (R.J.); 18260068857@163.com (H.K.); gexn@cau.edu.cn (X.G.); leosj@cau.edu.cn (L.Z.); hanx0158@cau.edu.cn (J.H.); zhangyongning@cau.edu.cn (Y.Z.); yanghanchun1@cau.edu.cn (H.Y.)
* Correspondence: guoxincau@cau.edu.cn; Tel.: +86-10-62732875
† These authors contributed equally to this work.

Abstract: Suid herpesvirus 1 (SuHV-1), known as pseudorabies virus (PRV), is one of the most devastating swine pathogens in China, particularly the sudden occurrence of PRV variants in 2011. The higher pathogenicity and cross-species transmission potential of the newly emerged variants caused not only colossal economic losses, but also threatened public health. To uncover the underlying pathogenesis of PRV variants, Tandem Mass Tag (TMT)-based proteomic analysis was performed to quantitatively screen the differentially expressed cellular proteins in PRV-infected Vero cells. A total of 7072 proteins were identified and 960 proteins were significantly regulated: specifically 89 upregulated and 871 downregulated. To make it more credible, the expression of XRCC5 and XRCC6 was verified by western blot and RT-qPCR, and the results dovetailed with the proteomic data. The differentially expressed proteins were involved in various biological processes and signaling pathways, such as chaperonin-containing T-complex, NIK/NF-κB signaling pathway, DNA damage response, and negative regulation of G2/M transition of mitotic cell cycle. Taken together, our data holistically outline the interactions between PRV and host cells, and our results may shed light on the pathogenesis of PRV variants and provide clues for pseudorabies prevention.

Keywords: pseudorabies virus; Vero cell; TMT-based proteomic analysis; differentially expressed proteins

Citation: Yang, X.; Xu, S.; Chen, D.; Jiang, R.; Kang, H.; Ge, X.; Zhou, L.; Han, J.; Zhang, Y.; Guo, X.; et al. Proteomic Analysis of Vero Cells Infected with Pseudorabies Virus. *Viruses* **2022**, *14*, 755. https://doi.org/10.3390/v14040755

Academic Editors: Yan-Dong Tang and Xiangdong Li

Received: 8 March 2022
Accepted: 31 March 2022
Published: 4 April 2022

Publisher's Note: MDPI stays neutral with regard to jurisdictional claims in published maps and institutional affiliations.

1. Introduction

Pseudorabies (PR), also known as Aujeszky's disease (AD), is one of the most notorious swine diseases and causes enormous economic losses to the pig-raising industry [1]. Typical clinical symptoms of PR include respiratory distress, nervous disorders, and reproductive failures in sows [2,3]. PR is caused by pseudorabies virus (PRV), also called suid herpesvirus 1 (SuHV-1), which belongs to the subfamily of *Alphaherpesvirus* in the family of *Herpesviridae*. The genome of PRV is about 175 kb in length and encodes over 70 viral proteins contributing to neuronal latent infection and immune modulation [4,5].

Since the first report of PR outbreak in the 1950s, PRV has spread through China over the past 70 years [6]. The intensive herd vaccination by attenuated live vaccine Bartha-K61 facilitates PR eradication, whereas strong immune pressure may accelerate the virus's evolution and pave the way for the emergence of variants. In 2011, large scale outbreaks of PR caused by PRV variants swept China [6,7]. Subsequent studies showed that the emerging variants had higher pathogenicity, and the typical vaccine Bartha-K61 only provided limited protection against PRV variants infections [7,8]. Despite Jianle Ren et al. reporting that glycoproteins C and D of PRV variant strain HB1201 contribute individually to the escape from Bartha-K61 vaccine-induced protection [9], the pathogenesis of PRV variants remains largely unclear.

PRV has a wide host range and is capable of infecting numerous animals. Increasing evidence suggested that the newly emerged variants from 2011 were the most prevalent genotypes worldwide and most frequently involved in cross-species transmission [10]. Next-generation sequencing and regular polymerase chain reaction (PCR) confirmed the presence of PRV genomes in cerebral spinal fluid from a 43-year-old patient [11]. In addition, a patient who presented with encephalitis and pulmonary infection also tested PRV positive in his cerebrospinal fluid and vitreous humor [12]. More severely, a PRV strain was isolated from an acute human encephalitis case in 2019, confirming the interspecies transmission between pigs and humans and the replication capacity of PRV in human [13]. Other animals, including bovine and wolf, were also reported to be infected by PRV [14,15]. Understanding the interactions in-depth between PRV infection and host may provide ideas for interspecies transmission prevention.

Innate immunity is the host's first line of defense against virus infection. When invading host cells, pathogens are recognized by specific pattern recognition receptors (PRRs) and then trigger immune responses [16]. To establish efficient infection, PRV has evolved various strategies to evade immune clearance. For example, PRV US3 degrades Bcl-2 associated transcription factor 1 to impair type I interferon production and benefit virus replication [17]; UL50 induces the degradation of type I interferon receptor via lysosomal pathway to antagonize interferon response [18]. Although NF-κB signaling pathway is activated during PRV infection, the expression of pro-inflammatory genes was inhibited [19]. Additionally, Wang et al. found that UL24 protein could abrogate tumor necrosis factor alpha (TNF-α)-mediated NF-κB activation [20]. We previously reported that PRV could dramatically enhance the dephosphorylation of eIF2α and thus promote host cell translation efficacy to facilitate its replication [21]. Higher pathogenesis and cross-species transmission ability of PRV may partly attribute to the enhanced immune evasion of PRV variants. Despite several decades of intensive study, the underlying mechanisms of PRV pathogenesis and immunomodulation still remain elusive. Hence, it is imperative to investigate the host factors involved in virus infection.

To date, proteomics is broadly applied to hunt for host factors relevant to virus infection [22]. Various animal viruses had been subjected to proteomic analysis to dissect the host factors involved in virus infection, such as porcine epidemic diarrhea virus (PEDV) [23], porcine reproductive and respiratory syndrome virus (PRRSV) [24] and porcine delta-coronavirus [25]. Tandem Mass Tag (TMT) technology, developed and launched by Thermo, is one of the most powerful quantitative methods for protein expression analysis with the highest throughput, the lowest systematic error, and the most powerful functions. In this study, TMT-based quantitative proteomics was employed to analyze protein profiles in mock- and PRV-infected Vero cells to gain insights into the virus-host interactions.

2. Materials and Methods

2.1. Cell Lines, Viruses, Chemicals, and Antibodies

African green monkey kidney cell (Vero), the immortalized porcine alveolar macrophage (CRL-2843), and porcine kidney cell (PK-15) were all cultured in Dulbecco's modified Eagle's medium (DMEM: Invitrogen, Carlsbad, CA, USA) containing 10% (*v/v*) fetal bovine serum (FBS, Thermo Fisher, Waltham, MA, USA) in a humidified 37 °C incubator with 5% CO_2 and stored in our lab. PRV HB1201 (GenBank accession number: KU057086.1) was a variant strain isolated from a pig in He Bei in China. 4′, 6′-diamidino-2-phenylindole (DAPI) and TMT 16Plex were purchaseded from Thermo Fisher Scientific (Waltham, MA, USA). The primary antibodies used in this study were specific for XRCC5 (16389-1-AP, Proteintech, Rosemont, IL, USA), XRCC6 (10723-1-AP, Proteintech, Rosemont, IL, USA), β-actin (66009-1-Ig, Proteintech, Rosemont, IL, USA), VP5 (prepared in our lab), and gB (prepared in our lab). The HRP-labeled secondary antibodies against rabbit (ZB2301) and mouse (ZB2305) were all purchased from ZSGB-BIO (Beijing, China).

2.2. Virus Inoculation and Protein Preparation

Vero cells were grown to monolayers in 10 cm cell culture dishes and then were inoculated with PRV HB1201 at 0.1 MOI for 1 h. Sustaining culture medium DMEM containing 2% FBS was added for another 24 h. Three independent experiments were conducted as biological replicates. The protein extraction procedure is as follows: at 24 h post-inoculation (h p.i.), the medium was removed and washed with 5 mL pre-cooling PBS twice; mock- or PRV-infected Vero cells were collected using a cell scraper and piped into 1.5 mL EP tubes; protein lysate (8 M urea, 1% SDS containing protease inhibitor) was added to lyse cell membrane and sonicated for 2 min to solubilize protein further; cell lysate was used to treat protein for another 30 min on ice and centrifuged (12,000 rpm for 15 min at 4 °C) to remove cellular debris. The protein concentration was analyzed by Bradford protein assay and SDS-PAGE was performed to evaluate the overall protein quality.

2.3. Reductive Alkylation and TMT Labeling

Protein reductive alkylation and TMT labeling procedures were conducted according to the instructions as follows. Briefly, 100 µg protein was treated with triethylammonium bicarbonate buffer (TEAB) to the final concentration of 100 mM, and then Tris (2-carboxyethyl) phosphine (TCEP) was added to make the final concentration 10 mM for 60 min at 37 °C; 40 mM iodoacetamide was added to the final concentration and reacted in a dark room for 40 min at room temperature (RT); ice-cold acetone was added (v:v = 6:1) and reacted for 4 h at -20 °C, and the liquid was removed after centrifugation at $10,000 \times g$ for 20 min; sediment was dissolved with 100 µL 100 mM TEAB and digested with trypsin (m:m = 1:50) fully overnight at 37 °C; finally, TMT was added to label proteins for 2 h at RT, followed by hydroxylamine treatment for another 30 min.

2.4. Immunofluorescence Assay (IFA)

The IFA was performed according to the protocol mentioned previously [21]. In brief, Vero cells seeded on coverslips in a six-well plate over 90% confluence were inoculated with 0.1 MOI PRV HB1201; then, the inoculated cells were fixed with 3.7% paraformaldehyde at indicated time points for 10 min and permeabilized with 2% bovine serum albumin (BSA) containing 0.1% Triton X-100 for 10 min; 2% BSA was used to block cells for 30 min and primary monoclonal antibody specific for gB with 1:1000 dilution incubated cells for 1 h at RT and then washed with PBS three times; secondary antibodies were added at RT for 1 h in a humid chamber; after one wash, nucleus were stained with DAPI (Molecular Probes) for 10 min and washed with PBS five times for 5 min each; finally, the coverslips were observed with a Nikon A1 microscope or laser confocal microscope.

2.5. RNA Extraction and Real-Time PCR Analysis

Total RNAs of mock- or PRV-infected Vero cells were extracted by TRIzol reagent (Biomed, Beijing, China). The culture medium was removed and the cells in six-well plates were lysed with 750 µL TRIzol for 5 min, then 250 µL chloroform was added to separate RNA. After centrifugation at 12,000 rpm at 4 °C for 10 min, the RNA fraction was transferred into a new tube and precipitated by 0.8 volumes of isopropanol. After centrifugation for 15 min at 12,000 rpm, RNA pellets were washed twice with 75% iced ethanol and resuspended in 20 µL RNase-free H_2O. The synthesis of cDNA was performed using Fast Quant RT Kit (With gDNase) (Tian Gen Biotech, Beijing, China) according to the manufacturer's instructions. The cDNA samples were quantified by SYBR Green RT-qPCR Master Mix (Vazyme, Nanjing, China) and repeated three times. All reactions were carried out by the Bio-Rad PCR system. All primers used in this study are listed in Table 1. The mRNA abundance of GAPDH, XRCC5, and XRCC6 were detected by RT-qPCR assay using specific primer sets GAPDHF/GAPDHR, XRCC5F/XRCC5R, and XRCC6F/XRCC6R respectively.

Table 1. Primers used in this study.

Primer Name	Primer Sequence
XRCC6-F	GCTCCTTGGTGGATGAGTTT
XRCC6-R	CTTGCTGATGTGGGTCTTCA
XRCC5-F	TGACTTCCTGGATGCACTAATCGT
XRCC5-R	TTGGAGCCAATGGTCAGTCG
GAPDH-F	CCTTCCGTGTCCCTACTGCCAAC
GAPDH-R	GACGCCTGCTTCACCACCTTCT

2.6. Western Blot Analysis

PRV HB1201-infected Vero, CRL-2843, and PK-15 cells were all harvested at 24 h p.i. The cells were lysed with radioimmunoprecipitation (RIPA) lysis buffer (Beyotime, Shanghai, China) containing protease inhibitor (1 mM PMSF) for 30 min, and the supernatant was transferred to a new tube after centrifugation. The protein concentration was determined with Pierce BCA Protein Assay Kit (Thermo Fisher, Waltham, MA, USA) and separated by sodium dodecyl sulfate-polyacrylamide gel electrophoresis (SDS-PAGE). Separated protein (10 μg each channel) was transferred onto polyvinylidene difluoride (PVDF) membrane (Millipore). PVDF membranes were blocked in 5% skimmed-milk-PBST at RT for 2 h, followed by incubation with primary antibodies at 4 °C overnight. Then, the PVDF membrane were washed three times with 0.05% PBST for 5 min each at a rotator and incubated with the HRP-conjugated secondary antibodies at 1:3000 dilution. After three washes, the membranes were incubated with ECL chemiluminescence detection kit (Pierce) for 2 min, and finally exposed to a chemiluminescence apparatus (Bio-Rad, Hercules, CA, USA).

2.7. Virus Titration

Viruses were serially diluted 10-fold with DMEM containing 2% FBS and inoculated into Vero cells at 90% confluence in 96-well culture plates. 72 h p.i. or later, the virus titers were calculated based on the cytopathic effects (CPE) according to the Reed-Muench method. Virus titers were determined from at least three independent experiments.

2.8. Data Analysis

All data were processed with GraphPad Prism 6 (GraphPad Software Inc., San Diego, CA, USA). The student's *t*-test or non-parametric test was used to analyze the difference between the values of two groups. A value of $p < 0.05$ was considered statistically significant.

3. Results

3.1. Kinetics of PRV HB1201 Replication in Vero Cells

Efficient viral infection and relatively mild cell collapse are critical factors for optimal sampling. PRV HB1201 could cause severe CPE and subsequently cell collapse on Vero cells, thus relative lower MOI (MOI = 0.1) was applied to infect Vero cells. To screen the optimal time points of sampling, the kinetics of PRV replication in Vero cells were determined at various time points by $TCID_{50}$. As shown in Figure 1B, the virus titers were up to 10^8 $TCID_{50}$/mL at 24 h p.i., similar to that at 30 to 48 h p.i., indicating PRV could propagate in Vero cells efficiently, and the virus titers reached a plateau at 24 h p.i. (Figure 1B). Furthermore, IFA results showed that gB positive cells increased as the infection progressed. Notably, most cells were infected at 24 h p.i., and the gB positive cells decreased after 30 h p.i. due to excessive cell collapse (Figure 1A). Meanwhile, the CPE was observed microscopically at various time points. Compared with mock-infected cells, PRV-Infected cells developed slightly visible CPE at 12 h p.i. and CPE were fairly apparent at 24 h p.i. (Figure 1A). Cell collapse soars from 30 h p.i., and many of the cells were detached and floated in the medium. In addition, the expression of viral capsid protein VP5 was detected by western blot. The level of VP5 increased gradually as infection progressed (Figure 1C). However, VP5 expression level decreased slightly at 30 h p.i. compared to that at 24 and 18 h p.i. This may result from cell detachment and virus release into the medium (Figure 1A).

Based on the results above, Vero cells infected with 0.1 MOI PRV for 24 h p.i. were regarded as optimal sampling time points and subjected to the following proteomic analysis.

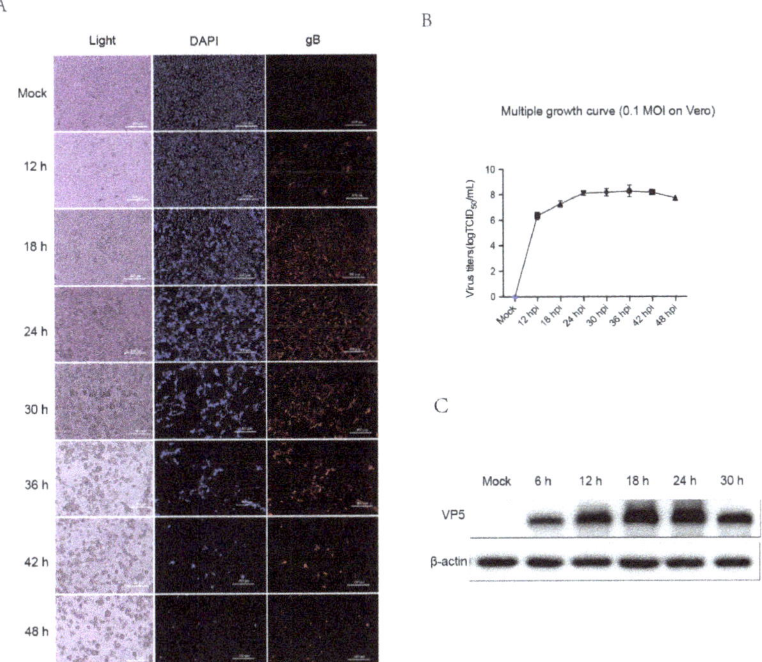

Figure 1. The replication of PRV variant HB1201 in Vero cells. (**A**) PRV at 0.1 MOI infects Vero cells and CPE was observed in microscopy at various time points. Meanwhile, IFA was also applied to view the efficiency of virus infection with antibody against gB protein. (**B**) The whole cells infected with 0.1 MOI virus were collected at indicated time points and tittered by $TCID_{50}$. (**C**) Vero cells infected with 0.1 MOI PRV were collected and the whole cell lysis was subjected to western blot analysis to detect the expression of VP5.

3.2. Protein Profiles Determined by TMT/MS Analysis

Proteomics is a systematic approach to study the virus-host interactions. To identify the differentially expressed proteins (DEPs) between mock- and PRV-infected cells, TMT-based quantitative proteomic analysis was performed, and the workflow is shown as Figure 2A. A total of 7072 cellular proteins were identified and quantified at 24 h p.i., among which 91 proteins were significantly upregulated and 879 proteins were downregulated compared to those in mock-infected Vero cells (Figure 2B) according to the criteria (p-value < 0.05 and fold change >1.5 or fold change <0.67). In addition, the top 20 upregulated and top 20 downregulated proteins are listed in Tables 2 and 3, respectively. Three technical replicates were carried out to improve the reliability of our data.

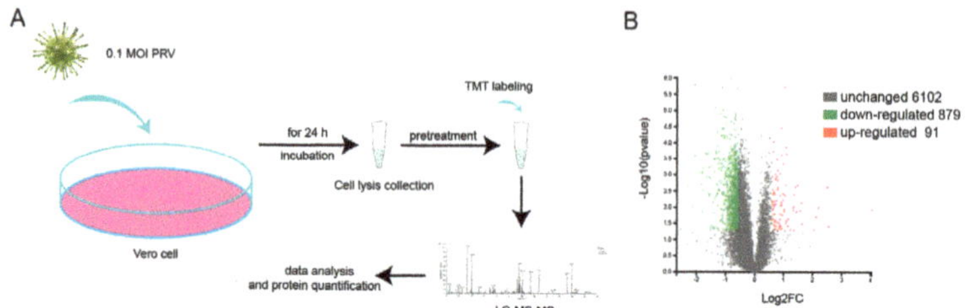

Figure 2. Overview of proteomic analysis procedure and DEPs. (**A**) The workflow of proteomic analysis. Vero cells were infected with 0.1 MOI PRV for 24 h and the whole cells lysis was collected after centrifugation at 12,000 rpm for 5 min. After reductive alkylation, the protein was labeled with TMT. Finally, the samples were subjected to liquid chromatography tandem mass spectrometry and bioinformatics analysis. (**B**) A total of 7022 proteins were identified, among which 89 proteins were markedly upregulated (red dots), 879 proteins were downregulated (green dots), and the remaining 6102 proteins stayed constant (gray dots). Proteins were considered significantly differently expressed when p value was less than 0.05 and fold change was less than 0.67 or more than 1.5 in this study.

3.3. Validation of TMT/MS Data by Western Blot and RT-qPCR

To verify TMT/MS data, X-ray repair cross-complementing protein 5 (XRCC5) and X-ray repair cross-complementing protein 6 (XRCC6) were analyzed by western blot in both mock- and PRV-infected Vero cells. The two proteins were selected for validation for the following reasons: they were downregulated significantly; they were closely related and involved in DNA repair process, which was a general cellular response during herpes virus infection [26]; and antibodies against them were commercially available. Western blot results showed that the protein level of XRCC5 and XRCC6 both decreased in PRV-infected Vero cells (Figure 3A). Then, Image J software was applied to quantify protein levels, and the ratios of the XRCC5 and XRCC6 between mock-and PRV-infected Vero cells coincided with proteomic data (Figure 3B). Moreover, the levels of XRCC5 and XRCC6 in PK-15 and CRL-2843 cells were also reduced (Figure 3E,F), indicating PRV-mediated XRCC5 and XRCC6 reduction was in a cell type-independent manner.

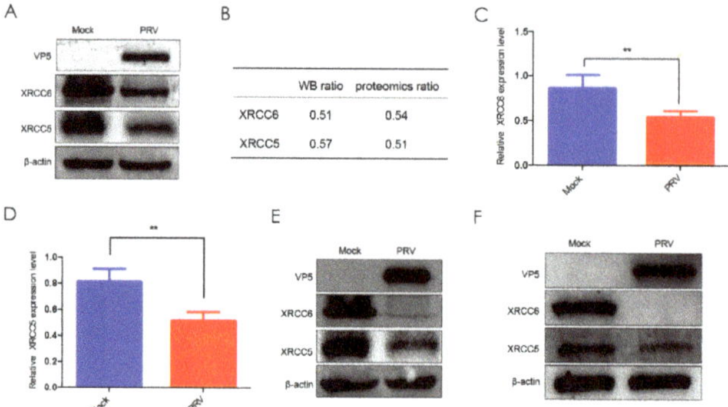

Figure 3. Validation of proteomics data by western blot and RT-qPCR. (**A**) Vero cells infected with PRV for 24 h were collected and western blot was performed to detect the expression of XRCC5 and XRCC6 with corresponding antibodies. (**B**) The western blot and proteomics ratio of XRCC5 and XRCC6. (**C**) Relative XRCC6 transcription in Vero cells. (**D**) Relative XRCC5 transcription in Vero cells. (**E**) The expression of XRCC5 and XRCC6 in PK-15 infected with PRV. (**F**) The expression of XRCC5 and XRCC6 in CRL-2843 infected with PRV. ** indicates significance at a 99% confidence interval ($p < 0.01$).

Table 2. Top 20 up-regulated proteins.

Accession	Description	FC (P_24h/M_24h)	*p* Value (P_24h/M_24h)	Significant
XP_007997295.1	4-hydroxybenzoate polyprenyltransferase, mitochondrial	16.440879	0.0111	Yes
XP_008000412.1	ATP synthase subunit gamma, mitochondrial isoform X2	5.82243	0.03774	Yes
XP_007964526.1	non-homologous end-joining factor 1	5.638689	0.002289	Yes
XP_007966188.1	transmembrane 7 superfamily member 3 isoform X1	4.570457	0.009848	Yes
XP_007985885.1	DNA-directed RNA polymerase III subunit RPC5 isoform X1	4.318305	0.01897	Yes
XP_008001057.1	proton myo-inositol cotransporter	4.10077	0.04733	Yes
XP_007978034.1	hemoglobin subunit alpha	3.092117	0.03053	Yes
XP_007959284.1	bromodomain-containing protein 9 isoform X1	3.054434	0.006955	Yes
XP_008010294.1	testis-expressed sequence 2 protein isoform X1	2.974047	0.01225	Yes
XP_007995408.1	relA-associated inhibitor isoform X1	2.802074	0.00199	Yes
XP_008014790.1	tropomodulin-2 isoform X1	2.732538	0.01336	Yes
XP_007965594.1	myeloid leukemia factor 2	2.647029	0.002251	Yes
XP_007958764.1	kinesin-like protein KIF16B isoform X1	2.550698	0.003898	Yes
XP_007958522.1	conserved oligomeric Golgi complex subunit 3	2.549788	0.04741	Yes
XP_007997053.1	serum albumin	2.544889	0.009266	Yes
XP_007977282.1	elongation of very long chain fatty acids protein 1	2.521127	0.01201	Yes
XP_008008965.1	vitronectin	2.339382	0.04719	Yes
XP_008012665.1	calcium signal-modulating cyclophilin ligand	2.218158	0.02212	Yes
XP_007995769.1	splicing factor, arginine/serine-rich 19 isoform X1	2.208965	0.037	Yes
XP_008014703.1	E3 ubiquitin-protein ligase NEDD4 isoform X3	2.12792	0.000212	Yes

In addition, the transcription level of XRCC5 and XRCC6 were also analyzed by RT-qPCR. Consistently, the mRNA level of XRCC5 and XRCC6 markedly decreased in virus-infected Vero cells compared with that in mock-infected cells (Figure 3C,D), suggesting that PRV-mediated XRCC5 and XRCC6 downregulation might result from transcription inhibition.

3.4. GO Analysis of The DEPs

GO annotation analysis could classify the tested proteins in three aspects: biological process (BP), cellular component (CC), and molecular function (MF). To dissect the function of DEPs, GO functional analysis revealed that 89 upregulated proteins and 871 downregulated proteins were involved in 12 biological processes (Figure 3A), including cellular processes, single-organism processes, metabolic processes, biological processes, regulation of biological processes, and so on; within the CC category, the DEPs were well distributed in different cell components, including cell parts, cells, organelle, organelle parts, and so on; in the MF category, the DEPs were involved in binding function.

Table 3. Top 20 down-regulated proteins.

Accession	Description	FC (P_24h/M_24h)	*p* Value (P_24h/M_24h)	Significant
XP_007995562.1	glioma tumor suppressor candidate region gene 2 protein	0.156304	0.007612	Yes
XP_008007884.1	UAP56-interacting factor isoform X1	0.171554	0.04785	Yes
XP_007975472.1	thioredoxin-interacting protein	0.202567	0.002175	Yes
XP_007971461.1	tripartite motif-containing protein 40	0.2282	0.009207	Yes
XP_008016505.1	general transcription factor II-I	0.237577	0.000002	Yes
XP_008016152.1	wolframin	0.239067	0.000248	Yes
XP_007972702.1	structural maintenance of chromosomes flexible hinge domain-containing protein 1 isoform X1	0.263824	0.000806	Yes
XP_007980315.1	epsilon-sarcoglycan isoform X1	0.280037	0.000878	Yes
XP_007966031.1	zinc finger protein AEBP2 isoform X1	0.280576	0.001929	Yes
XP_008013602.1	histone-lysine N-methyltransferase, H3 lysine-36 and H4 lysine-20 specific isoform X1	0.284503	0.01345	Yes
XP_007962440.1	regulator of G-protein signaling 10 isoform X1	0.287038	0.000822	Yes
XP_007982422.1	vasorin	0.289883	0.000601	Yes
XP_007977511.1	probable U3 small nucleolar RNA-associated protein 11 isoform X1	0.289894	0.002949	Yes
XP_008001366.1	bax inhibitor 1	0.311135	0.000205	Yes
XP_007962616.1	antigen KI-67 isoform X1	0.316703	0.001139	Yes
XP_007960624.1	ribosome biogenesis protein BMS1 homolog	0.323469	0.000212	Yes
XP_008016787.1	zinc finger and SCAN domain-containing protein 21 isoform X2	0.328314	0.003606	Yes
XP_008012144.1	alpha-protein kinase 2 isoform X1	0.332672	0.00059	Yes
XP_008013174.1	treacle protein isoform X1	0.334286	0.004324	Yes
XP_008007268.1	solute carrier organic anion transporter family member 2A1	0.338268	0.001218	Yes

In addition, GO enrichment analysis demonstrated that DEPs were mostly enriched in chaperonin-containing T-complex within the CC category. Furthermore, the majority of DEPs were enriched in the BP category, such as NIK/NF-κB signaling, Fc-epsilon receptor signaling pathway, negative regulation of G2/M transition of mitotic cell cycle, and innate immune response activating cell surface receptor signaling pathway (Figure 4B).

3.5. KEGG Functional Annotation of DEPs

KEGG pathway analysis was performed to further explore the underlying signaling pathways or functions among DEPs. As shown in Figure 5A, the 89 upregulated proteins participated in 32 pathways, and the top three were related to the immune system, signal transduction, and cancer. Meanwhile, the 871 downregulated proteins were involved in 44 pathways, and the top three were the "folding, sorting, and degradation of protein", signal transduction, and translation (Figure 5B).

KEGG enrichment analysis were also conducted to analyze the enriched signaling pathways in DEPs. Among all 970 DEPs, 20 pathways were significantly enriched, and the top three were proteasome, amino sugar and nucleotide sugar metabolism, and RNA polymerase (Figure 5C).

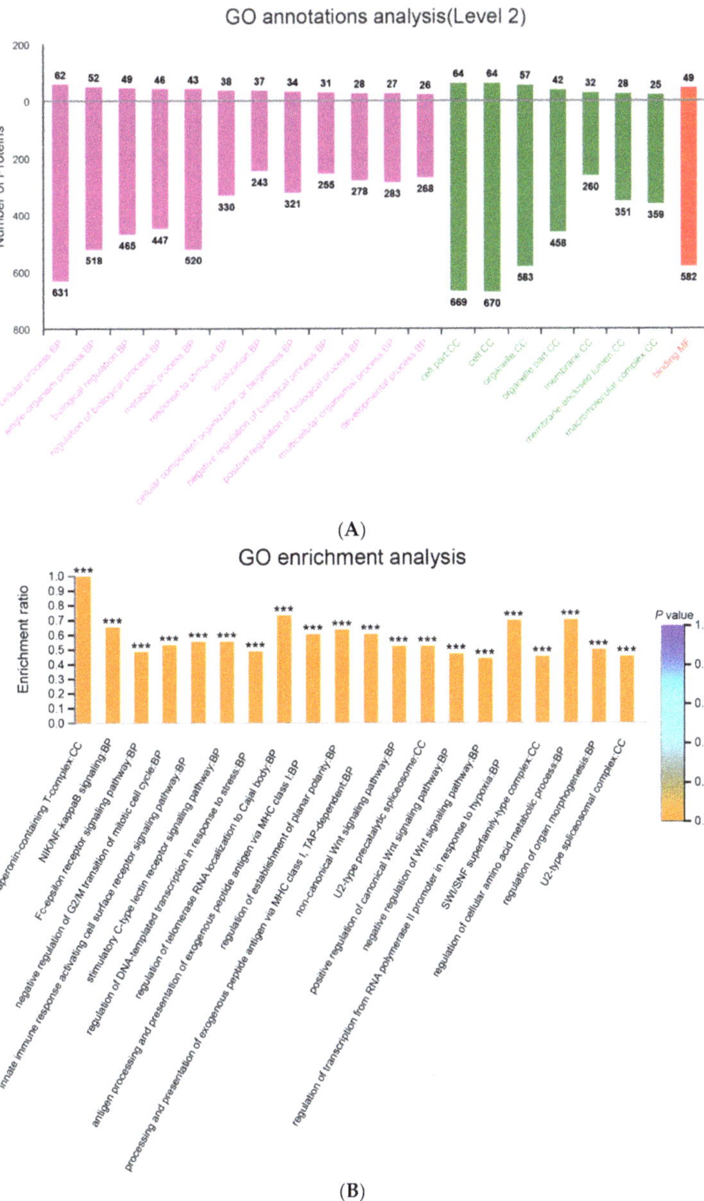

(**A**)

(**B**)

Figure 4. GO annotation and GO enrichment analysis of DEPs between mock- and PRV-infected Vero cells. (**A**) Up- and downregulated proteins are classified into three categories, respectively, by GO analysis: biological process (BP), cellular component (CC), and molecular function (MF). The x-axis represents the specific categories in BP, CC, and MF. The numbers on the y-axis indicate proteins in the category. (**B**) DEPs were subjected to GO enrichment analysis and the top 20 GO terms are listed on the x-axis. The y-axis indicates the enrichment ratio of DEPs and different colors represent different *p* values. *** indicates significance at a 99.9% confidence interval ($p < 0.001$).

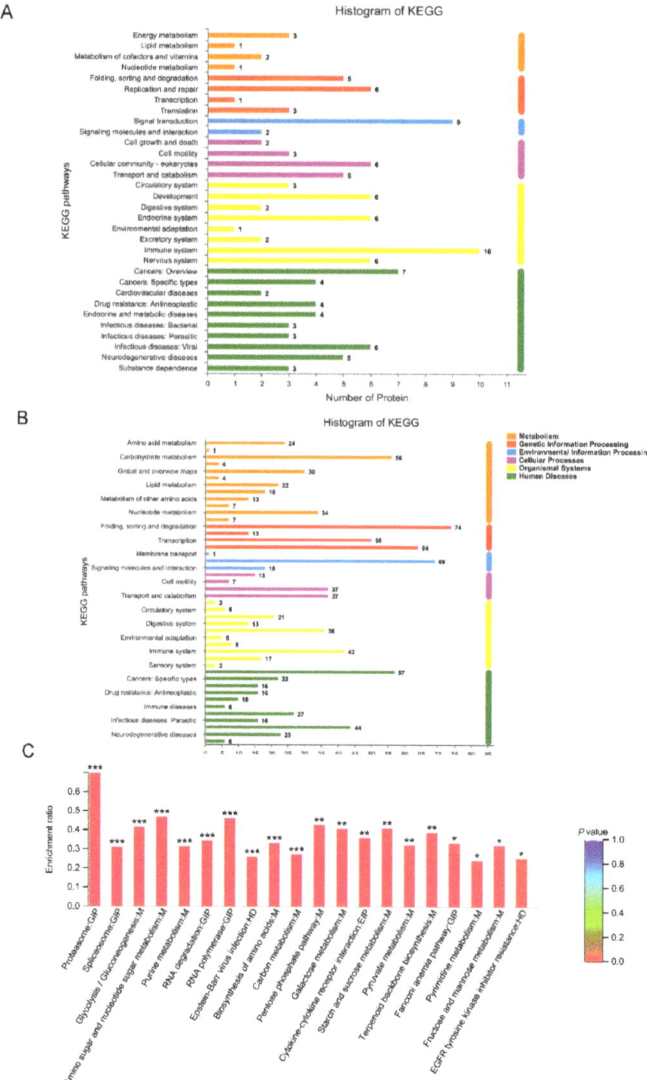

Figure 5. (**A**) The 89 upregulated proteins are classified into six main categories by KEGG analysis: metabolism, genetic information processing, environmental information processing, cellular processes, organismal systems, and human diseases. The x-axis indicates the numbers of proteins within particular categories. The y-axis indicates the specific pathways within six main categories. (**B**) The 871 downregulated proteins are classified in the same manner as Figure A: (**C**) KEGG pathway enrichment analysis of DEPs. The x-axis indicates the name of the KEGG pathways, the y-axis indicates the enrichment ratio (there were no apparent differences between 0.01 and 0.05, so the colors look similar and can't be distinguished by the naked eye). * indicates significance at a 95% confidence interval ($p < 0.05$), ** indicates significance at a 99% confidence interval ($p < 0.01$), *** indicates significance at a 99.9% confidence interval ($p < 0.001$).

3.6. COG Annotation of DEPs

The COG database is able to predicate the function of proteins based on protein sequence. To categorize the functions of DEPs, COG analysis was performed. As shown in

Figure 6 (left panel) 10 categories were involved in upregulated proteins. In particular, seven proteins were related to posttranslational modification, protein turnover, and chaperones; four proteins were classified into general function prediction only; three proteins were related to replication, recombination, and repair; two proteins were relevant to energy production and conversion, transcription, intracellular trafficking, secretion, vesicular transport, and so on. In addition, 22 categories were involved in 879 downregulated proteins: 61 proteins were related to posttranslational modification, protein turnover, and chaperones; 46 proteins were relevant to translation, ribosomal structure, and biogenesis; 39 proteins were classified into general function prediction only shown in Figure 6 (right panel). Further research is imperative to characterize the involvement of these categories during PRV infection.

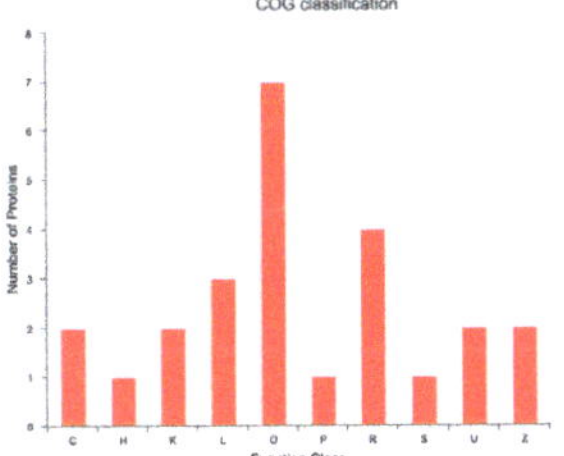
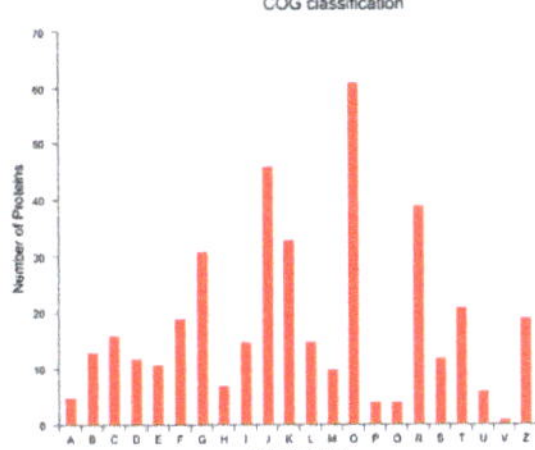

Figure 6. The COG annotation of significantly upregulated proteins (**left panel**) and downregulated proteins (**right panel**). The x-axis indicates the COG functional classification (presented with A to Z): (A) RNA processing and modification; (B) Chromatin structure and dynamics; (C) Energy production and conversion; (D) Cell cycle control, cell division, and chromosome partitioning; (E) Amino acid transport and metabolism; (F) Nucleotide transport and metabolism; (G) Carbohydrate transport and metabolism; (H) Coenzyme transport and metabolism; (I) Lipid transport and metabolism; (J) Translation, ribosomal structure, and biogenesis; (K) Transcription; (L) Replication, recombination, and repair; (M) Cell wall/membrane/envelope biogenesis; (O) Posttranslational modification, protein turnover, and chaperones; (P) Inorganic ion transport and metabolism; (Q) Secondary metabolites biosynthesis, transport, and catabolism; (R) General function prediction only; (S) Function unknown; (T) Signal transduction mechanisms; (U) Intracellular trafficking, secretion, and vesicular transport; (V) Defense mechanisms; (Z) Cytoskeleton. The y-axis indicates the protein number of particular functional classifications.

4. Discussion

PRV variant HB1201 exhibits higher pathogenicity, and its pathogenesis remains poorly defined. Nowadays, proteomics has been broadly used in profiling cellular protein expression patterns in virus-infected cells. In this paper, a TMT-based quantitative proteomics approach was applied, and we revealed striking protein profile shifts in PRV-infected Vero cells compared with those in mock-infected cells.

In the present study, a total of 7072 proteins were identified in whole Vero cells, among which 980 proteins were differentially expressed at 24 h p.i. Among the top 20 upregulated proteins, non-homologous end joining-1 (NHEJ-1) markedly induced an over five-fold change (Table 2), which is reported to be involved in DNA repair [27]. DNA viruses replicates their genomes in the nuclei of cells, and the mass accumulation of viral DNA genome in the nucleus may trigger host cell DNA damage responses. For example, intensive studies showed that herpes virus may engage components of DNA damage response to enhance its replication, while some of the DNA repair components are antiviral [28–30]. In our analysis, the expression of two DNA repair-related proteins, XRCC5 (Ku80) and XRCC6 (Ku70), were both shown to be reduced in PRV-infected cells. Moreover, western blot and RT-qPCR results supported the proteomic data at both protein and transcription level. XRCC5 and XRCC6 comprise the heterodimer, which recognizes and binds to double strand DNA break ends, and then promotes non-homologous end joining [30] or induces innate immune defenses against DNA virus infection [31,32]. Previous reports showed that XRCC6 not only modulated human T lymphotropic virus type 1 (HTLV-1) replication [33], but also regulated DNA virus-mediated innate immune response [34]. However, the expression of XRCC6 was significantly upregulated in HTLV-1-infected cells compared with PRV. We hold that PRV, with its larger genome, encodes more proteins and evolves more sophisticated strategies to evade host immune clearance by targeting XRCC5 and XRCC6. As compensation, many other tricks have been developed instead by HTLV-1. For example, HTLV-1 Tax could impair K63-linked ubiquitination of STING to evade host innate immunity [35]; HTLV-1 Tax blocks IRF3 phosphorylation through the interaction with and inhibition of TBK1 kinase [36]. These results above indicated that the DNA damage repair signaling pathway might be closely related to virus infection.

Our previous study showed that PRV infection induced the phosphorylation of PERK; however, the expression of GRP78 stayed unaltered [21], indicating that other host factors might alleviate the intensity of unfolded protein responses (UPR). The endoplasmic reticulum (ER) is a major factor of glycoprotein synthesis, and the excessive expression of glycoprotein may activate UPR [37]. According to proteomic data, the expression of seven-transmembrane superfamily member 3 (TM7SF3), engaged in the attenuation of cellular stress and the subsequent UPR [38], was significantly induced. TM7SF3 is a downstream target of p53 [38], which is involved in innate immune response regulation, cell cycling, DNA repair, and apoptosis [39,40]. Although Xun Li et al. reported that overexpression of p53 positively regulated PRV replication both in vivo and in vitro [41], many questions were still hanging in the air: for example, the expression level and activation status of p53 during PRV infection and its contributions to TM7SF3 overexpression. Most importantly, the biological significance of TM7SF3 on UPR and virus replication were imperative to be elucidated.

Innate immune response, particularly type I interferon production and inflammatory cytokines secretion, is the first line to fight against pathogen invasion. PRRs recognize pathogen-associated molecular patterns (PAMPs) and then trigger innate immune responses. PRV is a common pathogen in multiple animal species and has even been isolated from human patients [42], thus attention should also be paid to the protein profile shifts in Vero lines. Vero cells are type I interferon-deficient, so inflammatory responses are emphasized in this paper. Our results showed that the NIK/NF-κB signaling pathway was markedly enriched by GO enrichment analysis. It was reported that the virulent PRV variant induced substantial lethal inflammatory response by TRL2, while the attenuated live vaccine of PRV lost the ability to activate an inflammatory response [13,43]. The abnormal inflammatory responses mediated by PRV variants might contribute to its pathogenicity. KEGG enrichment analysis showed that DEPs were significantly enriched in proteasome (Figure 5C); in particular, 28 DEPs were relevant to proteasome. Proteasome was reported to shape innate immune response and regulate the production of inflammatory cytokines [44]. Therefore, we proposed that PRV could modulate inflammatory response via regulating 26s proteasome non-ATPase regulatory subunits and other proteasome-related proteins expression. During PRV propagation, there are amounts of viral proteins synthesized in cells. We hold that the proteasome-related proteins may also be involved in useless or damaged protein degradation to maintain cellular homeostasis. In addition, cells may recruit proteasome to degrade viral proteins by ubiquitinating them and achieve an antiviral during PRV infection. Furthermore, the PRRs regulator tripartite motif-containing protein 40 (TRIM40) was also significantly down-regulated, indicating PRV might subvert innate immune responses by inhibiting PRRs activation. Additionally, the enzymatic activity of cGAS is tightly regulated by XRCC5 and XRCC6 to maintain immune homeostasis [45,46]. PRV is enveloped, and multiple processes all require lipids, such as virus-cell membrane fusion and virus budding. Our results showed that the elongation of very long chain fatty acids protein 1 (ELOVL1), involved in unsaturated fatty acid biosynthetic process [47], was significantly upregulated during PRV infection. This suggests that cellular lipids metabolism may take part in PRV propagation and pathogenesis. In addition, E3 ubiquitin-protein ligase NEDD4 was also significantly upregulated in PRV-infected cells. NEDD4 is essential for neural development and homeostasis of neural circuit excitability during neuronal ER stress [48], indicating this may be a protective mechanism to maintain cell homeostasis and normal biological functions during PRV infection. Taken together, the interactions between PRV infection and innate immune responses are complex and need further investigation.

This study systematically analyzed the protein profiles of PRV-infected Vero cells using a TMT-based proteomic analysis method. Eighty-nine upregulated and 871 downregulated proteins were identified, and biological analysis demonstrated that various cellular processes were involved in PRV-infected cells, including cellular processes, single-organism processes, metabolic processes, biological processes, regulation of biological processes, and so on. Unfortunately, our analysis of DEPs remains only instructional, and the elucidation of their biological functions is required. This research will help to deepen the understanding of the virus pathogenesis and host immune responses.

Author Contributions: Conceptualization, X.Y., S.X. and X.G. (Xin Guo); methodology, X.Y., S.X.; software, S.X.; validation, X.Y., S.X. and X.G. (Xin Guo); formal analysis, S.X. and X.Y.; investigation, X.Y.; resources, X.G. (Xin Guo); data curation, S.X.; writing—original draft preparation, S.X., and X.Y.; writing—review and editing, X.Y., S.X., D.C., R.J., H.K., X.G. (Xin Guo), L.Z., J.H., Y.Z., H.Y. and X.G. (Xinna Ge); funding acquisition, X.G. (Xin Guo) All authors have read and agreed to the published version of the manuscript.

Funding: This research was funded by Natural Science Foundation of Beijing Municipal (Number: 6192014).

Institutional Review Board Statement: Not applicable for studies not involving humans or animals.

Informed Consent Statement: Not applicable.

Data Availability Statement: The proteomic data is available with the link: https://pan.baidu.com/s/12D_7pNG1VP79E_aNfgi-LQ (Password: 39ht, accessed on 1 March 2022).

Acknowledgments: This research was supported by Natural Science Foundation of Beijing Municipal (Number: 6192014).

Conflicts of Interest: The authors declare that there are no conflict of interest.

References

1. Freuling, C.M.; Muller, T.F.; Mettenleiter, T.C. Vaccines against pseudorabies virus (PrV). *Vet. Microbiol.* **2017**, *206*, 3–9. [CrossRef] [PubMed]
2. Ruiz-Fons, F.; Vidal, D.; Hofle, U.; Vicente, J.; Gortazar, C. Aujeszky's disease virus infection patterns in European wild boar. *Vet. Microbiol.* **2007**, *120*, 241–250. [CrossRef] [PubMed]
3. Muller, T.; Hahn, E.C.; Tottewitz, F.; Kramer, M.; Klupp, B.G.; Mettenleiter, T.C.; Freuling, C. Pseudorabies virus in wild swine: A global perspective. *Arch. Virol.* **2011**, *156*, 1691–1705. [CrossRef] [PubMed]
4. Pomeranz, L.E.; Reynolds, A.E.; Hengartner, C.J. Molecular Biology of Pseudorabies Virus: Impact on Neurovirology and Veterinary Medicine. *Microbiol. Mol. Biol. Rev.* **2005**, *69*, 462–500. [CrossRef]
5. Yan, K.; Liu, J.; Guan, X.; Yin, Y.X.; Peng, H.; Chen, H.C.; Liu, Z.F. The Carboxyl Terminus of Tegument Protein pUL21 Contributes to Pseudorabies Virus Neuroinvasion. *J. Virol.* **2019**, *93*, e02052-18. [CrossRef]
6. An, T.Q.; Peng, J.M.; Tian, Z.J.; Zhao, H.Y.; Li, N.; Liu, Y.M.; Chen, J.Z.; Leng, C.L.; Sun, Y.; Chang, D.; et al. Pseudorabies virus variant in Bartha-K61-vaccinated pigs, China, 2012. *Emerg. Infect. Dis.* **2013**, *19*, 1749–1755. [CrossRef]
7. Yu, X.; Zhou, Z.; Hu, D.; Zhang, Q.; Han, T.; Li, X.; Gu, X.; Yuan, L.; Zhang, S.; Wang, B.; et al. Pathogenic pseudorabies virus, China, 2012. *Emerg. Infect. Dis.* **2014**, *20*, 102–104. [CrossRef]
8. Tong, W.; Liu, F.; Zheng, H.; Liang, C.; Zhou, Y.J.; Jiang, Y.F.; Shan, T.L.; Gao, F.; Li, G.X.; Tong, G.Z. Emergence of a Pseudorabies virus variant with increased virulence to piglets. *Vet. Microbiol.* **2015**, *181*, 236–240. [CrossRef]
9. Ren, J.; Wang, H.; Zhou, L.; Ge, X.; Guo, X.; Han, J.; Yang, H. Glycoproteins C and D of PRV Strain HB1201 Contribute Individually to the Escape From Bartha-K61 Vaccine-Induced Immunity. *Front. Microbiol.* **2020**, *11*, 323. [CrossRef]
10. He, W.; Auclert, L.Z.; Zhai, X.; Wong, G.; Zhang, C.; Zhu, H.; Xing, G.; Wang, S.; He, W.; Li, K.; et al. Interspecies Transmission, Genetic Diversity, and Evolutionary Dynamics of Pseudorabies Virus. *J. Infect. Dis.* **2019**, *219*, 1705–1715. [CrossRef]
11. Yang, H.; Han, H.; Wang, H.; Cui, Y.; Liu, H.; Ding, S. A Case of Human Viral Encephalitis Caused by Pseudorabies Virus Infection in China. *Front. Neurol.* **2019**, *10*, 534. [CrossRef] [PubMed]

12. Wang, Y.; Nian, H.; Li, Z.; Wang, W.; Wang, X.; Cui, Y. Human encephalitis complicated with bilateral acute retinal necrosis associated with pseudorabies virus infection: A case report. *Int. J. Infect. Dis.* **2019**, *89*, 51–54. [CrossRef] [PubMed]

13. Liu, Q.; Wang, X.; Xie, C.; Ding, S.; Yang, H.; Guo, S.; Li, J.; Qin, L.; Ban, F.; Wang, D.; et al. A novel human acute encephalitis caused by pseudorabies virus variant strain. *Clin. Infect. Dis.* **2021**, *73*, e3690–e3700. [CrossRef]

14. Cheng, Z.; Kong, Z.; Liu, P.; Fu, Z.; Zhang, J.; Liu, M.; Shang, Y. Natural infection of a variant pseudorabies virus leads to bovine death in China. *Transbound Emerg Dis* **2020**, *67*, 518–522. [CrossRef] [PubMed]

15. Lian, K.; Zhang, M.; Zhou, L.; Song, Y.; Wang, G.; Wang, S. First report of a pseudorabies-virus-infected wolf (*Canis lupus*) in China. *Arch. Virol.* **2020**, *165*, 459–462. [CrossRef] [PubMed]

16. Nelemans, T.; Kikkert, M. Viral Innate Immune Evasion and the Pathogenesis of Emerging RNA Virus Infections. *Viruses* **2019**, *11*, 961. [CrossRef] [PubMed]

17. Qin, C.; Zhang, R.; Lang, Y.; Shao, A.; Xu, A.; Feng, W.; Han, J.; Wang, M.; He, W.; Yu, C.; et al. Bclaf1 critically regulates the type I interferon response and is degraded by alphaherpesvirus US3. *PLoS Pathog.* **2019**, *15*, e1007559. [CrossRef] [PubMed]

18. Zhang, R.; Xu, A.; Qin, C.; Zhang, Q.; Chen, S.; Lang, Y.; Wang, M.; Li, C.; Feng, W.; Zhang, R.; et al. Pseudorabies Virus dUTPase UL50 Induces Lysosomal Degradation of Type I Interferon Receptor 1 and Antagonizes the Alpha Interferon Response. *J. Virol.* **2017**, *91*, e01148-17. [CrossRef]

19. Romero, N.; Van Waesberghe, C.; Favoreel, H.W. Pseudorabies Virus Infection of Epithelial Cells Leads to Persistent but Aberrant Activation of the NF-κB Pathway, Inhibiting Hallmark NF-κB-Induced Proinflammatory Gene Expression. *J. Virol.* **2020**, *94*, e00196-20. [CrossRef]

20. Wang, T.Y.; Yang, Y.L.; Feng, C.; Sun, M.X.; Peng, J.M.; Tian, Z.J.; Tang, Y.D.; Cai, X.H. Pseudorabies Virus UL24 Abrogates Tumor Necrosis Factor Alpha-Induced NF-κB Activation by Degrading P65. *Viruses* **2020**, *12*, 51. [CrossRef]

21. Xu, S.; Chen, D.; Chen, D.; Hu, Q.; Zhou, L.; Ge, X.; Han, J.; Guo, X.; Yang, H. Pseudorabies virus infection inhibits stress granules formation via dephosphorylating eIF2α. *Vet. Microbiol.* **2020**, *247*, 108786. [CrossRef] [PubMed]

22. Munday, D.C.; Surtees, R.; Emmott, E.; Dove, B.K.; Digard, P.; Barr, J.N.; Whitehouse, A.; Matthews, D.; Hiscox, J.A. Using SILAC and quantitative proteomics to investigate the interactions between viral and host proteomes. *Proteomics* **2012**, *12*, 666–672. [CrossRef] [PubMed]

23. Zeng, S.; Zhang, H.; Ding, Z.; Luo, R.; An, K.; Liu, L.; Bi, J.; Chen, H.; Xiao, S.; Fang, L. Proteome analysis of porcine epidemic diarrhea virus (PEDV)-infected Vero cells. *Proteomics* **2015**, *15*, 1819–1828. [CrossRef] [PubMed]

24. Zhao, F.; Fang, L.; Wang, D.; Song, T.; Wang, T.; Xin, Y.; Chen, H.; Xiao, S. SILAC-based quantitative proteomic analysis of secretome of Marc-145 cells infected with porcine reproductive and respiratory syndrome virus. *Proteomics* **2016**, *16*, 2678–2687. [CrossRef] [PubMed]

25. Zhou, X.; Zhou, L.; Ge, X.; Guo, X.; Han, J.; Zhang, Y.; Yang, H. Quantitative Proteomic Analysis of Porcine Intestinal Epithelial Cells Infected with Porcine Deltacoronavirus Using iTRAQ-Coupled LC-MS/MS. *J. Proteome Res.* **2020**, *19*, 4470–4485. [CrossRef] [PubMed]

26. Volcy, K.; Fraser, N.W. DNA damage promotes herpes simplex virus-1 protein expression in a neuroblastoma cell line. *J. Neurovirol.* **2013**, *19*, 57–64. [CrossRef]

27. Chang, H.H.Y.; Pannunzio, N.R.; Adachi, N.; Lieber, M.R. Non-homologous DNA end joining and alternative pathways to double-strand break repair. *Nat. Rev. Mol. Cell Biol.* **2017**, *18*, 495–506. [CrossRef]

28. Full, F.; Ensser, A. Early Nuclear Events after Herpesviral Infection. *J. Clin. Med.* **2019**, *8*, 1408. [CrossRef]

29. Smith, S.; Weller, S.K. HSV-I and the cellular DNA damage response. *Future Virol.* **2015**, *10*, 383–397. [CrossRef]

30. Li, Z.; Pearlman, A.H.; Hsieh, P. DNA mismatch repair and the DNA damage response. *DNA Repair* **2016**, *38*, 94–101. [CrossRef]

31. Dempsey, A.; Bowie, A.G. Innate immune recognition of DNA: A recent history. *Virology* **2015**, *479–480*, 146–152. [CrossRef]

32. Ferguson, B.J.; Mansur, D.S.; Peters, N.E.; Ren, H.; Smith, G.L. DNA-PK is a DNA sensor for IRF-3-dependent innate immunity. *eLife* **2012**, *1*, e00047. [CrossRef] [PubMed]

33. Wang, J.; Kang, L.; Song, D.; Liu, L.; Yang, S.; Ma, L.; Guo, Z.; Ding, H.; Wang, H.; Yang, B. Ku70 Senses HTLV-1 DNA and Modulates HTLV-1 Replication. *J. Immunol.* **2017**, *199*, 2475–2482. [CrossRef] [PubMed]

34. Sui, H.; Zhou, M.; Imamichi, H.; Jiao, X.; Sherman, B.T.; Lane, H.C.; Imamichi, T. STING is an essential mediator of the Ku70-mediated production of IFN-λ1 in response to exogenous DNA. *Sci. Signal.* **2017**, *10*, eaah5054. [CrossRef] [PubMed]

35. Wang, J.; Yang, S.; Liu, L.; Wang, H.; Yang, B. HTLV-1 Tax impairs K63-linked ubiquitination of STING to evade host innate immunity. *Virus Res.* **2017**, *232*, 13–21. [CrossRef] [PubMed]

36. Yuen, C.K.; Chan, C.P.; Fung, S.Y.; Wang, P.H.; Wong, W.M.; Tang, H.V.; Yuen, K.S.; Chan, C.P.; Jin, D.Y.; Kok, K.H. Suppression of Type I Interferon Production by Human T-Cell Leukemia Virus Type 1 Oncoprotein Tax through Inhibition of IRF3 Phosphorylation. *J. Virol.* **2016**, *90*, 3902–3912. [CrossRef]

37. Johnston, B.P.; Pringle, E.S.; McCormick, C. KSHV activates unfolded protein response sensors but suppresses downstream transcriptional responses to support lytic replication. *PLoS Pathog.* **2019**, *15*, e1008185. [CrossRef]

38. Isaac, R.; Goldstein, I.; Furth, N.; Zilber, N.; Streim, S.; Boura-Halfon, S.; Elhanany, E.; Rotter, V.; Oren, M.; Zick, Y. TM7SF3, a novel p53-regulated homeostatic factor, attenuates cellular stress and the subsequent induction of the unfolded protein response. *Cell Death Differ.* **2017**, *24*, 132–143. [CrossRef]

39. Aloni-Grinstein, R.; Charni-Natan, M.; Solomon, H.; Rotter, V. p53 and the Viral Connection: Back into the Future. *Cancers* **2018**, *10*, 178. [CrossRef]

40. Mehrbod, P.; Ande, S.R.; Alizadeh, J.; Rahimizadeh, S.; Shariati, A.; Malek, H.; Hashemi, M.; Glover, K.K.M.; Sher, A.A.; Coombs, K.M.; et al. The roles of apoptosis, autophagy and unfolded protein response in arbovirus, influenza virus, and HIV infections. *Virulence* **2019**, *10*, 376–413. [CrossRef]
41. Li, X.; Zhang, W.; Liu, Y.; Xie, J.; Hu, C.; Wang, X. Role of p53 in pseudorabies virus replication, pathogenicity, and host immune responses. *Vet. Res.* **2019**, *50*, 9. [CrossRef] [PubMed]
42. Laval, K.; Vernejoul, J.B.; Van Cleemput, J.; Koyuncu, O.O.; Enquist, L.W. Virulent Pseudorabies Virus Infection Induces a Specific and Lethal Systemic Inflammatory Response in Mice. *J. Virol.* **2018**, *92*, e01614-18. [CrossRef] [PubMed]
43. Laval, K.; Van Cleemput, J.; Vernejoul, J.B.; Enquist, L.W. Alphaherpesvirus infection of mice primes PNS neurons to an inflammatory state regulated by TLR2 and type I IFN signaling. *PLoS Pathog.* **2019**, *15*, e1008087. [CrossRef] [PubMed]
44. Kammerl, I.E.; Meiners, S. Proteasome function shapes innate and adaptive immune responses. *Am. J. Physiol. Lung Cell. Mol. Physiol.* **2016**, *311*, L328–L336. [CrossRef] [PubMed]
45. Lu, M.; Qiu, S.; Zhang, L.; Sun, Y.; Bao, E.; Lv, Y. Pseudorabies virus glycoprotein gE suppresses interferon-β production via CREB-binding protein degradation. *Virus Res.* **2021**, *291*, 198220. [CrossRef] [PubMed]
46. Sun, X.; Liu, T.; Zhao, J.; Xia, H.; Xie, J.; Guo, Y.; Zhong, L.; Li, M.; Yang, Q.; Peng, C.; et al. DNA-PK deficiency potentiates cGAS-mediated antiviral innate immunity. *Nat. Commun.* **2020**, *11*, 6182. [CrossRef]
47. Jakobsson, A.; Westerberg, R.; Jacobsson, A. Fatty acid elongases in mammals: Their regulation and roles in metabolism. *Prog. Lipid Res.* **2006**, *45*, 237–249. [CrossRef]
48. Lodes, D.E.; Zhu, J.; Tsai, N.P. E3 ubiquitin ligase Nedd4-2 exerts neuroprotective effects during endoplasmic reticulum stress. *J. Neurochem.* **2022**, *160*, 613–624. [CrossRef]

Article

The Genetic Characterization of a Novel Natural Recombinant Pseudorabies Virus in China

Jianbo Huang [1], Wenjie Tang [2], Xvetao Wang [3], Jun Zhao [1], Kenan Peng [1], Xiangang Sun [1], Shuwei Li [2,3], Shengyao Kuang [2,3], Ling Zhu [1,4], Yuancheng Zhou [2,3,*] and Zhiwen Xu [1,4,*]

[1] College of Veterinary Medicine, Sichuan Agricultural University, Chengdu 611130, China; jianbohuang90@outlook.com (J.H.); zhaojunjoy@126.com (J.Z.); kenan-peng@outlook.com (K.P.); sun.xian.gang@163.com (X.S.); abtczl72@126.com (L.Z.)

[2] Livestock and Poultry Biological Products Key Laboratory of Sichuan Province, Animtech Bioengineering Co., Ltd., Chengdu 610299, China; wenhan28@126.com (W.T.); lishuwei84511614@126.com (S.L.); ksy_cd@163.com (S.K.)

[3] Veterinary Biologicals Engineering and Technology Research Center of Sichuan Province, Animtech Bioengineering Co., Ltd., Chengdu 610066, China; 15983342580@163.com

[4] Key Laboratory of Animal Diseases and Human Health of Sichuan Province, Chengdu 611130, China

* Correspondence: abtczyc@163.com (Y.Z.); abtcxzw@126.com (Z.X.); Tel.: +86-1822-7601-509 (Y.Z.); +86-1398-1604-765 (Z.X.)

Abstract: We sequenced the complete genome of the pseudorabies virus (PRV) FJ epidemic strain, and we studied the characteristics and the differences compared with the classical Chinese strain and that of other countries. Third-generation sequencing and second-generation sequencing technology were used to construct, sequence, and annotate an efficient, accurate PRV library. The complete FJ genome was 143,703 bp, the G+C content was 73.67%, and it encoded a total of 70 genes. The genetic evolution of the complete genome and some key gene sequences of the FJ strain and PRV reference strains were analyzed by the maximum likelihood (ML) method of MEGA 7.0 software. According to the ML tree based on the full-length genome sequences, PRV FJ strain was assigned to the branch of genotype II, and it showed a close evolutionary relationship with PRV epidemic variants isolated in China after 2011. The gB, gC, gD, gH, gL, gM, gN, TK, gI, and PK genes of the FJ strain were assigned to the same branch with other Chinese epidemic mutants; its gG gene was assigned to the same branch with the classic Chinese Fa and Ea strains; and its gE gene was assigned to a relatively independent branch. Potential recombination events were predicted by the RDP4 software, which showed that the predicted recombination sites were between 1694 and 1936 bp, 101,113 and 102,660 bp, and 107,964 and 111,481 bp in the non-coding region. This result broke the previously reported general rule that pseudorabies virus recombination events occur in the gene coding region. The major backbone strain of the recombination event was HLJ8 and the minor backbone strain was Ea. Our results allowed us to track and to grasp the recent molecular epidemiological changes of PRV. They also provide background materials for the development of new PRV vaccines, and they lay a foundation for further study of PRV.

Keywords: pseudorabies virus; complete genome sequencing; phylogenetic analysis; gene recombination

Citation: Huang, J.; Tang, W.; Wang, X.; Zhao, J.; Peng, K.; Sun, X.; Li, S.; Kuang, S.; Zhu, L.; Zhou, Y.; et al. The Genetic Characterization of a Novel Natural Recombinant Pseudorabies Virus in China. *Viruses* **2022**, *14*, 978. https://doi.org/10.3390/v14050978

Academic Editors: Yan-Dong Tang and Xiangdong Li

Received: 7 March 2022
Accepted: 2 May 2022
Published: 6 May 2022

Publisher's Note: MDPI stays neutral with regard to jurisdictional claims in published maps and institutional affiliations.

1. Introduction

Pseudorabies (PR), also known as Aujeszky's disease (AD), is an acute infectious disease caused by the pseudorabies virus (PRV) [1]. The disease can infect many livestock species and wild animals [2]. Pigs are the main vector for the virus. One of the main symptoms of diseased pigs is an elevated body temperature. In addition, newborn piglets mainly show neurological symptoms of encephalomyelitis, which can also invade the digestive system. Adult pigs often show recessive infection; pregnant sows can have miscarriages, stillbirths, and mummified fetuses; and boars show reproductive disorders and dyspnoea [3]. The disease is distributed all over the world, and it has been eradicated in the United States, Germany, the United Kingdom, Denmark, and the Netherlands. In other countries, it is one of the major diseases that greatly harms the swine industry [4,5]. The World Organization for Animal Health (OIE) lists pseudorabies as a notifiable infectious disease. In China, pseudorabies is classified as a second-class animal epidemic.

The PRV virus is a member of the Herpesviridae, Alphaherpesvirinae subfamily, varicella virus genus [6]. In terms of genome structure, PRV consists of a unique long region (UL), a unique short region (US), and a terminal repeat sequence (TR) and an internal repeat sequence (IR) at both ends of the US region [5]. At present, there is only one PRV serotype, and its genome is composed of double-stranded DNA, the length of which is approximately 143 kilobase pairs (kbp); the GC bases content can be as high as 74%. It contains at least 70 open reading frames (ORF) of which more than 50 proteins are structural; they can participate in the formation of viral capsid, tegument, and the envelope structure [7,8].

In the 1970s, the PRV Bartha-k61 vaccine strain was imported into China, and pseudorabies was well controlled for a time [9,10]. However, since 2012, outbreaks of porcine pseudorabies have been reported in many areas of China, and they have seriously endangered the development of the swine industry [11,12]. These outbreaks are due to the emergence of mutated PRV in various parts of China. The amino acid sequences of important glycoproteins such as gB, gC, gD, and gE have changed, and the existing Bartha-k61 vaccines can no longer elicit 100% protection against mutated PRV strains. It is imperative to develop new PRV vaccines. So, several years ago, PRV was divided into two distinct clusters with the gC gene used as the criterion, with Chinese strains classified as genotype II and PRVs isolated from Europe and North America classified as genotype I [13]. To track the genetic variation of PRV in China, we recently sequenced the complete genome of the PRV FJ strain isolated and identified from the brain tissue of suckling piglets in a pig farm in Fujian province. The genetic relationship between this strain and PRV strains in China and abroad were revealed by a series of bioinformatic analyses so as to provide data support for the development of a new genetically engineered PRV vaccine.

2. Materials and Methods

2.1. Isolation of PRV FJ Strain

Recently, pseudorabies broke out at an intensive pig farm in Fujian province, China. Some suckling piglets and weaned piglets had fever, lethargy, neurological symptoms, and they died. Sows gave birth to stillbirths and weak fetuses. The staff immunized pigs with PRV vaccine by the method of nasal drops six months ago. The incidence rate was approximately 23%, and the death rate of piglets was over 14% on the pig farm. We isolated and identified a PRV strain from 35 brain and tonsil samples of suckling piglets brought from the farm, and we named it the PRV FJ strain. After many rounds of virus multiplication and plaque purification in cell culture flasks, fluid virus samples were collected and frozen in a cryogenic refrigerator at $-80\ ^{\circ}\mathrm{C}$.

2.2. Concentration and Purification of the Virion

The PRV FJ strain was inoculated into a full monolayer of BHK-21 cells. After the cytopathic effect (CPE) reached 80−90%, the cell culture flask was placed in a cryogenic refrigerator at $-80\ ^{\circ}\mathrm{C}$. After freezing and thawing three times, the cell culture flask was separately packed into a 50 mL aseptic centrifuge tube, and the supernatant was centrifuged at 4 °C and 3500 rpm for 15 min. After the supernatant was sterilized and filtered by a 0.45 μm filter membrane (Millipore, Billerica, MA, USA), it was transferred to a 15 mL ultrafilter tube with a maximum cut-off of 100 kD (Millipore, Billerica, MA, USA). The tube was centrifuged at $4000 \times g$ for 30 min according to the manufacturer's instructions. After centrifugation, the concentrated liquid on the filter membrane was carefully aspirated with a 2–200 μL range pipette (Eppendorf, Hamburg, HAM, Germany).

2.3. DNA Extraction

DNA of the PRV FJ strain was extracted using the phenol–chloroform method:

(1)   200 µL of the 10% SDS solution, and 15 µL of the 10 mg/mL RNase A were added in the Eppendorf tubes and incubated at 60 °C for 30 min in the metal bath (Cole-Parmer, Chicago, IL, USA).

(2)   100 µL of the 10 mg/mL Proteinase K was added and incubated in a metal bath at 56 °C for 30 min.

(3)   The ddH$_2$O was added to make up the concentrated viral solution to 400 µL, then 600 µL of phenol: chloroform: isoamyl alcohol = 25:24:1 DNA extraction reagent was added, the Eppendorf tubes were carefully inverted and mixed, then the Eppendorf tubes were stood for 5 min to make the liquid stratified.

(4)   These Eppendorf tubes were centrifuged for 5 min at 12,000 r/min with a microcentrifuge (Thermo Scientific, Waltham, MA, USA), and they took the supernatant to avoid aspirating to the impurities in the middle layer.

(5)   Repeat steps (3) and (4).

(6)   An equal volume of isopropanol was added, mixed lightly, and precipitated for 1 h in a refrigerator at −20 °C.

(7)   These Eppendorf tubes were centrifuged at 12,000 r/min for 5 min with the microcentrifuge to discard the supernatant, 800 µL of anhydrous ice ethanol was added, then 1/10 volume of 3 mol/L NaAc was added, washed with light mixing, and left for 5 min.

(8)   These Eppendorf tubes were centrifuged at 12,000 r/min for 5 min with the microcentrifuge at 4 °C to discard the supernatant, the precipitate was placed in a biosafety cabinet, the exhaust air was turned on and blown until there was no smell of alcohol.

(9)   The precipitate was carefully dissolved in 100 µL TE solution.

Then the extracted DNA samples were stored in a refrigerator at −20 °C for further use.

2.4. Sequencing the Complete Genomes

The extracted PRV FJ DNA samples were sent to Wuhan BaiYi biotechnology company for complete genome sequencing. After the samples were qualified, the database was built with the PRV HLJ8 strain (National Center of Biotechnology Information [NCBI] accession number: KT824771.1) as the reference sequence. Third- and second-generation high-throughput sequencing was carried out using a PacBio RS II sequencing system and a MGISEQ-2000 sequencing system, respectively. For the PacBio RS II system, the Sequel Binding Kit 2.1, the Sequel Sequencing Kit 2.1, and the Sequel SMRT Cell 1mv2 (Pacific Biosciences, Menlo Park, CA, USA) were used for sequencing. The data were processed with the SMRT LINK 6.0 software. The read quality value in the original data was filtered. Based on the complete genome sequencing using the PacBio equipment, the obtained sequence was corrected using the MGISEQ-2000 s-generation sequencing platform. Finally, the complete PRV FJ genome sequence was assembled and annotated.

2.5. Genome and Related Gene Homology and Phylogenetic Analysis

Fifteen PRV genome sequences uploaded to NCBI (Table 1) were compared with the PRV FJ genome sequence and its major virulence, glycoprotein, and immunogenicity-related coding sequences(CDS): TK, PK, gB, gC, gD, gG, gH, gL, gM, gN, gI, and gE. A genetic evolution tree was drawn and analyzed using the MEGA 7.0 software (https://www.megasoftware.net, accessed on 25 October 2021).

2.6. Prediction of Potential Genome Recombination Events

The genome alignments from the 15 PRV reference strains and the FJ strain were analyzed with the Recombination Detection Program 4 (RDP4) software to screen for potential recombination events. Seven algorithms, including RDP, BootScan, GENECONV, Maxchi, SiScan, Chimera, and 3Seq were employed [14].

Table 1. Complete genome sequence information of pseudorabies viruses in the National Center of Biotechnology Information databases.

No.	Strain Name	Accession Number	Country	Isolation Date
1	Becker	JF797219.1	USA	1970
2	Bartha	JF797217.1	Hungary	1961
3	Ea	KU315430.1	China	1990
4	Fa	KM189913.1	China	2012
5	GD0304	MH582511.1	China	2015
6	HB1201	KU057086.1	China	2012
7	HeN1	KP098534.1	China	2012
8	HLJ8	KT824771.1	China	2012
9	HNB	KM189914.3	China	2012
10	HNX	KM189912.1	China	2012
11	Kaplan	JF797218.1	Hungary	1959
12	Kolchis	KT983811.1	Greece	2010
13	MY-1	AP018925.1	Japan	2015
14	ZJ-01	KM061380.1	China	2012
15	TJ	KJ789182.1	China	2012

2.7. Sequence Submission

The complete PRV FJ genome sequence was deposited in the GenBank database (http://www.ncbi.nim.nih.gov/genbank, accessed on 18 October 2021) under the accession numbers of MW286330.

3. Results

3.1. Complete Genome Sequence Analysis

After comparing the genome sequence assembly using the reference PRV sequences from NCBI databases, the PRV FJ complete genome length was 143,703 bp, the GC bases content was 73.67%, and it encoded a total of 70 genes without insertion and deletion of the rest of the coding sequences. The sequence was divided into four parts: UL, US, IRs, and TRs (Table 2). We annotated the linear map, gene arrangement, and distribution of the complete genome sequence using the Snapgene software (Figure 1), and we noted the annotations of each open reading frame (ORF) (Table 3).

Table 2. Nucleotide sequence coordinates and lengths are given relative to the genome sequence of PRV FJ strain.

Region	Location	Length (bp)
UL	1–101,012	101,012
IRs	101,013–117,681	16,669
US	117,682–127,034	9353
TRs	127,035–143,703	16,669

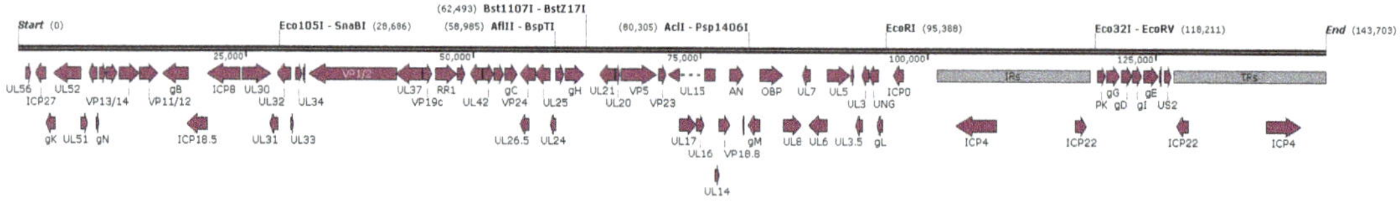

Figure 1. Sequences and distribution of the PRV FJ complete genome.

Table 3. Annotation of each open reading frame of from the PRV FJ complete genome.

Protein Name	Location of ORF (bp)	Length (aa)	Function
UL56	754–1377	207	Possibly vesicular trafficking
ICP27	1932–3017	361	Gene regulation; early protein
gK	3096–4034	312	Viral glycoprotein K; type III membrane protein
UL52	3989–6895	968	DNA replication; primase subunit of ULS/UL8/UL52 complex
UL51	6882–7613	243	Tegument protein
dUTPase	7812–8621	269	dUTPase
gN	8542–8841	99	Glycoprotein N; type I membrane protein; complexed with gM
VP22	8879–9619	246	Interacts with C-terminal domains of gE and gM; tegument protein
VP16	9683–10,924	413	Gene regulation (transactivator); egress (secondary envelopment); tegument protein
VP13/14	11,034–13,250	738	Viral egress (secondary envelopment); tegument protein
VP11/12	13,269–15,356	695	Possibly gene regulation; tegument protein
gB	15,905–18,649	914	Viral entry (fusion); cell–cell spread; glycoprotein B; type I membrane protein
ICP18.5	18,520–20,688	722	DNA cleavage and encapsulations (terminase); associated with UL15, UL33 and UL6
ICP8	20,836–24,378	1180	DNA replication-recombination; binds single-stranded DNA
UL30	24,677–27,823	1048	DNA replication; DNA polymerase subunit of UL30/UL42 complex
UL31	27,744–28,559	271	Viral egress (nuclear egress); primary virion tegument protein; interacts with UL34
UL32	28,552–29,967	471	DNA packaging; efficient localization of capsids to replication compartments
UL33	29,966–30,322	118	DNA cleavage and encapsidation; associated with UL28 and UL 15
UL34	30,494–31,279	261	Viral egress (nuclear egress); primary virion envelop protein tail-anchored type II nuclear membrane protein; interacts with UL31
VP26	31,334–31,645	103	Capsid protein
VP1/2	32,057–41,644	3195	Large tegument protein; interacts with UL37 and UL19
UL37	41,682–44,441	919	Tegument protein; interacts with UL36
VP19c	44,498–45,604	368	Capsid protein; forms triplexes together with ULl8
RR1	45,941–48,307	788	Nucleotide synthesis; large subunit of ribonucleotide reductase
RR2	48,317–49,228	303	Nucleotide synthesis; small subunit of ribonucleotide reductase
vhs	49,843–50,940	365	Gene regulation (inhibitor of gene expression); virion host cell shutoff
UL42	51,069–52,226	385	DNA replication; polymerase accessory subunit of UL30/UL42 complex
UL43	52,286–53,407	373	Unknown; type III membrane protein
gC	53,474–54,937	487	Viral entry (virion attachment); glycoprotein C; type I membrane protein; binds to heparan sulfate
UL26.5	55,233–56,093	286	Scaffold protein; substrate for UL26; required for capsid formation and maturation
VP24	55,233–56,831	532	Scaffold protein; proteinase; required for capsid formation and maturation
UL25	56,883–58,493	536	Capsid-associated protein; required for capsid assembly
UL24	58,592–59,107	171	Unknown; type III membrane protein
TK	59,100–60,062	320	Nucleotide synthesis; thymidine kinase
gH	60,198–62,255	685	Viral entry (fusion); cell–cell spread; glycoprotein H; type I membrane protein; complexed with gL
UL21	64,012–65,613	533	Capsid-associated protein
UL20	65,720–66,217	165	Viral egress; type III membrane protein
VP5	66,306–70,298	1330	Major capsid protein; forms hexons and pentons
VP23	70,473–71,363	296	Capsid protein; forms triplexes together with UL38
UL17	72,739–74,538	599	DNA cleavage and encapsidation
UL16	74,565–75,551	328	Possibly virion morphogenesis
UL15	71,546–72,687 75,572–76,691	753	DNA cleavage and encapsidation; terminase subunit; interacts with UL33 UL28, and UL6
UL14	76,690–77,169	159	Virion morphogenesis

Table 3. *Cont.*

Protein Name	Location of ORF (bp)	Length (aa)	Function
VP18.8	77,139–78,314	391	Protein-serine/threonine kinase
AN	78,280–79,731	483	DNA recombination; alkaline exonuclease
UL11	79,689–79,880	63	Viral egress (secondary envelopment); membrane-associated tegument protein
gM	80,309–81,490	393	Viral egress (secondary envelopment); glycoprotein M; type III membrane protein; C terminus interacts with UL49; inhibits membrane fusion in transient assays; complexed with gN
OBP	81,489–84,023	844	Sequence-specific ori-binding protein
UL8	84,020–86,086	688	DNA replication; part of ULS/UL8/UL52 helicase-primase complex
UL7	86,288–87,088	266	Virion morphogenesis
UL6	86,979–88,916	645	DNA packaging Capsid protein; portal protein; docking site for terminase
UL5	88,915–91,470	851	DNA replication; part of ULS/LJL8/UL52 helicase-primase complex; helicase motif
UL4	91,528–91,965	145	Nuclear protein
UL3.5	92,141–92,809	222	Possibly virion morphogenesis
UL3	92,806–93,540	244	Nuclear protein
UNG	93,597–94,568	323	Uracil-DNA glycosylase
gL	94,546–95,016	156	Viral entry; cell–cell spread; glycoprotein L; membrane-anchored via complex with gH
ICP0	96,248–97,348	366	Gene regulation (transactivator of viral and cellular genes); early protein
ICP4	103,130–107,544	1471	Gene regulation; immediate early protein
ICP22	116,146–117,339	397	Gene regulation
PK	118,467–119,471	334	Minor form of protein kinase (53-kDa mobility); viral egress (nuclear egress); major form of protein kinase (41-kDa mobility)
gG	119,531–121,030	499	Cell–cell spread; secreted; glycoprotein G
gD	121,214–122,422	402	Viral entry (cellular receptor binding protein); glycoprotein D
gI	122,446–123,543	465	Cell–cell spread; glycoprotein I; type I membrane protein; complexed with gE
gE	123,647–125,386	579	Cell–cell spread; glycoprotein E; type I membrane protein; complexed with gI; C terminus interacts with UL49
US9(11K)	125,444–125,740	98	Protein sorting in axons; type II tail-anchored membrane protein
US2(28K)	125,994–126,764	256	Possibly envelope associated

Abbreviations: aa, amino acids; ORF, open reading frame.

3.2. Genomic Genetic Evolution Analysis

Nucleotide homology comparison between the FJ strain and the reference strains using the MEGA 7.0 software showed that the FJ strain had the highest homology with Chinese PRV mutant strains isolated after 2011, with 99.9% and 99.7% homology with classical PRV Fa and Ea strains, respectively, isolated in the 20th century in China. The homology with the other country's MY-1, Bartha, Becker, Kaplan, and Kolchis strains was 99.0%, 95.7%, 95.7%, 96.0%, and 95.7%, respectively; these values are relatively low (Table 4). It is worth noting that the MY-1 strain is an Asian strain.

Table 4. Complete gene sequence nucleotide homology analysis.

Virus Strain	MY-1	FJ *	Bartha	Kaplan	Becker	TJ	ZJ01	HNX	Fa	HNB	HeN1	HLJ8	Kolchis	HB1201	Ea	GD0304
								Nucleotides Homology (%)								
MY-1																
FJ *	99.0															
Bartha	95.4	95.7														
Kaplan	95.6	96.0	99.4													
Becker	95.4	95.7	98.6	98.9												
TJ	99.0	100	95.7	96.0	95.7											
ZJ01	99.0	99.9	95.6	95.9	95.6	99.9										
HNX	99.0	100	95.7	96.0	95.7	100	99.9									
Fa	98.9	99.7	95.7	96.0	95.6	99.7	99.7	99.7								
HNB	99.0	100.0	95.7	96.0	95.7	100.0	99.9	100.0	99.7							
HeN1	99.0	100.0	95.7	96.0	95.7	100.0	99.9	100.0	99.7	100.0						
HLJ8	99.0	100.0	95.7	96.0	95.7	100.0	99.9	100.0	99.7	100.0	100.0					
Kolchis	95.4	95.7	99.1	99.5	99.0	95.7	95.6	95.7	95.7	95.7	95.7	95.7				
HB1201	99.0	100.0	95.7	96.0	95.7	100.0	99.9	100.0	99.7	100.0	100.0	100.0	95.7			
Ea	98.9	99.7	95.7	96.0	95.6	99.7	99.7	99.7	100.0	99.7	99.7	99.7	95.7	99.7		
GD0304	99.0	100.0	95.7	96.0	95.7	100.0	99.9	100.0	99.7	100.0	100.0	100.0	95.7	100.0	99.7	

Note. "*" indicates that this PRV strain is the target PRV strain.

We analyzed the complete genome sequence homology of the reference strains and the FJ strain with the online program mVista (http://genome.lbl.gov/vista/mvista/submit.shtml, accessed on 3 October 2021). Compared with the Bartha strain, the FJ strain had low homology in UL56, UL51, UL27, UL36, UL41, UL28.5, UL21, and LLT genes. Except for the HB1201 strain, the other Chinese reference strains and the MY-1 strain also showed homology differences in the above regions. The other country's Becker, Kolchis, and Kaplan reference strains only showed significant homology differences in UL27, UL36, UL21, and US1 gene regions. The homology difference between HB1201 and Bartha was the greatest, and there were large base deletions in the UL56, UL27, UL21, UL36, LLT, US1, US3, and IE180 gene regions. In summary, after homology comparison with the Bartha strain, the regions with lower homology between the FJ strain and the other reference strains were mainly distributed in the non-coding region (Figure 2).

We constructed and analyzed the genetic evolution tree of the complete genome sequence of PRV strains with the maximum likelihood (ML) method. All the strains were classified into two major branches. The other country's Bartha, Becker, Kolchis, and Kaplan strains were located in the European and the North American genotype (genotype I) branch, while the Chinese strains and the MY-1 strain were located in the Asian genotype (genotype II) branch, which was consistent with the results of other reported genetic evolution analyses. The FJ strain was still located in the genotype II branch, close to the GD0304 strain branch, and it had the lowest genetic relationship with the ZJ01 strain in the genotype II branch (Figure 3).

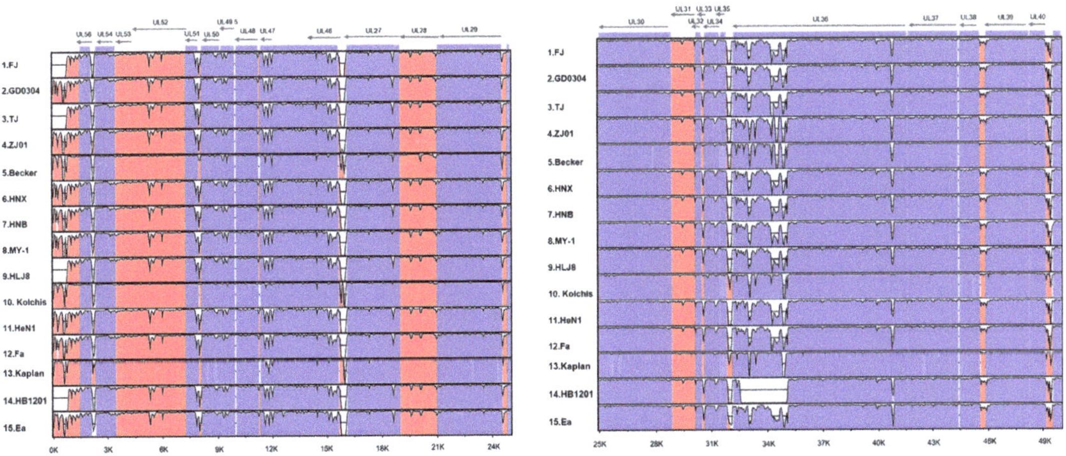

Figure 2. *Cont.*

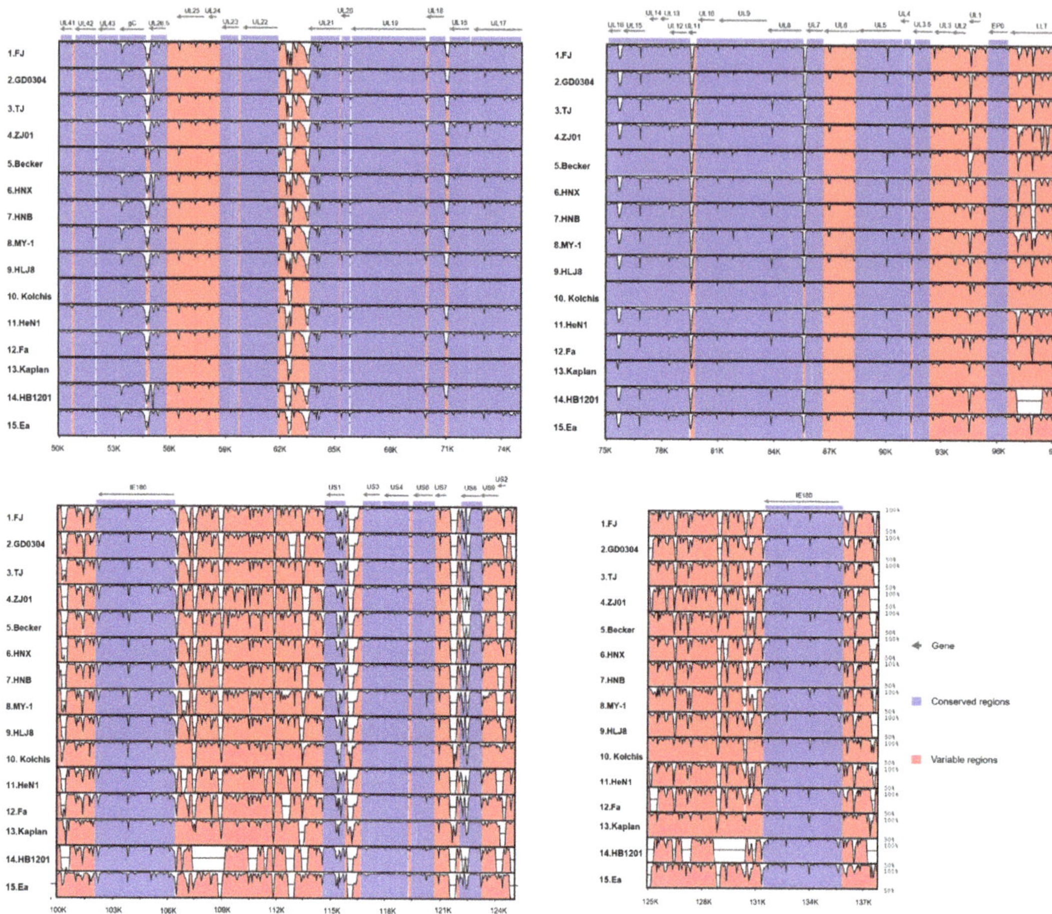

Figure 2. Genome organization and comparison of the PRV Bartha strain with the PRV FJ strain and the remaining 14 reference PRV strains. Comparison of the PRV genome shows conserved (blue) and variable (pink) regions. Open reading frames (ORFs) are represented by gray horizontal arrows across the top of each panel, and genome coordinates in kilobase pairs (kbp) are shown along the bottom.

3.3. Phylogenetic Analysis of Related Gene Sequences

We selected the coding sequences of 12 genes related to immunogenicity and virulence of PRV, including TK, PK, gB, gC, gD, gG, gH, gL, gM, gN, gI and gE genes and analyzed them using the ML method of the MEGA 7.0 software. Except for gL, the phylogenetic trees of all genes produced the typical genotype I and genotype II branches, while the gL evolutionary tree showed that the Chinese epidemic mutant HeN1 strain belonged to the European and the North American genotype I. The Becker strain belonged to genotype II. All the above genes of the FJ strain were located in the large genotype II branch, and its gB, gC, gD, gH, gL, gM, gN, TK, gI, and PK genes were in the same branch as other Chinese mutants. Its gG gene was assigned to the same branch with the classical Chinese PRV Fa and Ea strains' gG gene, while its gE gene was assigned to a relatively independent branch. All the genes of the MY-1 strain were located in the large branch of genotype II, except for TK, gL, gM, and gN; the other genes were located in a single branch compared with Chinese strains. The selected PRV FJ genes were far away from the other country's strains, and they were very close to the Chinese mutants; and, the above-mentioned immunogenicity and virulence-related genes were not significantly different from the previous PRV variants (Figure 4).

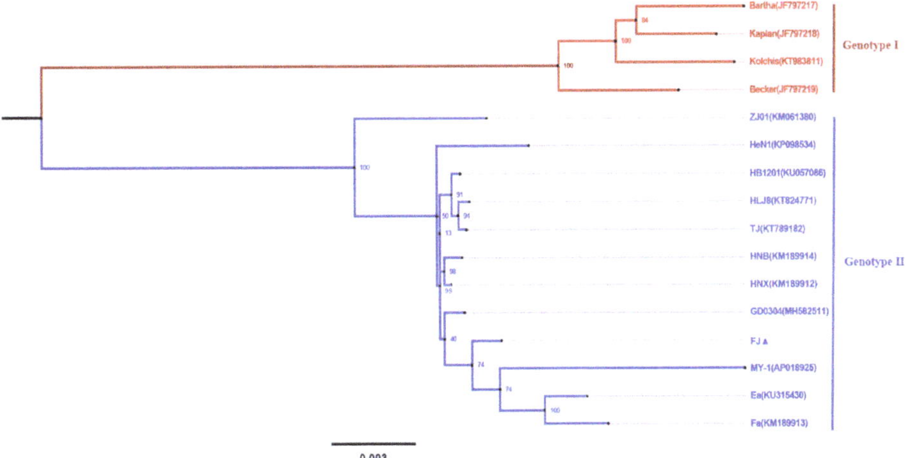

Figure 3. Phylogenetic analysis of the PRV complete genome sequences. The tree was constructed using the MEGA 7.0 software with the maximum likelihood method and 1000 bootstrap replicates. The bar and the number represent the genetic distance scale of these genes at this length is 0.003.

3.4. Recombination Analyses

We compared the FJ strain with the 15 PRV reference genome sequences using the RDP4 software (http://web.cbio.uct.ac.za/~darren/rdp.html, accessed on 14 October 2021); we predicted the recombination possibilities of the strain using Bootscan, LARD, 3seq, PhylPro, Maxchi, SiScan, and Chimaera algorithms. We detected several recombination signals for the FJ genome sequence (Figure 5). The major backbone of the FJ strain was the HLJ8 strain; the minor backbone was the Ea strain. We analyzed the potential recombination events of the FJ complete genome sequence using the above-mentioned algorithms; the p value of each algorithm was $<10^{-3}$. The predicted recombination sites were between 1694 and 1936 bp, between 101,113 and 102,660 bp, and between 107,964 and 11,148,1 bp; four algorithms supported the recombination events in each segment. Among them, two algorithms in the 1694–1936 bp section showed that the recombination event was credible; one algorithm in the 101,113–102,660 bp section showed that the recombination event was credible; and four algorithms in the 107,964–111,481 bp section showed that the recombination event was credible (Table 5).

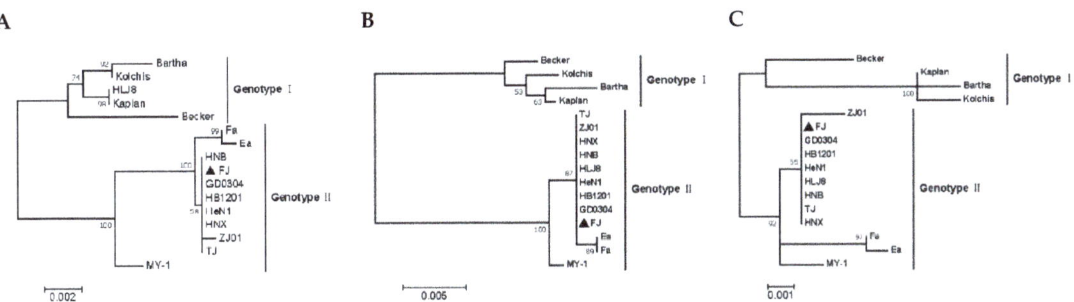

Figure 4. *Cont.*

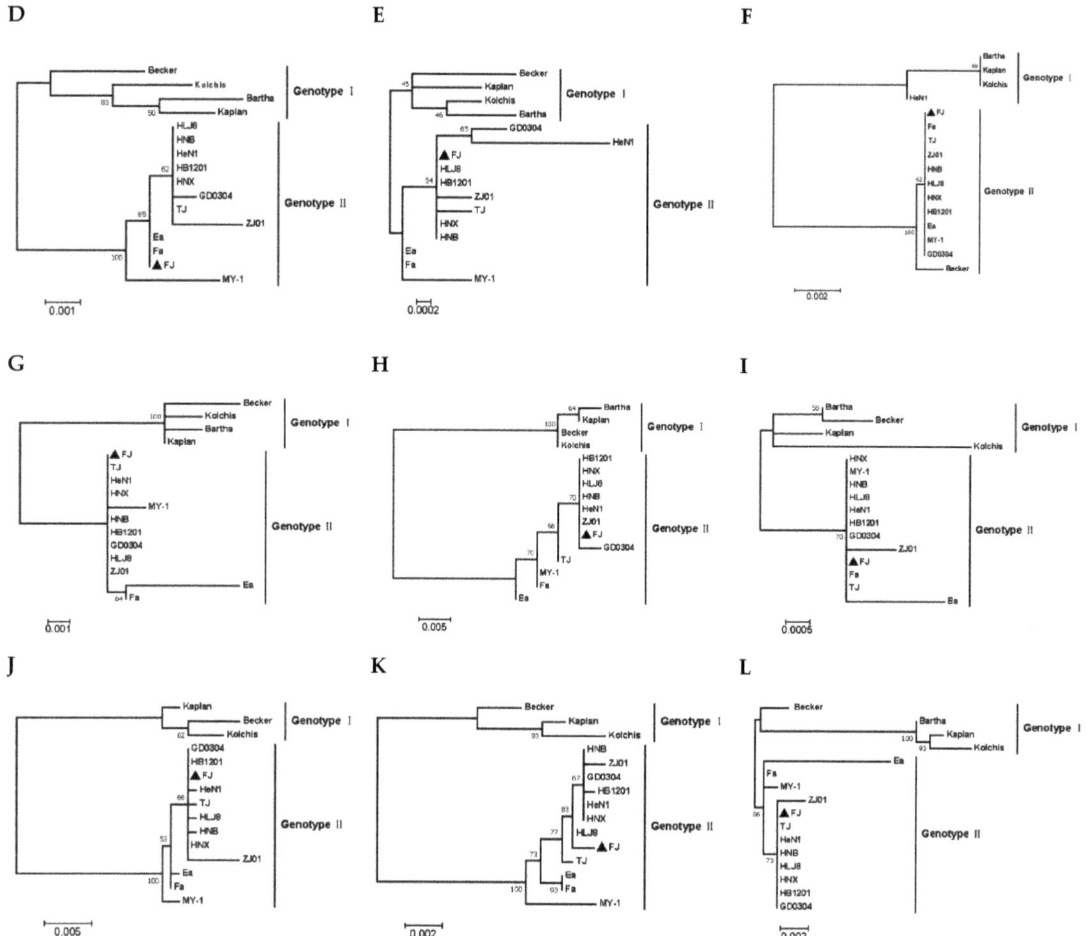

Figure 4. Phylogenetic analysis based on nucleotide sequences of PRV virulence-related and immunogenicity genes: (**A**) gB, (**B**) gC, (**C**) gD, (**D**) gG, (**E**) gH, (**F**) gL, (**G**) gM, (**H**) gN, (**I**) TK, (**J**) gI, (**K**) gE and (**L**) PK. The tree was constructed with the same method described in Figure 3.

Table 5. Analysis of PRV FJ recombination events with different algorithms; the numbers represent the *p* value of each algorithm.

Position (bp)	Method and *p* Value						
	Bootscan	Maxchi	Chimaera	SiScan	PhylPro	LARD	3Seq
1694–1936	2.31×10^{-11} *	3.50×10^{-9} *	NS	2.27×10^{-2}	1.05×10^{-2}	NS	NS
101,113–102,660	6.54×10^{-5} *	6.26×10^{-3}	NS	5.67×10^{-3}	1.82×10^{-3}	NS	NS
107,964–111,481	9.87×10^{-10} *	9.81×10^{-9} *	NS	1.93×10^{-5} *	1.68×10^{-4} *	NS	NS

Note. NS, does not support reorganization events; * $p < 10^{-3}$.

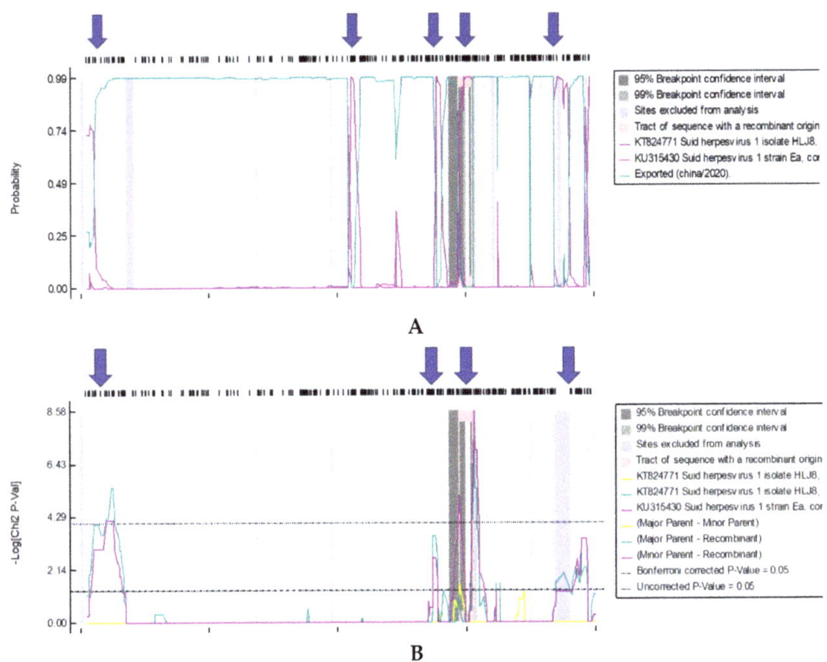

Figure 5. Putative recombination events in the PRV FJ strain complete genome. Recombination sites are marked by a blue arrow. The region with a pink background represents the potential recombination region. (**A**): BootScan method; (**B**): RDP method.

4. Discussion

Due to the large number of genes in the PRV complete genome, the high content of GC bases, and the presence of more than 900 nucleotide repeat sequences, it is relatively difficult to sequence the complete genome; research in the PRV gene function and comparative genomics had been somewhat restricted. In 2011, American researchers were the first to obtain and to publish the complete genome sequences of some representative PRV strains such as Bartha, Kaplan, and Becker using Illumina second-generation sequencing technology [3]. With the popularity of second-generation high-throughput sequencing around the world, the complete genome sequences of PRV isolates from various regions have been published in China since 2014 [15]. The advantage of second-generation sequencing is that segmented sequencing can be used to ensure the accuracy of sequencing results; the cost of sequencing at this stage is very low; and the DNA samples do not need to be of very high quality to sequence. However, there are several disadvantages, including that the sequencing time is very long; the content of GC bases in PRV genomes is very high so that it is hard to completely sequence the genomes at one time; every sequencing of a complete genome will generate many gap sequences, which need to be amplified and filled by multiple pairs of primers; and the sequencing technology still needs to be innovated. In recent years, with the advent of the third-generation PacBio RSII gene sequencer, long and complex sequencing has become very convenient.

In this study, we combined second- and third-generation sequencing, an approach that provides the benefits of third-generation sequencing efficiency, ultra-long reading length, short sequencing cycle, no base preference, and no gap sequences. In addition, this approach allows the sequencing of complex structures at one time, and it makes use of MGISEQ-2000 s generation sequencing to make up for the shortcomings of low third-generation sequencing flux and manual correction of sequencing errors. After sequencing, we assembled the PRV FJ complete genome quickly, accurately, and completely. The PRV FJ complete genome was 143,703 bp, had a G&C bases content of 73.67%, and encoded 70 ORFs. The length and the structural range of the complete genome sequence are consistent with the previously tested Chinese epidemic mutant HNX strain (full length = 142,294 bp, G&C bases content = 73.56%, encoding 70 ORFs) [16]; the HNB strain (full length = 142,255 bp, G&C bases content = 73.61%, encoding 70 ORFs) [17]; the TJ strain (full length = 143,642 bp, encoding 67 ORFs) [15]; and the HeN1 strain (full length = 141,803 bp, G&C bases bases content = 73.3%, encoding 69 ORF) [18]. Thus, our FJ strain sequencing results are reliable.

Among the PRV genes we selected for phylogenetic tree analysis, gB, gD, gH, gL, and gK are necessary to ensure their replication, growth, and proliferation in cells [19]. As PRV immunogenicity-related proteins, gB, gC, and gD can induce neutralizing antibody production [20]. The proteins expressed by gH and gL, gE and gI, and gM and gN genes can form heterodimers, which are related to virus infection and immune escape [21]. TK, PK, gE and gI are virulence-related genes. Single or multiple deletions or insertion mutations in these genes affect the virulence of PRV [22,23]. The gG gene encodes the only protein component released by PRV outside the virus; it was released out of the cell by protease hydrolysis, and it was disconnected from the virion after passing through the cell membrane [24]. Among the above-mentioned genes, only the gG gene of the PRV FJ strain was located in the subbranch of the classic Chinese strains identified before 2011. The reason is that the 244th base of the gG gene of the FJ strain, classic Chinese strain and other country's strains is 'T', while the corresponding base of the Chinese epidemic mutant is 'C'. The other genes were in the same subbranch with the mutants in China. Therefore, the results showed that the FJ strain has also been a common variant in China in recent years, and the genetic variation has been stable up to now.

Through the detection of the complete genome recombination events of the PRV FJ strain, we found recombinant signals in the 1694–1936, 101,113–102,660, and 107,964–111,481 bp regions, indicating that recombination occurred in the corresponding regions of the HLJ8 and the Ea strains. According to the location of coding sequence annotations in Table 3, we found the recombinant region was located in the non-coding region of UL and IRs of the FJ strain, so there is no recombination mutation event in the coding sequence. Non-coding regions in virus genomes have a variety of functions. For example, a non-coding region of Japanese encephalitis virus antagonizes the interferon response by blocking interferon regulatory factor 3 transport [25]; the replication of Marburg virus can be regulated by its non-coding region [26]. Influenza virus infection can be regulated by its non-coding region [27]. The non-coding region of the Epstein–Barr virus (EBV) plays an important role in the life cycle and the pathogenesis of EBV [28]. Natural recombination between different PRV strains has been common, as authors have reported, although the mechanism is unclear [29].

However, gene recombination in the non-coding region may affect the ability of PRV to induce interferon-beta promoter activity and regulate viral messenger RNA (mRNA) [30,31]. The major backbone strain for the recombination event was HLJ8, which is an epidemic variant strain that was isolated in China after 2011, while the minor backbone strain was Ea, a classic strain that was isolated in China in the 20th century. Hence, the FJ strain might have the ability to recombine with the Chinese epidemic variant strain and the classical strain. With regard to the cause of the natural recombination phenomenon, we speculate that the FJ wild type strain may have arisen due to natural recombination in pigs when they were immunized with the commercial vaccine with the Ea strain as the parent strain. During large-scale importation of breeding pigs in this pig farm, the cross-provincial transportation of breeding pigs caused some pigs with latent PRV infection in other areas, so it was difficult to show positive results accurately during PRV detection, while these pigs are still traded in the market. After the infection of the FJ strain, gene recombination might occur among PRV in the host.

Author Contributions: Conceptualization, Z.X. and Y.Z.; methodology, J.H.; software, W.T.; validation, L.Z. and K.P.; formal analysis, J.Z.; investigation, X.W.; resources, S.L.; data curation, Z.X.; writing—original draft preparation, J.H.; writing—review and editing, J.H.; visualization, S.K.; supervision, X.S.; project administration, Y.Z.; funding acquisition, L.Z. and Y.Z. All authors have read and agreed to the published version of the manuscript.

Funding: This research was funded by the Science and Technology Program of Sichuan province, grant number 2021ZDZX0010 and 20MZGC0024, and the Sichuan Veterinary Medicine and Drug Innovation Group of the China Agricultural Research System, grant number CARS-SVDIP.

Institutional Review Board Statement: Not applicable.

Informed Consent Statement: Not applicable.

Data Availability Statement: The complete genome of the virus described in detail here was deposited in GenBank under the following Accession Numbers: MW286330.

Conflicts of Interest: There is no financial and technical conflict of interest among the institutions in the article. The authors also declare no conflict of interest.

References

1. Aujeszky, A. *Über Eine Neue Infektionskrankheit bei Haustieren*; Fischer: Berlin, Germany, 1902; Volume 32.
2. Lee, J.; Wilson, M. A review of pseudorabies (Aujeszky's disease) in pigs. *Can. Vet. J.* **1979**, *20*, 65. [PubMed]
3. Szpara, M.L.; Tafuri, Y.R.; Parsons, L.; Shamim, S.R.; Verstrepen, K.J.; Legendre, M.; Enquist, L. A wide extent of inter-strain diversity in virulent and vaccine strains of alphaherpesviruses. *PLoS Pathog.* **2011**, *7*, e1002282. [CrossRef] [PubMed]
4. Müller, T.; Hahn, E.; Tottewitz, F.; Kramer, M.; Klupp, B.; Mettenleiter, T.; Freuling, C. Pseudorabies virus in wild swine: A global perspective. *Arch. Virol.* **2011**, *156*, 1691. [CrossRef] [PubMed]
5. Klupp, B.G.; Hengartner, C.J.; Mettenleiter, T.C.; Enquist, L.W. Complete, annotated sequence of the pseudorabies virus genome. *J. Virol.* **2004**, *78*, 424–440. [CrossRef] [PubMed]
6. King, A.M.; Lefkowitz, E.; Adams, M.J.; Carstens, E.B. *Virus Taxonomy: Ninth Report of the International Committee on Taxonomy of Viruses*; Elsevier: Amsterdam, The Netherlands, 2011; Volume 9.
7. Pomeranz, L.E.; Reynolds, A.E.; Hengartner, C.J. Molecular biology of pseudorabies virus: Impact on neurovirology and veterinary medicine. *Microbiol. Mol. Biol. Rev. MMBR* **2005**, *69*, 462–500. [CrossRef] [PubMed]
8. Nauwynck, H.; Glorieux, S.; Favoreel, H.; Pensaert, M. Cell biological and molecular characteristics of pseudorabies virus infections in cell cultures and in pigs with emphasis on the respiratory tract. *Vet. Res.* **2007**, *38*, 229–241. [CrossRef] [PubMed]
9. An, T.-Q.; Peng, J.-M.; Tian, Z.-J.; Zhao, H.-Y.; Li, N.; Liu, Y.-M.; Chen, J.-Z.; Leng, C.-L.; Sun, Y.; Chang, D. Pseudorabies virus variant in Bartha-K61–vaccinated pigs, China, 2012. *Emerg. Infect. Dis.* **2013**, *19*, 1749. [CrossRef]
10. Sun, Y.; Luo, Y.; Wang, C.-H.; Yuan, J.; Li, N.; Song, K.; Qiu, H.-J. Control of swine pseudorabies in China: Opportunities and limitations. *Vet. Microbiol.* **2016**, *183*, 119–124. [CrossRef]
11. Yu, X.; Zhou, Z.; Hu, D.; Zhang, Q.; Han, T.; Li, X.; Gu, X.; Yuan, L.; Zhang, S.; Wang, B. Pathogenic pseudorabies virus, China, 2012. *Emerg. Infect. Dis.* **2014**, *20*, 102. [CrossRef]
12. Yang, Q.-Y.; Sun, Z.; Tan, F.-F.; Guo, L.-H.; Wang, Y.-Z.; Wang, J.; Wang, Z.-Y.; Wang, L.-L.; Li, X.-D.; Xiao, Y. Pathogenicity of a currently circulating Chinese variant pseudorabies virus in pigs. *World J. Virol.* **2016**, *5*, 23. [CrossRef]
13. Ye, C.; Zhang, Q.-Z.; Tian, Z.-J.; Zheng, H.; Zhao, K.; Liu, F.; Guo, J.-C.; Tong, W.; Jiang, C.-G.; Wang, S.-J. Genomic characterization of emergent pseudorabies virus in China reveals marked sequence divergence: Evidence for the existence of two major genotypes. *Virology* **2015**, *483*, 32–43. [CrossRef] [PubMed]
14. Martin, D.P.; Murrell, B.; Golden, M.; Khoosal, A.; Muhire, B. RDP4: Detection and analysis of recombination patterns in virus genomes. *Virus Evol.* **2015**, *1*, vev003. [CrossRef] [PubMed]
15. Luo, Y.; Li, N.; Cong, X.; Wang, C.-H.; Du, M.; Li, L.; Zhao, B.; Yuan, J.; Liu, D.-D.; Li, S. Pathogenicity and genomic characterization of a pseudorabies virus variant isolated from Bartha-K61-vaccinated swine population in China. *Vet. Microbiol.* **2014**, *174*, 107–115. [CrossRef]
16. Yu, T.; Chen, F.; Ku, X.; Fan, J.; Zhu, Y.; Ma, H.; Li, S.; Wu, B.; He, Q. Growth characteristics and complete genomic sequence analysis of a novel pseudorabies virus in China. *Virus Genes* **2016**, *52*, 474–483. [CrossRef] [PubMed]
17. Yu, T.; Chen, F.; Ku, X.; Zhu, Y.; Ma, H.; Li, S.; He, Q. Complete genome sequence of novel pseudorabies virus strain HNB isolated in China. *Genome Announc.* **2016**, *4*, e01641-15. [CrossRef] [PubMed]
18. Ye, C.; Zhao, K.; Guo, J.; Jiang, C.; Chang, X.; Wang, S.; Wang, T.; Peng, J.; Cai, X.; Tian, Z. Genomic sequencing and genetic diversity analysis of a pseudorabies virus based on main infection or virulence genes. *Zhongguo Yufang Shouyi Xuebao/Chin. J. Prev. Vet. Med.* **2015**, *37*, 581–584.
19. Mettenleiter, T.C.; Lukàcs, N.; Thiel, H.-J.; Schreurs, C.; Rziha, H.-J. Location of the structural gene of pseudorabies virus glycoprotein complex gII. *Virology* **1986**, *152*, 66–75. [CrossRef]

20. Eloit, M.; Vannier, P.; Hutet, E.; Fournier, A. Correlation between gI, gII, gIII, and gp 50 antibodies and virus excretion in vaccinated pigs infected with pseudorabies virus. *Arch. Virol.* **1992**, *123*, 135–143. [CrossRef]
21. Jöns, A.; Dijkstra, J.M.; Mettenleiter, T.C. Glycoproteins M and N of pseudorabies virus form a disulfide-linked complex. *J. Virol.* **1998**, *72*, 550–557. [CrossRef]
22. Kit, S.; Kit, M.; Pirtle, E. Attenuated properties of thymidine kinase-negative deletion mutant of pseudorabies virus. *Am. J. Vet. Res.* **1985**, *46*, 1359–1367.
23. Lipowski, A. Evaluation of efficacy and safety of Aujeszky's. *Pol. J. Vet. Sci.* **2006**, *9*, 75–79. [PubMed]
24. Schwartz, J.A.; Brittle, E.E.; Reynolds, A.E.; Enquist, L.W.; Silverstein, S.J. UL54-null pseudorabies virus is attenuated in mice but productively infects cells in culture. *J. Virol.* **2006**, *80*, 769–784. [CrossRef] [PubMed]
25. Chang, R.-Y.; Hsu, T.-W.; Chen, Y.-L.; Liu, S.-F.; Tsai, Y.-J.; Lin, Y.-T.; Chen, Y.-S.; Fan, Y.-H. Japanese encephalitis virus non-coding RNA inhibits activation of interferon by blocking nuclear translocation of interferon regulatory factor 3. *Vet. Microbiol.* **2013**, *166*, 11–21. [CrossRef] [PubMed]
26. Alonso, J.A.; Patterson, J.L. Sequence variability in viral genome non-coding regions likely contribute to observed differences in viral replication amongst MARV strains. *Virology* **2013**, *440*, 51–63. [CrossRef]
27. Landeras-Bueno, S.; Ortin, J. Regulation of influenza virus infection by long non-coding RNAs. *Virus Res.* **2016**, *212*, 78–84. [CrossRef]
28. Iwakiri, D. Multifunctional non-coding Epstein–Barr virus encoded RNAs (EBERs) contribute to viral pathogenesis. *Virus Res.* **2016**, *212*, 30–38. [CrossRef]
29. Glazenburg, K.; Moormann, R.; Kimman, T.; Gielkens, A.; Peeters, B. Genetic recombination of pseudorabies virus: Evidence that homologous recombination between insert sequences is less frequent than between autologous sequences. *Arch. Virol.* **1995**, *140*, 671–685. [CrossRef]
30. Da Silva, L.F.; Jones, C. Small non-coding RNAs encoded within the herpes simplex virus type 1 latency associated transcript (LAT) cooperate with the retinoic acid inducible gene I (RIG-I) to induce beta-interferon promoter activity and promote cell survival. *Virus Res.* **2013**, *175*, 101–109. [CrossRef]
31. Umbach, J.L.; Kramer, M.F.; Jurak, I.; Karnowski, H.W.; Coen, D.M.; Cullen, B.R. MicroRNAs expressed by herpes simplex virus 1 during latent infection regulate viral mRNAs. *Nature* **2008**, *454*, 780–783. [CrossRef]

Article

Epidemiological Investigation and Genetic Analysis of Pseudorabies Virus in Yunnan Province of China from 2017 to 2021

Jun Yao [1,†], Juan Li [2,†], Lin Gao [1], Yuwen He [1], Jiarui Xie [1], Pei Zhu [1], Ying Zhang [3], Xue Zhang [2], Luoyan Duan [4], Shibiao Yang [1], Chunlian Song [3,*] and Xianghua Shu [2,*]

1   Yunnan Tropical and Subtropical Animal Virus Diseases Laboratory, Yunnan Animal Science & Veterinary Institute, Kunming 650224, China; yaojun_joshua@hotmail.com (J.Y.); 13987123280@163.com (L.G.); heyuwen1117@163.com (Y.H.); xjr1990123@sina.com (J.X.); zpcau@sina.com (P.Z.); yangsb3799@sina.com (S.Y.)
2   Yunnan Sino-Science Gene Technology Co., Ltd., Kunming 650501, China; ynndlj@126.com (J.L.); xuezhangx@outlook.com (X.Z.)
3   College of Animal Medicine, Yunnan Agricultural University, Kunming 650201, China; ynau_zhangying@163.com
4   College of Veterinary Medicine, China Agricultural University, Beijing 100193, China; sy20193050805@cau.edu.cn
*   Correspondence: 2011009@ynau.edu.cn (C.S.); ynsxh@aliyun.com (X.S.)
†   These authors contributed equally to this work.

Citation: Yao, J.; Li, J.; Gao, L.; He, Y.; Xie, J.; Zhu, P.; Zhang, Y.; Zhang, X.; Duan, L.; Yang, S.; et al. Epidemiological Investigation and Genetic Analysis of Pseudorabies Virus in Yunnan Province of China from 2017 to 2021. *Viruses* 2022, 14, 895. https://doi.org/10.3390/v14050895

Academic Editors: Yan-Dong Tang and Xiangdong Li

Received: 24 March 2022
Accepted: 15 April 2022
Published: 25 April 2022

Publisher's Note: MDPI stays neutral with regard to jurisdictional claims in published maps and institutional affiliations.

Abstract: In recent years, the prevalence of pseudorabies virus (PRV) has caused huge economic losses to the Chinese pig industry. Meanwhile, PRV infection in humans also sounded the alarm about its cross-species transmission from pigs to humans. To study the regional PRV epidemic, serological and epidemiological investigations of PRV in pig populations from Yunnan Province during 2017–2021 were performed. The results showed that 31.37% (6324/20,158, 95% CI 30.73–32.01) of serum samples were positive for PRV glycoprotein E (gE)-specific antibodies via enzyme-linked immunosorbent assay (ELISA). The risk factors, including the breeding scale and development stage, were significantly associated with PRV seroprevalence among pigs in Yunnan Province. Of the 416 tissue samples collected from PRV-suspected pigs in Yunnan Province, 43 (10.33%, 95% CI 7.41–13.26) samples were positive for PRV-*gE* nucleic acid in which 15 novel PRV strains from these PRV-positive samples were isolated, whose *gC* and *gE* sequences were analyzed. Phylogenetic analysis showed that all 15 isolates obtained in this study belonged to the genotype II. Additionally, the *gC* gene of one isolate (YuN-YL-2017) was genetically closer to variant PRV strains compared with others, while the *gE* gene was in the same clade with other classical PRV strains, indicating that this isolate might be a recombinant strain generated from the classical and variant strains. The results revealed the severe PRV epidemic in Yunnan Province and indicated that PRV variants are the major genotypes threatening the pig industry development.

Keywords: pseudorabies virus; seroprevalence; epidemiology; phylogenetic analysis; variants

1. Introduction

Pseudorabies (PR) is a devastating infectious disease that poses a huge threat to the development of the pig industry worldwide [1]. The causative agent of PR, pseudorabies virus (PRV) or Suid herpesvirus (SuHV-1), is an enveloped double-stranded DNA virus that belongs to the subfamily *Alphaherpesvirinae* of the family of *Herpesviridae* [2]. Pigs are known as the natural host and reservoir for PRV. The clinical symptoms of pigs infected with PRV vary depending on the growth stages: in newborn piglets, PRV infection causes severe diarrhea, vomiting, and neurological symptoms, resulting in high morbidity; in pregnant sows, PRV infection leads to reproductive failure [2,3]. Moreover, PRV has an intensive cross-species transmission capacity, which can infect a wide variety of animals, such as pigs, ruminants, carnivores, bears, etc. [4]. Notably, PRV transmission from pigs to humans has raised worldwide concerns since Chinese researchers recently have successfully isolated a variant PRV strain from an acute human encephalitis case [5].

Since the first detection of PRV in the United States, the disease caused by this pathogen has been observed in many countries, including Canada, China, and Hungary [6]. PR has been successfully controlled or eradicated in some countries or regions, such as Canada and Mexico, due to the application of multiple diagnosis approaches and glycoprotein E (gE)-deleted live or attenuated PRV vaccines [2]. However, this infectious disease remains widely prevalent in Chinese populations. Since late 2011 especially, PRs caused by PRV variants have frequently erupted in some Bartha-K61-immuized pig farms in China [7,8]. Subsequent experiments showed that the Bartha-K61 vaccine could not provide complete protection against these variants [8].

Currently, PRV strains are composed of two genotypes (genotype I and genotype II). PRV strains from Europe and USA belong to the genotype I, while most of genotype II PRV strains are isolated from Asian countries, mainly in China [2]. Moreover, the genotype II strains can be further divided into two sub-genotypes (classical PRV strains and variant PRV strains) [2]. According to the genetic characteristics among different PRV genotype strains, several amino acid (aa) insertions and deletions were observed, for example, the PRV genotype II strains have a 3-aa continuous deletion (^{75}VPG^{79}) in the UL27 gene and a 7-aa continuous insertion (^{63}AASTPAA^{69}) in the UL44 gene compared with PRV genotype I strains [9].

An investigation of the prevalence of PRV is required to build up strategies to control and even eradicate PR and minimize the risk of humans contacting this infectious pathogen. Though the prevalence and genetic characteristics of PRV have been documented in several regions or provinces of China [2,3,10,11], the relevant information in Yunnan Province in recent years is still not available. To fill in this gap, 20,158 pig serum samples were collected from 2017 to 2021 to investigate the epidemiology of PRV in Yunnan Province. Furthermore, the genetic characteristics of 15 newly isolated PRV strains were analyzed based on their *gC* and *gE* sequences.

2. Materials and Methods

2.1. Samples Collection

A total of 20,158 pig serum specimens were collected from 573 pig farms between March 2017 and December 2021, which nearly covered the entire Yunnan Province, China. The sampled pigs were chosen according to the breeding scale and breeding model. In brief, approximately equal numbers of specimens were collected from different growth stages (sucking piglets, nursery pigs, fattening pigs, sows, and gilts). Meanwhile, approximately equal sampling frequency was applied; 10, 25~30, and 50~60 serum samples were collected from each small (<100 sows), medium (100~500 sows), and large-scaled pig farm (>500 sows), respectively. In addition, tissue samples (such as brain, lymph node, lung, and kidney) were collected from 416 PRV infection-suspected pigs in 107 farms; the clinical symptoms of these diseased pigs mainly included encephalitis, diarrhea, fever, etc. The specimens were collected with standard procedures and delivered to Yunnan Animal Science and Veterinary Institute in a cold environment. Detailed information of each sample was documented.

2.2. Serological Detection of Anti PRV-gE Antibodies

Anti-gE antibodies in each serum sample were detected with Pseudorabies Virus (PRV)-gE antibody ELISA Kits (Cat: CP144, IDEXX Laboratories, Westrook, ME, USA) following the manufacturer's instructions, which could be used to differentiate the vaccine strain or field strain-infected pigs.

2.3. Virus Detection and Isolation

Viral DNA were extracted from the tissue samples using a DNA Isolation Kit (Genenode Biotech Co.Ltd., Beijing, China) according to the manufacturer's instructions. PCR was performed targeting the partial PRV-*gE* gene, with primers gE-F/R (gE-F: 5′-CCCAACGAC ACGGGCCTCTA-3′; gE-R:5′-GCACAGCACGCAGAGCCAGA-3′). The virus was isolated from PRV-positive tissue samples for subsequent experiments. Briefly, the tissue samples were homogenized and subjected to three freeze–thaw cycles. The supernatants, containing PRV virus, were filtered through a 0.22 μm filter after centrifugation and inoculated into a monolayer of BHK-21 or ST cells, which were cultured in a 5% CO_2 incubator at 37 °C. The supernatants and cells with obvious cytopathic effects (CPE) were harvested for plaque purification assays [3] and molecular identification by real-time PCR assays. Viral titers were determined by the Reed–Muench method in ST cells and the 50% lethal dose (LD_{50}) of which in mice models were calculated as described by Luo et al. [12].

2.4. Sequencing and Genetic Analysis

PCR was performed to amplify the complete sequences of *gE* and *gC* of 15 novel PRV strains as described previously [2]. The positive PCR products were purified and cloned into the pUCm-T vector. The plasmid carrying either the *gE* or *gC* gene was sequenced in duplicate. The full-length of *gE* or *gC* sequences of 15 newly isolated PRV strains and reference strains were compared using the DNAStar version 7.10 software. The phylogenetic tree based on the *gE* or *gC* gene was generated using the neighbor-joining (NJ) method in MEGA X software, with 1000 bootstrap replicates [13]. Detailed information of 15 novel PRV isolates and reference strains were available in the NCBI database as shown in Table 1.

Table 1. Detailed information of PRV strains identified in this study and reference strains, including strain name, collection year, isolation region, viral titer, the median lethal doses (LD_{50}) to mice, and GenBank accession numbers.

Strains	Collection Year	Isolation Region	Pig Farm Size	Tissue Type	$TCID_{50}$/ 0.1 mL	LD_{50}	GenBank Accession
YuN-YL-2017	2017	Yunan, China	Small	Lung, fattening pig	$10^{5.25}$	$10^{3.5}$	OM982597(*gC*), ON012780 (*gE*)
YuN-KD-2017	2017	Yunan, China	Large	Aborted fetus	$10^{6.58}$	$10^{2.65}$	OM982598 (*gC*), ON012781 (*gE*)
YuN-XN-2017	2017	Yunan, China	Large	Aborted fetus	$10^{5.75}$	$10^{2.85}$	OM982599 (*gC*), ON012782 (*gE*)
YuN-FL-2017	2017	Yunan, China	Medium	Aborted fetus	$10^{6.083}$	$10^{2.63}$	OM982600 (*gC*), ON012783 (*gE*)
YuN-QJ-2018	2018	Yunan, China	Medium	Aborted fetus	$10^{6.5}$	$10^{2.5}$	OM982601 (*gC*), ON012784 (*gE*)
YuN-LL-2018	2018	Yunan, China	Large	Aborted fetus	$10^{6.875}$	$10^{2.08}$	OM982602 (*gC*), ON012785 (*gE*)
YuN-KM-2018	2018	Yunan, China	Small	Aborted fetus	$10^{7.0}$	$10^{2.85}$	OM982603 (*gC*), ON012786 (*gE*)
YuN-YX-2019	2019	Yunan, China	Medium	Aborted fetus	$10^{6.0}$	$10^{1.80}$	OM982604 (*gC*), ON012787 (*gE*)
YuN-KM-2019	2019	Yunan, China	Large	Aborted fetus	$10^{6.38}$	$10^{2.0}$	OM982605 (*gC*), ON012788 (*gE*)
YuN-QJ-2019	2019	Yunan, China	Small	Aborted fetus	$10^{6.59}$	$10^{2.5}$	OM982606 (*gC*), ON012789 (*gE*)

Table 1. *Cont.*

Strains	Collection Year	Isolation Region	Pig Farm Size	Tissue Type	TCID$_{50}$/ 0.1 mL	LD$_{50}$	GenBank Accession
YuN-FY-2020	2020	Yunan, China	Large	Aborted fetus	$10^{7.12}$	$10^{2.43}$	OM982607 (*gC*), ON012790 (*gE*)
YuN-QJ-2020	2020	Yunan, China	Small	Aborted fetus	$10^{6.0}$	$10^{2.63}$	OM982608 (*gC*), ON012791 (*gE*)
YuN-ST-2020	2020	Yunan, China	Large	Aborted fetus	$10^{7.0}$	$10^{2.5}$	OM982609 (*gC*), ON012792 (*gE*)
YuN-DH-2021	2021	Yunan, China	Medium	Aborted fetus	$10^{6.67}$	$10^{2.38}$	OM982610 (*gC*), ON012793 (*gE*)
YuN-KM-2021	2021	Yunan, China	Large	Aborted fetus	$10^{6.25}$	$10^{2.43}$	OM982611 (*gC*), ON012794 (*gE*)
hSD-1	2019	Shandong, China			-	-	MT468550
JXCH2-16	2016	Jiangxi, China			-	-	MK806387
SD-18	2020	China			-	-	MT949536
HN1201	2012	Henan, China			-	-	KP722022
ZJ01	2012	Zhejiang, China			-	-	KM061380
SC	1986	Sichuan, China			-	-	KT809429
HLJ-2013	2013	Heilongjiang, China			-	-	MK080279
HeN1	2012	Henan, China			-	-	KP098534
Ea	1993	Hubei, China			-	-	KX423960
HuB17	2020	Hubei, China			-	-	MT949537
Fa	2012	Fujian, China			-	-	KM189913
JS-2012	2012	Jiangsu, China			-	-	KP257591
Bartha	-	Hungary			-	-	JF797217
Kaplan	-	Hungary			-	-	KJ717942
Kolchis	2010	Greece			-	-	KT983811

2.5. Data Analyses

The seroprevalence of PRV in pigs was presented as the minimum infection rate (MIR) with 95% confidence intervals (CIs). The statistical significance of PRV-gE seroprevalence among different groups was analyzed using a Chi-square test in SPSS 21.0 software (SPSS Inc., Chicago, IL, USA). A difference with a *p*-value < 0.05 was considered statistically significant.

3. Results

3.1. Seroprevalence of PRV-gE in Yunnan Province during 2017–2021

In total, 573 pig farms were included in this survey, where nearly all sampled pigs had been immunized with an attenuated PRV vaccine (Bartha-K61 or HB-98 strain) or inactivated PRV vaccine. Of the collected serum samples, 6324 out of 20,158 samples were seropositive for PRV-gE specific antibodies, contributing to the overall positive rate of 31.37% (95% CI 30.73–32.01). The seroprevalence rates of PRV-gE from March 2017 to Augest 2018, September 2018 to January 2020, and April 2020 to December 2021 were 29.25% (2355/8051), 41.48% (2449/5904), and 24.50% (1520/6203), respectively (Table 2) (*p* < 0.01).

Table 2. Seroprevalence of PRV-gE among pigs in Yunnan province with different risk factors.

	Category	No. Sample	No. Positive	% (95% CI)	*p*-Value
	March 2017 to August 2018	8051	2355	29.25 (28.26–30.24)	<0.001
Period	September 2018 to January 2020	5904	2449	41.48 (40.22–42.74)	<0.001
	April 2020 to December 2021	6203	1520	24.50 (23.43–25.57)	Reference
	Piglets	2644	442	16.72 (15.29–18.14)	Reference
	Nursery pigs	5304	1467	27.66 (26.45–28.86)	<0.001
Pig herd	Fattening pigs	5621	2301	40.94 (39.65–42.22)	<0.001
	Sows	4039	1245	30.82 (29.40–32.24)	<0.001
	Gilts	2123	761	35.84 (33.81–37.89)	<0.001
	Boars	427	108	25.29 (21.17–29.42)	<0.001
	Small	3438	1273	37.03 (35.41–39.64)	<0.001
Pig farm size	Medium	6273	1556	24.80 (23.74–25.87)	Reference
	Large	10,447	3495	33.45 (32.55–34.36)	<0.001
		20,158	6324	31.37 (30.73–32.01)	

In terms of pig herds, the average PRV-gE seroprevalence rate in piglets (16.72%, 442/2644) was significantly lower than these of other development stages of pigs (25.29~40.94%) ($p < 0.01$) (Table 2). Moreover, we further investigated the seroprevalence of PRV in pig farms with different breeding scales, which showed that the lowest seroprevalence was observed in medium scale farms (24.80%, 1556/6273), followed by small-scale farms and large-scale farms at 33.45% (3495/10,447) and 37.03% (1273/3438), respectively ($p < 0.01$) (Table 2).

3.2. PRV Detection and Viral Isolation

As shown in Table 3, of the 416 tissue samples collected from PR-suspected pigs, 43 (10.33%, 95% CI 7.41–13.26) samples were positive for PRV-gE nucleic acids. The detection rate of PRV among collected samples from March 2017 to August 2018, September 2018 to January 2020, and April 2020 to December 2021 were 9.04% (16/177), 14.56% (15/103), and 8.82% (12/136), respectively ($p > 0.05$). In terms of tissue samples from the pigs with different clinical symptoms, the positive rates of PRV infection among aborted fetuses (13.89%, 15/108) and piglets with neurological symptoms (18.07%, 15/83) were higher than other samples (5.78%, 13/225) ($p < 0.01$).

To further investigate the genetic features of PRV strains prevalent in Yunnan Province in recent years, 15 PRV strains were successfully isolated from the PRV-positive samples, purified via plaque purification, and further validated by PCR. The viral titers of these PRV strains were determined via the Reed–Muench method in ST cells, varying from $\sim 10^{5.25}$ to $10^{7.4}$ TCID$_{50}$/0.1 mL (Table 1). The subsequent animal experiments showed that the LD$_{50}$ of 15 novel PRV strains to six-week-old female Kunming-mice ranged from $\sim 10^{2.0}$ to $10^{3.5}$ TCID$_{50}$ (Table 1).

Table 3. The PRV-gE DNA positive rates among pigs with different risk factors.

	Category	No. Sample	No. Positive	% (95% CI)	*p*-Value
	March 2017 to August 2018	177	16	9.04 (4.82–13.26)	0.947
Period	September 2018 to January 2020	103	15	14.56 (7.75–21.38)	0.165
	April 2020 to December 2021	136	12	8.82 (4.06–13.59)	Reference
	Aborted fetus	108	15	13.89 (7.37–20.41)	< 0.01
Samples	Piglets with neurological symptoms	83	15	18.07 (9.79–26.35)	< 0.01
	Others	225	13	5.78 (2.73–8.83)	Reference
		416	43	10.33 (7.41–13.26)	

3.3. Phylogenetic Analysis

PRV *gE* and *gC* of the newly identified 15 PRV stains were amplified by PCR and cloned into a pUCm-T vector for sequencing [2]. According to the phylogenetic analysis based on PRV *gE* or *gC* sequences, all PRV strains, including 15 novel PRV strains and reference strains, were divided into two genotypes: genotype I and genotype II (Figure 1A,B). In agreement with a previous study [14], most of the isolates from China were clustered as genotype II, which could be further divided into the classical (before 2012) and variant (after 2012) sub-genotypes, while PRV strains from other parts, such as Europe and the U.S., belonged to genotype I. Notably, all 15 PRV isolates obtained in this study belonged to genotype II. Importantly, the *gE* phylogenetic tree showed that one isolate from Yunnan Province in 2017 (designed as YuN-YL-2017) was genetically closer to classical PRV strains compared with others (Figure 1A), while the *gC* gene was in the same clade with other PRV variants (Figure 1B).

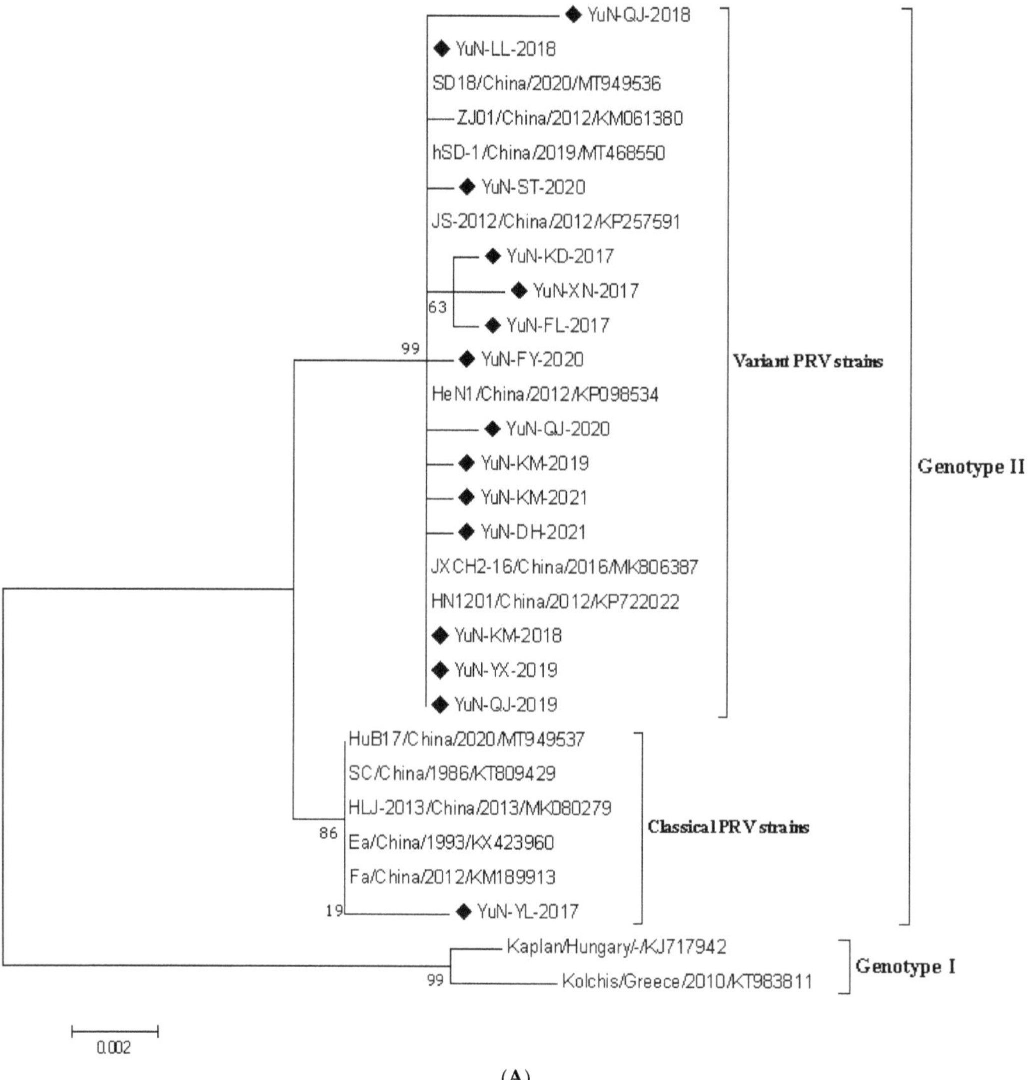

(**A**)

Figure 1. *Cont.*

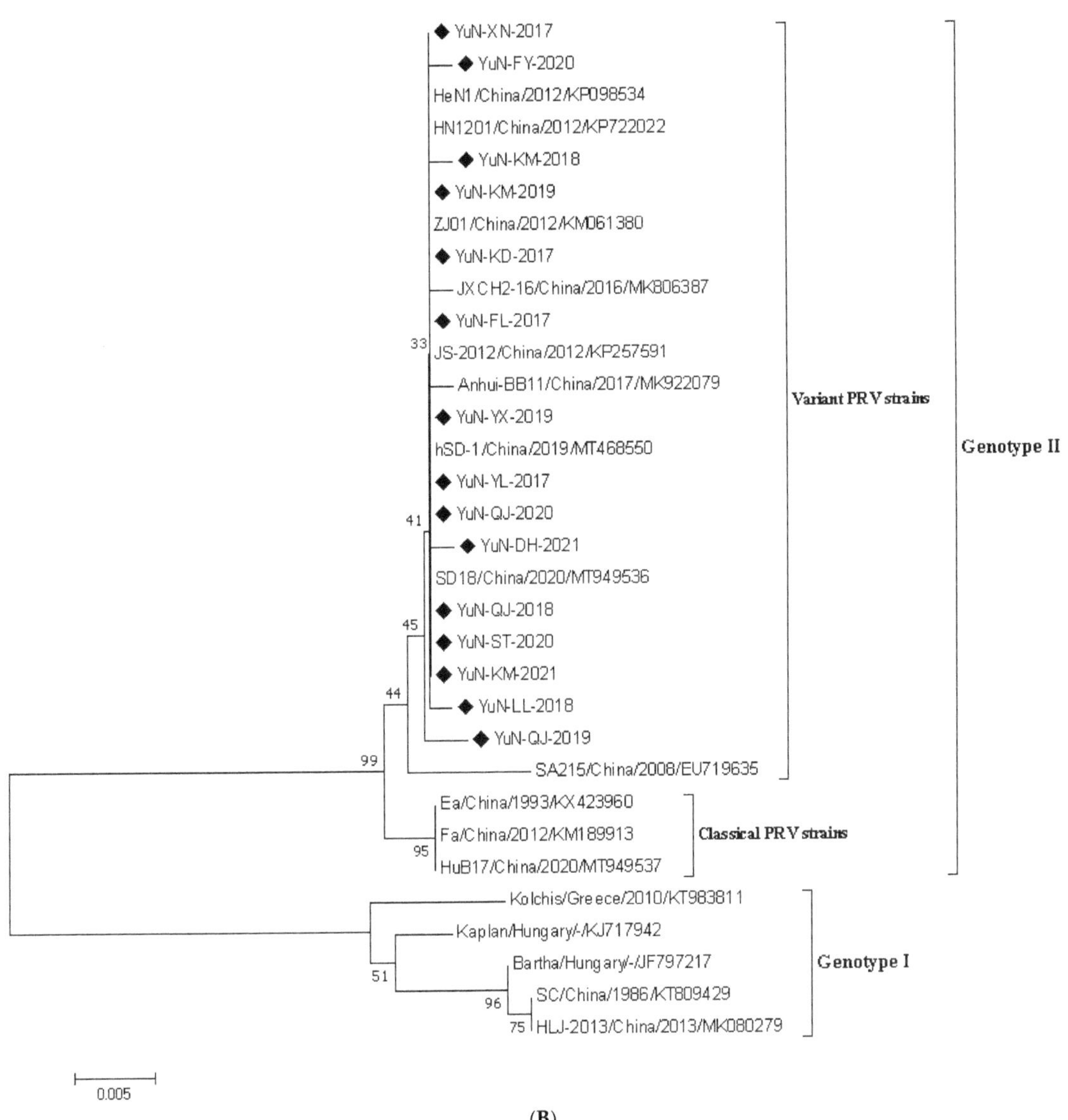

Figure 1. Phylogenetic analysis based on the nucleotide sequences of *gE* (**A**) and *gC* (**B**) genes of the 15 novel PRV isolates obtained in this study and other reference strains. A phylogenetic tree was generated using the neighbor-joining method with 1000 bootstrap replicates in MEGA X software. The black triangle represents the 15 PRV isolates.

3.4. Analysis of PRV gC and gE

The nucleotide and the corresponding amino acid sequence variations for *gC* (1464 bp) and *gE* (1734~1740 bp) genes of 15 novel PRV strains within the isolates were 0.0~0.3%, 0.0~0.7%, and 0.0~0.8%, 0.0~1.7%, respectively (Table 4). Moreover, compared with PRV variants and classical PRV strains, these 15 PRV strains exhibited a 99.6~100.0%, 99.1~99.4% nucleotide and 99.4~100.0%, 98.3~98.7% amino acid sequence identity in the gC gene and a 99.3~100.0%, 99.2~99.8% nucleotide and 98.6~100.0%, 98.8~99.5% amino acid sequence identity in the gE gene (Table 4), respectively. Remarkable, the *gE* gene of the YuN-YL-2017 strain showed a higher sequence homology with classical PRV strains (such as Ea and Fa), while its *gC* gene was highly homologous to the variants (such as HeN1 and ZJ01).

Table 4. Sequence similarity analysis of the gC and gE sequences of PRV strains identified in this study.

Selected Strains	Nucleotide Sequences (%)		Amino Acid Sequences (%)	
	gC	*gE*	gC	gE
15 PRV strains obtained in this study	99.7~100.0	99.3~100.0	99.2~100.0	98.3~100.0
Compared with PRV variants	99.6~100.0	99.3~100.0	99.4~100.0	98.6~100.0
Compared with classical PRV strains	99.1~99.4	99.2~99.8	98.3~98.7	98.8~99.5
Compared with PRV strains in genotype I	94.2~96.1	97.4~97.8	89.2~96.5	95.3~96.0

PRV gC and gE proteins sequences among the PRV strains were further aligned. The results revealed that there was no amino acid insertions or deletions, but several mutations were observed among gC proteins of 15 PRV strains when compared with other PRV variants. Except for some amino acid mutations among gE proteins, compared with the PRV variants, the YuN-YL-2017 strain had two amino acid deletions at site 48 (D) and 498 (D).

4. Discussion

Since the emergence of variant PRV strains in China in 2011, the disease caused by PRV variants has been considered a major factor contributing to huge economic losses to the swine industry. Recently, the cross-species transmission events of PRV from pigs to humans have also attracted increasing attention [15,16]. Great efforts have been made for the control of PR; particularly, this disease was listed in the "Mid- and Long-term Animal Disease Prevention and Control Program in China (2012–2020)". Nevertheless, PR remains widely spread in Chinese pig populations and pose a challenge for other animals breeding in China, such as fox and mink. Thus, obtaining accurate data on the epidemiological characteristics of PRV is beneficial for formulating control or eradication measures.

The present results showed that the average PRV-gE seropositive rate was 31.37% among 20,158 serum samples from Yunnan Province from 2017 to 2021. Further analysis showed that the PRV seroprevalence in Yunnan Province between September 2018 to January 2020 (41.48%, 2449/5904) was higher than these during March 2017–August 2018 (29.25%, 2355/8051) and April 2020–December 2021 (24.50%, 1520/6203), and a similar epidemiological trend was also observed in the pathogen detection section in this study. Since the outbreaks of African swine fever (ASF) and its rapid spread since August 2018 contributed to the substantial reduction of the sow population in China, many PRV-positive sows might have been introduced into pig farms to keep the breeding scale, which contributed to the high seroprevalence of PRV in some regions of China [2]. Owing to the fact that the prevalence of ASF has been controlled in 2020 [17] and the excessive pig production in China recently, many pig farms subsequently focused on the prevention or eradication of other infectious diseases, including PR, classical swine fever, etc.

Two factors, "pig herd" and "breeding scale", were significantly associated with the seroprevalence of PRV of pigs in Yunnan Province. The seroprevalence of PRV in fattening pigs, sows, and gilts was higher than these in piglets, nursery pigs, and boars; similar results were also observed in previous research [18]. On one hand, the occurrences of PR in fattening pigs are often neglected since they only display mild symptoms. On the other hand, fattening pigs are not immunized with PRV vaccines in some pig farms. Meanwhile, long-term feeding increases the probability of PRV infection among sows and gilts. Moreover, as reported in Lin's study [2], we also found that a lower PRV-gE seropositive rate among pigs was detected from medium-sized farms compared with those in large and small ones, which might suggest that the medium-density feeding mode is more suitable for infectious diseases control.

The gC protein participates in viral abortion on a host cellular surface; meanwhile, this protein is an important target for neutralizing an antibody [19]. The gE protein is mainly involved in viral virulence [4]. Phylogenetic analysis based on the *gE* or *gC* gene revealed that PRV strains prevalent worldwide can be divided into two genotypes (namely, genotype I and genotype II), and most PRV strains circulated in China belong to the genotype II [1,14]. In line with these, 15 novel PRV strains obtained in this study formed one large clade with Chinese PRV variants (after 2012) and Chinese classical PRV strains (before 2012) and belonged to the genotype II, which showed a distinct relationship to genotype I strains, such as Bartha and Backer (Figure 1A,B). Remarkably, one isolate, namely, YuN-YL-2017, was identified as a PRV variant according to the genetic analysis of *gC* gene, which belonged to the classical strains according to the *gE* gene. These results indicated this strain might be a recombinant variant strain. Further analysis showed that the LD_{50} of YuN-YL-2017 to mice was higher than those of other PRV variants ($10^{3.5}$ $TCID_{50}$ VS $10^{2.0-2.8}$ $TCID_{50}$), suggesting that the recombinant event in the genome of YuN-YL-2017 decreased its virulence to mice, and the underlying mechanisms will be explored in the future.

5. Conclusions

In conclusion, this study comprehensively investigated the prevalence and genetic features of PRV in Yunnan Province from 2017 to 2021, showing that PR remains highly prevalent among pig populations in Yunnan Province, China. Phylogenetic analysis showed that all 15 PRV strains isolated in this study belonged to the genotype II, displaying a distinct evolutionary relationship with the Bartha strain in genotype I, which might partly explain the immune failure of the PRV Bartha-K61 vaccine in pigs challenged by PRV variants, and further suggesting that novel vaccines should be developed for the control of PR in this region. In addition, the results above also highlighted the importance of continuous monitoring the molecular epidemiology of such recombinant PRV strains in the future.

Author Contributions: Conceptualization, J.Y., C.S. and J.L.; methodology, J.Y. and X.S.; software, L.G. and Y.H.; formal analysis, C.S., J.L., L.G., S.Y. and Y.H.; investigation, J.Y., Y.H., J.X., P.Z., Y.Z., X.Z. and L.D.; resources, S.Y., X.S., C.S. and J.L.; original draft writing, J.Y.; review and revising, J.Y., C.S., J.L. and X.S.; funding acquisition, J.Y. and X.S. All authors participated in editing the article and approved the final manuscript. All authors have read and agreed to the published version of the manuscript.

Funding: This work was supported by the National Key Research and Development Program of China (No. 2017YFD0501800) and the Major Specialized Projects of Yunnan Science and Technology (No. 202102AE090007).

Institutional Review Board Statement: All animal experiments were conducted according to the rules of Animal Ethics Committee of the Yunnan Agricultural University, Kunming, China (YNAU2019llwyh022).

Informed Consent Statement: Not applicable.

Data Availability Statement: All data presented in the present study can be found in online repositories.

Conflicts of Interest: The authors declare no conflict of interest.

References

1. Sun, Y.; Luo, Y.; Wang, C.-H.; Yuan, J.; Li, N.; Song, K.; Qiu, H.-J. Control of swine pseudorabies in China: Opportunities and limitations. *Vet. Microbiol.* **2016**, *183*, 119–124. [CrossRef] [PubMed]
2. Lin, Y.; Tan, L.; Wang, C.; He, S.; Fang, L.; Wang, Z.; Zhong, Y.; Zhang, K.; Liu, D.; Yang, Q.; et al. Serological Investigation and Genetic Characteristics of Pseudorabies Virus in Hunan Province of China From 2016 to 2020. *Front. Vet. Sci.* **2021**, *8*, 1503. [CrossRef] [PubMed]
3. Zheng, H.-H.; Jin, Y.; Hou, C.-Y.; Li, X.-S.; Zhao, L.; Wang, Z.-Y.; Chen, H.-Y. Seroprevalence investigation and genetic analysis of pseudorabies virus within pig populations in Henan province of China during 2018–2019. *Infect. Genet. Evol.* **2021**, *92*, 104835. [CrossRef] [PubMed]

4. Tan, L.; Shu, X.; Xu, K.; Liao, F.; Song, C.; Duan, D.; Yang, S.; Yao, J.; Wang, A. Homologous recombination technology generated recombinant pseudorabies virus expressing EGFP facilitates to evaluate its susceptibility to different cells and screen antiviral compounds. *Res. Vet. Sci.* **2022**, *145*, 125–134. [CrossRef] [PubMed]

5. Liu, Q.; Wang, X.; Xie, C.; Ding, S.; Yang, H.; Guo, S.; Li, J.; Qin, L.; Ban, F.; Wang, D.; et al. A Novel Human Acute Encephalitis Caused by Pseudorabies Virus Variant Strain. *Clin. Infect. Dis.* **2021**, *73*, E3690–E3700. [CrossRef] [PubMed]

6. Mueller, T.; Hahn, E.C.; Tottewitz, F.; Kramer, M.; Klupp, B.G.; Mettenleiter, T.C.; Freuling, C. Pseudorabies virus in wild swine: A global perspective. *Arch. Virol.* **2011**, *156*, 1691–1705. [CrossRef]

7. An, T.-Q.; Peng, J.-M.; Tian, Z.-J.; Zhao, H.-Y.; Li, N.; Liu, Y.-M.; Chen, J.-Z.; Leng, C.-L.; Sun, Y.; Chang, D.; et al. Pseudorabies Virus Variant in Bartha-K61-Vaccinated Pigs, China, 2012. *Emerg. Infect. Dis.* **2013**, *19*, 1749–1755. [CrossRef]

8. Tong, W.; Liu, F.; Zheng, H.; Liang, C.; Zhou, Y.-J.; Jiang, Y.-F.; Shan, T.-L.; Gao, F.; Li, G.-X.; Tong, G.-Z. Emergence of a Pseudorabies virus variant with increased virulence to piglets. *Vet. Microbiol.* **2015**, *181*, 236–240. [CrossRef]

9. Tan, L.; Yao, J.; Yang, Y.-D.; Luo, W.; Yuan, X.-M.; Yang, L.-C.; Wang, A.-B. Current Status and Challenge of Pseudorabies Virus Infection in China. *Virol. Sin.* **2021**, *36*, 588–607. [CrossRef] [PubMed]

10. Gu, J.; Hu, D.; Peng, T.; Wang, Y.; Ma, Z.; Liu, Z.; Meng, F.; Shang, Y.; Liu, S.; Xiao, Y. Epidemiological investigation of pseudorabies in Shandong Province from 2013 to 2016. *Transbound. Emerg. Dis.* **2018**, *65*, 890–898. [CrossRef] [PubMed]

11. Xia, L.; Sun, Q.; Wang, J.; Chen, Q.; Liu, P.; Shen, C.; Sun, J.; Tu, Y.; Shen, S.; Zhu, J.; et al. Epidemiology of pseudorabies in intensive pig farms in Shanghai, China: Herd-level prevalence and risk factors. *Prev. Vet. Med.* **2018**, *159*, 51–56. [CrossRef] [PubMed]

12. Luo, Y.; Li, N.; Cong, X.; Wang, C.-H.; Du, M.; Li, L.; Zhao, B.; Yuan, J.; Liu, D.-D.; Li, S.; et al. Pathogenicity and genomic characterization of a pseudorabies virus variant isolated from Bartha-K61-vaccinated swine population in China. *Vet. Microbiol.* **2014**, *174*, 107–115. [CrossRef] [PubMed]

13. Kumar, S.; Stecher, G.; Li, M.; Knyaz, C.; Tamura, K. MEGA X: Molecular Evolutionary Genetics Analysis across Computing Platforms. *Mol. Biol. Evol.* **2018**, *35*, 1547–1549. [CrossRef] [PubMed]

14. Ye, C.; Zhang, Q.-Z.; Tian, Z.-J.; Zheng, H.; Zhao, K.; Liu, F.; Guo, J.-C.; Tong, W.; Jiang, C.-G.; Wang, S.-J.; et al. Genomic characterization of emergent pseudorabies virus in China reveals marked sequence divergence: Evidence for the existence of two major genotypes. *Virology* **2015**, *483*, 32–43. [CrossRef] [PubMed]

15. Wong, G.; Lu, J.; Zhang, W.; Gao, G.F. Pseudorabies virus: A neglected zoonotic pathogen in humans? *Emerg. Microbes Infect.* **2019**, *8*, 150–154. [CrossRef] [PubMed]

16. Guo, Z.; Chen, X.-X.; Zhang, G. Human PRV Infection in China: An Alarm to Accelerate Eradication of PRV in Domestic Pigs. *Virol. Sin.* **2021**, *36*, 823–828. [CrossRef]

17. Liu, J.; Liu, B.; Shan, B.; Wei, S.; An, T.; Shen, G.; Chen, Z. Prevalence of African Swine Fever in China, 2018–2019. *J. Med. Virol.* **2020**, *92*, 1023–1034. [CrossRef]

18. Zhou, H.; Pan, Y.; Liu, M.; Han, Z. Prevalence of Porcine Pseudorabies Virus and Its Coinfection Rate in Heilongjiang Province in China from 2013 to 2018. *Viral Immunol.* **2020**, *33*, 550–554. [CrossRef]

19. Gerdts, V.; Jons, A.; Mettenleiter, T.C. Potency of an experimental DNA vaccine against Aujeszky's disease in pigs. *Vet. Microbiol.* **1999**, *66*, 1–13. [CrossRef]

*viruses*

MDPI

Article

Serological Investigation and Genetic Characteristics of Pseudorabies Virus between 2019 and 2021 in Henan Province of China

Ximeng Chen [1,†], Hongxuan Li [1,†], Qianlei Zhu [2,†], Hongying Chen [1,*], Zhenya Wang [3,*], Lanlan Zheng [1], Fang Liu [1] and Zhanyong Wei [1]

[1] International Joint Research Center of National Animal Immunology, College of Veterinary Medicine, Henan Agricultural University, Zhengdong New District Longzi Lake 15#, Zhengzhou 450046, China; simonarchimonde@126.com (X.C.); li17836997254@126.com (H.L.); zhll2000@sohu.com (L.Z.); liufang.vet@henau.edu.cn (F.L.); weizhanyong@henau.edu.cn (Z.W.)

[2] Henan Center for Animal Disease Control and Prevention, Zhengzhou 450002, China; zhuqianlei123@126.com

[3] Key Laboratory of "Runliang" Antiviral Medicines Research and Development, Institute of Drug Discovery & Development, Zhengzhou University, Zhengzhou 450001, China

[*] Correspondence: chhy927@163.com (H.C.); zhenyawang@zzu.edu.cn (Z.W.); Tel./Fax: +86-371-55369208 (H.C.)

[†] These authors contributed equally to this work.

Abstract: In late 2011, severe pseudorabies (PR) outbreaks occurred among swine herds vaccinated with the Bartha-K61 vaccine in many provinces of China, causing enormous economic losses for the pork industry. To understand the epidemic profile and genetic characteristics of the pseudorabies virus (PRV), a total of 35,796 serum samples were collected from 1090 pig farms of different breeding scales between 2019 and 2021 in the Henan province where swine had been immunized with the Bartha-K61 vaccine, and PRV glycoprotein E (gE)-specific antibodies were detected using an enzyme-linked immunosorbent assay (ELISA). The results reveal that the overall positive rate for PRV gE antibodies was 20.33% (7276/35,796), which decreased from 25.00% (2596/10,385) in 2019 to 16.69% (2222/13,315) in 2021, demonstrating that PR still existed widely in pig herds in the Henan province but displayed a decreasing trend. Further analysis suggested that the PRV-seropositive rate may be associated with farm size, farm category, quarter, region and the cross-regional transportation of livestock. Moreover, the gE gene complete sequences of 18 PRV isolates were obtained, and they shared a high identity (97.1–100.0%) with reference strains at the nucleotide level. Interestingly, the phylogenetic analysis based on the gE complete sequences found that there were both classical strains and variant strains in pig herds. The deduced amino acid sequence analysis of the gE gene showed that there were unique amino acids in the classical strains, the variant strains and genotype II strains. This study provides epidemiological data that could be useful in the prevention of pseudorabies in Henan, China, and this finding contributed to our understanding of the epidemiology and evolution of PRV.

Keywords: Aujeszky's disease; pseudorabies virus; epidemiological characteristics; phylogenetic analysis

Citation: Chen, X.; Li, H.; Zhu, Q.; Chen, H.; Wang, Z.; Zheng, L.; Liu, F.; Wei, Z. Serological Investigation and Genetic Characteristics of Pseudorabies Virus between 2019 and 2021 in Henan Province of China. *Viruses* 2022, 14, 1685. https://doi.org/10.3390/v14081685

Academic Editors: Yan-Dong Tang and Xiangdong Li

Received: 12 June 2022
Accepted: 27 July 2022
Published: 30 July 2022

Publisher's Note: MDPI stays neutral with regard to jurisdictional claims in published maps and institutional affiliations.

1. Introduction

Porcine pseudorabies (PR), also called Aujeszky's disease, is an acute infectious disease with high morbidity and mortality which has caused great harm to the global animal husbandry industry [1]. The causative agent, suid herpesvirus type 1 (SuHV-1, syn. Aujeszky's disease virus or pseudorabies virus (PRV)), is an enveloped and double-stranded linear DNA virus which is taxonomically classified into the family *Herpesviridae*, subfamily *Alphaherpesvirinae*, genus *Varicellovirus* [2]. Members of the family Suidae (true pigs), as unique natural hosts and reservoirs of PRV, are susceptible to PRV at all ages [3], with clinical symptoms including lethal encephalitis and neurological symptoms in neonatal piglets, respiratory disease in finishing pigs and reproductive failure in infected sows [4,5]. Without specific host tropism, PRV was previously deemed to infect a wide range of mammals including ruminants, carnivores and rodents, with the exception of higher-order primates and humans [6–8]. Nonetheless, multiple reports have shown that PRV can cross the livestock-to-human species barrier, invade the human central nervous system (CNS) and induce human encephalitis [9–12], in the case of a PRV strain isolated from acute human encephalitis, suggesting that humans may be a potential host for PRV [13]. Besides, there are currently no effective drugs to prevent and treat the disease, which could pose a potential threat to public health.

In China, PR was first reported in the 1950s, and was well-controlled between 1990 and 2011 due to the widespread use of glycoprotein E (gE)-negative vaccines based on the Hungarian strain Bartha-K61 [14]. However, at the end of 2011, PR outbreaks took place on many Chinese farms where swine had been immunized with the Bartha-K61 vaccine, and rapidly spread to many regions of China, causing a significant impact on the pig-farming industry [15–18]. Studies indicated that the re-emerging PR was caused by PRV variants, and Bartha-K61 vaccines only confer partial protection on piglets against these new variants [15,17,19,20]. It is reported that about 2600 newborn piglets and 200 sows died from the PRV variant infections in one swine herd in the Henan province, resulting in a direct economic loss of at least one million Chinese yuan (CNY 156,000) [21].

To better prevent and control the disease, serological and molecular epidemiology investigations of PRV infection were carried out extensively in the Henan province of China before 2019 [18,19,22], but its epidemic profile in recent years remains unclear. Hence, the current study was designed to investigate the seroprevalence of PR from 2019 to 2021 in the Henan province of China and to analyze the gE gene of PRV strains isolated in this study.

2. Materials and Methods

2.1. Sample Collection

From January 2019 to December 2021, a total of 35,796 serum samples were collected from 35,796 pigs on 1090 farms in the Henan province of China (Figure 1). The 1090 pig farms consisted of three different categories: slaughterhouses, commercial pig farms and breeding farms. Owing to various breeding scales of stock farms, 15~29, 30 and 31~200 samples were collected from each small- (<500 pigs), medium- (500~2000 pigs) and large-scale farms (>2000 pigs), respectively. Each batch of serum samples was collected on the day after blood collection and stored at −80 °C, and the specific detection was completed the next day. The results from different cities will be aggregated into the database monthly. Additionally, 389 tissue samples containing brains, lymph nodes and lungs were collected from 389 diseased pigs during outbreaks of PR. These tissue samples were homogenized, diluted at a ratio of 1:5 in phosphate-buffered saline (PBS) and then stored at −80 °C until use.

2.2. Serological Detection

Serological analysis was performed to detect anti-gE antibodies (Abs) using commercial blocking enzyme-linked immunosorbent assay (ELISA) kits (Pseudorabies Virus gE Antibody Test Kit) (IDEXX Laboratories, Westbrook, ME, USA), according to the manufacturer's procedure. Results were calculated by dividing the absorbance at 650 nm (A (650)) of the tested sample by the mean A (650) of the negative control, resulting in a sample/negative (S/N) value that was used to differentiate infected from vaccinated animals. The S/N value was inversely proportional to the quantity of Abs. Therefore, S/N ≤ 0.60 was considered as positive, S/N > 0.70 was regarded as negative, and 0.60 < S/N ≤ 0.70 was judged to be suspicious, necessitating that the test be retested or retested over time.

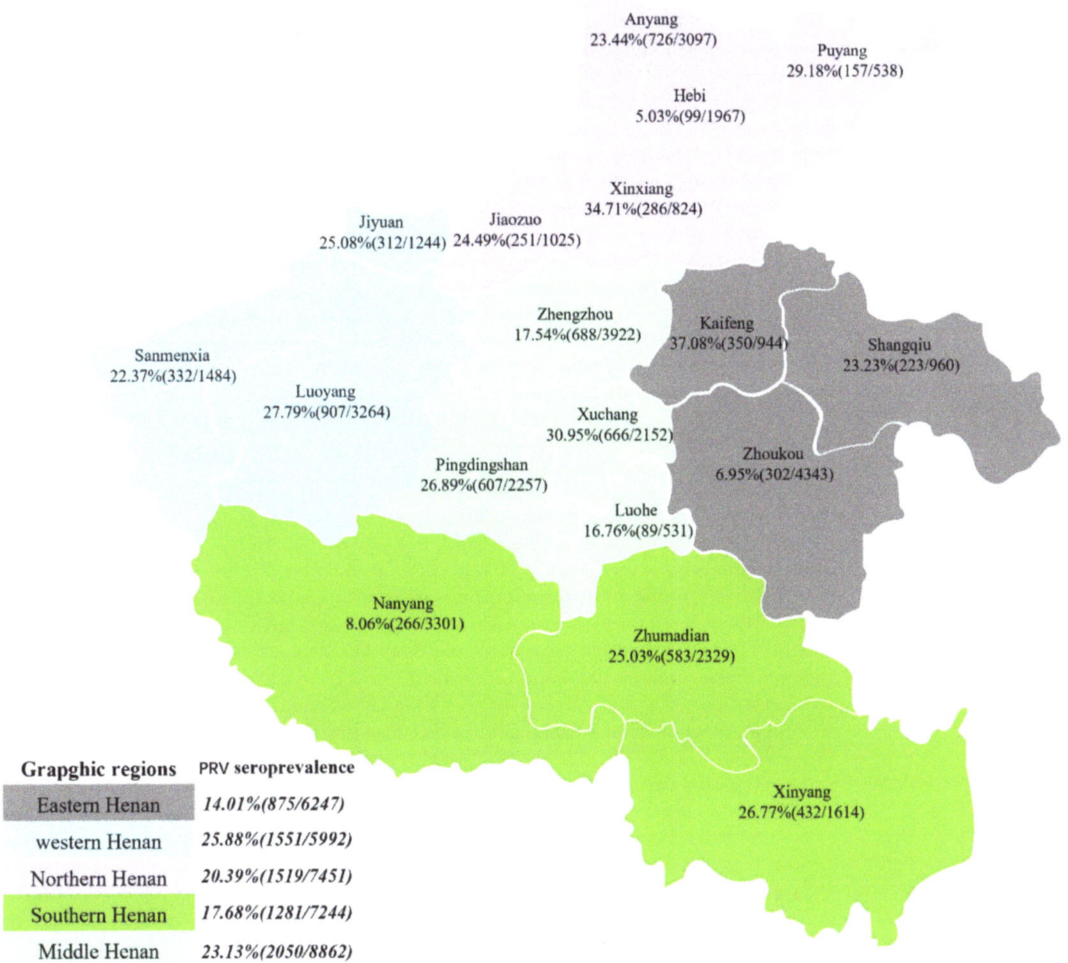

Figure 1. Seroprevalence of pseudorabies virus (PRV) in different geographical regions in the Henan province of China during 2019–2021.

2.3. Virus Detection and Isolation

The tissue samples were freeze–thawed three times to release the virus, and the homogenates were centrifuged at $8000\times g$ for 5 min. Viral DNA was extracted from the tissue samples using the Nucleic Acid Extraction and Purification Kit (Omega Bio-Tek, Inc., Norcross, GA, USA) according to the manufacturer's instructions. The presence of PRV nucleic acids was screened by polymerase chain reaction (PCR) as described previously [22]. PRV-positive tissue supernatants were filtered through a 0.22 μm filter (EMD Millipore, Billerica, MA, USA), and then inoculated into swine testis (ST) cell monolayers for 2 h in a 37 °C incubator supplemented with 5% CO_2. The ST cells were maintained for 72 h to produce a cytopathic effect (CPE) in Dulbecco's modified Eagle's medium (DMEM) (Gibco, Billings, MT, USA) supplemented with 2% heat-inactivated fetal bovine serum (FBS) (Gibco). When obvious CPE appeared, cells were collected. The collected viruses were plaque-purified in 2 mL of DMEM containing 1% (w/v) low-melting-point agarose and 2% FBS, and their identity was validated by PCR as described previously [22].

2.4. Sequencing and Phylogenetic Analysis

The full-length gE gene was amplified by PCR from viral DNA extracted from the isolates as described previously [23]. The PCR product was purified using the V-ELUTE Gel Mini Purification Kit (Beijing Zoman Biotechnology Co., Ltd., Beijing, China), then ligated into the pMD^{TM} 18 T Vector Cloning Kit (Takara, Dalian, China), and finally transformed into Trelief^{TM} 5α Chemically Competent Cells (Tsingke, Beijing, China). The positive monoclonal clones were verified by PCR and sequenced by Wuhan AuGCT DNA-SYN Biotechnology Co., Ltd. (Wuhan, China), in triplicate.

Twenty-eight PRV strains were retrieved from the NCBI database and served as the reference strains. The nucleotide sequences and the corresponding amino acid variations in the gE gene between PRV isolates sequenced in our study and the reference strains were analyzed using DNASTAR Lasergene.v7.1 (DNASTAR, Inc., Madison, WI, USA). The phylogenetic tree was constructed by the neighbor-joining method using the MEGA 7.0 software (www.megasoftware.net) with a bootstrap of 1000 replicates [24].

3. Results

3.1. Seroprevalence of PRV in Henan Province

In the present study, a total of 35,796 serum samples were collected in the Henan province, including 10,385 in 2019, 12,096 in 2020 and 13,315 in 2021. Meanwhile, 1090 involved farms were inoculated, including 324 in 2019, 385 in 2020 and 381 in 2021, from which sera were collected. Data obtained from the ELISA assay demonstrated that the positive rate at the serum sample level was 25.00% in 2019, 20.32% in 2020 and 16.69% in 2021 (Table 1), respectively, while at the farm level it was 50.31% in 2019, 50.65% in 2020 and 49.87% in 2021 (Table 2). Of the 1090 farms, 2278 serum samples were obtained from 98 small farms, 26,430 from 881 medium farms and 7088 from 111 large farms. The positive rate at small farms, medium farms and large farms was 25.72%, 22.15% and 11.78% at the serum sample level, and 69.39%, 48.81% and 45.05% at the farm level (Table 2). The seroprevalence of slaughterhouses, commercial pig farms and breeding farms was 25.83%, 23.80% and 8.14% (Table 1).

Regional variation, seasonal variation and other factors with a potential association with PRV were evaluated. From 2019 to 2021, the total seroprevalence was 14.01% in eastern Henan, 25.88% in western Henan, 17.68% in southern Henan, 20.39% in northern Henan and 23.13% in middle Henan (Figure 2a), revealing that PRV infection rates varied from region to region. From 2019 to 2021, the total seroprevalence rate was 21.70% in the first quarter (Q1), 24.16% in Q2, 21.38% in Q3 and 15.29% in Q4, implying PRV infection rates varied by quarter. In detail, the peak of PRV infection was 31.59% in Q3 of 2019 but 28.15% in Q1 of 2020 and 21.48% in Q2 of 2021, indicating that the peak time of PRV infection each year was inconsistent, while the lowest number of infections usually occurred in Q4 (Figure 2b). The performance in different distributions was more pronounced. During 2019 to 2021, the peak of PRV infection was 41.24% and 26.36% in Q1 in western Henan and northern Henan, compared to 30.36% in Q2 in middle Henan, and 19.52% and 26.87% in Q3 in eastern Henan and southern Henan (Figure 2c). Nevertheless, the lowest number of infections occurred in Q4, with 8.59% in the east, 22.32% in the west, 12.27% in the south, 12.04% in the north and 20.52% in the middle. Last but not least, the specific changes in the seroprevalence rates by quarter and by region are shown in Figure 2d, and the annual infection rate for each city is illustrated by Figure S1a–c.

Table 1. The detection results for pseudorabies virus in different regions and different pig farms in the Henan province of China during 2019–2021.

Time	The Regions of Henan Province, China					The Categories of Pig Farms			Total
	Eastern Henan	Western Henan	Southern Henan	Northern Henan	Middle Henan	Slaughterhouse	Commercial Farm Households	Breeding Stock Farm	
January–March in 2019 (Q1)	0.18% (1/570)	61.92% (161/260)	11.40% (107/939)	18.78% (74/394)	22.13% (52/235)	62.50% (50/80)	27.50% (327/1189)	1.59% (18/1129)	16.47% (395/2398)
April–June in 2019 (Q2)	16.13% (121/750)	25.45% (112/440)	31.22% (153/490)	44.03% (295/670)	44.25% (200/452)	65.00% (195/300)	29.90% (572/1913)	19.35% (114/589)	31.44% (881/2802)
July–September in 2019(Q3)	17.79% (71/399)	30.11% (137/455)	39.21% (149/380)	44.13% (124/281)	31.12% (164/527)	46.86% (112/239)	27.96% (402/1438)	35.89% (131/365)	31.59% (645/2042)
October–December in 2019 (Q4)	15.22% (35/230)	24.67% (111/450)	12.69% (82/646)	17.76% (114/642)	28.34% (333/1175)	19.22% (69/359)	28.48% (508/1784)	9.80% (98/1000)	21.48% (675/3143)
Total for 2019	11.70% (228/1949)	32.46% (521/1605)	20.00% (491/2455)	30.55% (607/1987)	31.35% (749/2389)	43.56% (426/978)	28.61% (1809/6324)	11.71% (361/3083)	25.00% (2596/10,385)
January–March in 2020 (Q1)	24.42% (138/565)	52.94% (63/119)	26.96% (86/319)	29.97% (205/684)	24.11% (102/423)	33.47% (159/475)	29.45% (415/1409)	19.06% (61/320)	28.15% (594/2110)
April–June in 2020 (Q2)	2.56% (10/390)	20.05% (149/743)	14.53% (100/688)	17.59% (57/324)	40.00% (212/530)	14.76% (31/210)	25.76% (448/1739)	6.75% (49/726)	19.74% (528/2675)
July–September in 2020(Q3)	14.63% (109/745)	26.06% (135/518)	22.00% (187/850)	17.01% (272/1599)	20.64% (219/1061)	18.97% (184/970)	22.87% (705/3082)	4.58% (33/721)	19.32% (922/4773)
October–December in 2020 (Q4)	8.46% (33/390)	29.82% (201/674)	14.65% (103/703)	1.98% (7/353)	16.75% (70/418)	17.78% (16/90)	17.27% (328/1899)	12.75% (70/549)	16.31% (414/2538)
Total for 2020	13.88% (290/2090)	26.68% (548/2054)	18.59% (476/2560)	18.28% (541/2960)	24.79% (603/2432)	22.33% (383/1715)	23.09% (1862/8065)	9.20% (213/2316)	20.32% (2458/12,096)
January–March in 2021 (Q1)	0.00% (0/330)	18.67% (56/300)	81.82% (90/110)	27.89% (94/337)	17.20% (65/378)	18.64% (11/59)	21.88% (258/1179)	16.59% (36/217)	20.96% (305/1455)
April–June in 2021 (Q2)	49.17% (118/240)	29.12% (166/570)	11.71% (41/350)	13.16% (109/828)	20.91% (252/1205)	14.11% (59/418)	25.28% (565/2235)	11.48% (62/540)	21.48% (686/3193)
July–September in 2021(Q3)	26.29% (168/639)	21.65% (176/813)	14.04% (99/705)	13.84% (62/448)	16.82% (215/1278)	18.67% (84/450)	23.90% (625/2615)	1.34% (11/818)	18.54% (720/3883)
October–December in 2021 (Q4)	7.11% (71/999)	12.92% (84/650)	7.89% (84/1064)	11.90% (106/891)	14.07% (166/1180)	20.42% (106/519)	13.95% (398/2854)	0.61% (7/1141)	10.68% (511/4784)
Total for 2021	16.17% (357/2208)	20.66% (482/2333)	14.09% (314/2229)	18.28% (371/2504)	17.27% (698/4041)	17.98% (260/1446)	20.78% (1846/8883)	3.88% (116/2986)	16.69% (2222/13,315)
Total for 2019–2021	14.01% (875/6247)	25.88% (1551/5992)	17.68% (1281/7244)	20.39% (1519/7451)	23.13% (2050/8862)	25.83% (1069/4139)	23.80% (5517/23,177)	8.14% (690/8480)	20.33% (7276/35,796)

Note: Q represents quarter.

Table 2. The prevalence of pseudorabies virus in different-scale pig herds in the Henan province during 2019–2021.

Year	Prevalence on Pig Farms of Different Sizes				Prevalence of Serum Samples on Pig Farms of Different Sizes			
	Small (15~29)	Medium (30)	Large (31~200)	Total	Small	Medium	Large	Total
2019	73.33% (22/30)	48.81% (123/252)	42.86% (18/42)	50.31% (163/324)	31.52% (197/625)	26.57% (2009/7560)	17.73% (390/2200)	25.00% (2596/10,385)
2020	69.23% (27/39)	49.20% (154/313)	42.42% (14/33)	50.65% (195/385)	24.49% (226/923)	21.84% (2051/9390)	10.15% (181/1783)	20.32% (2458/12,096)
2021	65.52% (19/29)	48.42% (153/316)	50.00% (18/36)	49.87% (190/381)	22.33% (163/730)	18.93% (1795/9480)	8.50% (264/3105)	16.69% (2222/13,315)
Total	69.39% (68/98)	48.81% (430/881)	45.05% (50/111)	50.28% (548/1090)	25.72% (586/2278)	22.15% (5855/26,430)	11.78% (835/7088)	20.33% (7276/35,796)

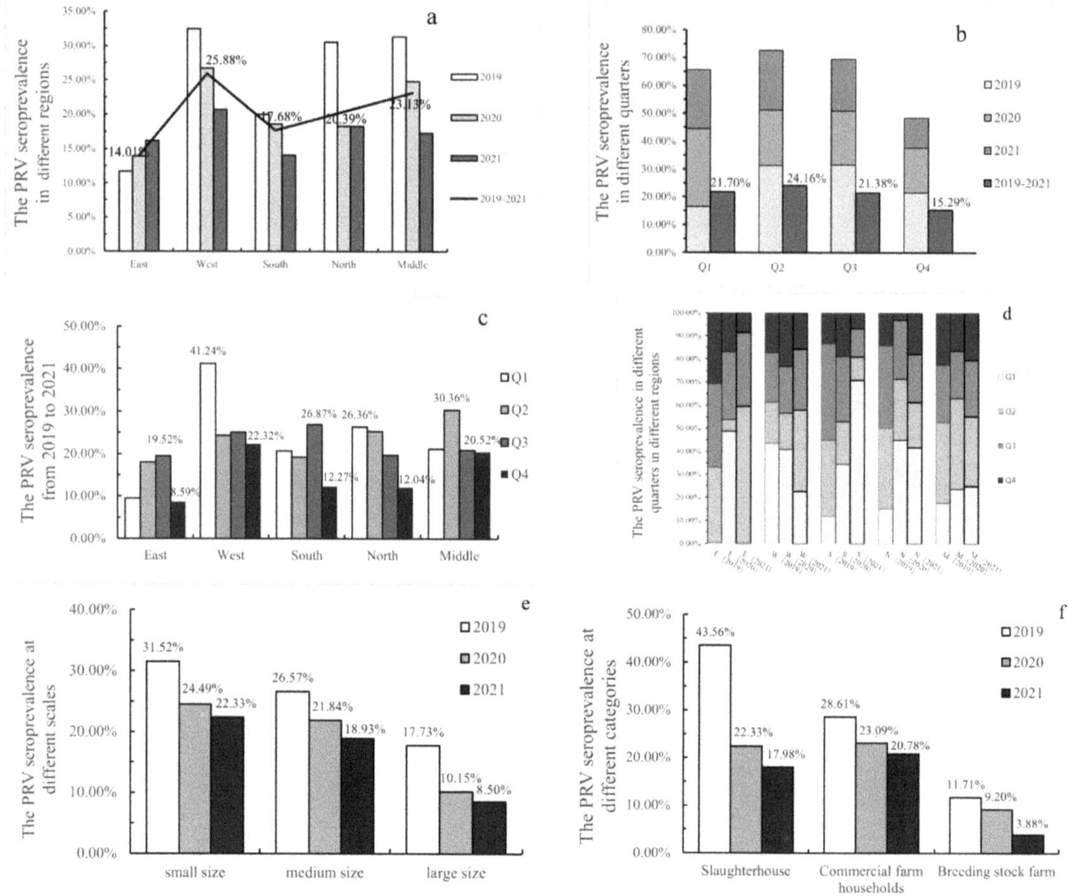

Figure 2. Seroprevalence rate of PRV-gE in the Henan province, China, 2019–2021, from different perspectives. (**a**) Seroprevalence rate of PRV-gE in different regions. (**b**) Seroprevalence rate of PRV-gE in different quarters. (**c,d**) Seroprevalence rate of PRV-gE in different quarters in different regions. (**e**) The PRV-gE seropositive rate of farms of different scales. (**f**) The PRV-gE seropositive rate of farms of different categories.

3.2. The Results of PRV Isolation

The PRV gE gene was detected by PCR in 52 of the 389 (13.37%) clinical case samples. PCR-positive tissue samples were inoculated into ST cells, and distinct CPEs appeared after three to four blind passages on ST cells. PCR detection results confirmed that 18 PRV isolates were obtained, and these were named as HD-1, HD-2, PY-1, SC1, SC2, HN-CM, HN-CY, HN-GM, HN-HY, HN-LL, HN-LH, HN-LY, HN-YH, HN-HX, HN-MY, HN-WZ, HN-XT and HN-YY.

3.3. Genetic Analysis Results Based on gE

The full-length gE genes of 18 PRV isolates were cloned and sequenced. These sequences were submitted to GenBank under the accession numbers listed in Table S1. Of the 18 full gE sequences examined, 12 isolates were 1737 nucleotides (nt) in length, encoding a protein of 578 amino acids (aas), and the other 6 isolates (HN-YH, PY-1, HD1, HD2, SC1 and SC2) were 1740 nt-long (579 aas). The nt sequencing analysis showed that the 18 isolates shared 98.6–99.9% nucleotide identity with each other, and these isolates shared 97.1% to 100.0% nucleotide sequence identity with 28 PRV reference strains (Table S1). The phylogenetic analysis indicated that PRV strains could be divided into two genotypes (Figure 3), and these observations were corroborated by gE nt sequence identities of 18 PRV isolates in this study and 28 PRV reference strains. Genotype I included 10 PRV reference strains from Europe and America, and 18 PRV isolates in this study displayed 97.1% to 98.2% nucleotide sequence identity with the 10 PRV reference strains. All 18 PRV isolates in this study were clustered in genotype II, together with 12 Chinese PRV strains (after 2012), 3 early Chinese PRV strains (before 2012) and 3 Asian PRV strains, including the South Korean strain Yangsan, Japanese strain RC1 and Malaysian strain P-PrV, with nt sequence identities ranging from 98.7% to 99.9% between 18 PRV isolates in this study and the 18 reference strains. Phylogenetically, 6 of 18 PRV isolates (HN-YH, PY-1, HD1, HD2, SC1 and SC2) in this study and 10 Chinese variant PRV reference strains (after 2012) were distributed within the variant PRV cluster in genotype II and had 99.6–100.0% nucleotide sequence identity. The remaining 12 isolates, which represent the current PRV classical strains from a high-positive city at different times in each region in the Henan province, were located in the classical cluster with 2 early Chinese classical PRV reference strains, Ea and SC (before 2012), and 2 Chinese PRV reference strains, HuN-YY and HuB17 (after 2012), with nt sequence identity ranging from 99.1% to 99.9%. These results reveal that among the 18 isolates, HN-YH, PY-1, HD1, HD2, SC1 and SC2 belonged to PRV variants of genotype II, and the other 12 isolates were grouped as the classical PRV strains. In addition, combined with amino acid sequence analysis, unique aa variations in gE protein were used as molecular markers for differentiating gE clade divisions in genetic evolution analysis (Figure 3).

The aa sequences of gE protein in the 46 PRV strains were aligned, and all aa insertions and aa substitutions are depicted in Figure 4. Compared with genotype I strains, 36 genotype II strains had an aa insertion at position 48 (34D/ 2no insertion) and 18 aa substitutions at positions 59 (36D/Y→36N), 63 (36N→35D/1N), 106 (36V→36L), 122 (36A→35S/1A), 149 (36R→36M), 179 (36T→36S), 181 (36R/Q→36L), 215 (36L→36A), 216 (36A→36D), 472 (36G→36R), 474 (36R→36H), 504 (36A→35I/1V), 509 (36S→36A), 522 (36V→35A/1T), 526 (36A→36P), 573 (36S→35N/1S), 577 (36N→33M /2N/1H) and 578 (36A→33S/2A/1V).

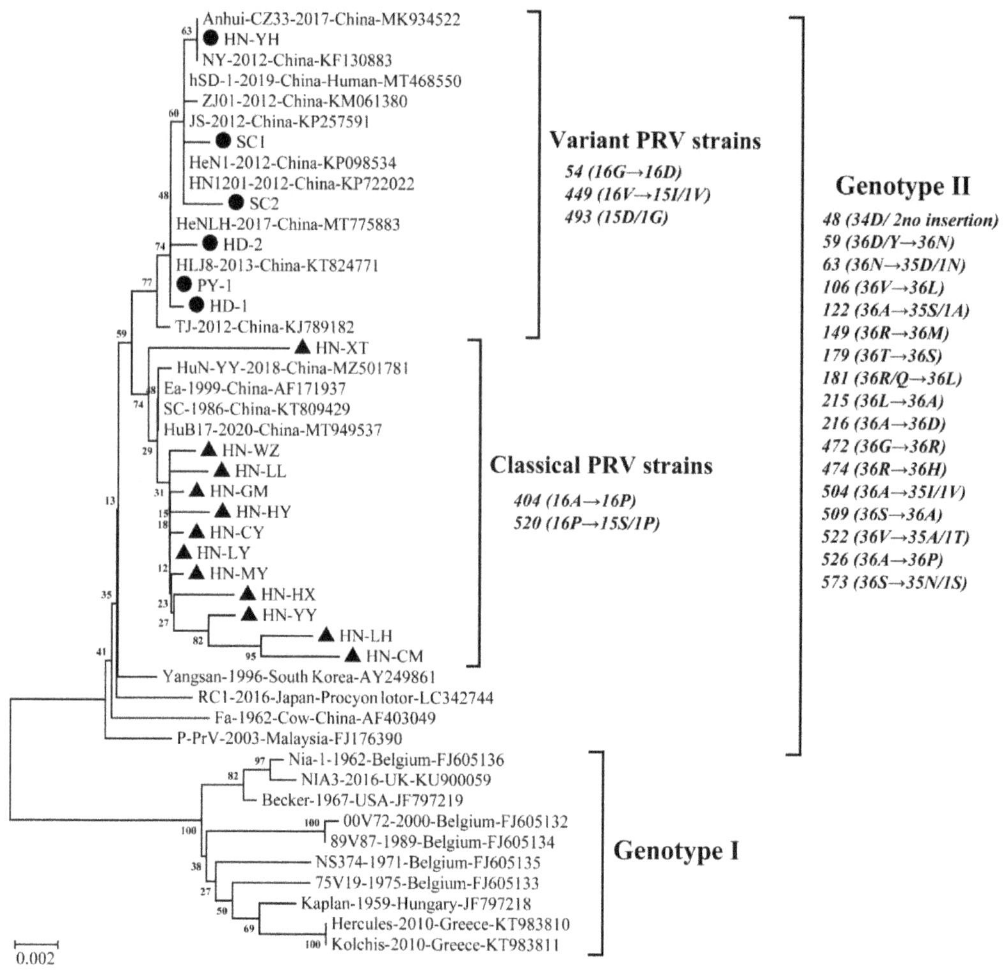

Figure 3. Phylogenetic tree based on the nucleotide sequences of gE gene of 18 PRV isolates determined in this study and 28 reference strains. Neighbor-joining trees were constructed with p-distance model and bootstrapping at 1000 replicates. Black solid circles (•) and black solid triangles (▲) represent variant strains and classical strains in this study, respectively.

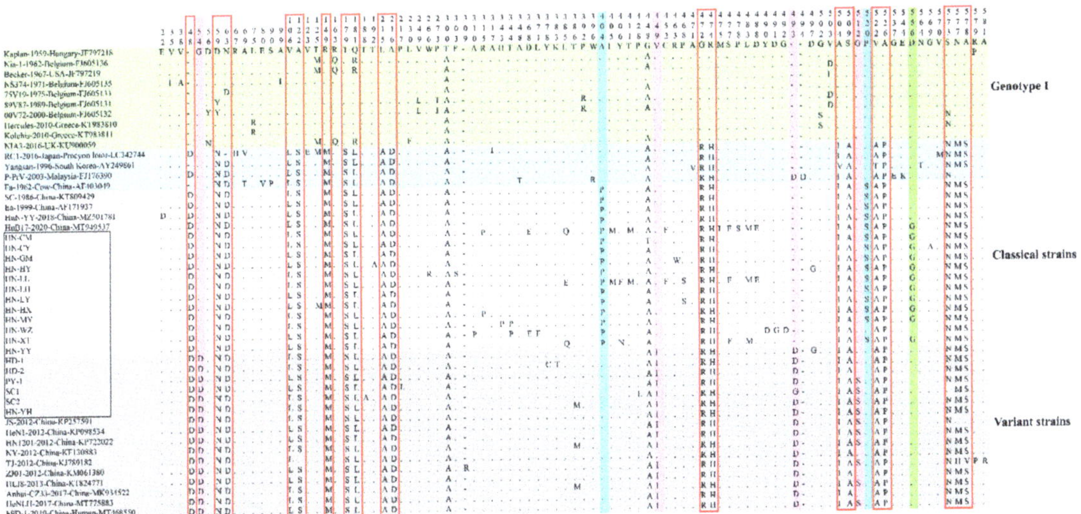

Figure 4. All amino acid mutation sites of gE protein of 18 PRV isolates sequenced in this study and 28 reference strains. All the strains were clustered into two genotypes, and genotype I only contained European–American PRV strains (mint green), while genotype II included Chinese variant strains (cloud gray), Chinese classical strains (pearl river gray) and Asian strains (Maya blue). The PRV isolates acquired in this study are shown in the hollow black rectangle. Compared with genotype I strains, unique amino acids substitutions of genotype II strains are shown in the hollow red rectangles. Amino acid mutations in variant strains are highlighted in lavender violet. The arctic blue areas represent unique amino acids substitutions in classical PRV strains. The screamin' green area represents unique amino acid substitutions in classical PRV strains from the Henan province.

4. Discussion

Due to dense swine populations, notifiable PR remains one of the most important diseases in regions of South America and some regions in Europe and Asia, especially in China, which is the largest producer of pork products in the world [3]. PR is recognized to have caused devastating damage to the Chinese swine industry and was listed in the "Middle-Long-term Animal Disease Prevention and Control Program in China (2012–2020)" [25], which aims to eradicate PR in pig farms in China by the end of 2020 (The State Council of the People's Republic of China 2012). As a result, the prevalence of virulent PRV has declined significantly in China since 2012, but the eradication target has not yet been reached. As almost all pig farms in China are still using the gE-deleted PRV vaccine (Bartha K61 strain), detection of gE-specific antibodies by ELISA remains a rapid and effective method for differentiating between infected and vaccinated animals (DIVA) [26]. In this study, the overall positive rate for PRV gE antibodies decreased from 25.00% in 2019 to 16.69% in 2021, being significantly lower than 94.2% (49/52) in 2011 [18] and 30.14% (1419/4708) in 2018 [22]. It can be seen that the PR epidemic situation in the Henan province has exhibited an obvious downward trend, which is in agreement with the tendency of eradication. Recent epidemiological studies showed that the seroprevalence of PRV decreased from 38.20% in 2018 to 18.12% in 2020 in the Shandong province [27] from 62.40% in 2013 to 51.59% in 2018 in Tianjin [28] and from 20.9% in 2013 to 11.6% in 2018 in the Heilongjiang province [29], but increased from 19.91% in 2016 to 25.46% in 2020 in the Hunan province [30]. These data reveal that the positive rate of PRV infection in pigs differed in different geographical regions of China, which may have been caused by different situations of co-infection, vaccine protection, feeding management, introduction and quarantine, etc. The positive rate in farms in the Henan province was 50.31% (163/324) in 2019, 50.65% (195/385) in 2020 and 49.87% (190/381) in 2021 (Table 2). Associated epidemiological studies showed that PRV gE-antibody-positive rates of farms were 58.2% (124/213) in 23 regions of China [31], 49.76% (516/1037) from 2010 to 2018 in Tianjin [28] and 43.19% (349/808) in the Hunan province during 2016–2020 [30]. A possible reason for why the Henan province was more affected by the variant PR outbreaks is because some sows that were negative but could have transmitted PRV were not culled. Compared with other provinces, it took longer for the Henan province to complete the evolution of PRV. All data demonstrate that PRV is still circulating in Chinese pig herds.

In this study, the seroprevalence in small farms, medium farms and large farms was 25.72%, 22.15% and 11.78% at the serum sample level. This suggests the larger-scale pig farms may be more likely to benefit from stricter biosecurity measures, such as compulsory pig vaccination campaigns and sufficient regulations to reduce PR incidence and prevalence. In addition, the seroprevalence of serum samples from pig farms of different sizes (Figure 2e) and pig farms of different categories, including slaughterhouses, commercial pig farms and breeding farms (Figure 2f), declined year by year, implying that PRV was effectively controlled. It may be that the enhanced immune quality of PRV and the development of PR eradication have contributed to the effective control of pseudorabies. The immune quality of PRV, including the vaccine quality and vaccination density, has been improved, thereby increasing the resistance of pigs to the disease, which would be an effective factor for controlling the epidemic of PRV [26,32]. Regional prevention strategies should be adopted [32]. At the end of 2020, the statistics released by the Henan Survey Organization National Bureau of Statistics Information Network showed that the pig population exhibited a clear tendency toward south > west > central > north > east. Southern and western herds were larger but had greatly lower PRV-gE positivity, especially in Nanyang, at 8.06% (266/3301), in the south and Zhoukou, at 6.95% (302/4343), in the west, suggesting that swine farms in these regions may have developed or may be developing eradication of pseudorabies.

PRV seroprevalence was reported to be higher in autumn than in other seasons [22,30]. In this study, the highest seroprevalence rate was 24.16% (2095/8670) in Q2. This difference may be due to differences in sample size, time span or regional scope, resulting in insufficient data to determine the underlying causes of sudden fluctuations in a given location compared to a country. It was usually be speculated to be latent infection of the virus [8], co-infection with other pathogens [33] and cross-species transmission, etc. Furthermore, the seroprevalence rate in eastern Henan rose sharply from 0.00% (0/330) in Q1 to 49.17% (118/240) in Q2 in 2021, compared with the increase from 0.18% (1/570) in Q1 to 16.13% (121/750) in Q2 in 2019, which may be associated with the rate of 81.82% (90/110) seen in Q1 in 2021 in southern Henan. This suggests that the seasonal or seasonal infection may be influenced by the input and output of pigs and is easily misjudged as a temperature factor or others. Benefiting from sufficiently detailed data in the present study, the three sets of data in Table 1 may objectively explain this issue by the spread of the virus with the cross-regional transportation of live pigs. Additionally, seasonal and regional variations in PRV infection rates were evident which are in agreement with Sun's study [16].

Nonetheless, no province has taken the lead in completing the eradication of PR. Regardless of near full vaccination rates, the virus, similar to other herpesviruses, can establish latent infection in the host's peripheral nervous system via PRV in the field and reactivate after natural stimuli or stress factors, which is recognized as the most critical source of infection [15,34]. The currently available vaccines only provide partial protection against virulent PRV infection. Recombination of PRV strains in Suidae may result in changes in antigenicity, virulence and thus immune failure, which could be the source of continuing epidemics in China [35]. PRV may have spread from domestic pigs to dogs, and then to wild boars, in which the virus established itself and continues to circulate [36,37], and PRV may also cause bovine death through interspecies infection [38]. In particular, gE is a virulence factor of PRV infection in pigs and determines tropism for the central nervous system [39], which is frequently used to investigate the epidemiology and evolution of the virus [19,40]. Therefore, a serosurvey to detect specific antibodies against PRV and the etiological as well as genetic characteristics of PRV isolates would contribute to the rational use of vaccines and other novel viral inhibitors.

The phylogenetic analysis indicated that 10 PRV strains from Europe and America were identified as genotype I, whereas 18 PRV isolates in this study together with 12 Chinese PRV strains (after 2012), 3 early Chinese PRV strains (before 2012) and 3 Asian PRV strains, including South Korean strain Yangsan, Japanese strain RC1 and Malaysian strain P-PrV, were clustered in genotype II, which is in accordance with a previous report that most PRV isolates from China and other countries in Asia were classified into genotype II, whereas PRV strains from Europe and America were assigned to genotype I [41]. Interestingly, all 12 PRV isolates in this study and 1 PRV isolate reported in the Hunan province [30] belonged to classical PRV strains, meaning that the currently popular PRV strains could possibly recover to the rates of the classical strains, which requires further investigation. As previously shown, the nt insertions at 138–140 resulted in an aa insertion at position 48 (D) [16,20,22]. In this study, the aa insertion at position 48 (34D/ 2no insertion), which occurred in all genotype II strains except the South Korean strain P-PrV from a pig and the Chinese strain Fa from a cow might be a characterization of the pseudorabies gE gene circulating in Asian pig herds after the first global outbreak. This finding confirms that PRV strains circulating in China evolved independently of strains isolated in Europe and America [42]. Furthermore, a cluster of 16 classical PRV strains has unique aa substitutions at positions 404 (16A→16P) and 520 (16P→15S/1P) compared with other clades. Surprisingly, the classical strains from the Henan province have a unique aa substitution at position 556 (12D→11G/1D) compared with other strains. Whether these mutations affect the virulence of PRV isolates currently circulating in China warrants further investigation.

Both an aa insertion at position 493 (15D/1G) and two aa mutations at positions 54 (16G→16D) and 449 (16V→15I/1V) were shared among the 16 PRV variants, which differ from other strains. A total of 10 out of 16 PRV variants (including 3 isolates in this study and 7 reference strains) had one substitution at position 512 (G→S). The nt insertions at positions 1472–1474 resulted in one aa insertion at position 493 (15D/1G), as also reported in previous studies [16,20,22], which represents a unique feature of variant PRV strains appearing after the pseudorabies outbreak in China in 2011. This supports the view that PRV strains in China may have evolved independently, leading to the emergence of variant strains [42]. Another feature found in the cluster of PRV variants was that nt substitutions at positions 161, 228 and 1345 resulted in aa substitutions at positions 54 (16G→16D), 449 (16V→15I/1V) and 512 (16G→10S/6G). Notably, the mutation at position 512 occurred in some of the variant strains, which indicated that PRV may still be mutating slowly. Due to the epitopes of gE protein localized at the aa positions 52–238 [43], the aa at position 54 (G→D) may be responsible for immune evasion. Previous studies showed novel PRV variants exhibited enhanced pathogenicity [35], and the variant strains HN1201, TJ, JS-2012 and hSD-1/2019 were more virulent and neurotropic to mice or pigs than the classical strains [13,17,44,45]. Whether the virulence of these isolates acquired in this study is enhanced remains to be further studied, involving proliferation characteristics in different types of macrophages, immune responses, pathogenicity to mice and pigs, etc.

In summary, the serological results demonstrate that that PR was still endemic at high levels in intensive pig herds in the Henan province, China, but displayed an obvious decreasing trend. Phylogenetic analysis based on the complete gE sequence found that both classical strains and variant strains existed in pig herds. Our finding of PR transmission through pigs transported across regions points to an inadequacy of PR prevention, and ongoing monitoring of PRV should be implemented for prevention and control.

Supplementary Materials: The supplementary material for this article can be found online: https://www.mdpi.com/article/10.3390/v14081685/s1. Figure S1. Seroprevalence of pseudorabies virus (PRV) in geographical distributions in Henan province of China from January to December. (a) Seroprevalence in 2019. (b) Seroprevalence in 2020. (c) Seroprevalence in 2021. Table S1. The information of PRV strains isolated in this study and reference strains for sequence alignment and phylogenetic analysis.

Author Contributions: Data curation, X.C.; Formal analysis, H.L.; Investigation, Q.Z.; Methodology, Z.W. (Zhanyong Wei) and F.L.; Software, Z.W. (Zhenya Wang); Writing—original draft, X.C.; Writing—review and editing, L.Z. and H.C. All authors have read and agreed to the published version of the manuscript.

Funding: This work was supported by the Department of Science and Technology of Henan province (No. 222102310081), the Special Project of scientific and Technological Research and development of Henan Province (No. 222102110053), Zhongyuan High Level Talents Special Support Plan (No. 204200510015), the Henan open competition mechanism to select the best candidate program (No. 211110111000) and Fang's family (Hong Kong) foundation.

Institutional Review Board Statement: All experimental procedures were reviewed and approved by the Henan Agriculture University Animal Care and Use Committee (license number SCXK (Henan) 2013–0001).

Informed Consent Statement: Not applicable.

Data Availability Statement: The accession numbers presented in this study can be found in online repositories (https://www.ncbi.nlm.nih.gov/) accessed on 24 May 2022.

Acknowledgments: We appreciate Liuhui Zhang, Youyi Zhao, Tong Xu, Chengyao Hou, Huihua Zheng, Yuanyuan Han, Yi Ai and Linlin Chen for their help. As the first author, I sincerely thank Xiaoqing Zeng, Heng Zhou, Bingqian Wang, Qixiao Ye, Zhongyi Ou, Jian Huang, Gang Huang, and Xinxin Yao for their encouragement and support.

Conflicts of Interest: The authors declare that they have no conflict of interest.

References

1. Nauwynck, H.J. Functional aspects of Aujeszky's disease (pseudorabies) viral proteins with relation to invasion, virulence and immunogenicity. *Vet. Microbiol.* **1997**, *55*, 3–11. [CrossRef]
2. Mettenleiter, T.C. Aujeszky's disease (pseudorabies) virus: The virus and molecular pathogenesis—State of the art, June 1999. *Vet. Res.* **2000**, *31*, 99–115. [CrossRef]
3. Müller, T.; Hahn, E.C.; Tottewitz, F.; Kramer, M.; Klupp, B.G.; Mettenleiter, T.C.; Freuling, C. Pseudorabies virus in wild swine: A global perspective. *Arch. Virol.* **2011**, *156*, 1691–1705. [CrossRef] [PubMed]
4. Hahn, E.C.; Fadl-Alla, B.; Lichtensteiger, C.A. Variation of Aujeszky's disease viruses in wild swine in USA. *Vet. Microbiol.* **2010**, *143*, 45–51. [CrossRef] [PubMed]
5. Pomeranz, L.E.; Reynolds, A.E.; Hengartner, C.J. Molecular biology of pseudorabies virus: Impact on neurovirology and veterinary medicine. *Microbiol. Mol. Biol. Rev.* **2005**, *69*, 462–500. [CrossRef] [PubMed]
6. Fenner, F.; Bachman, P.A.; Gibbs, E.J.P. Pseudorabies. In *Veterinary Virology*; Fenner, F., Bachman, P.A., Gibbs, E.J.P., Eds.; Academic Press, Inc.: San Diego, CA, USA, 1987; pp. 353–356.
7. Marcaccini, A.; López Peña, M.; Quiroga, M.I.; Bermúdez, R.; Nieto, J.M.; Alemañ, N. Pseudorabies virus infection in mink: A host-specific pathogenesis. *Vet. Immunol. Immunopathol.* **2008**, *124*, 264–273. [CrossRef] [PubMed]
8. Pensaert, M.B.; Kluge, P. Pseudorabies Virus (Aujeszky's Disease). In *Virus Infections of Porcines*; Pensaert, M.B., Ed.; Elsevier Science Publishers: New York, NY, USA, 1989; pp. 39–64.
9. Ai, J.W.; Weng, S.S.; Cheng, Q.; Cui, P.; Li, Y.J.; Wu, H.L.; Zhu, Y.M.; Xu, B.; Zhang, W.H. Human Endophthalmitis Caused by Pseudorabies Virus Infection, China, 2017. *Emerg. Infect. Dis.* **2018**, *24*, 1087–1090. [CrossRef] [PubMed]
10. Meng, X.J.; Lindsay, D.S.; Sriranganathan, N. Wild boars as sources for infectious diseases in livestock and humans. *Philos. Trans. R. Soc. Lond. B Biol. Sci.* **2009**, *364*, 2697–2707. [CrossRef] [PubMed]
11. Wang, D.; Tao, X.; Fei, M.; Chen, J.; Guo, W.; Li, P.; Wang, J. Human encephalitis caused by pseudorabies virus infection: A case report. *J. Neurovirol.* **2020**, *26*, 442–448. [CrossRef] [PubMed]
12. Yang, H.; Han, H.; Wang, H.; Cui, Y.; Liu, H.; Ding, S. A Case of Human Viral Encephalitis Caused by Pseudorabies Virus Infection in China. *Front. Neurol.* **2019**, *10*, 534. [CrossRef] [PubMed]
13. Liu, Q.; Wang, X.; Xie, C.; Ding, S.; Yang, H.; Guo, S.; Li, J.; Qin, L.; Ban, F.; Wang, D.; et al. A Novel Human Acute Encephalitis Caused by Pseudorabies Virus Variant Strain. *Clin. Infect. Dis. Off. Publ. Infect. Dis. Soc. Am.* **2021**, *73*, e3690–e3700. [CrossRef] [PubMed]
14. Tong, G.; Chen, H.C. Pseudorabies epidemic status and control measures in China. *Chin. J. Vet. Sci.* **1999**, *19*, 1–2.
15. An, T.Q.; Peng, J.M.; Tian, Z.J.; Zhao, H.Y.; Li, N.; Liu, Y.M.; Chen, J.Z.; Leng, C.L.; Sun, Y.; Chang, D.; et al. Pseudorabies virus variant in Bartha-K61-vaccinated pigs, China, 2012. *Emerg. Infect. Dis.* **2013**, *19*, 1749–1755. [CrossRef] [PubMed]
16. Sun, Y.; Liang, W.; Liu, Q.; Zhao, T.; Zhu, H.; Hua, L.; Peng, Z.; Tang, X.; Stratton, C.W.; Zhou, D.; et al. Epidemiological and genetic characteristics of swine pseudorabies virus in mainland China between 2012 and 2017. *PeerJ* **2018**, *6*, e5785. [CrossRef]

17. Tong, W.; Liu, F.; Zheng, H.; Liang, C.; Zhou, Y.J.; Jiang, Y.F.; Shan, T.L.; Gao, F.; Li, G.X.; Tong, G.Z. Emergence of a Pseudorabies virus variant with increased virulence to piglets. *Vet. Microbiol.* **2015**, *181*, 236–240. [CrossRef] [PubMed]
18. Wu, R.; Bai, C.; Sun, J.; Chang, S.; Zhang, X. Emergence of virulent pseudorabies virus infection in northern China. *J. Vet. Sci.* **2013**, *14*, 363–365. [CrossRef] [PubMed]
19. Wang, Y.; Qiao, S.; Li, X.; Xie, W.; Guo, J.; Li, Q.; Liu, X.; Hou, J.; Xu, Y.; Wang, L.; et al. Molecular epidemiology of outbreak-associated pseudorabies virus (PRV) strains in central China. *Virus Genes* **2015**, *50*, 401–409. [CrossRef]
20. Yu, X.; Zhou, Z.; Hu, D.; Zhang, Q.; Han, T.; Li, X.; Gu, X.; Yuan, L.; Zhang, S.; Wang, B.; et al. Pathogenic pseudorabies virus, China, 2012. *Emerg. Infect. Dis.* **2014**, *20*, 102–104. [CrossRef]
21. Ye, P.G. New trends of the pseudorabies infection in China. *Swine Indus* **2013**, *6*, 23.
22. Zheng, H.H.; Jin, Y.; Hou, C.Y.; Li, X.S.; Zhao, L.; Wang, Z.Y.; Chen, H.Y. Seroprevalence investigation and genetic analysis of pseudorabies virus within pig populations in Henan province of China during 2018–2019. *Infect. Genet. Evol. J. Mol. Epidemiol. Evol. Genet. Infect. Dis.* **2021**, *92*, 104835. [CrossRef]
23. Zhao, Y.; Wang, L.Q.; Zheng, H.H.; Yang, Y.R.; Liu, F.; Zheng, L.L.; Jin, Y.; Chen, H.Y. Construction and immunogenicity of a gE/gI/TK-deleted PRV based on porcine pseudorabies virus variant. *Mol. Cell. Probes* **2020**, *53*, 101605. [CrossRef]
24. Kumar, S.; Stecher, G.; Tamura, K. MEGA7: Molecular Evolutionary Genetics Analysis Version 7.0 for Bigger Datasets. *Mol. Biol. Evol.* **2016**, *33*, 1870–1874. [CrossRef]
25. Sun, Y.; Luo, Y.; Wang, C.H.; Yuan, J.; Li, N.; Song, K.; Qiu, H.J. Control of swine pseudorabies in China: Opportunities and limitations. *Vet. Microbiol.* **2016**, *183*, 119–124. [CrossRef] [PubMed]
26. van Oirschot, J.T. Diva vaccines that reduce virus transmission. *J. Biotechnol.* **1999**, *73*, 195–205. [CrossRef]
27. Ren, Q.; Ren, H.; Gu, J.; Wang, J.; Jiang, L.; Gao, S. The Epidemiological Analysis of Pseudorabies Virus and Pathogenicity of the Variant Strain in Shandong Province. *Front. Vet. Sci.* **2022**, *9*, 806824. [CrossRef]
28. Zhang, L.; Ren, W.; Chi, J.; Lu, C.; Li, X.; Li, C.; Jiang, S.; Tian, X.; Li, F.; Wang, L.; et al. Epidemiology of Porcine Pseudorabies from 2010 to 2018 in Tianjin, China. *Viral Immunol.* **2021**, *34*, 714–721. [CrossRef] [PubMed]
29. Zhou, H.; Pan, Y.; Liu, M.; Han, Z. Prevalence of Porcine Pseudorabies Virus and Its Coinfection Rate in Heilongjiang Province in China from 2013 to 2018. *Viral Immunol.* **2020**, *33*, 550–554. [CrossRef] [PubMed]
30. Lin, Y.; Tan, L.; Wang, C.; He, S.; Fang, L.; Wang, Z.; Zhong, Y.; Zhang, K.; Liu, D.; Yang, Q.; et al. Serological Investigation and Genetic Characteristics of Pseudorabies Virus in Hunan Province of China From 2016 to 2020. *Front. Vet. Sci.* **2021**, *8*, 762326. [CrossRef]
31. Yang, H.C. Epidemiological situation of swine diseases in 2014 and the epidemiological trend and control strategies in 2015. *Swine Ind. Sci.* **2015**, *32*, 38–40. (In Chinese)
32. Liu, Y.; Zhang, S.; Xu, Q.; Wu, J.; Zhai, X.; Li, S.; Wang, J.; Ni, J.; Yuan, L.; Song, X.; et al. Investigation on pseudorabies prevalence in Chinese swine breeding farms in 2013-2016. *Trop. Anim. Health Prod.* **2018**, *50*, 1279–1285. [CrossRef]
33. Ma, Z.; Liu, M.; Liu, Z.; Meng, F.; Wang, H.; Cao, L.; Li, Y.; Jiao, Q.; Han, Z.; Liu, S. Epidemiological investigation of porcine circovirus type 2 and its coinfection rate in Shandong province in China from 2015 to 2018. *BMC Vet. Res.* **2021**, *17*, 17. [CrossRef] [PubMed]
34. Sabó, A. Analysis of reactivation of latent pseudorabies virus infection in tonsils and Gasserian ganglia of pigs. *Acta Virolog.* **1985**, *29*, 393–402.
35. Zhou, Q.; Zhang, L.; Liu, H.; Ye, G.; Huang, L.; Weng, C. Isolation and Characterization of Two Pseudorabies Virus and Evaluation of Their Effects on Host Natural Immune Responses and Pathogenicity. *Viruses* **2022**, *14*, 712. [CrossRef] [PubMed]
36. Ferrara, G.; Longobardi, C.; D'Ambrosi, F.; Amoroso, M.G.; D'Alessio, N.; Damiano, S.; Ciarcia, R.; Iovane, V.; Iovane, G.; Pagnini, U.; et al. Aujeszky's Disease in South-Italian Wild Boars (Sus scrofa): A Serological Survey. *Animals* **2021**, *11*, 3298. [CrossRef] [PubMed]
37. Moreno, A.; Sozzi, E.; Grilli, G.; Gibelli, L.R.; Gelmetti, D.; Lelli, D.; Chiari, M.; Prati, P.; Alborali, G.L.; Boniotti, M.B.; et al. Detection and molecular analysis of Pseudorabies virus strains isolated from dogs and a wild boar in Italy. *Vet. Microbiol.* **2015**, *177*, 359–365. [CrossRef] [PubMed]
38. Cheng, Z.; Kong, Z.; Liu, P.; Fu, Z.; Zhang, J.; Liu, M.; Shang, Y. Natural infection of a variant pseudorabies virus leads to bovine death in China. *Transbound. Emerg. Dis.* **2020**, *67*, 518–522. [CrossRef] [PubMed]
39. Klupp, B.G.; Hengartner, C.J.; Mettenleiter, T.C.; Enquist, L.W. Complete, annotated sequence of the pseudorabies virus genome. *J. Virol.* **2004**, *78*, 424–440. [CrossRef]
40. Kimman, T.G.; de Wind, N.; Oei-Lie, N.; Pol, J.M.; Berns, A.J.; Gielkens, A.L. Contribution of single genes within the unique short region of Aujeszky's disease virus (suid herpesvirus type 1) to virulence, pathogenesis and immunogenicity. *J. Gen. Virol.* **1992**, *73 Pt 2*, 243–251. [CrossRef] [PubMed]
41. Tan, L.; Yao, J.; Yang, Y.; Luo, W.; Yuan, X.; Yang, L.; Wang, A. Current Status and Challenge of Pseudorabies Virus Infection in China. *Virolog. Sinic.* **2021**, *36*, 588–607. [CrossRef]
42. Wang, X.; Wu, C.X.; Song, X.R.; Chen, H.C.; Liu, Z.F. Comparison of pseudorabies virus China reference strain with emerging variants reveals independent virus evolution within specific geographic regions. *Virology* **2017**, *506*, 92–98. [CrossRef] [PubMed]
43. Jacobs, L.; Meloen, R.H.; Gielkens, A.L.; van Oirschot, J.T. Epitope analysis of glycoprotein I of pseudorabies virus. *J. Gen. Virol.* **1990**, *71 Pt 4*, 881–887. [CrossRef] [PubMed]

44. Luo, Y.; Li, N.; Cong, X.; Wang, C.H.; Du, M.; Li, L.; Zhao, B.; Yuan, J.; Liu, D.D.; Li, S.; et al. Pathogenicity and genomic characterization of a pseudorabies virus variant isolated from Bartha-K61-vaccinated swine population in China. *Vet. Microbiol.* **2014**, *174*, 107–115. [CrossRef] [PubMed]
45. Yang, Q.Y.; Sun, Z.; Tan, F.F.; Guo, L.H.; Wang, Y.Z.; Wang, J.; Wang, Z.Y.; Wang, L.L.; Li, X.D.; Xiao, Y.; et al. Pathogenicity of a currently circulating Chinese variant pseudorabies virus in pigs. *World J. Virol.* **2016**, *5*, 23–30. [CrossRef] [PubMed]

MDPI

St. Alban-Anlage 66

4052 Basel

Switzerland

Tel. +41 61 683 77 34

Fax +41 61 302 89 18

www.mdpi.com

Viruses Editorial Office

E-mail: viruses@mdpi.com

www.mdpi.com/journal/viruses

www.ingramcontent.com/pod-product-compliance
Lightning Source LLC
LaVergne TN
LVHW071518170726
843515LV00008B/2007